Introduction to the Biology of Marine Life

Jones and Bartlett Titles in Biological Science

Introduction to the Biology of Marine Life

EIGHTH EDITION

James L. Sumich
Grossmont College

John F. Morrissey
Hofstra University

JONES AND BARTLETT PUBLISHERS

Sudbury, Massachusetts

BOSTON   TORONTO   LONDON   SINGAPORE

World Headquarters
Jones and Bartlett Publishers
40 Tall Pine Drive
Sudbury, MA 01776
978-443-5000
info@jbpub.com
www.jbpub.com

Jones and Bartlett Publishers Canada
2406 Nikanna Road
Mississauga, ON L5C 2W6
CANADA

Jones and Bartlett Publishers International
Barb House, Barb Mews
London W6 7PA
UK

Copyright © 2004 by Jones and Bartlett Publishers, Inc.

PRODUCTION CREDITS
Chief Executive Officer: Clayton Jones
Chief Operating Officer: Don W. Jones, Jr.
Executive V.P. & Publisher: Robert W. Holland, Jr.
V.P., Design and Production: Anne Spencer
V.P., Sales and Marketing: William Kane
V.P., Manufacturing and Inventory Control: Therese Bräuer
Executive Editor, Science: Stephen L. Weaver
Managing Editor, Science: Dean W. DeChambeau
Associate Editor, Science: Rebecca Seastrong
Senior Production Editor: Louis C. Bruno, Jr.
Marketing Manager: Matthew Bennett
Associate Marketing Manager: Matthew Payne
Text and Cover Design: Anne Spencer
Composition and Illustrations: Precision Graphics
Printing and Binding: Courier Kendallville
Cover Printing: John Pow Co.
Cover image: © Stephen Frink Collection/Alamy Images

Library of Congress Cataloging-in-Publication Data
Sumich, James L.
   Introduction to the biology of marine life/ James L. Sumich, John F. Morrissey.-8th ed.
      p   cm.
   Includes bibliographical references and index.
   ISBN 0-7637-3313-X (alk. paper)
      1. Marine biology. I.Morrissey, John F. (John Francis), 1960- II. Title
   QH91.S95 2004
   578.77-dc22
                                                      200443812

Printed in the United States of America
08 07 06 05 04   10 9 8 7 6 5 4 3 2 1

Brief Contents

Contents

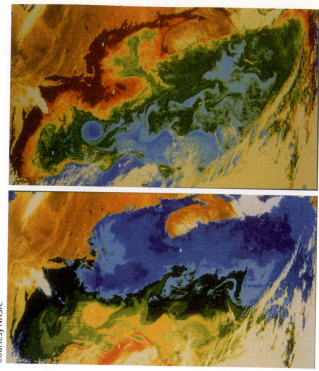

Courtesy NASA.

Michael Van Wort/ORA/NESDIS/NOAA.

© Photodisc.

4   Marine Plants 98

5   Marine Protozoans and Invertebrates 124

6   Marine Vertebrates 156

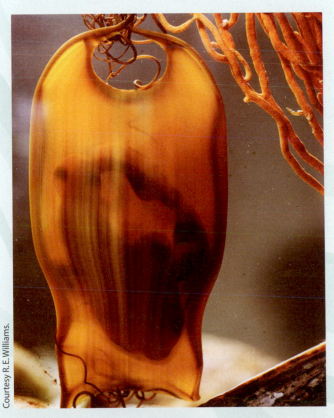
Courtesy R. E. Williams.

Courtesy M.Snyderman.

Louise Kane/NOAA Restoration Center

Preface

An Introduction to the Biology of Marine Life was written to engage introductory, college-level students in the excitement and challenge of understanding marine organisms and the environments in which they live. No previous knowledge of marine biology is assumed; however, some exposure to the fundamental concepts of biology is helpful. Selected groups of marine organisms are used to develop an understanding of biological principles and processes that are basic to all forms of life in the sea. To build on these basics, information dealing with several aspects of taxonomy, evolution, ecology, behavior, and physiology of selected groups of marine organisms is presented. We have intentionally avoided adopting any one of these major subdivisions as the framework of this text. *Biology* is an inclusive term, and we hope that a student's initial venture into this field provides some flavor of the mix of disciplines that constitute modern biological sciences.

Like most textbooks, the sequence of topics in this one is somewhat arbitrary and is intended to be flexible. While this text includes more material than might be covered in one semester, instructors can select and mold the material to match their teaching styles and time limitations. With judicious use of outside supplementary readings, this text can easily provide the structure for a two-quarter or two-semester course.

Changes in the Eighth Edition

The widespread and positive reception of the previous editions of this text continues to be encouraging. This edition represents our continuing effort to meet better the needs of those who use the text. The many welcome suggestions and comments from readers and reviewers have been considered in the changes that were made, and those changes have been considerable.

Each chapter contains many new color photographs, line-art drawings, graphs, and charts to assist students in visualizing the concepts presented. End-of-chapter summaries, questions for discussion, and supplementary reading lists, which encourage further in-depth exploration of covered topics, have been revised and updated. Additional updated references, listed at the end of the text, will be useful to students who have the enthusiasm and intuitive skills necessary to cope with the challenges of reading original literature.

Reorganization

This eighth edition has been completely reorganized. Chapter 1 serves as an introduction to *The Ocean as a Habitat*, stressing the tremendous differences between terrestrial biology and life in the sea. Chapter 2 provides a necessary review of basic *Patterns of Association* in biology for the beginning student and emphasizes evolutionary, taxonomic, and trophic relationships among marine organisms. Chapters 3 through 6 summarize all life in the sea, with two chapters dedicated to autotrophic producers followed by two chapters of heterotrophic consumers. Chapters 3 and 4 have been entirely reorganized so that considerations of primary productivity in the sea are discussed concurrently with presentations of the important marine primary producers, including *Phytoplankton* and *Marine Plants*. Chapter 5 summarizes our current knowledge of *Marine Protozoans and Invertebrates*, including important updates of their biology, interrelationships, and diversity. Chapter 6 includes expanded coverage of *Marine Vertebrates*, including marine reptiles, birds, and mammals. Chapters 7 through 12 are organized around major marine habitats, with a completely new chapter (12) on polar seas. Chapter 7 describes the ecology of *Estuaries*, their biologic, economic, esthetic, and recreational importance, as well as their unusual vulnerability to pollution. *Temperate Coastal Seas* are the subject of Chapter 8, which includes all intertidal communities, the best known, most studied, and most easily disturbed marine habitats. The unique set of marine habitats found in *Tropical and Subtropical Shallow Seas* is the subject of Chapter 9. Although coral reefs are again the emphasis of this chapter, it is greatly expanded with updates on worldwide coral mortality, the mating systems of reef fishes, and discussions of seagrass meadows and their tetrapod inhabitants. *The Open Sea* is presented in Chapter 10, complete with major pelagic players, such as zooplankton and vertebrates, as well as unique phenomena of this realm, including vertical migration, buoyancy strategies, methods of locomotion, and orientation mechanisms. Chapter 11 describes perhaps the largest, yet least understood, habitat on Earth,

The Deep-Sea Floor. Chapter 12, a totally new chapter for this eighth edition, summarizes the biology of *Marine Birds and Mammals in Polar Seas.* And finally, Chapter 13 describes our history of and prognosis for *Harvesting Living Marine Resources,* with updated discussions of fishing down the food chain as well as an entirely new presentation of the impact of trade of marine ornamental species.

Research in Progress Boxes

Marine biology, like all sciences, is a dynamic and active field. Each year, recent discoveries about the sea are published in thousands of new scientific papers. In this edition, we strive to present our current understanding of marine biology as a work in progress. In each chapter, contemporary research issues, recent technological advances, and current topics of interest are presented as *Research in Progress* boxes, with selected references to lead interested students to additional background information. Through these boxes we hope to show the process of science as well as suggest to our readers that marine biology is a vibrant field of study, yearning for their future contributions. The *Research in Progress* boxes, in chapter order are

Powers of 10 and the Metric System
Metabolic Rates of Large Mammals
Oceanography from Space
Mangrove Trees: Connections between Cattle Egrets and Yellow-spotted Stingrays
Can Transplanted Sea Urchins Aid Reef Corals in Their Recovery?
Serendipity in Science—The Discovery of Living Coelacanths
A Surprising Result of Wetland Loss
War: The Forgotten Source of Oil Spills
Threats to the Diversity of Living Sea Turtles
Dolphin Swimming
Deep Submersibles for Seafloor Studies
Marine Sanctuaries
Using Science to Detect Misleading Science

Ancillaries

Instructor's Tool Kit—CD-ROM Compatible with Windows and Macintosh platforms, this CD-ROM provides instructors with the following traditional ancillaries.

The Instructor's Manual, provided as a text file, includes chapter outlines, summaries, and objectives; key con-cepts and terms; teaching tips, insights, and anecdotes relevant to each chapter; lists of the key genera discussed within each chapter; essay questions; and additional critical thinking questions to test the students' level of comprehension of more difficult topics. To answer, the student must understand the facts as well as the concepts and bring them together.

The Test Bank is available as text files and as part of the Diploma™ Test Generator software included on the CD. The Test Generator software enables you to choose an appropriate variety of questions, create multiple versions of tests, even administer and grade tests on-line.

The PowerPoint™ Lecture Outline Slides presentation package provides lecture notes and images for each chapter of *An Introduction to the Biology of Marine Life.* A PowerPoint™ viewer is provided on the CD. Instructors with the Microsoft PowerPoint™ software can customize the outlines, art, and order of presentation.

The Kaleidoscope Media Viewer provides a library of all the art, tables, and photographs in the text to which Jones and Bartlett Publishers holds the copyright or that are in the public domain. The Kaleidoscope Media Viewer uses your browser—Internet Explorer or Netscape Navigator—so you may project images from the text in the classroom, insert images into PowerPoint presentations, or print your own acetates.

Animations of selected illustrations in the text expand the reach of the figures beyond the static page. Because certain concepts in marine biology involve motion, these animated illustrations are a valuable tool for helping students understand the concepts.

Additional public-domain movies and animations from the National Oceanographic and Atmospheric Administration (NOAA) are provided as a courtesy.

 Marine Biology On-Line This text's web site, **www.bioscience.jbpub.com/ marinebiology**, provides additional resources to expand the scope of the textbook and make sure students have access to the most up-to-date information in marine biology. Students can find a variety of study aids in the eLearning area, such as chapter outlines, flash cards, and review questions. Carefully chosen links to relevant Web sites enable students to explore specific topics in marine biology in more detail. A brief description of each link places the site in context *before* the student connects to it.

Acknowledgments

Much credit for the ongoing development of this text goes to students and instructors who have used previous editions and have offered valuable comments and criticisms. We thank our instructors of the past and colleagues of the present for their contributions to this book. Special thanks also go to the many colleagues and institutions that graciously permitted use of their exceptional photographs. Finally and especially, we thank our present and former students for their interest and enthusiasm in discovering rewarding methods of communicating this information.

Many colleagues at numerous institutions reviewed drafts of various editions and collectively improved the text. We thank the following reviewers who have been generous with their time and comments most recently:

Holly Ahern, Adirondack Community College; William G. Ambrose, Bates College; Paul A. Billeter, Charles County Community College; Brenda Blackwelder, Central Piedmont Community College; James L. Campbell, Los Angeles Valley College; Gregory M. Capelli, College of William & Mary; Sneed Collard, University of West Florida; Harold N. Cones, Christopher Newport University; Susan Cormier, University of Louisville; J. Nicholas Ehringer, Hillsborough Community College; Gina Erickson, Highline Community College; Paul E. Fell, Connecticut College; Susan Flanagan, Nunez Community College; Robert T. Galbraith, Crafton Hills College; Dominic Gregorio, Cypress College; Lynn Hansen, Modesto Jr. College; Marty L. Harvill, Bowling Green State University; Richard Heard, Gulf Coast Research Lab; Mary Katherine Wicksten Texas A&M University; Susan Keys, Springfield College; Gil Bane, Kodiak College; Matthew Landau, Stockton State College; Nan Ho, Las Positas College; Cynthia Lewis, San Diego State University; Vicky J. Martin, University of Notre Dame; Jeremy Montague, Barry University; Donald Munson, Washington College; Valerie Pennington, Southwestern College; Richard A. Roller, University of Wisconsin-Stevens Point; Mary Beth Saffo, University of California–Santa Cruz; L. Scott Quackenbush, Los Angeles Valley College; Robert E Shields, U.S. Coast Guard Academy; Cynthia C. Strong, Bowling Green State University; Doug Tupper, Southwestern College; Jefferson T. Turner, Southeastern Massachusetts University; Jacqueline Webb, New York State College of Veterinary Medicine, Cornell University; John T. Weser, Scottsdale Community College; Robert Whitlatch, University of Connecticut; and Richard B. Winn, Duke University Marine Laboratory.

We also acknowledge the fine editorial and production team at Jones and Bartlett Publishers. Stephen Weaver, Executive Editor, Science, had faith in signing us up; Dean DeChambeau contributed his experience and textbook development skills to the project; Anne Spencer provided stellar interior and cover designs; Lou Bruno deftly supervised the production process; Pam Thomson, the copy editor, made sure our grammar was correct and our style consistent; the artists and typesetters at Precision Graphics provided first-class artwork and page makeup; Kimberly Potvin found many of the photos that enliven these pages; and Rebecca Seastrong coordinated the flow of material and the ancillaries.

To Students Using This Text:

We encourage you to immerse yourself in this material as much as possible and in as many ways as you can invent. Spend time at the seashore just messing about. Walk along a beach after high tide and examine the bits of plant and animal debris the sea left behind. Wade into a rocky tide pool, stand still, and watch the action around you. If you can swim, snorkel or dive for a closer look. If you don't swim, learn. Pick up the less fragile organisms for a closer look. If you dislike touching cold and wet creatures, work around your reluctance to change what you dislike. Watch how young children observe things, and mimic their enthusiasm. Take a day trip on a fishing boat. Volunteer at a local aquarium even if you think you don't yet know enough to contribute; you'll learn. Mostly, it is a matter of investing time, time in the field and seat time in classroom. You will get to experience the fun stuff only if you are there.

James L. Sumich
Grossmont College

John F. Morrissey
Hofstra University

CHAPTER 1

A synthetic view of the North Pacific Ocean.

The Ocean as a Habitat

Earth's oceans are home to an extraordinary variety of living organisms adapted to the special conditions of the sea. The characteristics of these organisms and the variety of marine life itself are consequences of the many properties of the ocean habitat. This chapter will provide a survey of the developmental history and present structure of the ocean basins and a general discussion of some properties of seawater and of ocean circulation processes. Adaptations to these properties and processes have molded the character of the ocean's inhabitants through their very long history of evolutionary development.

As students of the Earth's oceans, an appropriate perspective is needed. We naturally tend to see the world from a human point of view, with human scales of time and distance, and with land under our feet and air surrounding us. To begin to understand the marine environment of our home planet and how it and its inhabitants evolved to their present forms, we must broaden our perspective to include very different time and distance scales. Terms such as "young" and "old" or "large" and "small" have limited meaning unless placed in some useful context. FIGURE 1.1 compares size and time scales for a few common oceanic features and inhabitants. Throughout this book, these scales will be revisited and others will be introduced to help you develop a practical sense of the time and space scales experienced by marine organisms.

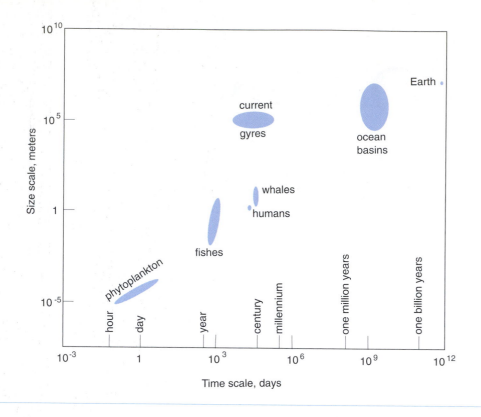

figure 1.1

Time and size scales for a range of major marine features.

1.1 The Changing Marine Environment

Our solar system, including Earth, is thought to have been formed about 5 billion years ago. Modern concepts of the origin of our solar system indicate that the planets aggregated from a vast cloud of cold gas and dust particles into clusters of solid matter. These clumps continued to grow as gravity attracted them together. As Earth grew in this manner, pressure from the outer layers compressed and heated the Earth's center. Aided by heat from decay of radioactive elements, the planet's interior melted. Iron, nickel, and other heavy metals settled to the core, whereas the lighter materials floated to the surface and cooled to form a density-layered planet with a relatively thin and rigid crust (FIG. 1.2).

Early in Earth's history, volcanic vents poked through the crust and tapped the upper mantle for liquid material and gases that were then spewed out over the surface of the young Earth, and a primitive atmosphere developed. Water vapor was certainly present. As it condensed, it fell as rain, accumulated in low places on the Earth's surface, and formed primitive oceans. Additional water may have arrived as "snowballs" from space in the form of comets col-

liding with the young Earth. Atmospheric gases dissolved into accumulating seawater, and other chemicals, dissolved from rocks and carried to the seas by rivers, added to the mixture, eventually creating that complex brew of water and chemicals that we call seawater.

Since their initial formation, ocean basins have experienced considerable change. New material derived from the Earth's mantle has extended the continents so that they are now larger and stand higher than at any time in the past. The oceans have kept pace, getting deeper with accumulations of new water from volcanic gases and from the chemical breakdown of rock. Earth's early life forms (represented by fossils older than about 3 billion years) also had a significant impact on the character of their physical environment. Whether the earliest life forms originated at hot seeps on the deep-sea floor or in warm pools at the sea's edges is a matter of continuing speculation and research. What is clear, however, is that life on this planet requires water; in fact, living organisms are mostly water.

Early in life's history on Earth, free oxygen (O_2) began to be produced in increasing amounts by microscopic photosynthetic cells. The O_2 content of the

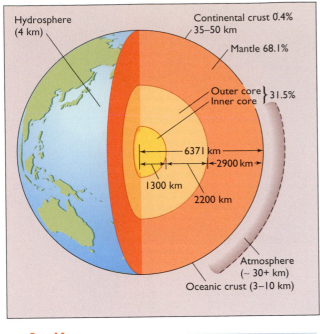

figure 1.2

A section through the Earth, showing its density-layered interior structure and the thickness of each layer.

Hydrosphere (4 km)
Continental crust 0.4% 35–50 km
Mantle 68.1%
Outer core / Inner core } 31.5%
6371 km
2900 km
1300 km
2200 km
Atmosphere (~ 30+ km)
Oceanic crust (3–10 km)

atmosphere 600 million years ago was probably about 1% of its present concentration. It was not much, but it was an important turning point, the time when organisms that could take advantage of O_2 in aerobic respiration became dominant and organisms not using O_2 (anaerobes) became less prevalent (more on this in Chapter 2).

The evolution of more complex life forms using increasingly efficient methods of energy utilization set the stage for an explosion of marine species. By 500 million years ago, most major groups of marine organisms had made their appearance. Worms, sponges, corals, and the distant ancestors of terrestrial animals and plants were abundant. But life at that time could exist only in the sea, where a protective blanket of seawater shielded it from intense solar radiation.

As O_2 became more abundant in the upper atmosphere, some of it was converted to **ozone** (O_3). The process of forming ozone absorbed much of the lethal ultraviolet radiation coming from the sun and prevented the radiation from reaching Earth's surface. The O_2 concentration of the atmosphere 400 million years ago is estimated to have reached 10% of its present level and achieved its current concentration in the Mesozoic Era, about 200 million years ago. The additional ozone screened out enough ultraviolet radiation to permit a few life forms to

abandon their sheltered marine home and colonize the land. Only recently have we become aware that industrialized society's increasing use of aerosols, refrigerants, and other atmospheric pollutants is gradually depleting this protective layer of ozone. **FIGURE 1.3** provides a general timeline for a few of the major events in the early development of life on Earth.

Charting the Deep

People must have explored their local coastal environments very early in their history, but few of their discoveries were recorded. By 325 B.C., Pytheas, a Greek explorer, had sailed to Iceland and developed a method for determining **latitude** (**FIG. 1.4**). About a century later, Eratosthenes of Alexandria, Egypt provided the earliest recorded estimate of Earth's size, its first dimension. His calculated circumference of 45,000 km was only about 12% greater than today's accepted value of 40,000 km. During the Middle Ages, Vikings, Arabians, Chinese, and Polynesians sailed over major portions of Earth's oceans. By the 15th century, all the major inhabitable land areas were occupied; only Antarctica remained unknown to and untouched by humans. Even so, precise charting of the ocean basins had to await several more voyages of discovery.

Between 1768 and 1779, James Cook, an English navigator, conducted three exploratory voyages, mostly in the Southern Hemisphere. He was the first to cross the Antarctic Circle and to understand and conquer scurvy (a disease caused by a deficiency of vitamin C). He is best remembered as the first global explorer to make extensive use of the marine chronometer developed by John Harrison, a British inventor. The chronometer, a very accurate shipboard clock, was necessary to establish the **longitude** of any fixed point on the Earth's surface. Together with Pytheas's 2000-year-old technique for fixing latitude, reasonably accurate positions of geographic features anywhere on the globe could be established for the first time, and our two-dimensional view of Earth's surface was essentially complete. Today, coastal LORAN stations and satellite-based global positioning systems enable individuals to determine their position to within a few meters anywhere on the Earth.

In 1882, one century after Cook's voyages, the first truly interdisciplinary global voyage for scientific exploration of the seas departed from England. The H.M.S. *Challenger* was converted expressly for this voyage. The voyage lasted over 3 years, sailed almost 125,000 km in a circumnavigation of the globe

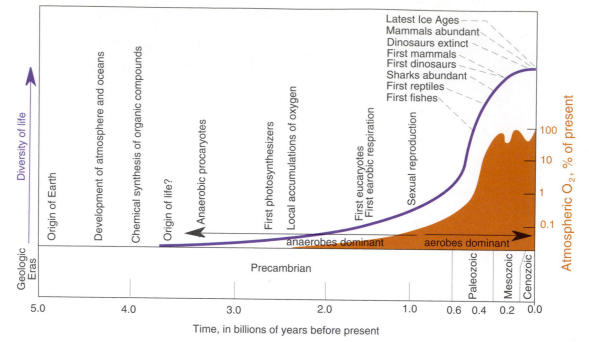

A summary of some biologic and physical milestones in the early development of life on Earth. The blue curve represents the relative diversity of life; the orange curve represents the O_2 concentration of the atmosphere. Several of the terms used here are defined in Chapter 2.

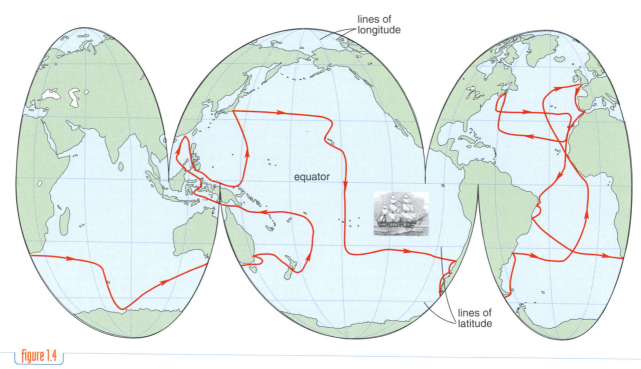

An "orange peel" projection of the Earth's surface, with latitude and longitude lines at 30-degree intervals. The red track traces the voyage of H.M.S. *Challenger* (inset).

(Fig. 1.4), and returned with such a wealth of information that 15 years and 50 large volumes were required to publish the findings. During the voyage, 492 depth soundings were made. These soundings traced the outlines of the Mid-Atlantic Ridge under 2 km of ocean water, plumbed the Mariannas Trench to a depth of 8185 m, and filled in rough outlines of the third dimension of the world ocean, its depth.

The scale of the world ocean is immense, and dealing with the large numbers needed to describe some of the ocean's features can be a confusing process. One simple solution is to express large numbers as powers of the number 10. This approach relies on multiplying or dividing numbers by some multiple of 10 (something we can do mentally very well) and expressing them as exponential powers of 10. Here are some examples of numbers and decimals as powers of 10:

Powers of 10 and the Metric System

Number Notation		Exponential Notation		Decimal Notation		Exponential Notation
1	=	1×10^0	=	1.0	=	1×10^0
10	=	1×10^1	=	0.1	=	1×10^{-1}
100	=	1×10^2	=	0.01	=	1×10^{-2}
1000	=	1×10^3	=	0.001	=	1×10^{-3}
1,000,000	=	1×10^6	=	0.000001	=	1×10^{-6}

In this manner, large numbers such as the area of the Pacific Ocean, listed in Table 1.1, of 165,200,000 km², can be written as 165.2×10^6 or 1.652×10^8 km².

To add or subtract exponential numbers, they must first be converted so that they have the same power of 10.

Example: $1.7 \times 10^2 + 3.4 \times 10^3 = 1.7 \times 10^2 + 34 \times 10^2 = 35.7 \times 10^2$

To multiply numbers expressed as exponents, the numbers are multiplied and the exponents are then added.

Example: $1.7 \times 10^2 \times 3.4 \times 10^3 = (1.7 \times 3.4) \times 10^{(2+3)} = 5.78 \times 10^5$

To divide numbers expressed as exponents, the numbers are divided and the exponents are subtracted.

Example: $(1.7 \times 10^2)/(3.4 \times 10^3)$
$= (1.7/3.4) \times 10^{(2-3)}$
$= 0.5 \times 10^{-1}$

The Systéme International d'Unitès (SI), commonly known as the metric system, is internationally accepted as the system of measurement units. Only Brunei, Myanmar, and the United States have not converted to the metric system for common usage, although in the United States it is used for reporting scientific and engineering data, including the measurements and units used in this book. The metric system is widely used because of its simplicity and ease of conversion between related units.

The SI is built on two simple concepts. First, fundamental units are defined. Next, units larger or smaller than these fundamental units are created by multiplying or dividing them by powers of 10. Unlike the English system that we commonly use, the basic SI units for distance, area, volume, mass, and heat energy are closely interrelated. Each basic

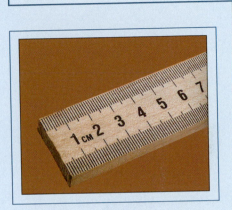

unit is related to the others by a simple equality, making it relatively easy to convert from one unit to another. Although several types of measurements exist in the metric system, all use the same prefix to refer to a particular power of 10 used as a multiplier:

Common SI Prefixes	Decimal Notation	Exponential Notation
Milli- (m) = one thousandth	0.001	1×10^{-3}
Centi- (c) = one hundredth	0.01	1×10^{-3}
Deci- (d) = one tenth	0.1	1×10^{-1}
Basic unit = one	1	1×10^0
Deca- (dk) = ten	10	1×10^1
Hecto- (h) = one hundred	100	1×10^2
Kilo- (k) = one thousand	1000	1×10^3

The basic unit of distance measure is the **meter (m);** the meter is slightly longer than a yard. Like all other basic units of the SI, the meter is subdivided into smaller units by factors of 10. The next smaller distance unit is the **decimeter (dm),** or 0.1 m. Each decimeter is further subdivided into 10 **centimeters (cm).** Further subdivisions or multiples of the meter provide additional distance units:

(continued on next page)

RESEARCH
in progress

Powers of 10 and the Metric System

Common units for distance
Millimeter (mm) = 0.001 m
Centimeter (cm) = 0.01 m
Decimeter (dm) = 0.1 m
Meter (m) = 1 m
Kilometer (km) = 1000 m

Two types of SI units are used to measure area. Some are simply squared distance measures, such as **square meters (m²)** and **square centimeters (cm²)**. Other SI units, such as the hectare, are exclusively for area measure:

Common units for area
Square millimeter (mm²)
    = 0.000001 m² = 1×10^{-6} m²
Square centimeter (cm²)
    = 0.0001 m² = 1×10^{-3} m²
Square kilometer (km²)
    = 1,000,000 m² = 1×10^6 m²
Are (a) = 100 m² = 1×10^2 m²
Hectare (ha) = 10,000 m² = 1×10^4 m²

Two separate, but related, systems of units are also used for volume measure. Some are based on the cube of distance measures, others on the **liter (L)** and its subdivisions. Some of the more common volume units are as follows:

Common units for volume
Milliliter (ml) = 0.001 liters = 1 cm³
Liter = 1 liter = 1×10^3 ml
Cubic centimeter (cm³) = 1 ml = 1 $\times 10^{-3}$ liters
Cubic meter (m³) = 1×10^6 cm³ = 1 $\times 10^6$ ml = 1×10^3 L

The basis for all SI units of mass is the **gram (g).** One gram is defined as the mass of pure water occupying 1 ml or 1 cm³ at 4°C. Other mass units derived from multiples or subdivisions of the gram are as follows:

Common units for mass
Milligram (mg) = 0.001g
Gram (g) = 1g
Kilogram (kg)   = 1000 g
Metric ton (MT) = 1×10^6 g

Heat energy is measured in calories. One calorie is defined as the amount of heat energy needed to change the temperature of 1 g (or 1 ml or 1 cm³) of pure water 1°C:

Common units for heat energy
Calorie (cal.) = 1 cal.
Kilocalorie (kcal) = 1000 cal. = 1×10^3 cal.
Dietary calorie (Cal.) = 1 kcal = 1 $\times 10^3$ cal.

Some approximate English system equivalents of the more commonly used SI units are listed below. Until using the SI becomes second nature, it may be helpful to learn a few SI–English unit equivalents so that you may quickly develop a mental concept of the magnitude of the units you are considering.

Some approximate SI–English unit equivalents
1 meter is a bit more than one yard (1.1 yard).
1 centimeter is a bit less than one-half inch (0.4 inch).
1 kilometer is a bit more than one-half mile (0.6 mile).
1 kilogram is a bit more than two pounds (2.2 pounds).
1 liter is a bit more than one quart (1.06 quarts).

Additional Reading
Hebra A. *Measure for Measure: The Story of Imperial, Metric, and Other Units.* Baltimore: Johns Hopkins University Press, 2003.
Morrison P. *Powers of Ten: About the Relative Size of Things in the Universe and the Effect of Adding Another Zero.* W.H. Freeman, 1985.

www.bioscience.jbpub.com/marinebiology
For an excellent visualization of powers of 10, go to this book's website at http://www.bioscience.jbpub.com/marinebiology and look for the Powers of Ten link in Chapter 1.

A Different View of the Ocean Floor

Early in the 20th century, Alfred Wegener proposed that the oceans were changing in other ways. Wegener developed a detailed hypothesis of **continental drift** to explain several global geologic features, including the remarkable jigsaw-puzzle fit of some continents (especially the west coast of Africa and the east coast of South America). He proposed that our present continental masses had drifted apart after the breakup of a single supercontinent, **Pangaea.** His evidence was ambiguous, though, and most scientists at the time remained unconvinced. It was not until the early 1960s that new evidence compelled two geophysicists to independently, and almost simultaneously, propose the closely related concepts of **seafloor spreading** and **plate tectonics.** In hindsight, these related concepts

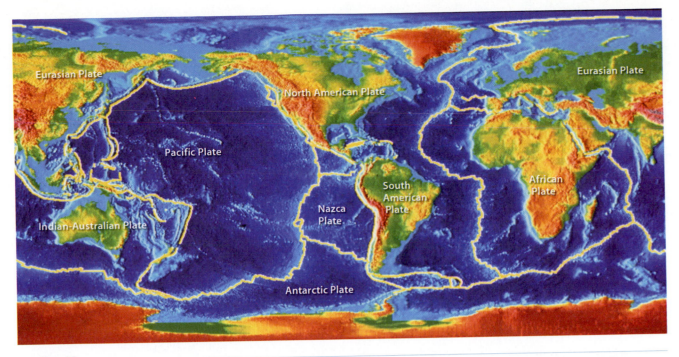

figure 1.5

The major plates of the Earth's crust. Compare the features of this map with those of Figure 1.11.

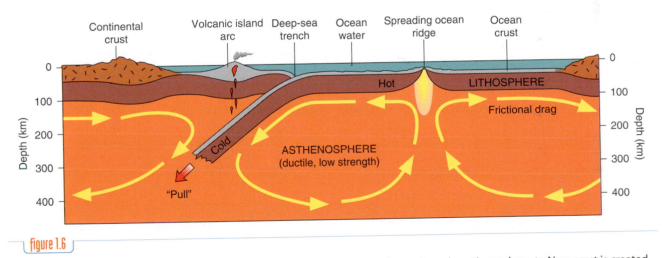

figure 1.6

Cross-section of a spreading ocean floor, illustrating the relative motions of oceanic and continental crusts. New crust is created at the ridge axis, and old crust is lost in trenches.

seem completely obvious—that the Earth's crust is divided into giant irregular plates (FIG. 1.5). These rigid crustal plates float on the denser and slightly more plastic mantle material. Each plate edge is defined by oceanic trench or ridge systems, and some plates include both oceanic and continental crusts. New oceanic crustal material is formed continually along the axes of oceanic ridges and rises. As crustal plates grow on either side of the ridge, they move away from the ridge axis in opposite directions, carrying bottom sediments and attached continental masses with them (FIG. 1.6).

In 1968, a new and unusual ship, the *Glomar Challenger,* was launched to probe time, the fourth dimension of the oceans. Equipped with a deck-mounted drilling rig, the *Glomar Challenger* was capable of drilling into the seafloor in water over 7000 m deep (FIG. 1.7). Within 2 years, the *Glomar Challenger* recovered vertical sediment core samples from enough sites on both sides of the Mid-Atlantic Ridge to finally and firmly confirm the concept of seafloor spreading and continental drift. Before being decommissioned in 1983, the *Glomar Challenger* traveled almost 700,000 km and drilled 318,461 m of

figure 1.7

The ocean drill ship, *Glomar Challenger.*

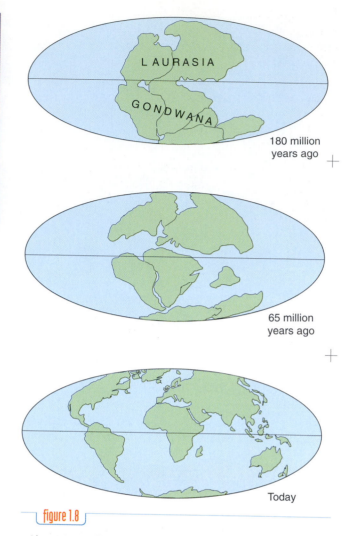

LAURASIA

GONDWANA

180 million years ago

65 million years ago

Today

figure 1.8

About 200 million years ago, the megacontinent, Pangaea, separated into two large continental blocks, Laurasia and Gondwana. Since then, they have fragmented into smaller continents and continue to drift apart. These maps outline the changing past positions of the continents and ocean basins. (Adapted from Dietz and Holden 1970.)

seafloor in 1092 drill holes at 624 sites in all ocean basins. Subsequent analyses of microscopic marine fossils recovered from this tremendous store of marine sediment samples have led to refined estimates of the ages and patterns of evolution of all the major ocean basins.

Scientists working in the deep-diving research submersible *Alvin* in 1977 made a remarkable discovery of previously unknown marine animal communities associated with seafloor hot water vents in the eastern equatorial Pacific Ocean. Several other vents and their associated animal communities found elsewhere during the past quarter century associated with ridge or rise axes are discussed in Chapter 11.

The changes that seafloor spreading and plate tectonics have wrought on the shapes and sizes of the oceans have been impressive. Currently, the African continent is drifting northward on a collision course with Europe, relentlessly closing the Mediterranean Sea. The Atlantic Ocean is becoming wider at the expense of the Pacific Ocean. Australia and India continue to creep northward, slowly changing the shapes of the ocean basins they border. Occasional violent earthquakes are only incidental tremors in this monumental collision of crustal plates. The rates of seafloor spreading have been determined for some oceans, and they vary widely. The South Atlantic is widening about 3 cm each year (or approximately your height in your lifetime). The Pacific Ocean is shrinking somewhat faster.

The breakup of the megacontinent, Pangaea, produced ocean basins where none existed before. The seas that existed 200 million years ago have changed size or have disappeared altogether. Some of the past positions of the continents and ocean basins, based on our present understanding of the processes involved, are reconstructed in FIGURE 1.8. Excess crust produced by seafloor spreading folds into mountain ranges (the Himalayas are a dramatic example) or slips down into the mantle and remelts (Fig. 1.6). Consequently, most marine fossils older than about 200 million years can never be studied; they too have been carried to destruction by the "conveyor belt" of subduction, the sinking of seafloor crust at trench locations to be remelted in the mantle. Ironically, the only fossil evidence we can find for the first 90% of the evolutionary history of marine life is found in landforms that were once ancient seabeds.

On much shorter time scales, other processes have been at work to alter the shapes and sizes of ocean basins. During the past 200,000 years, our planet has experienced two major episodes of global cooling associated with extensive continental glaciation. Just 18,000 years ago, northern reaches of Europe, Asia, and North America were frozen under the grip of the most recent ice age, or the **last glacial maximum.** The massive amount of water contained in those glaciers lowered sea level about 150 m below its present (and also its preglacial) level. Between 18,000 and 10,000 years ago, melting and shrinking of these continental glaciers were accompanied by a 150-m rise in global sea level and the flooding of land exposed during the last glacial maximum. Coral reefs, estuaries, and other shallow coastal habitats were modified extensively during this flooding; these topics are discussed in Chapters 7 and 9. At the present time, warmer summer temperatures around Antarctica are creating some concern about the potential for melting glaciers to cause another rise in global sea levels of a few meters (http://earthobservatory.nasa.gov/Study/ArcticIce/).

1.2 The World Ocean

Currently, Earth is the only planet in our solar system that has liquid water at its surface. Unlike the faces of any planet we can see from Earth, from space our Blue Planet stands apart from all others in our solar system, because 70% of our planet's solid face is hidden by water too deep for light to penetrate. Our world ocean has an average depth of about 3800 m (3.8 km). This may seem like a lot of water, but when compared with Earth's diameter of 13,250 km, the world ocean is actually a relatively thin film of water filling the low places of Earth's crustal surface. On the scale of the view of Earth from space shown on page 1, the average ocean water depth is represented by a distance of about 0.04 mm (about one one-thousandth of an inch). Although shallow on a planetary scale, these water depths of several thousand meters easily dwarf the largest of the plants or animals living there.

Visualizing the World Ocean

One can visualize the Earth's marine environment as a single large interconnected ocean system as shown in FIGURE 1.9. The Antarctic continent is surrounded by a "Southern Ocean," which has three large embayments extending northward. These three oceanic extensions, partially separated by continental barriers, are the Atlantic, Pacific, and Indian oceans. Other smaller oceans and seas, such as the Arctic Ocean and the

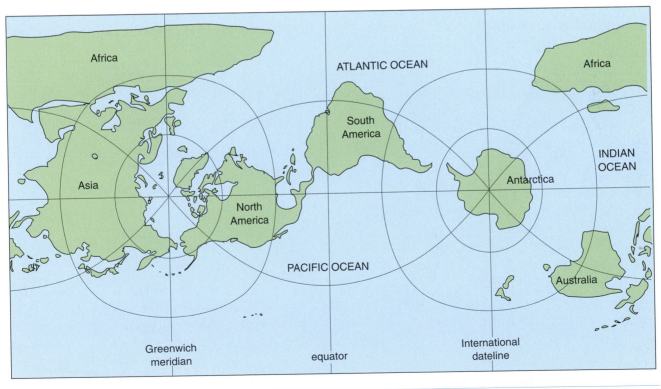

figure 1.9

A modified polar view of the world ocean, emphasizing the extensive oceanic connections between major ocean basins in the Southern Hemisphere.

Mediterranean Sea, project from the margins of the larger ocean basins. These broad connections between major ocean basins permit exchange of both seawater and the organisms living in it, reducing and smoothing out differences between adjacent ocean basins.

FIGURE 1.10 represents a more conventional view of the world ocean, with the world ocean separated into four major ocean basins, the Atlantic, Pacific, Indian, and Arctic, and without the emphasis on the extensive southern connections apparent in Figure 1.9. The format of Figure 1.10 is often more useful because our interest in the marine environment has been focused in the temperate and tropical regions of Earth.

The equator also is a very real physical boundary extending across the tropical center of the large ocean basins. The curvature of the Earth's surface causes areas near the equator to receive more radiant energy from the sun than equal-sized areas in polar regions. The resultant heat gradient from warm tropical to cold polar regions establishes the basic patterns of atmospheric and oceanic circulation. Surface ocean current patterns display a nearly mirror-image symmetry in the northern and southern halves of the Pacific and Atlantic oceans. This symmetry establishes it as a natural, although intangible, focus for the graphic representation of these features and of the life zones they define.

Nearly two thirds of our planet's land area is located in the Northern Hemisphere. The Southern Hemisphere is an oceanic hemisphere, with 80% of its surface covered by water. The Pacific Ocean alone accounts for nearly one half of the total ocean area. A few descriptive measurements for features of the six largest marine basins are listed in TABLE 1.1.

Maximum oceanic depths extend to over 11,000 m, but most of the ocean floor lies between 3000 and 6000 m below the sea surface. A synthetic image of the northern and central parts of the Atlantic Ocean (FIG. 1.11) illustrates some of the larger scale features of the ocean floor. The **continental shelf**, which extends seaward from the shoreline and is actually a structural part of the continental landmass, would not be considered an oceanic feature if sea level were lowered by as little as 5% of its present average depth. The width of continental shelves varies, from almost nothing off southern Florida to over 800 km wide in the Arctic Ocean north of Siberia. Continental shelves account for 8% of the ocean's surface area; this is equivalent to about one sixth of the Earth's total land area.

Most continental shelves are relatively smooth and slope gently seaward. The outer edge of the shelf, called the **shelf break**, is a vaguely defined feature that usually occurs at depths between 120 and 200 m. Beyond the shelf break, the bottom steepens slightly to become the **continental slope**.

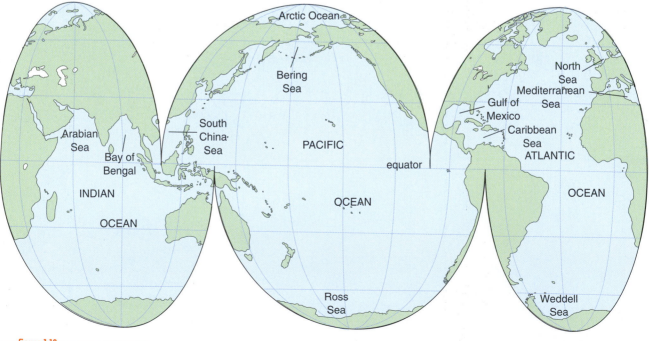

figure 1.10

An equatorial view of the world ocean.

table 1.1

Some Comparative Features of the Major Ocean Basins

Ocean	Area × 10⁶ km²	Volume × 10⁶ km³	Average depth (m)	Maximum depth (m)
Pacific	165.2	707.6	4282	11,033
Atlantic	82.4	323.6	3926	9200
Indian	73.4	291.0	3963	7460
Arctic	14.1	17.0	1205	4300
Caribbean	4.3	9.6	2216	7200
Mediterranean	3.0	4.2	1429	4600
Other	18.7	17.3		
Totals (average)	361.1	1370.3	(3795)	

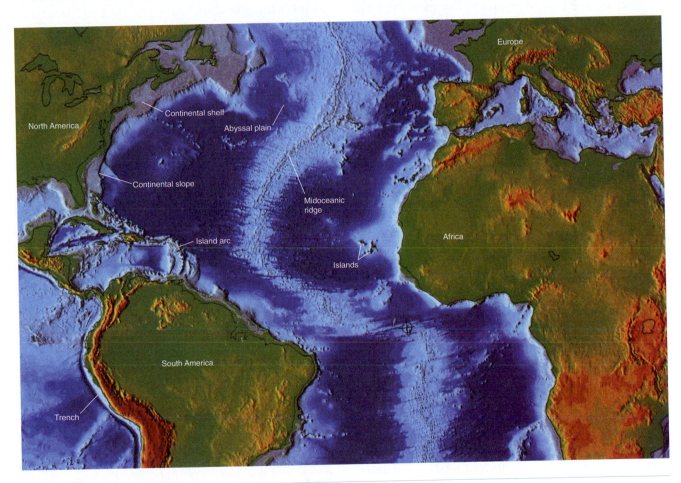

figure 1.11

Some large-scale features of the North Atlantic seafloor.

Continental slopes reach depths of 3000 to 4000 m and form the boundaries between continental masses and the deep ocean basins.

A large portion of deep ocean basins consists of flat sediment-covered areas called **abyssal plains**. These plains have almost imperceptible slopes, much like that of eastern Colorado. The sediment blanket over abyssal plains often completely buries smaller crustal elevations, called abyssal hills. Most abyssal plains are situated near the margins of the ocean basins at depths between 3000 and 5000 m.

Oceanic **ridge** and **rise systems**, such as the Mid-Atlantic Ridge and East Pacific Rise, occupy over 30% of the ocean basin area. The ridge and rise systems are rugged more or less linear features that form a continuous underwater mountain chain encircling the Earth. The Mid-Atlantic Ridge actually resembles a 20- to 30-km-wide canyon like the East African Rift

Valley, sitting atop a broadly elevated seafloor aligned down the center of the Atlantic Ocean. The top of the Mid-Atlantic Ridge may extend 4 km above the surrounding abyssal hills, and isolated peaks occasionally extend above sea level to form islands such as Iceland and Ascension Island. The East Pacific Rise is much lower and broader, with a barely perceptible rift about 2 km wide.

Trenches are distinctive ocean floor features with depths usually extending deeper than 6000 m. Most trenches, including the five deepest, are located along the margins of the Pacific Ocean. The bottom of the Challenger Deep, in the Mariana Trench of the western North Pacific, is 11,033 m deep, the greatest ocean depth found anywhere. (This is as far below sea level as commercial jets typically fly above the sea surface.) The enormous depths of trenches impose extreme conditions of high water pressure and low temperature on their inhabitants. Although trenches account for less than 2% of the ocean bottom area, they are integral parts in the processes of seafloor spreading and plate tectonics, acting as sites where crustal plates tens or hundreds of millions of years old are finally subducted back into the mantle to be remelted.

Most oceanic **islands, seamounts,** and abyssal hills have been formed by volcanic action. Oceanic islands are volcanic mountains that extend above sea level; seamounts are volcanic mountains whose tops remain below the sea surface. Most of these features are located in the Pacific Ocean. Islands in tropical areas are often submerged and capped by coral atolls or fringed by coral reefs. These reefs, examined in Chapter 9, form some of the most beautiful and complex animal communities found anywhere.

Seeing in the Dark

Humans are visual beings; we continually examine our surroundings through the clarity of air. In the oceans, however, the opacity of seawater has long thwarted our ability to examine its depths in detail. Fortunately, the opacity of seawater to light is compensated by its transparency to sound. The early ancestors of dolphins capitalized on this feature of water about 20 million years ago by evolving sophisticated biosonar systems capable of high-resolution target discrimination. Biosonar is based on the production of sharp sounds, detecting the echo as it bounces off a target, and measuring the time delay between sound production and echo return (more details on dolphin biosonar can be found in Chapter 10). In a technologic sense, we are catching up with dolphins with the development of various types of electronic sonar (<u>so</u>und <u>na</u>vigating <u>r</u>anging) systems. Without going into the history of the development of sonar, it is sufficient to indicate that several different types of sonar systems with different resolutions have been central to obtaining detailed pictures of the deep ocean floor.

The most widespread and familiar type of sonar is operated from surface ships. You may be familiar with commercial versions of these as fish finders or depth finders on personal boats. At sea, surface sonar systems operate from a single ship-based transmitter to produce a single line or track of sequential depth measurements below the ship as it is underway. This system provides an acoustic view of the seafloor in two dimensions, with poor resolution of small structures or fine detail.

Multiple beam sonar systems also are operated from surface vessels, but with an array of several sound transmitters and receivers arranged from bow to stern. The numerous overlapping sound beams return much more information about the seafloor, information that would be a useless jumble of data without computers to decipher it. Multiple beam sonar can map a swath of ocean bottom several kilometers wide in a single pass and with much higher resolution than single transmitter systems. The result is the acoustic equivalent of a continuous strip of aerial photographs of the ocean floor (**FIG. 1.12**). Sidescan sonar systems are variants of the multiple beam technique. As its name implies, sidescan sonar directs sound beams to the sides of the ship's track. Images obtained with sidescan sonar show in rich detail the fine texture of the seafloor, equivalent to close-up photographs of the ocean floor.

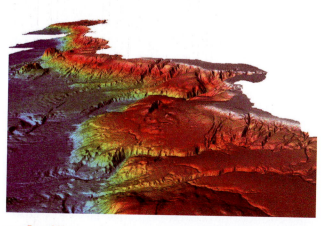

figure 1.12

A multiple beam SONAR image of the coastal margin of southern California.

1.3 Properties of Seawater

Several common properties of seawater are crucial to the survival and well-being of the ocean's inhabitants. Water accounts for 80–90% of the volume of most marine organisms. It provides buoyancy and body support for swimming and floating organisms and reduces the need for heavy skeletal structures. Water is also the medium for most chemical reactions needed to sustain life. The life processes of marine organisms in turn alter many fundamental physical and chemical properties of seawater, including its transparency and chemical makeup, making organisms an integral part of the total marine environment. Understanding the interactions between organisms and their marine environment requires a brief examination of some of the more important physical and chemical attributes of seawater. The characteristics of pure water and seawater differ in some respects, so we consider first the basic properties of pure water and then examine how those properties differ in seawater.

Pure Water

Water is a common yet very remarkable substance. It is the only substance on Earth that is abundant as a liquid (mostly in oceans), with substantial quantities left over as a gas in the atmosphere and as a solid in the form of ice and snow. Individual water molecules have a simple structure, represented by its molecular formula, H_2O. Yet the collective properties of many water molecules interacting with each other in a liquid are quite complex. Each water molecule has one atom of oxygen (O) and two atoms of hydrogen (H), which together form water (H_2O). Some characteristics of these and other biologically important elements are listed in Appendix B.

The many unusual properties of water stem from its molecular shape: a four-cornered tetrahedron with the two hydrogen atoms forming angles of about 105 degrees with the oxygen atom. This molecular shape is simplified to two dimensions in FIGURE 1.13. This atomic configuration creates an asymmetric water molecule, with the oxygen atom dominating one end of the molecule and the hydrogen atoms dominating the other end. The **covalent bond** between each hydrogen and the oxygen atom is formed by the sharing of two negatively charged electrons. The oxygen atom attracts the electron pair of each bond, causing the oxygen end of each water molecule to assume a slight negative charge. The hydrogen end of the molecule, by giving up part of its electron complement, is left with a small positive charge. The resulting electrical polarization of water molecules, one end with a positive charge and the other end with a negative charge, has profound consequences for liquid water. When water is in its liquid form, each end of one water molecule attracts the oppositely charged end of other water molecules. This attractive force creates a weak bond, a **hydrogen bond** or **H-bond**, between adjacent water molecules (FIG. 1.14). These bonds are much weaker and less stable than the covalent bonds within a single water molecule and are continually breaking and reforming as they change partners with other water molecules millions of times a second.

The H-bonds between water molecules require that water must be warmed to a much higher temperature to boil than that needed for other substances with similar molecular sizes, such as O_2 or CO_2. Water's high freezing point of 0°C and boiling point at 100°C causes most water at the Earth's surface to exist as a liquid, making life as we know it possible. Hydrogen bonding also accounts for several other unique and important properties of water. TABLE 1.2 lists and the following paragraphs discuss some of these properties.

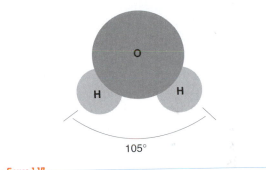

figure 1.13

The arrangement of H and O atoms in a molecule of water (H_2O). The oxygen end has a slight net negative charge, whereas the hydrogen end has a slight net positive charge.

figure 1.14

Hydrogen bonding between adjacent molecules of liquid water. Blue dashed lines represent hydrogen bonds.

Table 1.2

Some Biologically Important Physical Properties of Water

Properties	Comparison with other substances	Importance in biologic processes
Boiling point	High (100°C) for molecular size	Causes most water to exist as a liquid at earth surface temperatures
Freezing point	High (0°C) for molecular size	Causes most water to exist as a liquid at earth surface temperatures
Surface tension	Highest of all liquids	Critical to position maintenance of sea-surface organisms
Density of solid	Unique among common natural substances	Causes ice to float and inhibits complete freezing of large bodies of water
Latent heat of vaporization	Highest of all common natural substances (540 cal/g)	Moderates sea-surface temperatures by transferring large quantities of heat to the atmosphere through evaporation
		Inhibits large-scale freezing oceans
Latent heat of fusion	Highest of all common natural substances (80 cal/g)	Maintains a large variety of substances in solution, enhancing a variety of chemical reactions
Solvent power	Dissolves more substances in greater amounts than any other liquid	Moderates daily and seasonal temperature changes
Heat capacity	High (1 cal/g/°C) for molecular size	Stabilizes body temperatures of organisms

Viscosity and Surface Tension

Hydrogen bonding between adjacent water molecules within a mass of liquid water creates a slight "stickiness" between these molecules. This property, known as **viscosity**, has a marked effect on all marine organisms. The viscosity of water reduces the sinking tendency of some organisms by increasing the frictional resistance between themselves and nearby water molecules. At the same time, viscosity magnifies problems of frictional drag that actively swimming animals must overcome.

At the surface of a water mass (such as the air–sea boundary), the mutual attraction of water molecules creates a flexible molecular "skin" over the water surface. This, the **surface tension** of water, is sufficiently strong to support the full weight of a water strider (FIG. 1.15). Both surface tension and viscosity are temperature dependent, increasing as the temperature decreases.

Density–Temperature Relationships

Most liquids contract and become more dense as they cool. The solid form of these substances is denser than the liquid form. Over most of the temperature range at which pure water is liquid, it behaves like other liquids. At 4°C or above, the **density** decreases with increasing temperature. Below 4°C, however, water behaves differently: The density–temperature pattern of pure water reverses so that the density begins to decrease as temperatures fall below 4°C. One model used to explain this unique behavior of water proposes that, at near-freezing temperatures, less-dense icelike clusters consisting of several water molecules form and disintegrate very rapidly within the body of liquid water. As liquid water continues to cool, more clusters form and the clusters survive longer. Eventually, at 0°C, all the water molecules become locked into a rigid solid crystal lattice of ice. The water ice formed is about 8% less dense than liquid water at the same temperature, so ice always floats on liquid water. This is an unusual, but very fortunate, property of water. Without this unique density–temperature relationship, ice would sink as it formed, and lakes, oceans, and other bodies of water would freeze solid from the bottom up. Winter survival for organisms living in such an environment would be much more difficult.[1]

Heat Capacity

Heat is a form of energy, the energy of molecular motion. The sun is the source of almost all energy entering the Earth's surface heat budget. At the surface of the sea, some of the sun's radiant energy is converted to heat energy that is then transferred from place to place primarily by **convection** (mixing) and secondarily by **conduction** (the exchange of heat energy between adjacent molecules). Heat energy is measured in **calories**.[2]

---

[1]The maximum density of pure water is used to define the fundamental metric measure of mass, the gram, which is defined as the mass of pure water at 4°C contained in the volume of one cubic centimeter. Thus, the density, the ratio of mass to volume, of pure water is 1.000 g/cm^3.

[2]A calorie is a unit of heat energy, defined as the quantity of heat needed to elevate the temperature of 1 g of pure water 1°C.

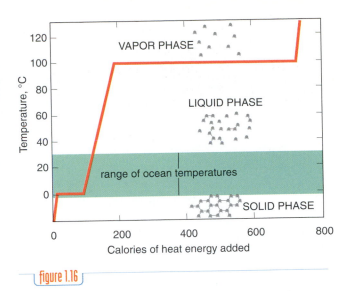

figure 1.16

The heat energy necessary to cause temperature and phase changes in water.

figure 1.15

A water strider (*Halobates*), one of the few completely marine insects, is supported by the surface tension of seawater.

Water has the ability to absorb or give up heat without experiencing much of a temperature change. To illustrate the high **heat capacity** of water, imagine a 1-g block of ice at −20°C on a heater that provides heat at a constant rate. Heating the ice from −20°C to 0°C requires 10 calories, or 0.5 calories per degree of temperature increase (the heat capacity of ice). However, converting 1 g of ice at 0°C to liquid at 0°C requires 80 calories. Conversely, 80 calories of heat must be removed from 1 g of liquid water at 0°C to freeze it to ice at the same temperature. (This is referred to as water's **latent heat of fusion**.) Continued heating of the 1-g water sample from 0°C requires one calorie of heat energy for each 1° change in temperature (the heat capacity of liquid water) until the boiling point (100°C) is reached. At this point, further temperature increase stops until all the water is converted to water vapor. For this conversion, 540 calories of heat energy are necessary (water's **latent heat of vaporization**). FIGURE 1.16 summarizes the energy requirements for these changes in water

temperature. The high heat capacity and the large amount of heat required for evaporation allow large bodies of water to resist extreme temperature fluctuations. Heat energy is absorbed slowly by water when the air above is warmer and is gradually given up when the air is colder, providing a crucial global-scale temperature-moderating mechanism for marine environments and adjacent land areas.

Solvent Action

The small size and polar charges of each water molecule enable it to interact with and dissolve most naturally occurring substances, especially salts, which are composed of atoms or simple molecules, called **ions**, that carry an electrical charge. Salts held together by **ionic bonds** (bonds between oppositely charged adjacent ions) are particularly susceptible to the solvent action of water. FIGURE 1.17 illustrates the process of a salt crystal dissolving in water. Initially, several water molecules form weak H-bonds with each Na⁺ and Cl⁻ ion, and they eventually overcome the mutual attraction of those ions that previously bound them together in the crystalline structure. As more Na⁺ and Cl⁻ ions are removed in this way, the solid crystal structure disintegrates and the salt dissolves.

Water is not a good universal solvent, however, for some large organic molecules like waxes and oils or for small molecules that lack electrical charges. A notable and biologically critical example of the second category is O_2, for water can only dissolve a few parts per million of O_2.

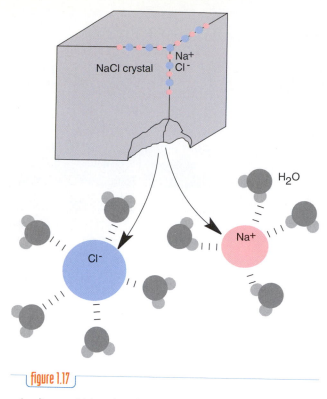

figure 1.17

A salt crystal (above) and the action of charged water molecules in dissolving the crystal to dissociated Na⁺ and Cl⁻ ions.

Seawater

Seawater is the accumulated product of several billion years of water condensing as rain from the atmosphere, eroding rocks and soil, and washing it all to the sea. About 3.5% of seawater is composed of dissolved compounds from these sources. The other 96.5% is pure water. Traces of all naturally occurring substances probably exist in the ocean and can be separated into three general categories: (1) inorganic substances, usually referred to as salts, including nutrients necessary for plant growth; (2) dissolved gases like N_2, O_2, and CO_2; and (3) organic compounds derived from living organisms. Organic compounds dissolved in seawater include fats, oils, carbohydrates, vitamins, amino acids, proteins, and other substances. Some compounds are valuable sources of nutrition for marine bacteria and some other organisms. Current research indicates that other organic compounds, especially synthetically created ones such as polychlorinated biphenyls and other chlorinated hydrocarbons, have accumulated in marine food chains and have had serious negative impacts on the development and reproduction of some forms of marine life.

Dissolved Salts

Salts account for most dissolved substances in seawater. The total amount of dissolved salts in seawater is referred to as its **salinity** and is measured in parts per thousand (‰) rather than in parts per hundred (%). Salinity values range from nearly zero at river mouths to over 40‰ in some areas of the Red Sea. Yet, in open-ocean areas away from coastal influences, salinity averages approximately 35‰ and varies only slightly over large distances (FIG. 1.18).

Salinity is altered by processes that add or remove salts or water from the sea. The primary mechanisms of salt and water addition or removal are evaporation, precipitation, river runoff, and the freezing and thawing of sea ice. When evaporation exceeds precipitation, it removes water from the sea surface, thereby concentrating the remaining salts and increasing the salinity. Excess precipitation decreases salinity by diluting the sea salts. Freshwater runoff from rivers has the same effect. FIGURE 1.19 illustrates the average annual north–south variation of sea-surface evaporation and precipitation. Areas with more evaporation than precipitation (Fig. 1.19, gray-shaded areas) generally correspond to the high surface salinity regions shown in Figure 1.18 and to the latitudes with most of the great land deserts of the world. In polar regions where seawater can freeze, only the water molecules are incorporated into the developing ice crystal. The dissolved ions are excluded from the growing ice crystal, causing the salinity of the remaining liquid seawater to increase. The process is reversed when ice melts. Freezing and thawing of seawater are usually seasonal phenomena, resulting in little long-term salinity differences.

When salts dissolve in water, they release both positively and negatively charged ions. The more common ions found in seawater are listed in TABLE 1.3 and are grouped as major or minor constituents according to their abundance. The major ions account for over 99% of the total salt concentration in seawater. In relation to each other, concentrations of these major ions remain remarkably constant even though their total abundance may differ from place to place. Most of the more abundant ions enumerated in Table 1.3 are important components of marine organisms. Magnesium, calcium, bicarbonate, and silica are important components of the hard skeletal parts of marine organisms. Plants need nitrate and phosphate for the synthesis of organic material.

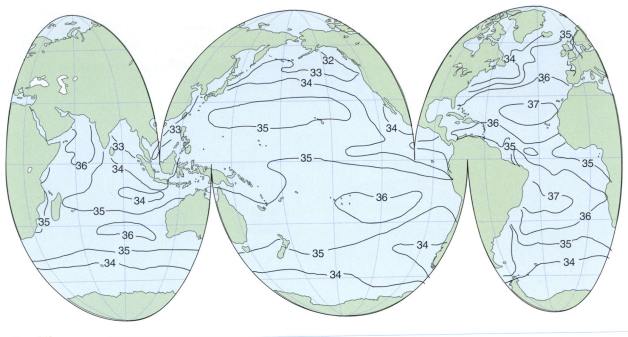

figure 1.18

Geographic variations of surface ocean salinities, expressed in ‰.

Salt and Water Balance

The well-being of living organisms requires that they maintain reasonably constant internal environmental conditions. **Homeostasis** is the term used to describe the tendency of living organisms to control or regulate fluctuations of their internal environment. Homeostasis is the result of coordinated biologic processes that regulate conditions such as body temperature or blood ion concentrations. When working properly, these processes result in a dynamic regulation of conditions that vary within definite and tolerable limits. This section describes those processes that affect the homeostasis of salt and water exchange between the body fluids of an organism and its seawater environment. Other homeostatic processes will be encountered in later chapters.

The body fluids of marine organisms are separated from seawater by boundary membranes that participate in several vital exchange processes, including absorption of oxygen and nutrients and excretion of waste materials. Small molecules, such as water, easily pass through some of these membranes, but the passage of larger molecules and the ions abundant in seawater is blocked. Such membranes are **selectively permeable**; they allow only small molecules and ions to pass through while blocking the passage of larger molecules and ions. When substances are free to move, as they are when dissolved in seawater, they move along a gradient from regions where they exist in high concentrations to regions of lower concentrations. This type of molecular or ionic motion is known as **diffusion.** Diffusion causes both water molecules and dissolved substances to move along concentration gradients within living organisms and sometimes across selectively permeable membranes between organisms and their surrounding seawater.

To illustrate the basic process of salt and water balance in marine organisms, let us examine two somewhat idealized animal examples: a sea cucumber and a salmon. A sea cucumber avoids problems of salt and water imbalance by maintaining an internal fluid medium chemically similar to seawater (about 35‰ dissolved substances). It can easily maintain this balance as long as the salt concentrations of the fluids on either side of its boundary membranes are equal and no concentration gradient exists. This is known as an **isosmotic** condition. A state of equilibrium is maintained as long as water diffuses out of the sea cucumber as rapidly as it enters and the salt content of the internal fluids remains equal to that of the seawater outside (FIG. 1.20, top).

However, if the sea cucumber is removed from the sea and placed in a freshwater lake, the salt concentration is then greater inside the animal (still 35‰)

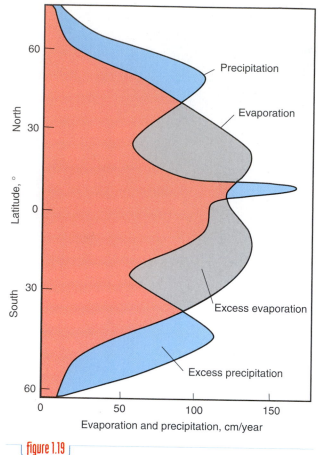

figure 1.19

Average north–south variation of sea surface evaporation and precipitation. (Redrawn by permission of G. Dietrich, 1963. *General Oceanography*. New York: Interscience Publishers.)

Major and Minor Ions in Seawater of 35‰ Salinity

table 1.3

Ion	Chemical formula	Concentration (‰)	
Chloride	Cl^-	19.3	
Sodium	Na^+	10.6	
Sulfate	SO_4^{2-}	2.7	
Magnesium	Mg^{2+}	1.3	Major
Calcium	Ca^{2+}	0.4	
Potassium	K^+	0.4	
Bicarbonate	HCO_3^-	0.1	
Bromide	Br^-	0.066	
Borate	$B(OH)_4^-$	0.027	
Strontium	Sr^{2+}	0.013	Minor
Fluoride	F^-	0.001	
Silicate	SiO_4^{4-}	0.001	

Plus traces of other naturally occurring elements

than outside (the body fluids are now **hyperosmotic** to the lake water), and the internal water concentration (965‰) is correspondingly less than the concentration of the lake water (1000‰). Water molecules, following their concentration gradient, diffuse across the selectively permeable boundary membranes into the sea cucumber. The movement of water across such a membrane is a special type of diffusion known as **osmosis.** The dissolved ions, now more concentrated within the animal than outside, cannot diffuse out of the sea cucumber because this movement is blocked by the impermeability of the membranes to the ions. The net result is an increase in the amount of water inside the sea cucumber. The additional water creates an internal **osmotic pressure** that is potentially damaging because the animal is incapable of expelling the excess water, and so it swells. Many other marine animals are similarly incapable of countering such os-

motic stresses. As a consequence, these organisms are restricted to regions of the ocean where salinity fluctuations are small.

In contrast to the lack of control that sea cucumbers have over their osmotic situation, bony fishes and some other marine animals and most marine plants possess well-developed osmoregulatory mechanisms. As a result, some of these organisms are free to occupy estuaries and other regions of varying salinities unhindered by osmotic upsets (estuarine organisms are discussed in Chapter 7). Salmon, which spend their early life in freshwater rivers and then grow to maturity at sea, are an example of how some organisms maintain a homeostatic internal medium regardless of external environmental conditions (Fig. 1.20, bottom).

The salt concentration of a salmon's body fluids, like those of most other bony fishes, is midway between the concentrations found in fresh water and in seawater (about 18‰). As such, the body fluids are hyperosmotic to fresh water and **hypoosmotic** to seawater. So these fish never achieve an osmotic balance with their external environment. Instead, they constantly expend energy to maintain a stable internal osmotic condition different from either river or ocean water. In seawater, salmon lose body water by osmosis and are constantly fighting dehydration. To counter these losses, they drink large amounts of seawater that are absorbed through their digestive tracts and into their bloodstreams. The water is retained in the body tissues, and excess salts taken in with the

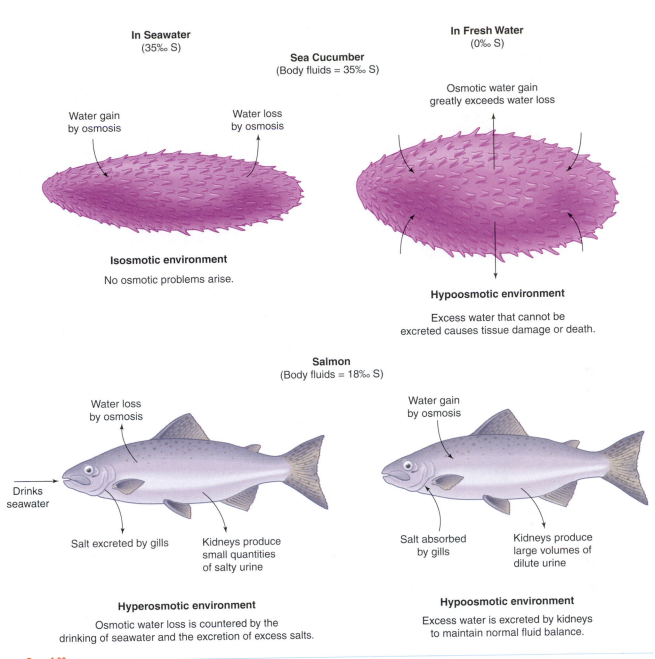

In Seawater
(35‰ S)

Sea Cucumber
(Body fluids = 35‰ S)

In Fresh Water
(0‰ S)

Water gain
by osmosis

Water loss
by osmosis

Osmotic water gain
greatly exceeds water loss

Isosmotic environment

No osmotic problems arise.

Hypoosmotic environment

Excess water that cannot be
excreted causes tissue damage or death.

Salmon
(Body fluids = 18‰ S)

Water loss
by osmosis

Drinks
seawater

Salt excreted by gills

Kidneys produce
small quantities
of salty urine

Water gain
by osmosis

Salt absorbed
by gills

Kidneys produce
large volumes of
dilute urine

Hyperosmotic environment

Osmotic water loss is countered by the
drinking of seawater and the excretion of excess salts.

Hypoosmotic environment

Excess water is excreted by kidneys
to maintain normal fluid balance.

figure 1.20

A comparison of the osmotic conditions of a sea cucumber and a salmon in seawater and in fresh water.

water are actively excreted by special **chloride cells** located in the gills. Because the kidneys of salmon are unable to produce urine with a salt concentration higher than that of their body fluids, their kidneys cannot get rid of the excess salts.

The osmotic challenges of salmon are completely reversed when, as adults, they return from the sea to spawn in freshwater rivers and lakes. Now the problems are excessive osmotic water gain across the gill and digestive membranes and a steady loss of salts to the surrounding water. Here, the salmon drink very little fresh water. To balance the osmotic water gain, the kidneys produce large amounts of dilute urine after effectively recovering most of the salts from that urine. Needed salts are obtained from food and are also actively scavenged from the surrounding water through other specialized cells in the gills. Thus, at a considerable expense of energy, salmon maintain a homeostatic internal fluid environment in either river or ocean water.

Light and Temperature in the Sea

Most marine organisms living in the upper portions of the sea use light energy from the sun for one of two functions, vision or photosynthesis. The amount of energy reaching the sea surface through the atmosphere depends on the presence of dust, clouds, and gases that absorb, reflect, and scatter a portion of the incoming solar radiation (FIG. 1.21). On an average day, about 65% of the sun's radiation arriving at the outer edge of our atmosphere reaches the Earth's surface. The intensity of incoming solar radiation is reduced when the angle of the sun is low, as it is in winter or at high latitudes, and a portion of the light that does make it through the atmosphere is reflected back into space by the sea surface.

Of the broad spectrum of the sun's electromagnetic radiation that makes it to sea level (FIG. 1.22, top), most marine animals can visually detect only a very narrow band near the center of the spectrum. Our eyes visually respond only to the portion labeled **visible** light in Figure 1.22 (violet through red), and most other animals with eyes, whether they see in color or not, respond visually to approximately the same portion of the electromagnetic radiation spectrum.

The band of light energy used by animals for vision broadly overlaps that used in photosynthesis. Photosynthetic organisms must remain in the upper region of the ocean (the **photic zone**) where solar energy sufficient to support photosynthesis will reach them. The depth of the photic zone is determined by how rapidly seawater absorbs light and converts it to heat energy. Dissolved substances, suspended sediments, and even plankton populations diminish the amount of light available for photosynthetic activity and cause the depth of light penetration to differ dramatically between coastal and oceanic water (Fig. 1.21).

As sunlight travels through our atmosphere and into the sea, its color characteristics are altered as seawater rapidly absorbs the violet and the orange-red por-

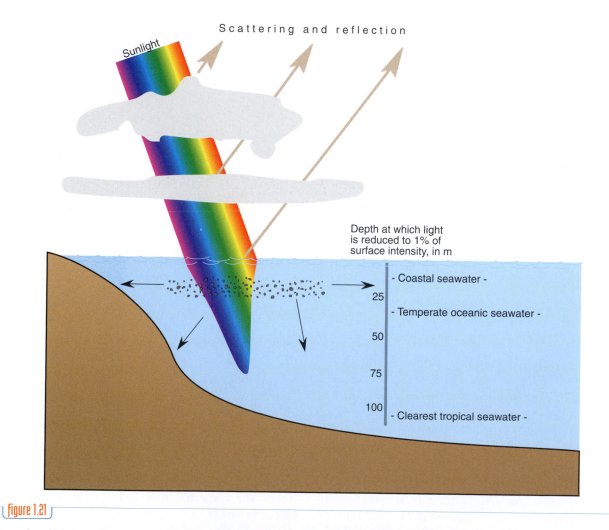

Scattering and reflection

Sunlight

Depth at which light is reduced to 1% of surface intensity, in m

- Coastal seawater -

25

- Temperate oceanic seawater -

50

75

100 - Clearest tropical seawater -

figure 1.21

Fate of sunlight as it enters seawater. The violet and red ends of the visible spectrum are absorbed first.

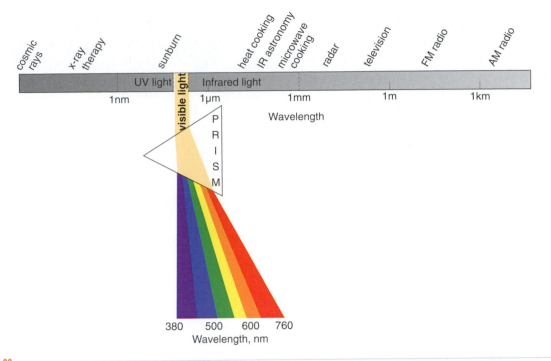

figure 1.22

The electromagnetic radiation spectrum. The small portion known as visible light is passed through a prism to separate the light into its component colors.

tions of the visible spectrum, leaving the green and blue wavelengths to penetrate deeper. Even in the clearest tropical waters, almost all red light is absorbed in the upper 10 m. Clear seawater is most transparent to the blue and green portions of the spectrum (450–550 nm); 10% of the blue light penetrates to depths of 100 m or more. However, even this light is eventually absorbed or scattered (FIG. 1.23). The deeper penetration and eventual back-scattering of blue light account for the characteristic blue color of clear tropical seawater. Coastal waters are commonly more turbid, with a greater load of suspended sediments and dissolved substances derived from land runoff. Here, there is a shift in the relative penetration of light energy, with green light penetrating deepest. In many coastal regions, green light is reduced to 1% of its surface intensity in less than 30 m. Adaptations to these different light regimes by photosynthetic organisms are described in Chapter 3.

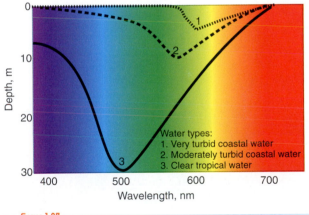

figure 1.23

Penetration of various wavelengths of light in three different water types: (1) very turbid coastal water, (2) moderately turbid coastal water, and (3) clear tropical water. Note the shift to shorter wavelengths (bluer light) in clearer water.

Temperature Effects

When sunlight is absorbed by water molecules, it is converted to heat energy, and the motion of the water molecules increases. **Temperature**, commonly reported in almost all countries of the world as degrees Celsius (°C), is the way we measure and describe that change in molecular motion. Temperature is a universal factor governing the existence and behavior of living organisms. Life processes cease to function above the boiling point of water, when protein structures are irreversibly altered (as when you cook an egg), or at subfreezing temperatures, when the formation of ice crystals damages cellular structures. But between these absolute temperature limits, life flourishes.

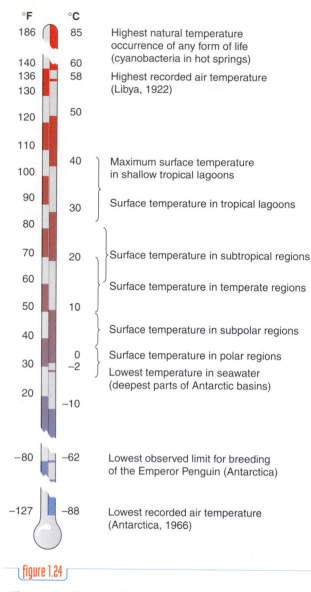

°F	°C	
186	85	Highest natural temperature occurrence of any form of life (cyanobacteria in hot springs)
140	60	
136	58	Highest recorded air temperature (Libya, 1922)
130		
120	50	
110		
100	40	Maximum surface temperature in shallow tropical lagoons
90	30	Surface temperature in tropical lagoons
80		
70	20	Surface temperature in subtropical regions
60		Surface temperature in temperate regions
50	10	
40		Surface temperature in subpolar regions
30	0	Surface temperature in polar regions
	−2	Lowest temperature in seawater (deepest parts of Antarctic basins)
20	−10	
−80	−62	Lowest observed limit for breeding of the Emperor Penguin (Antarctica)
−127	−88	Lowest recorded air temperature (Antarctica, 1966)

figure 1.24

The range of biologically important temperatures at the Earth's surface. The temperature ranges of marine climatic regions in Figure 1.27 are included.

figure 1.25

Emperor penguins are the only penguins to lay eggs and rear their young on ice during the Antarctic winter.

The high heat capacity of water limits marine temperatures to a much narrower range than air temperatures over land (FIG. 1.24). Some marine organisms survive in coastal tropical lagoons at temperatures as high as 40°C. Bacteria associated with deep-sea hydrothermal vents sometimes experience water temperatures above 60°C. Other deep-sea animals spend their lives in water perpetually within one or two degrees of 0°C. Penguins and a few other birds and mammals adapted to extreme cold commonly tolerate air temperatures far below 0°C in polar regions. Emperor penguins even manage to incubate and hatch eggs under these conditions (FIG. 1.25). But these are exceptions; most ma-

rine species typically experience water temperatures between 5 and 30°C.

Individual activity, cell growth, oxygen consumption, and other physiologic functions, collectively termed **metabolism**, proceed at temperature-regulated rates. Most animals lack mechanisms for body temperature regulation. These are **poikilotherms** (often inappropriately described as cold-blooded). These organisms are also referred to as **ectotherms**; their body temperatures vary with, and are largely controlled by, outside environmental temperatures. The terms *poikilotherm* and *ectotherm*, often used interchangeably, refer to distinct aspects of body temperature control. Poikilotherms do not regulate their body temperatures; external conditions govern the body temperatures of ectotherms. In the sea, the temperature-moderating properties of water restrict fluctuations of temperatures experienced by marine ectotherms. Most marine organisms are simultaneously ectothermic and poikilothermic.

For marine ectotherms, water temperature is a principal factor controlling metabolic rates. Marine ectotherms generally have fairly narrow optimal temperature ranges, bracketed on either side by wider and less optimal, but still tolerable, ranges. Within these tolerable temperature limits, the metabolic rate of many poikilotherms is roughly doubled by a 10°C temperature increase. This, however, is only a general rule of thumb; some processes may accelerate sixfold with a 10°C temperature increase, whereas other processes may change very little. The actual effect of water temperature on the feeding rate of a

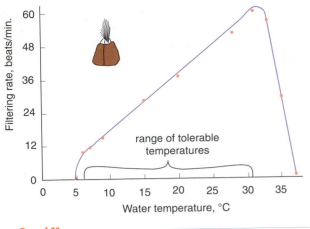

figure 1.26

Filtering rate of an intertidal barnacle as a function of water temperature. Only within the range of tolerable temperatures is the filtering rate proportional to the water temperature. (Adapted from Southward 1964.)

typical marine ectotherm, a submerged barnacle, is shown in FIGURE 1.26.

Only birds and mammals maintain nearly constant body temperatures. They are known as **homeotherms**. Their normal core body temperatures are maintained between 37 and 40°C by the production of heat by internal tissues and organs. Thus, they are also **endotherms**. Endothermic homeotherms are less restricted by environmental temperatures than

are their poikilothermic neighbors. As a result, they range widely to exploit resources over all thermal regimes available in the sea.

A few large tunas, billfishes, and sharks exhibit a thermoregulatory condition intermediate to the two just discussed. These fishes are poikilothermic, so their body temperatures fluctuate with that of the surrounding seawater. Even so, they are unlike most other poikilotherms because they retain some of the heat produced by their swimming muscles. These animals are endothermic, yet they lack the constant body temperatures characteristic of birds and mammals. The special mechanisms used for this heat retention are described in Chapter 10.

The distribution of various forms of marine life is closely associated with geographic differences in seawater temperatures. Surface ocean temperatures are highest near the equator and decrease toward both poles. This temperature gradient establishes several east–west-trending marine climatic zones (FIG. 1.27). The approximate temperature range of each zone is included in Figure 1.24. These marine climatic zones serve as an important framework for organizing the remainder of this book.

Our Planetary Greenhouse

The average temperature of Earth's surface is maintained at its present temperature by a finely tuned global heat engine. About half of the solar energy

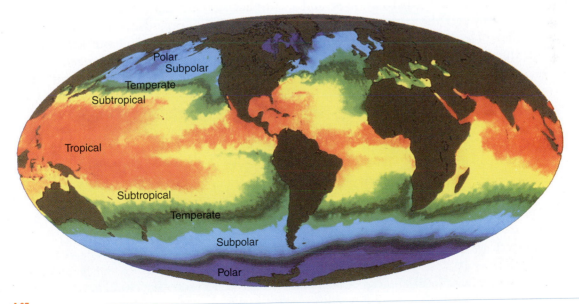

figure 1.27

Earth's sea surface temperatures obtained from 2 weeks of satellite infrared observations July 1984. Temperatures are color coded, with red being warmest and decreasing through oranges, yellows, greens, and blues. The temperature ranges of the labeled marine climatic zones are listed in Figure 1.24 and are shifted slightly northward during the Northern Hemisphere summer.

hitting our upper atmosphere penetrates to Earth's surface, where it is converted to heat energy as it is absorbed by water, vegetation, soil, and human-made structures. If the average temperature of Earth's surface is to remain stable, an equal amount of heat energy must be re-radiated back into space. Heat energy, however, radiates at longer wavelengths than does incoming visible light, and some atmospheric gases are more transparent to visible light than they are to radiated heat. These atmospheric greenhouse gases (especially water vapor, carbon dioxide, methane, and ozone) serve as a natural part of the global heat budget system by trapping heat near the Earth's surface and keeping most of our solar-powered planet well above the freezing temperature of water.

We have, since the beginning of the Industrial Revolution, enhanced the greenhouse effect by substantially increasing the concentrations of natural greenhouse gases such as CO_2 in our atmosphere (FIG. 1.28). Burning of fossil fuels and devegetation of land surfaces (especially burning of tropical rain forests, clear-cutting of temperate forests, and urban development) appear to be the main sources of the excess CO_2. Within the next century, a doubling of preindustrial levels of this gas is a virtual certainty.

What is not yet certain is what effect CO_2 will have on planetary temperatures and ultimately on climate. The most likely eventual response of our climate to increased concentrations of greenhouse gases is a general long-term global warming, causing deserts to expand, continental ice caps to partially melt, and the sea level to rise. However, the magnitude and geographic distribution of such a warming trend cannot be predicted with confidence using current climate prediction models.

One reason for the uncertainty of predictions from the best available climate models is the role of the world ocean in both global heat and carbon budgets. Seawater has a high heat capacity, and the water temperature of deep-ocean basins is substantially lower than surface-water temperatures. One difficulty in understanding the processes governing heat transfer from air to water is the difference in time scales; days or weeks for the atmosphere are years or centuries for the ocean. Any temperature response by the ocean to increased atmospheric temperatures may be too slow to moderate rising surface temperatures. Other complications also exist. As water warms, it expands. This thermal expansion of the world ocean could cause average sea level to rise as much as 1 m for each 1°C increase in temperature and cause significant worldwide changes in the present pattern of human occupation of low-lying coastal plains.

The world ocean is also a carbon dioxide sink; it already contains more than 50 times as much CO_2 as our atmosphere and 20 times as much CO_2 as the Earth's total biosphere. Several features of the ocean govern its capacity to absorb and hold CO_2. First, the rate of vertical circulation and mixing limits the direct exchange of CO_2 between the atmosphere and the ocean. Once the CO_2 is in seawater, its solubility is much greater than that of other atmospheric gases because it reacts with water to form carbonate and bicarbonate ions (this process is discussed later in the chapter). In addition, there exists a biologic "carbon pump" of organisms that absorb carbonate in surface waters and make hard skeletons of it. These skeletons rapidly sink to the deep seafloor, transporting massive amounts of CO_2 with them. These

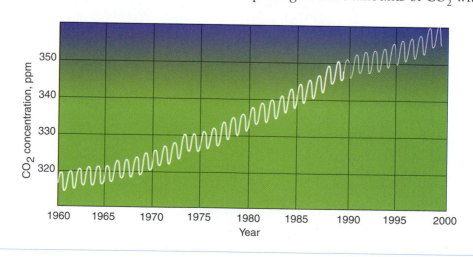

figure 1.28

Pattern of atmospheric CO_2 increase over four decades. The slight annual variations are due to seasonal CO_2 uptake and release by land plants.

features make it exceptionally difficult to predict the behavior of CO_2 exchange between the atmosphere and the world ocean accurately.

Salinity–Temperature–Density Relationships

Seawater density is a function of both temperature and salinity, increasing with either a temperature decrease or a salinity increase. Under typical oceanic conditions, temperature fluctuations exert a greater influence on seawater density because the range of marine temperatures is much greater (−2 to 30°C) than the range of open-ocean salinities. FIGURE 1.29 graphically demonstrates the relationships between the temperature, salinity, and density of water.

Seawater sinks when its density increases. So the densest seawater is naturally found near the sea bottom; however, the physical processes that create this dense water (evaporation, freezing, cooling) occur only at the ocean surface. Consequently, dense water on the sea bottom originally must have sunk from the ocean surface. This sinking process is the only mechanism available to drive circulation of water in the deep portions of ocean basins. An obvious feature in most oceans is a **thermocline**, a subsurface zone of rapid temperature decrease with depth (about 1°C/m). The temperature drop that exists across the thermocline creates a zone of comparable density increase known as a **pycnocline** (FIG. 1.30). The large

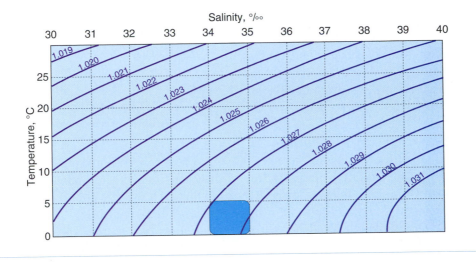

figure 1.29

Temperature–salinity–density diagram for seawater. Blue curved lines represent density values (in g/cm³) resulting from the combined effects of temperature and salinity. Three fourths of the volume of the ocean is remarkably uniform, with salinity, temperature, and density characteristics defined by the dark blue area.

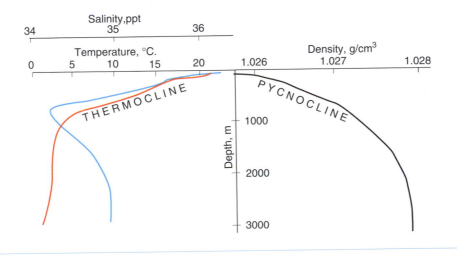

figure 1.30

Variations in water temperature (orange curve) and salinity (blue curve) at a GEOSECS station in the western South Atlantic Ocean. The resulting density profile is shown at right (black curve).

density differences on either side of the thermocline effectively separate the oceans into a two-layered system: a thin well-mixed surface layer above the thermocline overlying a heavier, cold, thick, stable zone below. The thermocline and resulting pycnocline inhibit mixing and the exchange of gases, nutrients, and sometimes even organisms between the two layers. In temperate and polar regions, the thermocline is a seasonal feature. During the winter, the surface water is cooled to the same low temperature as the deeper water. This cooling causes the thermocline to disappear and results in wintertime mixing between the two layers. Warmer marine climates of the tropics and subtropics are more often characterized by well-developed permanent thermoclines and their associated pycnoclines.

Water Pressure

Organisms living below the sea surface constantly experience the pressure created by the weight of the overlying water. At sea level, pressure from the weight of the Earth's envelope of air is about 1 kg/cm², or 15 lb/in², or 1 **atmosphere (atm)**. Pressure in the sea increases another 1 atm for every 10-m increase in depth to more than 1000 atm in the deepest trenches. Most marine organisms can tolerate the pressure changes that accompany moderate changes in depth, and some even thrive in the constant high-pressure environment of the deep sea. However, fishes with swim bladders or whales with lungs that collapse and expand with depth (and pressure) changes have evolved some sophisticated solutions to the problems associated with large and rapid pressure changes. These adaptations are described in Chapter 6.

Dissolved Gases and Acid-Base Buffering

The solubility of gases in seawater is influenced by temperature, with greater solubility occurring at lower temperatures. Nitrogen, carbon dioxide, and oxygen are the most abundant gases dissolved in seawater. Although molecular nitrogen (N_2) accounts for 78% of our atmosphere, it is comparatively nonreactive and therefore is not used in the basic life processes of most organisms. (Notable exceptions are some N_2-fixing bacteria and the occasional careless scuba diver who dives too deep for too long.) Carbon dioxide and oxygen, on the other hand, are metabolically very active. Carbon dioxide and water are used in photosynthesis to produce oxygen and high-energy organic compounds such as carbohydrates and fats. Respiration reverses the results of the photosynthetic process by releasing the usable energy incorporated

in the organic components of an organism's food. In contrast to photosynthesis, oxygen is used in respiration, and carbon dioxide is given off.

Carbon dioxide is abundant in most regions of the sea, and concentrations too low to support plant growth seldom occur. Seawater has an unusually large capacity to absorb CO_2 because most dissolved CO_2 does not remain as a gas. Rather, much of the CO_2 combines with water to produce a weak acid, carbonic acid (H_2CO_3). Typically, carbonic acid dissociates to form a hydrogen ion (H^+) and a bicarbonate ion (HCO_3^-) or two H^+ ions and a carbonate ion (CO_3^{2-}). These reactions can be summarized in the following chemical equations:

$$CO_2 + H_2O \leftrightarrow H_2CO_3 \qquad (1)$$
carbon dioxide / water / carbonic acid

$$H_2CO_3 \leftrightarrow H^+ + HCO_3^- \qquad (2)$$
carbonic acid / hydrogen ion / bicarbonate ion

$$HCO_3^- \leftrightarrow H^+ + CO_3^{2-} \qquad (3)$$
bicarbonate ion / hydrogen ion / carbonate ion

The arrows pointing in both directions indicate that each reaction is reversible, either producing or removing H^+ ions. The abundance of hydrogen ions in water solutions controls the acidity or alkalinity of that solution and is measured on a scale of 0 to 14 pH units (FIG. 1.31). The pH units are a measure of the hydrogen ion concentration. Water with a low pH is very acidic because it has a high H^+ ion concentration. A pH of 14 is very basic (or alkaline) and denotes low H^+ ion concentrations. Neutral pH (the pH of pure water) is 7 on the pH scale. The carbonic acid–bicarbonate–carbonate system in seawater functions to **buffer** or to limit changes of seawater pH. If excess hydrogen ions are present, the reactions described above proceed to the left and the excess hydrogen ions are removed from solution. Otherwise, the solution would become more acidic. If too few hydrogen ions are present, more are made available by the conversion of carbonic acid to bicarbonate and bicarbonate to carbonate. In open-ocean conditions, this buffering system is very effective, limiting ocean water pH fluctuations to a narrow range between 7.5 and 8.4. This buffering system also functions as a crucial component of our planet's ability to accommodate the increasing concentrations of atmospheric CO_2 resulting from our industrial and cultural practices on land.

Oxygen in the form of O_2 is necessary for the survival of most organisms. (The major exceptions

	pH	$[H^+]$	$[OH^-]$	
Acidic Solutions	1	10^{-1}	10^{-13}	Hydrochloric acid
	2	10^{-2}	10^{-12}	Lime juice
	3	10^{-3}	10^{-11}	Acetic acid
	4	10^{-4}	10^{-10}	Tomato juice
	5	10^{-5}	10^{-9}	Black coffee
	6	10^{-6}	10^{-8}	Milk
Neutral	7	10^{-7}	10^{-7}	Pure water
	8	10^{-8}	10^{-6}	Seawater
	9	10^{-9}	10^{-5}	Borax solution
Basic Solutions	10	10^{-10}	10^{-4}	
	11	10^{-11}	10^{-3}	Milk of magnesia
	12	10^{-12}	10^{-2}	Household ammonia
	13	10^{-13}	10^{-1}	Lye
	14	10^{-14}	10^{-0}	Sodium hydroxide

figure 1.31

The pH scale, showing the concentration of H^+ ions at each pH value. Note that the concentration scale is exponential.

are some anaerobic microorganisms.) Water, however, is not a good solvent for O_2. The concentration of O_2 in seawater generally remains between 0 and 8 parts per million, not much for active oxygen-hungry animals. Oxygen is used by organisms in all areas of the marine environment, including the deepest trenches. However, the transfer of oxygen from the atmosphere to seawater and the production of excess oxygen by photosynthetic marine organisms are the only global-scale processes available to introduce oxygen into seawater. Both of these processes occur only at or near the surface of the ocean. Oxygen consumed near the bottom can only be replaced by oxygen from the surface. If replenishment is not rapid enough, available oxygen supplies will be reduced or removed completely. Oxygen replenishment occurs by very slow diffusion processes from the oxygen-rich surface layers downward and also by density-driven sinking that carries oxygen-enriched waters to deep-ocean basins. At depths of about 1000 m, animal respiration and bacterial decomposition use O_2 as fast as it is replaced, creating an **oxygen minimum zone**. FIGURE 1.32 illustrates a typical vertical profile of dissolved oxygen concentration from the surface to the bottom of the sea.

Dissolved Nutrients

Nitrate (NO_3^-) and phosphate (PO_4^{3-}) dissolved in seawater are the critical fertilizers of the sea. Nitrate is the most common reactive form of nitrogen in sea-

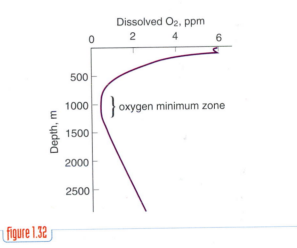

figure 1.32

Vertical distribution of dissolved O_2 in the North Pacific (150° W, 47° N) during winter. (Data from Barkley 1968.)

water. These and smaller amounts of other nutrients are used by photosynthetic organisms living in the near-surface waters. These same nutrients eventually are excreted back into the water at all depths as waste products of the organisms that had consumed and digested the photosynthetic material. This sinking of once-living material eventually removes nutrients from near-surface waters and increases their concentrations in deeper waters. (More details concerning these patterns of nutrient distribution appear in Chapters 3 and 4.)

The vertical distribution of dissolved nutrients is usually opposite that of dissolved oxygen. This

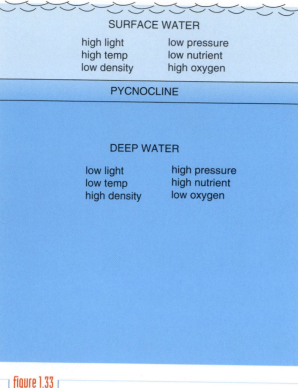

figure 1.33

Contrasting features of shallow and deep ocean water resulting in a two-layer system separated by a pycnocline.

SURFACE WATER

high light   low pressure
high temp   low nutrient
low density   high oxygen

PYCNOCLINE

DEEP WATER

low light   high pressure
low temp   high nutrient
high density   low oxygen

opposing pattern of vertical oxygen and nutrient distribution reflect the contrasting biologic processes that influence their concentrations in seawater. Oxygen is normally produced by near-surface photosynthesizers and consumed by animals and bacteria at all depths, whereas nutrients are consumed by photosynthesizers near the surface and are excreted by organisms at all depths. These contrasting vertical patterns of nutrient and oxygen abundance reinforce the temperature and density-based stratification of much of the ocean into a two-layer system (FIG. 1.33), a warm, low-density, oxygen-rich, nutrient poor surface layer of ocean water a few tens of meters deep overlying a much thicker, cold, high-density, oxygen-poor, nutrient-rich layer below.

1.4 The Ocean in Motion

Ocean water is constantly in motion, moving and mixing nutrients, oxygen, and heat. Wave action, tides, currents, and density-driven vertical water movements enhance mixing and reduce variations in salinity and temperature. Oceanic circulation processes also serve to disperse swimming and floating organisms and their eggs, spores, and larvae. Toxic body wastes are carried away, and food, nutrients, and essential elements are replenished.

Wind Waves

The curvature of the sphere of the Earth causes the sun to warm latitudes of the Earth's surface and overlying air near the equator more than at the poles. This differential warming of the atmosphere by the sun drives patterns of winds that blow across the ocean surface, creating **waves** and **surface currents**. Surface ocean waves are periodic vertical disturbances of the sea surface. They typically travel in a repeating series of alternating wave crests and troughs. The size and energy of waves are dependent on the wind's velocity, duration, and **fetch** (the distance over which the wind blows in contact with the sea surface). Ocean waves range in height from very small capillary waves a few millimeters high to monster storm waves surpassing 30 m in height (FIG. 1.34). Waves are commonly characterized by their height, wavelength (FIG. 1.35), and **period** (the time required for two successive wave crests to pass a fixed point). Regardless of their size, these general features of waves apply to all ocean waves.

Once created by surface winds, ocean waves travel away from their area of formation. Only the wave shape advances, however, transmitting the energy forward. The water particles themselves do not advance in the direction of the wave. Instead, their paths approximate vertical circles with little net forward motion (Fig. 1.35). Waves provide an important mechanism to mix the near-surface layer of the sea. The depth to which waves produce noticeable motion is about one half the wavelength. As wavelengths seldom exceed 100 m in any ocean, the effective depth of mixing by wind-driven waves is generally no greater than 50 m.

Waves entering shallow water behave differently than open-ocean waves. When the water depth is less than one half the wavelength, bottom friction begins to slow the forward speed of the waves. This slowing causes the waves to become higher and steeper. At the point where the wave height becomes greater than about one seventh of the wavelength, the wave top becomes unstable as it outruns its base. It then pitches forward and breaks. The energy released on shorelines (and on the organisms living there) by breaking waves can be enormous and is a major force in shaping the physical and biologic character of most coastlines.

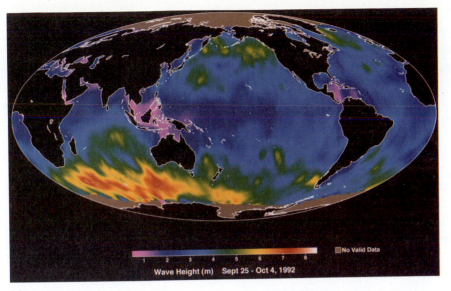

(a)

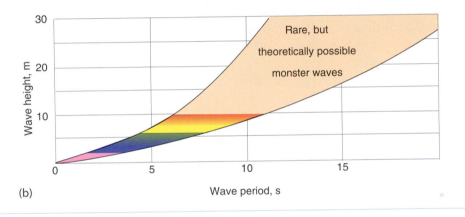

(b)

figure 1.34

(a) Range of ocean surface wave heights, determined over a 10-day period by the TOPEX/Poseidon satellite. (b) Wave periods and their corresponding range of heights.

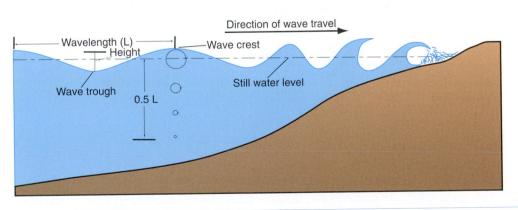

figure 1.35

Wave form and pattern of water motion in a deep-water wave as it moves to the right in to shore. Circles indicate orbits of water particles diminishing with depth. There is little water motion below a depth of one half of the wavelength.

Surface Currents

Ocean surface currents occur when winds blow over the ocean with a constancy of direction and velocity for a sufficient period of time to transfer momentum to the water through friction. Unlike ocean waves, surface currents do represent the horizontal movement of water molecules. The momentum imparted to the sea by these winds drives regular patterns of broad, slow, relatively shallow ocean surface currents. Some currents transport more than 100 times the volume of water carried by all the Earth's rivers combined. Currents of such magnitude greatly affect the distribution of marine organisms and the rate of heat transport from tropical to polar regions.

These currents are driven by stable patterns of winds at the ocean surface. Three major wind belts occur in the Northern Hemisphere. The **trade winds**, near 15° N latitude, blow from northeast to southwest. The **westerlies**, in the middle latitudes, blow primarily from the west-southwest. And the **polar easterlies**, at very high latitudes, blow from east to west. Each of these wind belts has its mirror-image counterpart in the Southern Hemisphere. As the surface layer of water is moved horizontally by these broad belts of surface winds, momentum is transferred first to the sea surface and then downward. The speed of the deeper water steadily diminishes as momentum is lost to overcome the viscosity of the water.

Eventually, at depths generally less than 200 m, the speed of wind-driven currents becomes negligible.

The surface water moved by the wind does not flow parallel to the wind direction but experiences an appreciable deflection, known as the **Coriolis effect**. The Coriolis effect influences moving air and water masses by causing a deflection to the right in the Northern Hemisphere and to the left in the Southern Hemisphere. As successively deeper water layers are set into motion by the water above them, they undergo a further Coriolis deflection away from the direction of the water just above to produce a spiral of current directions from the surface downward (**FIG. 1.36**). The magnitude of the Coriolis deflection of wind-driven currents varies from about 15 degrees in shallow coastal regions to nearly 45 degrees in the open ocean. The net Coriolis deflection from the wind headings creates a pattern of wind-forced ocean-surface currents that flow primarily in east to west or west to east directions.

Except for the region just north of Antarctica, the continuous flow of these wind-driven east–west currents is obstructed by continents. This causes water transported by currents from one side of the ocean to pile up against continental margins on the other side. The surface of the equatorial Pacific Ocean, for example, is about 2 m higher on the west side than it is on the east side. The opposite is true in the middle latitudes of both hemispheres, where the east

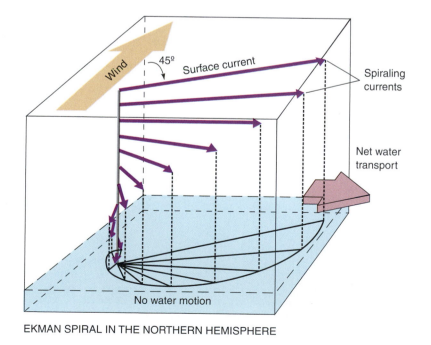

EKMAN SPIRAL IN THE NORTHERN HEMISPHERE

figure 1.36

A spiral of current directions, indicating greater deflection to the right (in the Northern Hemisphere), which increases with depth due to the Coriolis effect. Arrow length indicates relative current speed.

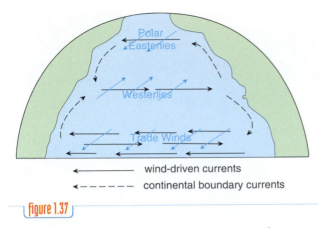

figure 1.37

Generalized surface-current flow in the North Pacific Ocean. Blue arrows indicate general directions of ocean-surface winds.

been removed. Both these current patterns exist, but they are particularly clear in the North Pacific Ocean (FIG. 1.37). An east-flowing Equatorial Countercurrent divides the west-flowing North Pacific Equatorial Current. The north–south-flowing continental boundary currents merge into the east–west currents to produce large circular current patterns, or **gyres**. Similar current patterns are found in the other major ocean basins (FIG. 1.38).

Maps similar to Figure 1.38 are useful for describing long-term average patterns of surface ocean circulation. However, they tend to hide the complexity and even the beauty that exist in these currents at any moment in time. Current maps are analogous to the blurred images taken of a night freeway scene when the camera shutter is held open for hours. The pattern of traffic flow is obvious, yet the details of vehicles' slowing, accelerating, and changing lanes are completely lost. The continued development of satellite monitoring of ocean surface phenomena provides an improved approach for visualizing and understanding global-scale surface current patterns. FIGURE 1.39 is a satellite image of a portion

side is higher. Eventually, the water must flow away from these areas of accumulation. Either it flows directly back against the established current, producing a **countercurrent**, or it flows as a **continental boundary current** parallel to a continental margin from areas of accumulation to areas where water has

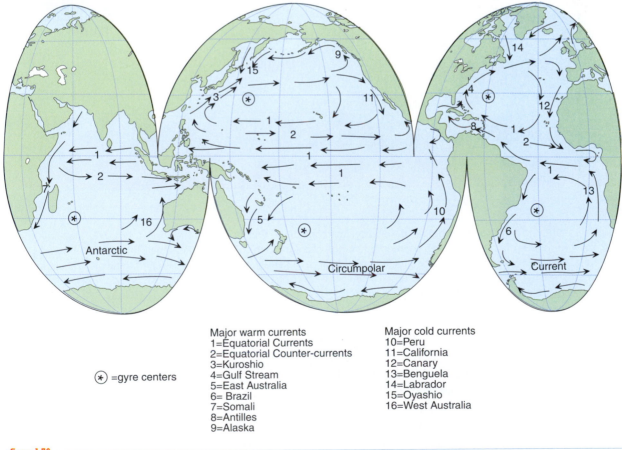

Major warm currents
1=Equatorial Currents
2=Equatorial Counter-currents
3=Kuroshio
4=Gulf Stream
5=East Australia
6= Brazil
7=Somali
8=Antilles
9=Alaska

Major cold currents
10=Peru
11=California
12=Canary
13=Benguela
14=Labrador
15=Oyashio
16=West Australia

(✳) =gyre centers

figure 1.38

The major surface currents of the world ocean. (Adapted from Pickard and Emory, 1982.)

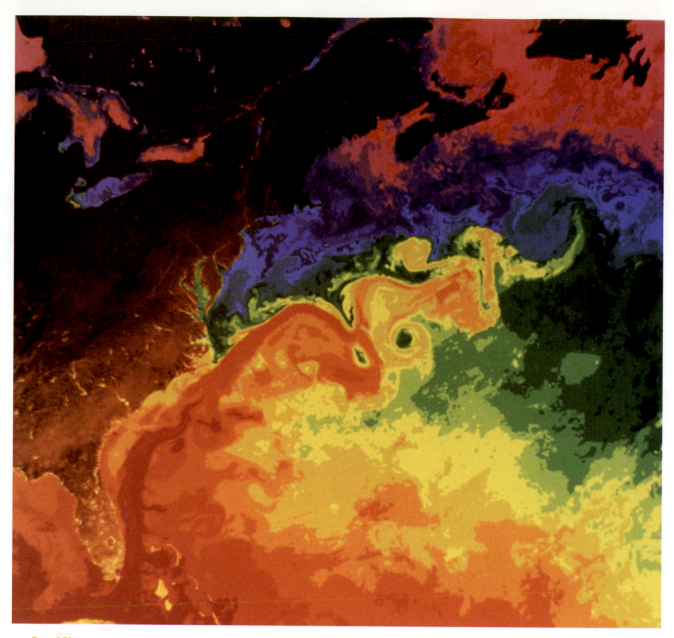

figure 1.39

National Oceanic and Atmospheric Administration satellite image of the Gulf Stream off the US East Coast during early April 1984. Water surface temperatures are represented by a range of colors, from cold (−2 to 9°C, violet) through blue, green, yellow, orange, and red (very warm, 26–28° C). The Gulf Stream appears as a narrow band of warm water off the Florida coast, and then cools gradually as it moves northeast and releases heat to the atmosphere. North of Cape Hatteras, the current begins to meander and form isolated rings typical of strong surface ocean currents.

of the North Atlantic Ocean, including the Gulf Stream. This image emphasizes ocean-surface temperature differences and reveals remarkable meanders, constrictions, and nearly detached rings of Gulf Stream water as the current flows north and east along the path shown in Figure 1.38.

Ocean Tides

Tides are ocean surface phenomena familiar to anyone who has spent time on a seashore. They are actually very-long-period waves that are usually imperceptible in the open ocean and only become noticeable near the shoreline, where they can be observed as a periodic rise and fall of the sea surface. The maximum elevation of the tide, known as **high tide**, is followed by a fall in sea level to a minimum elevation, or **low tide**. On most coastlines, two high tides and two low tides occur each day. The vertical difference between consecutive high tides and low tides is the **tidal range**, which varies from a just few centimeters in the Mediterranean Sea to more than 15 m in the long narrow Bay of Fundy between Nova Scotia and New Brunswick. The global tidal range averages about 2 m.

In 1687, in his *Principia Mathematica,* Sir Isaac Newton explained ocean tides as the consequence of the gravitational attraction of the moon and sun on the oceans of the Earth. According to Newton's law of universal gravitation, our moon, because of its closeness to the Earth, exerts about twice as much tide-generating force as does the more distant but much larger sun. The constantly changing positions of the Earth relative to the moon and sun nicely account for the timing of ocean tides, yet Newton's equilibrium model of ocean tides is seriously deficient in its ability to explain real tidal patterns in real oceans. Those of you interested in Newton's model and other models of ocean tides should consult a current oceanography text. For our purposes, a description of the results of tide-producing forces, rather than their causes, will suffice.

The moon completes one orbit around the Earth each lunar month (27.5 days). Hypothetically, if the Earth were completely covered with water, two bulges of water, or lunar tides, would occur: one on the side of the Earth facing the moon and the other on the opposite side of the globe (FIG. 1.40). As the Earth

makes a complete rotation every 24 hours, a point on the Earth's surface (indicated by the marker in Fig. 1.40) would first experience a high tide (a), then a low tide (b), another high tide (c), another low tide (d), and finally another high tide (e). During that rotation, however, the moon advances in its own orbit so that an additional 50 minutes of the Earth's rotation is required to bring our reference point directly in line with the moon again. Thus, the reference point experiences only two equal high and two equal low tides every 24 hours and 50 minutes (a **lunar day**).

In a similar manner, the sun–Earth system also generates tide-producing forces that yield a solar tide about one half as large as the lunar tide. The solar tide is expressed as a variation on the basic lunar tidal pattern, not as a separate set of tides. When the sun, the moon, and the Earth are in approximate alignment (at the time of the new moon and full moon, FIG. 1.41), the solar tide has an additive effect on the lunar tide, creating several days of extra-high high tides and very-low low tides known as **spring tides**. One week later, when the sun and moon are at right angles to each other relative to the Earth, the solar tide partially cancels the effects of the lunar tide to produce moderate tides known as **neap tides**. During each lunar month, two sets of spring tides and two sets of neap tides occur.

So far, only the effects of tide-producing forces in a not very realistic ocean covering a hypothetical planet without continents have been considered. What happens when continental landmasses are taken into consideration? The continents block the westward passage of the tidal bulges as the Earth rotates

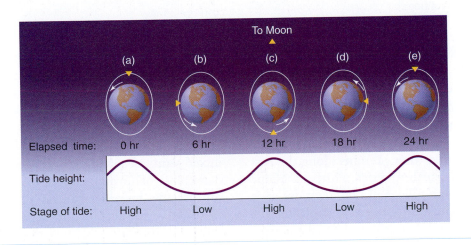

figure 1.40

Each day as the Earth rotates, a point on its surface (indicated by the marker) experiences high tides when under tidal bulges and low tides when at right angles to tidal bulges.

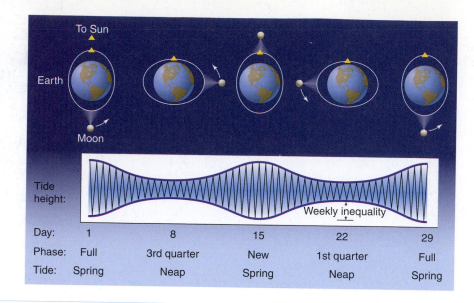

figure 1.41

Weekly tidal variations caused by changes in the relative positions of the Earth, moon, and sun.

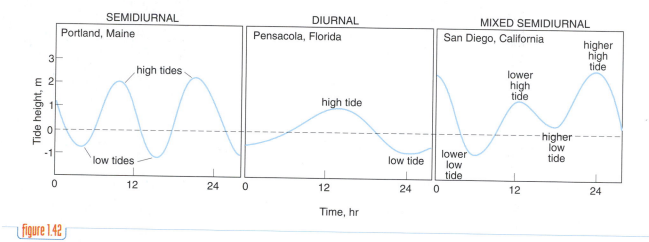

figure 1.42

Three common types of tides.

under them. Unable to move freely around the globe, these tidal impulses establish complex patterns within each ocean basin that may differ markedly from the tidal patterns of adjacent ocean basins or even other regions of the same ocean basin.

FIGURE 1.42 shows some regional variations in the daily tidal configuration at three stations along the east and west coasts of North America. Portland, Maine experiences two high tides and two low tides each lunar day. The two high tides are quite similar to each other, as are the two low tides. Such tidal patterns, referred to as **semidiurnal** (semidaily) tides, are characteristic of much of the East Coast of the United States. The tidal pattern at Pensacola, Florida, on the Gulf Coast, consists of one high tide and one low tide

each lunar day. This is a **diurnal**, or daily, tide. Different yet is the daily tidal pattern at San Diego, California. There, two high tides and two low tides occur each day, but successive high tides are quite different from each other. This type of tidal pattern, characteristic of the West Coast of North America, is a **mixed semidiurnal** tide. FIGURE 1.43 outlines the geographic occurrence of diurnal, semidiurnal, and mixed semidiurnal tides for coastal areas.

Tidal conditions for any day on a selected coastline can be predicted because the periodic nature of tides is easily observed and recorded. For the most part, prediction of the timing and amplitude of future tides is based on the astronomical positions of the sun and moon relative to the Earth and on his-

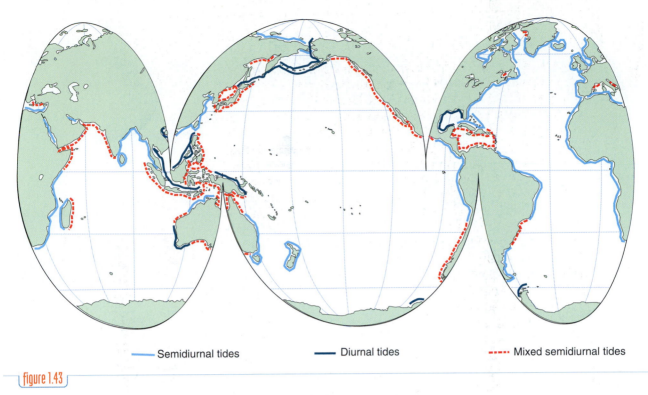

—— Semidiurnal tides   —— Diurnal tides   ---- Mixed semidiurnal tides

figure 1.43

The geographic occurrence of the three types of tides described in Figure 1.42.

torical observations of actual tidal occurrences at tide-recording stations along coastlines and in harbors around the world. The National Ocean Survey of the US Department of Commerce uses information from these records to compile and publish annual "Tide Tables of High and Low Water Predictions" for principal ports along most coastlines of the world.

Vertical Water Movements

Vertical movements of ocean water are produced by upwelling and sinking processes. These processes tend to break down the vertical stratification established by the pycnocline (Fig. 1.33). Localized areas of upwelling are created by several oceanic processes that bring deeper nutrient-rich waters to the surface. These processes are considered in Chapter 3. The continuous availability of deep-water nutrients that can be used by photosynthetic organisms accounts for the high productivity so characteristic of upwelling regions. Several of the world's most important fisheries are based in upwelling areas.

The physical processes that increase seawater density are strictly surface features that affect water temperature or salinity, and the resulting vertical circulation is referred to as **thermohaline** circulation. Seawater sinking from the surface is usually highly oxygenated because it has been in contact with the

atmosphere, so it transports dissolved oxygen to deep areas of the ocean basins that would otherwise be **anoxic** (lacking oxygen). The chief areas of sinking are located in the colder latitudes, where sea surface temperatures are low. After sinking, this dense water continues to spread and flow horizontally as very slow and ill-defined deep ocean currents. Time spans of a few hundred to a thousand years are required for water that sinks in the North Atlantic to reach the surface again in the Southern Hemisphere. FIGURE 1.44 outlines the general patterns of large-scale deep-ocean thermohaline circulation.

On a somewhat smaller scale, the Mediterranean and Black Seas provide two contrasting examples of density-driven thermohaline circulation. In the arid climate of the Mediterranean Sea, particularly at its eastern end, evaporation from the sea surface greatly exceeds precipitation and runoff. The resulting high-salinity and high-density water sinks and fills the deeper parts of the Mediterranean basin. The sinking of surface water provides substantial mixing and O_2 replenishment for the deep water of the Mediterranean and is similar to the deep circulation of the open ocean. Part of this deep dense water eventually flows out of the Mediterranean over a shallow sill at Gibraltar and down into the Atlantic Ocean. To compensate for the outflow and losses due to evaporation, nearly two million cubic meters of Atlantic surface water flow into

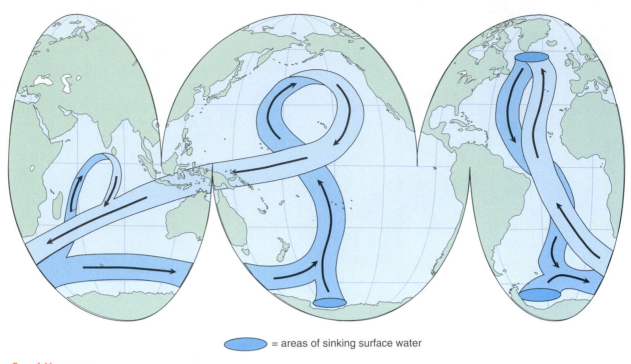

= areas of sinking surface water

figure 1.44

The general pattern of deep-ocean circulation in the major ocean basins. Light blue represents flow at intermediate depths, whereas darker blue represents deeper flow. Sinking of cold surface water occurs at high latitudes indicated by ovals. (Adapted from Broecker *et al*. 1985.)

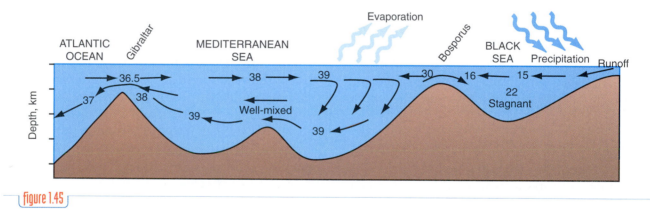

figure 1.45

A comparison of the deep-ocean circulation patterns of two marginal seas, the Mediterranean Sea and the Black Sea. The numbers represent salinity in ‰.

the Mediterranean each second. The currents at Gibraltar can be compared with two large rivers flowing in opposite directions, one above the other (FIG. 1.45).

Like the Mediterranean, the Black Sea is isolated by a shallow sill (at the Bosporus). In contrast to the Mediterranean Sea, however, the Black Sea is characterized by a large excess of precipitation and river runoff. In this sense, the circulation of the Black Sea resembles that of some semienclosed fjords of Scandinavia and the west coast of Canada. The dilute surface waters of the Black Sea form a shallow low-density layer that does not mix with the higher salinity denser water below. Instead, it flows into the Mediterranean Sea through the Bosporus (Fig. 1.45). Low-salinity oxygen-rich surface water does not sink, so the more common oxygen-dependent forms of marine life are restricted to the uppermost layer. Below 150 m, the Black Sea is stagnant and anoxic. Yet these anoxic deep waters of the Black Sea (more than 80% of its volume) are by no means lifeless. The rain of organic material from above accumulates and provides abundant nourishment for several types of anaerobic bacteria.

1.5 Classification of the Marine Environment

The large size and enormous complexity of the marine environment make it a difficult system to classify. Many systems of classification have been proposed, each reflecting the interest and bias of the classifier. The system presented here is a slightly modified version of a widely accepted scheme proposed by Hedgpeth over a half-century ago. The terms used in FIGURE 1.46 designate particular zones of the marine environment; these terms are easily confused with the names of groups of organisms that normally inhabit these zones (These will be introduced in the next chapter.) The boundaries of these zones are defined on the basis of physical characteristics such as water temperature, water depth, and available light.

Working downward from the ocean surface, the limits of **intertidal** zones are defined by tidal fluctuations of sea level along the shoreline. These zones and their inhabitants are examined in detail in Chapter 8. The splash, intertidal, and inner shelf zones occur in the photic (lighted) zone, where the light intensity is great enough to accommodate photosynthesis. The depth of the photic zone depends on conditions that affect light penetration in water, extending much deeper (up to 200 m) in clear tropical waters than in murky coastal waters of temperate or polar seas (sometimes less than 5 m). The rest of the ocean volume is the perpetually dark **aphotic** (unlighted) **zone**, where the absence of sunlight prohibits photosynthesis.

The **benthic division** refers to the environment of the sea bottom. The inner shelf includes the seafloor from the low-tide line to the bottom of the photic zone. Beyond that, to the edge of the continental shelf, is the outer shelf. The bathyal zone is approximately equivalent to the continental slope areas. The abyssal zone refers to abyssal plains and other ocean-bottom areas between 3000 and 6000 m in depth. The upper boundary of this zone is sometimes defined as the region where the water temperature never exceeds 4°C. The hadal zone is that part of the ocean bottom below 6000 m, primarily the trench areas.

The **pelagic division** includes the entire water mass of the ocean. For our purposes, it will be sufficient to separate the pelagic region into two provinces: the **neritic province**, which includes the water over the continental shelves, and the **oceanic province**, the water of the deep ocean basins. Each of these subdivisions of the ocean environment is inhabited by characteristic assemblages of marine organisms. It is these organisms and their interactions with their immediate surroundings that are the subject of the remaining chapters of this book.

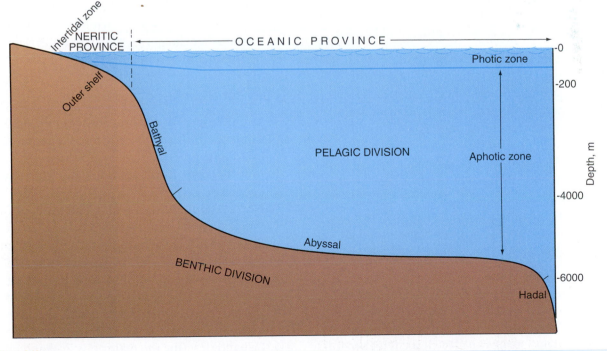

figure 1.46

A system for classifying the marine environment. (Adapted from Hedgpeth 1957.)

Summary Points

The Ocean as a Habitat

- True understanding of the sea requires a shift in perspective, away from human-based concepts of time and distance, toward a "geological perspective" that allows for our five-billion-year history and vast distances greater than several thousand kilometers.

- Terrestrial habitats often are discussed in terms of acreage, square miles, or some other two-dimensional measurement. The sea must be viewed as a three-dimensional environment, with depth distributions of organisms and processes being as biologically significant as descriptions based on longitude and latitude.

The Changing Marine Environment

- Earth's surface is dominated by our world ocean, a single interconnected volume of seawater partially separated by land masses into a number of ocean basins and coastal seas. The sizes and shapes of the ocean basins, as well as the composition of seawater itself, are continually changing. Billions of years of geological activity, such as volcanism, seafloor spreading, and plate tectonics, as well as persistent erosion and recurrent episodes of glaciation all contribute to this constant metamorphosis.

Properties of Seawater

- The unique characteristics of pure water determine most of the basic properties of salt water. The differential affinity of oxygen and hydrogen for electrons creates an electrical charge separation within water molecules that forms hydrogen bonds between adjacent water molecules. Hydrogen bonding, in turn, affects water's basic properties, including viscosity, surface tension, heat capacity, solvent capability, density-temperature relationships, and its stability as a liquid.

- Salt water contains a great variety of dissolved salts, gases, and other inorganic and organic substances. These dissolved molecules and compounds affect many characteristics of seawater, including its density, osmotic properties, buffering capacity, and other biologically significant features.

The Ocean in Motion

- The sea is constantly moving, both horizontally and vertically. Winds, waves, tides, currents, sinking water masses, and upwelling all contribute to the remarkable homogeneity of the world ocean.

Classification of the Marine Environment

- Energy from the sun warms the sea's surface and creates winds that result in a two-layered world ocean, with a shallow, well-mixed, warm, sunlit layer overlaying a much deeper, cold, dark, high-pressure layer of slowly moving water below.

- The three-dimensional marine environment can be separated into two broad divisions, the benthic realm of the sea floor and the pelagic water column. These in turn may be subdivided into smaller categories based on water depth, light availability, and ambient temperature.

Topics for Discussion and Review

1. Volcanic activity, earthquakes, and the distribution of continents on Earth today are all due to a single process. State what it is and describe how it is responsible for the above phenomena.

2. Label a diagrammatic cross-section of the North Atlantic Ocean with all benthic features (such as trenches and shelf breaks) presented in this chapter. Then discuss whether or not this is the best way to categorize the sea floor.

3. What is sonar and why is it perhaps the best way to determine the dimensions of ocean basins?

4. List three properties of pure water that are highest of all common natural substances.

5. List and describe the major physical and chemical features of seawater that change markedly from the sea surface downward. How do these

same features change along the sea surface as one proceeds from the equator north or south to either pole?

6. Compare and contrast the osmotic changes that a Portuguese man-of-war will experience while being blown from the open sea into the mouth of a large river.

7. Why are the terms "cold-blooded" and "warm-blooded" often inappropriate when used to describe the thermal strategies of marine animals?

8. What is the Greenhouse Effect and why is it so difficult to predict its future influence on our climate?

9. Describe the buffering mechanism at work in the sea, then compare it to the buffering of your own blood.

10. Define spring and neap tides, and then explain why both occur twice each month.

Suggestions for Further Reading

Bonati, E. 1987. The rifting of continents. *Scientific American* 256:97–103.

Broad, W. J. 1997. *The Universe Below: Discovering the Secrets of the Deep Sea.* Simon & Schuster, New York.

Broecker, W. S. 1983. The ocean. *Scientific American* 249:146–160.

Bryan, K. 1978. The ocean heat balance. *Oceanus* 21:18–26.

Cloud, P. 1989. *Oasis in space: Earth history from the beginning.* W.W. Norton, New York.

Duxbury, A. C., and A. B. Duxbury. 1991. *An Introduction to he World's Oceans.* Wm. C. Brown Publishing, Iowa.

Edmond, J. M. 1982. Ocean hot springs. *Oceanus* 25:22–27.

Ellis, R. 2000. *Encyclopedia of the Sea.* Knopf, New York.

Hogg, N. 1992. The Gulf Stream and its recirculation. *Oceanus* 35:28–37.

Jenkyns, H. C. 1994. Early history of the oceans. *Oceanus* 36:49–52.

Kunzig, R. 1999. *The Restless Sea: Exploring the World Beneath the Waves.* W. W. Norton, New York.

McDonald, J. E. 1970. The Coriolis effect. *Scientific American* 222:72–76.

McDonald, K. C., and P. J. Fox. 1990. The mid-ocean ridge. *Scientific American* 262:72–79.

Melville, W. K., and P. Matusov. 2002. Distribution of breaking waves at the ocean surface. *Nature* 417:58–63.

Monahan, D., and M. Leier. 2001. *World Atlas of the Oceans: More than 200 Maps and Charts of the Ocean Floor.* Firefly Books, Spain.

National Geographic. 2001. *Atlas of the Ocean: The Deep Frontier.* Washington, DC.

Ross, D. A. 1988. *Introduction to Oceanography,* 4th edition. Princeton Hall, New Jersey.

Vink, G. E., W. Morgan, and P. Vogt. 1985. The earth's hot spots. *Scientific American* 252:50–57.

Webster, P. J. 1981. Monsoons. *Scientific American* 244:108–118.

Weller, R. A., and D. M. Farmer. 1992. Dynamics of the ocean's mixed layer. *Oceanus* 35:46–55.

Whitworth, T. III. 1988. The Antarctic circumpolar current. *Oceanus* 31:53–58.

CHAPTER 2

A rock tide pool on the Oregon coast.

Patterns
of Associations

Living organisms require space in which to live, material and
energy from their surroundings, and an uninterrupted ancestral lineage to provide the genetic heritage that defines
what they are and how they function. To satisfy these requirements, living organisms organize themselves into complex patterns of associations that are often difficult to consider
in their entirety. To cope with this complexity, we tend to divide these complex systems into smaller more manageable
subunits and then organize these subunits by relating them
to the whole system on the basis of certain characteristics. The
classification of the marine environment (see Fig. 1.46) is a
simple example of this approach for a nonliving system.

To be useful, any classification scheme must present the
information in a generally accepted manner. Doing so requires an orderly framework to classify the available information so that it becomes more meaningful or useful and,
most importantly, more understandable. In this chapter, three
different broadly overlapping systems that we commonly use
to describe the manner by which marine organisms associate and interact with each other are introduced. These are
based on the physical space in the oceans they occupy, on
their individual evolutionary histories, and on the feeding
relationships they have evolved to obtain their nourishment.

2.1 Spatial Distribution

A simple way to classify marine organisms is according to where they live (FIG. 2.1). At first glance, this figure may look like Figure 1.46 used in Chapter 1 to classify the marine environment into broad categories of habitats. The categories indicated in Figure 2.1, however, refer to large and general groups of organisms based on what part of the marine environment they occupy. The **benthos** includes all organisms living on the sea bottom (the **epifauna**) or in the sediment of the seafloor (the **infauna**). This definition is often extended to include those fishes and other swimming animals that are closely associated with the ocean bottom. Benthic plants, because of their need for sunlight, are restricted to intertidal areas and shallow inner shelves of the ocean's margins where the seafloor lies within the photic zone. Below the photic zone, plants disappear and animals, bacteria, and fungi survive almost entirely on the rain of organic material sinking from the sunlit waters above.

The large actively swimming marine animals found in the pelagic division are the **nekton**. This group includes a large variety of vertebrates (fishes, reptiles, birds, and mammals) and a few types of invertebrates such as squid and some large crustaceans. In contrast, pelagic **plankton** (derived from the Greek word *planktos*, which means "to wander") are defined by their movements and their small size. Carried about by water currents, plankton have little or no ability to control their geographic distribution, although some have reasonable abilities to swim ver-

tically. Plankton are usually small mostly microscopic organisms; however, some jellyfish have tentacles over 15 m long and a bell 2 m across. Photosynthetic members of the plankton are termed **phytoplankton**. They are nearly all microscopic, either a single cell or loose aggregates of a few cells, and, like plants, are restricted to the sunlit photic zone of the pelagic environment. **Zooplankton** are the nonphotosynthetic plankton, ranging in size and complexity from microscopic single-celled organisms to large multicellular animals. Zooplankton are distributed throughout the pelagic division of the marine environment. Sometimes, distinctions between these major groups are not clearcut. Many fishes, for example, begin life as tiny zooplankton and then gradually develop into nektonic animals as their size increases and their swimming abilities improve. Despite some fuzziness around the boundaries of these definitions, they serve as a convenient way of referring to major associations of marine organisms living under similar environmental conditions.

Spanning the major categories outlined in Figure 2.1, different life zones occur along gradients of latitude, reflecting changes in water temperature and light availability, and with distances from continental shorelines. These life zone differences are reflected in the organization of this text, with Chapters 7 to 9 focusing on coastal seas and Chapters 10 to 12 emphasizing the oceanic realm with individual chapters reflecting differences between marine temperate, tropical, and polar living conditions.

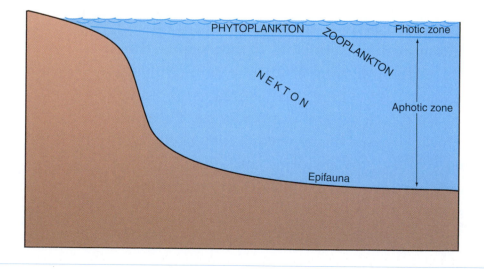

figure 2.1

A spatial classification of marine organisms. Compare this figure with the classification scheme for the marine environment in Figure 1.46.

2.2 Evolutionary Relationships and Taxonomic Classification

All living organisms exhibit varying capabilities for both **ecologic** and **evolutionary adaptations** to changing conditions. Ecologic adaptations are adjustments made by individuals during their lifetimes to changing environmental conditions. In contrast, evolutionary adaptations are products of the changing response of a population of individuals over many generations. The combined results of ecologic and evolutionary adaptations determine whether individuals will obtain sufficient resources to survive until they successfully reproduce. By the simplest of definitions, to reproduce successfully means only that an organism must replace itself with an offspring also capable of reproducing successfully (by the same definition).

Evolutionary Adaptations

Natural populations are characterized by reproductive potentials that exceed those needed to maintain the population size and that are in excess of the number their habitat can support. Eventually, expanding populations outgrow their necessary resources, and competition between individual members of the population intensifies. An individual's ability to survive and reproduce depends on its genetic and physical uniqueness. Natural conditions cause many to perish before reaching sexual maturity. Only those equipped to compete and survive in their current local environment succeed in passing on their genetic traits to future generations.

The offspring inherit characteristics that, in turn, provide a similar ability to compete and survive. Superior fitness may result from a resistance to disease, starvation, or climatic variations, or it may be a capacity to reproduce quickly. This competition and differential survival are summarized in the overworked phrase "survival of the fittest." However, the rules and conditions for survival change continuously and unpredictably. The selection factor for one generation might be a food shortage and for the next generation, disease. As a result, "survival of the fitter" might be a more appropriate phrase, for organisms never evolve to fit their total environment perfectly.

The basic biologic units of evolutionary adaptation, then, are populations. Evolutionary adaptation occurs in populations, never in individuals. Individuals perish regardless of whether or not their populations continue to exist and evolve. Of the millions of types of organisms that have evolved in more than 3 billion years of life's history on Earth, only a tiny fraction exist today. These are the temporary winners.

In the sea, these current winners continually confront variations in temperature, salinity, available oxygen, light, and food, as well as attempts by their neighbors to crowd them out or consume them. In day-to-day ecologic time, these stresses mold the structures of communities and ecosystems. Over much longer periods of time, they shape the evolutionary path of populations through time. Collectively, these processes working over time spans measured in millions of years have resulted in an extraordinary diversity of organisms, a diversity estimated to be between 10 and 30 million distinct species of organisms currently inhabiting this planet.

What is the source of all this fantastic diversity? If life originated only once on our planet, why are the seas populated with so many kinds of organisms? How do we explain the presence of more than 10,000 species of annelid worms or more than 100,000 species of mollusks if each descended from a single ancestral worm or mollusk?

The diversity of life forms that we see today represents the physical expression of their genetically coded information that has been passed on in an uninterrupted lineage from generation to generation. Every cell of each living organism has thousands of different genes incorporated into chromosomes. It is the genes located on these chromosomes that govern the regulatory and structural features of cells. In this way, each organism maintains its own coded repository for all the required information needed to develop, function, and behave in its own unique manner. When it reproduces, it passes copies of its genetic information to its offspring.

We often refer to reproduction as the process by which we replicate ourselves from one generation to the next. In fact, reproduction might better be described as the process by which sets of genes are transferred through generations of organisms. Organisms that reproduce asexually first copy their chromosomes and then themselves. The products of their reproductive efforts are genetically identical to each other and to their parent (FIG. 2.2).

Sexual reproduction, however, is fundamentally different. The whole point of sexual reproduction is to provide a mechanism whereby diverse, rather than identical, offspring can be produced by the same parents. In its most basic form, sexual reproduction involves the unpairing of paired chromosomes in a

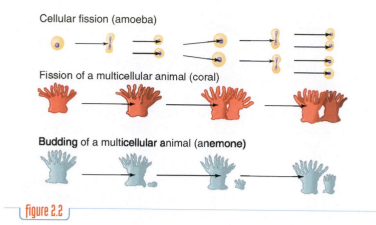

Cellular fission (amoeba)

Fission of a multicellular animal (coral)

Budding of a multicellular animal (anemone)

figure 2.2

Three approaches to asexual reproduction.

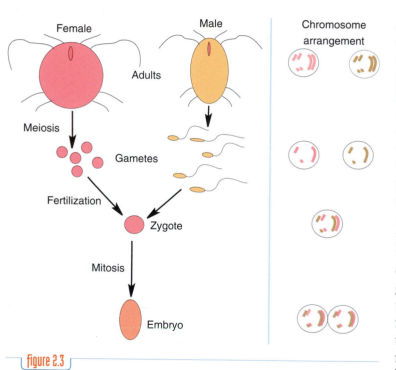

Female

Male

Chromosome arrangement

Adults

Meiosis

Gametes

Fertilization

Zygote

Mitosis

Embryo

figure 2.3

The basic components of sexual reproduction. The chromosome arrangement of each cell is shown to the right.

production, because gamete production is followed by fertilization, the remaining obligatory part of sexual reproduction. In fertilization, the chromosomes carried by the sperm cell are combined with those of the egg to form a **zygote** with double the chromosome number of either of the gametes. This double, or **diploid**, set of chromosomes is carried by all cells in the development to sexual maturity. Because the sequence of events in plant sexual reproduction is so different from that of animals, these patterns are examined separately in Chapter 4 for plants and Chapters 5 and 6 for animals.

The genetic diversity expressed by the offspring of sexually reproducing parents stems from two sources, mutations and sexual reproduction itself. Mutations are random structural alterations of the information coded in a gene. They may result from errors in the process of copying existing genetic codes, or they may be induced by external factors such as radiation. Although some mutations bestow survival advantages on those who carry them, much more often mutations are deleterious and are usually fatal. Even so, it has been the slow accumulation over time of nonfatal mutations that has provided the genetic diversity, the raw material on which sexual reproduction has operated. With parents of slightly different genetic histories available, sexual reproduction (in essence, meiosis followed by fertilization) provides the mechanism to repackage the genetic diversity of parents into large numbers of different offspring. One pair of human parents, for example, each with 23 pairs of chromosomes, hypothetically could produce 2^{23}, or 8,388,608 genetically different offspring. When other chromosomal recombination maneuvers at meiosis are considered, the amount of diversity generated by the sexual reproduction of two individuals is almost limitless.

Sexual reproduction creates genetic diversity, but it is not responsible for evolutionary changes in the occurrence of particular genes in a population over time (evolutionary adaptation). Evolutionary adaptations are the expressions of the changing responses of a population of individuals to changes or variations in their environment over many generations. In natural populations, the major mechanisms for evolutionary change over time are gene flow and natural selection.

process known as **meiosis** and then recombining them at **fertilization** to form totally new chromosome pairings. When mature, sexually reproducing adults produce **gametes** (either eggs or sperm) by meiosis. Meiosis is a cell division process in which the chromosomes of the gametes produced include one of each of the pairs of chromosomes characteristic of the other cells of the adult individuals (FIG. 2.3). This is termed a **haploid** chromosome condition. Halving of chromosome numbers in the formation of gametes is a necessary component of sexual re-

Gene flow can occur when individuals from one population migrate into and interbreed with another population of the same species. The impact of gene flow is highly variable, depending on the degree of genetic difference between the two populations as well as the number of immigrants from one population to the other. If either of these factors is large, genetic differences in subsequent generations can occur and spread rapidly through the affected population.

Natural selection in populations occurs over generations because of differential rates of reproduction and mortality in different gene lineages within a population. Over long periods of time, biologic mechanisms that create genetic variations and work on those variations to create change through time have given rise to genetically isolated populations of organisms. Populations can become reproductively isolated from other closely related populations by any of several extrinsic or intrinsic factors. Common extrinsic mechanisms include geographic and climatic barriers to migration and, consequently, to gene flow. Intrinsic isolating mechanisms include behavioral, structural, or ecologic differences that might split a population into two or more subgroups and prevent their interbreeding.

A clear example of intrinsic isolating mechanisms exists in two different populations of Pacific Northwest killer whales, which appear to be well into the process of becoming separate species. A "resident" population feeds almost exclusively on fishes, whereas a "transient" population preys principally on marine mammals, especially harbor seals and sea lions. These two populations of killer whales use different foraging strategies, even when in close proximity to each other. These strategies reflect not only different prey types, but also different pod traditions. Transient groups commonly encircle their prey, often using repeated tail slaps or ramming actions to kill their prey before consuming it (FIG. 2.4). Differences in the foraging behaviors of the two overlapping populations in the Pacific Northwest persist through time, with each group expressing different vocalizations and different social group sizes as well as different prey preferences and foraging behaviors.

figure 2.4

Tail slap of killer whale directed at its sea lion prey.

Table 2.1

A Comparison of Foraging-related Differences between Transient and Resident Killer Whales of the Pacific Northwest

Character	Residents	Transients
Group size	Large (3–80)	Small (1–15)
Dive pattern	Short and consistent	Long and variable
Temporal occurrence	With salmon runs	Unpredictable
Foraging areas	Deep water	Shallow water
Vocalize when hunting?	Frequently	Less frequently
Prey type	Salmon and other fish	Mammals
Relative prey size	Small	Large
Prey sharing	Usually no	Usually yes

Modified from Baird *et al.* 1992.

TABLE 2.1 summarizes some important differences between these two populations.

Reproductively isolated populations experience different environmental (and therefore different selective) pressures, and given sufficient time, they must diverge from each other as they evolve. Eventually, isolated populations evolving in different directions achieve enough uniqueness to be accorded the status of species. It is this repetitive pattern of reproductive isolation and genetic divergence that, through more than 3 billion years of Earth's history, has been the source of the tremendous variety of species that we see today.

Taxonomy and Classification

Biologists estimate that between 10 and 30 million different **species** of organisms exist on Earth today. Of these, only about 1.9 million have been identified and formally described, and we know very little about the vast majority of those. Although most species that have been described are land-dwelling insects, the diversity of life in the sea is immense. Because of the evolutionary processes that started in the sea and have operated there for the past 3 to 4 billion years of Earth's history, each of these species exhibits some genetic relationship to all other species.

How do we organize and manipulate all the information we have accumulated regarding the evolutionary histories and relationships of millions of species? Sometimes, evolutionary relationships between organisms are obvious, for example, porpoises and dolphins. At other times, however, such relationships are more obscure. The **taxonomic system of classification** provides a means to deal with this vast and often confusing array of diversity by reflecting these evolutionary, or **phylogenetic**, relationships of organisms. The process of taxonomic classification consists of three basic steps. First, closely related groups of individual organisms must be recognized and described. Next, these groups, called **taxa** (singular, taxon), are assigned Latin (or latinized Greek) names according to procedures established by international conventions. Finally, the described and labeled groups are fitted into a hierarchy of larger more inclusive taxa.

The discovery and description of species and the recognition of the patterns of relationships among them is based on the processes of biologic evolution. Patterns of relationships among species are based on changes in the features or **characters** of an organism. Characters are the varied inherited characteristics of organisms that include DNA makeup, anatomic structures, physiologic features, and behavioral traits. Evolution of a character may be recognized as a change from a preexisting or **ancestral** character state to a new **derived** character state. A fundamental underpinning of phylogeny is the concept of **homology**, the similarity of features resulting from common ancestry. Two or more features are homologous if their common ancestor possessed the same feature. For example, the flipper of a seal and the flipper of a walrus are homologous as flippers because their common ancestor had flippers.

The basic tenet of **phylogenetic taxonomy**, or **cladistics** (from the Greek word *klados,* meaning "branch"), is that shared derived character states provide strong evidence that two or more species possessing these features share a common ancestry. In other words, the shared derived features represent unique evolutionary events that may be used to link two or more species together in a common evolutionary history. Thus, by sequentially linking species together based on their common possession of derived shared characteristics, the evolutionary history of those taxa can be inferred.

Relationships among taxonomic groups (such as a species) are commonly represented in the form of a **cladogram**, or **phylogenetic tree**, a branching diagram that represents our current best estimate of phylogeny. The lines of the cladogram are known as **lineages** or **clades.** Lineages represent the sequence of descendant populations through time. Branching of the lineages at **nodes** on the cladogram represents **speciation events**, a splitting of a lineage resulting in the formation of two species from one common ancestor. The two Pacific Northwest killer whale populations described earlier are thought to be at such a speciation node. A more extensive example of a cladogram traces the evolution of manatees, dugongs, and elephants from their common ancestor (FIG. 2.5).

The fundamental unit of taxonomic classification, then, is the species. A species is a group of closely related individuals that are similar in appearance and that can and normally do interbreed and produce fertile offspring. The free exchange of genetic information between individuals of such groups connects each individual to a common gene pool and steers them along a common evolutionary path, with whole populations adapting to environmental influences over long periods of time.

This widely accepted definition of a species, however, poses special problems for the classification of marine organisms. Because of the environmental extremes occupied by many marine organisms, they are quite often difficult, or even impossible, to study alive, and little is known of their reproductive habits. In such cases, another somewhat circuitous definition is used: A species is a group of closely related individuals classified as a species by a competent taxonomist on the basis of anatomy, physiology, and other characteristics (including genetic comparisons where possible). Whichever definition is used, the species must be regarded as a functional biologic unit that can be identified and studied.

Assigning names to species or larger groups of organisms is a process more regimented than merely recognizing and describing the species. Common

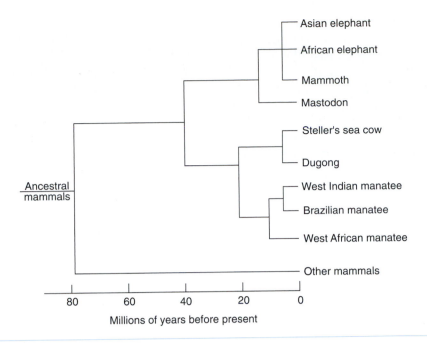

figure 2.5

A cladogram illustrating the relationships between Sirenians and elephants and their close relatives based on differences in mitochondrial DNA. Time scale for speciation events estimated from rates of DNA change.

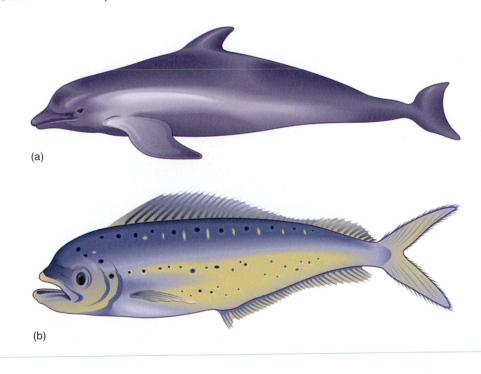

(a)

(b)

figure 2.6

(a) Common dolphin, *Delphinus*. (b) Dolphinfish, *Coryphaena*, also known as a dorado or mahi-mahi.

names are often used in localized areas, but the lack of standardization in the use of common names detracts from their widespread usefulness and acceptance. To some people, the name *dolphin* refers to an air-breathing porpoise-like marine mammal (FIG. 2.6A). To others, a dolphin is a prized game fish (Fig. 2.6b) that is also known as mahi-mahi in Hawaii and as dorado in Spanish-speaking countries. The confusion created by these common names is eliminated when species and other taxonomic groups are assigned names that are accepted by international agreement as standard group names.

Evolutionary Relationships and Taxonomic Classification   **49**

Following the scheme first introduced by the Swedish botanist Linnaeus over two centuries ago, taxonomic names of a species consist of two terms. The first is the genus name followed by the species name. The genus name is always capitalized; the species name is not. Both are either italicized or underlined. Each taxonomic name is unique and represents only one species. Thus, there can be no confusion that *Delphinus delphis,* the common dolphin, is not the same animal as the dolphinfish, *Coryphaena hippurus.*

The naming of a species does not complete the taxonomic classification process. The species is only part of a larger classification scheme that consists of a hierarchy of taxonomic categories:

> Kingdom
> > Phylum (Division for photosynthetic groups)
> > > Class
> > > > Order
> > > > > Family
> > > > > > Genus
> > > > > > > Species

Each category is constructed so that it encompasses one or more categories from the next lower level. All categories above the species level lack the type of natural definition we afford the species.

These groups are not completely arbitrary, however. Each group reflects the evolutionary relationships known or assumed to exist between its component taxa on the basis of its anatomy, embryology, and biochemistry. It is because much of the evolutionary history of some lineages is not known in detail that a classification based on little information may not accurately reflect the lineage's actual relationships with other closely related groups. Ideally, each genus is composed of a group of very closely related, but genetically isolated, species. Families include related genera that have many features in common. Orders include related families based on generalized characteristics. Classes, phyla (singular, phylum), and kingdoms are increasingly inclusive categories based on even more general features. The term *division* is used in place of phylum for photosynthetic organisms and fungi.

TABLE 2.2 summarizes the taxonomy of a few organisms mentioned so far in this book. Dolphins and blue whales are more closely related to each other than to the other organisms listed in Table 2.2, so they are placed in the same order, Cetacea, that includes other whales but excludes all other species of organisms. Copepods do not resemble whales or dolphins, yet their evolutionary connections are closer to both dolphins and blue whales (organisms in the same kingdom) than they are to mangroves in the kingdom Plantae. In this way, the taxonomic system of classification serves as a framework to support our understanding of the evolutionary relationships that exist between groups of organisms.

In FIGURE 2.7, the major phyla and divisions of marine organisms are arranged to illustrate the presumed evolutionary relationships of each group. Only phyla or divisions with several free-living nonparasitic marine species are included. These phyla and divisions are then grouped into the five kingdoms outlined in Figure 2.7. The kingdom-level classification system used in this text is slightly modified from a widely accepted system proposed by Whittaker in 1969, which groups living organisms into five kingdoms (Fig. 2.7) based on their cellular structures, their modes of nutrition, and their deduced patterns of evolutionary relationships.

In this system, three divisions, Bacteria, Cyanobacteria, and Archeobacteria, are placed in the kingdom Monera. (Another classification system gaining acceptance separates Archeobacteria into a unique sixth kingdom.) These small organisms lack much of the complex subcellular structure found in other modern cells, a condition described as procaryotic (FIG. 2.8).

Taxonomic Classification of Some Marine Organisms

table 2.2

Organism	Kingdom	Phylum/Division	Class	Order	Family	Genus	Species
Blue whale	Animalia	Chordata	Mammalia	Cetacea	Balaenopteridae	*Balaenoptera*	*musculus*
Dolphin	Animalia	Chordata	Mammalia	Cetacea	Delphinidae	*Delphinus*	*delphis*
Dolphinfish	Animalia	Chordata	Osteichthyes	Teleostei	Coryphaenidae	*Coryphaena*	*hippurus*
Copepod	Animalia	Arthropoda	Crustacea	Calanoida	Calanidae	*Calanus*	*finmarchicus*
Mangrove	Plantae	Anthophyta	Dicotyledones	Laminales	Avicenniaceae	*Avicennia*	*germinans*
Tintinnid	Protista	Ciliophora	Ciliata	Spirotricha	Tintinnidae	*Halteria*	*grandinella*

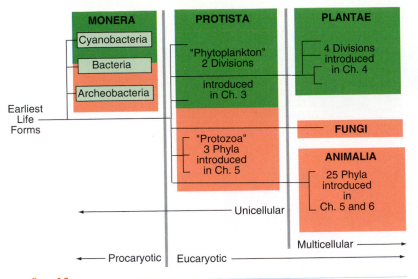

figure 2.7

A phylogenetic tree illustrates the evolutionary relationships of the major groups of marine organisms. The five kingdoms are shaded in color, green for photosynthetic groups, salmon for heterotrophic ones. The three Moneran division names are listed here. Other division and phylum names will be added in later chapters as they are introduced.

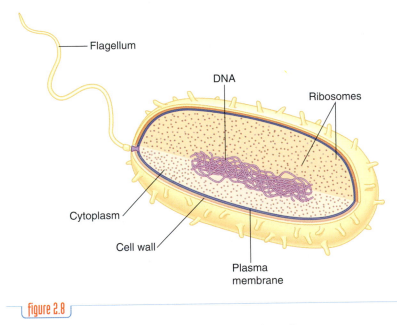

figure 2.8

A simplified diagram of a procaryotic bacterial cell, *Bacillus*.

A **cell wall** provides form and mechanical support for the cell. Inside the cell wall, a selectively permeable **plasma membrane** separates the internal fluid environment (the **cytoplasm**) from the exterior environment of the cell and regulates exchange between the cell and its external medium. In some species, limited movement is provided by a whiplike **flagellum**. Internally, the genetic information is coded and

stored in a single circular strand of DNA. Small ribosomes use that information to direct the synthesis of enzymes. The enzymes, in turn, control and regulate all other chemical reactions that occur in living cells.

Bacteria are the most prolific organisms on Earth. They represent most of the described species of Monerans. The phylogenetic patterns of relationships between bacterial species are not well understood, and for our convenience these important decomposers are grouped in a single division, the Bacteria.

Despite the name, the lineage of Archeobacteria ("ancient bacteria") actually is not as old as that of Bacteria. Some genetic comparisons indicate that Archeobacteria are sufficiently different from Bacteria to warrant a separate kingdom status. More research on the origins of this recently discovered group is expected to clarify their taxonomic status. For our purposes, they will be treated as a division of the kingdom Monera. Marine Archeobacteria include several diverse groups of "extremophiles" with descriptive names, including thermophiles ("heat lovers") that live in extremely hot water around deep-sea volcanic vents, halophiles ("salt lovers") in high salinity coastal ponds and lagoons, and barophiles ("pressure lovers") found at extreme ocean depths.

All Cyanobacteria are photosynthetic, and many are important in marine phytoplankton communities. Fossil remains over 3 billion years old of simple cells very much like modern cyanobacteria have been reported from scattered sites in Africa and Australia. These were likely the first photosynthesizers in Earth's early oceans. Some bacteria are also photosynthetic, but most can be found as decomposers in all marine habitats.

The procaryotic Monerans have been eclipsed in most environments by groups of relatively large and ecologically dominant organisms, the eucaryotes. The complexity and diversity of eucaryotic cells are responsible for much of the immense variety of life forms on Earth today. These larger and structurally more

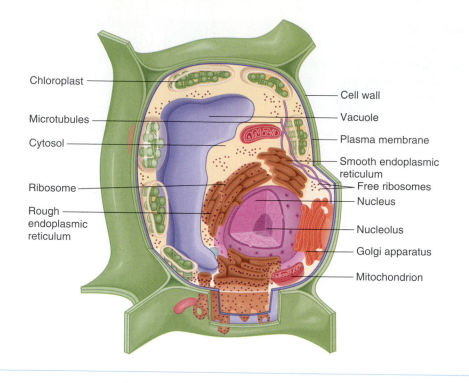

Chloroplast

Microtubules

Cytosol

Ribosome

Rough endoplasmic reticulum

Cell wall

Vacuole

Plasma membrane

Smooth endoplasmic reticulum

Free ribosomes

Nucleus

Nucleolus

Golgi apparatus

Mitochondrion

figure 2.9

Simplified diagram of a eucaryotic cell. In addition to several types of subcellular organelles common to most eucaryotes, photosynthetic cells contain chloroplasts and vacuoles, and are typically supported externally by rigid cell walls.

complicated eucaryotic cells possess a nucleus and a variety of other membrane-bound structures not found in procaryotes (FIG. 2.9). The chromosomes and their surrounding **nuclear membrane** form a central structure, the nucleus. The enzymes involved in respiration and energy release are associated with numerous small **mitochondria**. Many of the enzyme-synthesizing ribosomes are free in the cytoplasm; others are arranged on a membranous **endoplasmic reticulum**. Food particles are ingested and stored in vacuoles within the cell. Other subcellular structures are involved in excretion of wastes, osmotic balance, and other cellular chores. In addition to these cellular structures common to all eucaryotes, photosynthetic eucaryotes typically possess chloroplasts and cell walls. **Chloroplasts** serve as the sites of photosynthesis, an important process described in the next section. Cell walls provide shape and support in a manner similar to those of Monerans. Cell walls also provide structural resistance to the stresses internal osmotic pressures place on fragile cell membranes. Often, cell walls alone are sufficient to deal with pressures generated by ion imbalances across cell membranes.

The eucaryotic organisms that are single celled or consist of simple aggregations of similar cells are placed in the kingdom Protista. Most nonparasitic protists are aquatic or live in soil. This kingdom in-

cludes the rest of the major phytoplankton (described in Chapter 3), a few terrestrial funguslike groups, and several nonphotosynthetic phyla sometimes referred to as protozoans. The marine protozoan groups are described in Chapter 5.

From the basic module of the eucaryotic cell, multicellular organisms of even greater size and organizational complexity have evolved. Fungi form a third kingdom of living organisms. Fungi are complex organisms whose bodies are composed of **hyphae**, which are fine threadlike tubes containing numerous haploid nuclei. The hyphae have cell walls made of chitin (similar to the material of arthropod exoskeletons, p. 144) and often lack cross walls, so true cells of fungi are difficult to define. Diffuse masses of hyphae form a mycelium, the nonreproductive part of a fungus. Fungi are much more abundant and diverse on land, where their wind-blown spores effectively disperse them over long distances. Only a few hundred species of mostly inconspicuous fungi live in the sea. However, along with bacteria, marine fungi function as decomposers in most benthic environments, and a few, with their photosynthetic algal symbionts, exist as lichens on intertidal rocks (p. 229).

The last two kingdoms, Plantae and Animalia, include groups of organisms familiar to most people.

Both kingdoms consist of multicellular and therefore usually larger organisms. Plants are photosynthetic nonmotile organisms with cellulose cell walls and life cycles that include alternating gametophyte and sporophyte generations (see Chapter 4). Most members of the kingdom Plantae live on land, yet several thrive in marine communities. Marine plants and their life cycles are described in Chapter 4. Animals lack cell walls and have some muscle-contracting and nerve-conducting capabilities. These attributes provide flexibility and mobility, two of the evolutionary hallmarks of animals. Chapters 5 and 6 include a survey of the common phyla of marine animals.

2.3 Trophic Relationships

Some relationships between different organisms can be described by their **trophic** associations. This approach involves determining what an organism eats and what eats it. It is the trophic connections that these organisms forge with others to obtain food that weaves much of the fabric of communities of organisms in all ecosystems. In this section we examine some of the more basic or obvious characteristics of these trophic interactions.

Harvesting Energy

Living organisms require two fundamental things from their nourishment, matter and energy. Matter is necessary for individual growth and for reproduction. Energy is needed to maintain the structured chemical state that distinguishes living organisms from nonliving masses of similar material. Almost all living organisms on Earth satisfy their energy needs either by photosynthesis or by cellular respiration, using molecules of **adenosine triphosphate (ATP)** as their currency of energy exchange.

Photosynthesis is a biochemical process that uses **chlorophyll** pigments to absorb some of the abundant energy of the sun's rays. In all photosynthetic organisms except Cyanobacteria, the photosynthetic pigments and enzymes are contained in the chloroplasts of the cell. In this process, ATP and other high-energy substances are made and then used to synthesize sugars, amino acids, and lipids from CO_2 and H_2O. For the present, photosynthesis can be summarized by the following general equation (more detail will be added in Chapter 3):

$$6CO_2 + 12H_2O \rightarrow C_6H_{12}O_6 + 6H_2O + 6O_2$$

carbon   water chlorophyll   sugar   water   oxygen
dioxide   sunlight

Most nonphotosynthetic organisms on Earth rely either directly or indirectly on the energy-rich organic substances produced by photosynthetic organisms. In environments with limited amounts of free O_2 (such as anoxic basins or deep-ocean bottom muds) and abundant supplies of organic material, **anaerobic respiration** (respiration without O_2) provides a mechanism to obtain energy for use in cellular processes. Several variations of anaerobic respiration are exhibited by plants and animals, yet all release energy from organic substances without using O_2. In **alcoholic fermentation**, for example, sugar is degraded, or broken down, to alcohol and CO_2. Energy is released in the form of ATP:

$$C_6H_{12}O_6 \rightarrow 2(C_2H_5OH) + 2CO_2 + energy$$

sugar   respiratory   alcohol   carbon   (equivalent
      enzymes   dioxide   to 2 ATP)

In most eucaryotic organisms, respiratory enzymes more complex than that of anaerobic respiration housed in the mitochondria completely oxidize high-energy compounds such as sugar to carbon dioxide and water and, in the process, release energy:

$$C_6H_{12}O_6 + 6O_2 \rightarrow 6CO_2 + 6H_2O + energy$$

sugar   oxygen   respira-   carbon   water   (equivalent
                 tory   dioxide   to 37 ATP)
                 enzymes

This process uses oxygen and is therefore referred to as **aerobic respiration**. In aerobic respiration, each molecule of sugar yields 18 times as much energy as it would if used in anaerobic respiration. Consequently, organisms that metabolize food with oxygen in this manner obtain a tremendous energetic advantage over their anaerobic competitors.

The transfer of matter and energy for use in metabolic processes has shaped the evolution of a close interdependence of three major categories of marine organisms: autotrophic primary **producers** and heterotrophic **consumers** and **decomposers**. Autotrophic means self-nourishing; these organisms use photosynthesis to build high-energy organic carbohydrates or lipids and other cell components from water, carbon dioxide, and small amounts of inorganic nutrients (primarily nitrate and phosphate) found in seawater. They are the **primary producers** of marine ecosystems and are placed in the first **trophic level**. Some bacterial autotrophs extract energy from inorganic compounds to build high-energy organic molecules. These autotrophs are **chemosynthetic**, and they are discussed more in Chapter 11.

Consumers and decomposers are unable to synthesize their own food from inorganic substances and

Metabolic Rates of Large Mammals

Oxygen is so important to most life on Earth that measuring its use by animals has been a standard tool for estimating their metabolic rates. Metabolic rates are the rates at which energy-using processes occur within cells. As the equation for cellular respiration on page 53 suggests, the rate at which O_2 is used in a cell (or a whole animal) is proportional to the rate at which energy is released, so it serves as an effective monitor of overall metabolic rates.

For many marine animals, the process of measuring their O_2 consumption rates is as easy as placing them in a sealed container in water of known O_2 content, waiting 1 hour, and then again measuring the O_2 content of the water. The difference in O_2 concentrations of the water represents the amount of O_2 used during that 1-hour period. For more active animals with larger O_2 requirements, this approach can be modified to allow a flow of water through the test chamber to replace O_2 as it is used while monitoring the O_2 content of the inflow and outflow water streams to determine the amount used.

Air-breathing marine animals, especially large ones, present a different challenge: How to measure O_2 consumption of an animal too large to restrain or to

place in a closed system for monitoring. One approach for large whales relies on the knowledge that their O_2 consumption rates are determined by three features of their breathing behavior: their breathing rates, their lung volumes, and their O_2 uptake rates.

Breathing rates by themselves have sometimes been used to estimate metabolic rates of large free-swimming whales, because they are roughly proportional to O_2 consumption rates. Typical ventilatory patterns of whales consist of an extended period of submergence interrupted by very abrupt exhalations (lasting usually less than 1 second) immediately followed by

Figure B2.1   Balloon collection of exhaled lung air from JJ.

equally short-duration inhalations. When a whale surfaces to breathe, the blow is usually quite visible, and counting breaths is a relatively easy task.

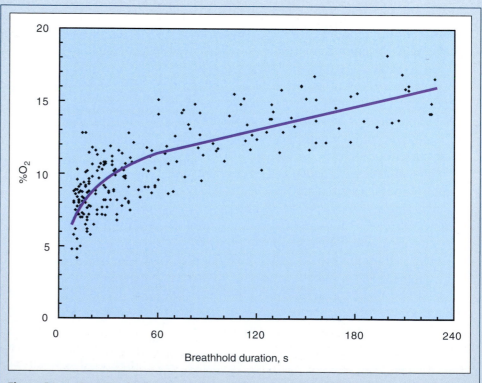

Figure B2.2   Pattern of oxygen uptake as a function of JJ's breathhold duration.

Figure B2.3 Large flowmeter in place to measure JJ's tidal lung volume.

Metabolic Rates of Large Mammals

Tidal lung volume (V_T) is the amount of air that is inhaled and then exhaled with every breath. The inhaled air always contains 21% O_2; the exhaled air contains somewhat less. The difference in O_2 content between inhaled and exhaled air (%O_2) is the O_2 that was assimilated from that tidal lungfull. When V_T and %O_2 can be measured or estimated and combined with breath-ing rates, O_2 consumption rates can be calculated:

$$V_T \text{ (L/breath)} \times \%O_2 \times \text{breathing rate (breaths/min)} = O_2 \text{ consumption (L/min)}$$

A rehabilitating gray whale calf ("JJ") that stranded in southern California in 1997 provided an excellent opportunity to explore the relation-ships among breathing rates, tidal lung volume, and oxygen extraction of a young and growing baleen whale un-der relatively controlled conditions. As JJ swam freely about her pool, sam-ples of expired lung air were collected with latex weather balloons (Figure B2.1), sealed, and then ana-lyzed with a gas analyzer system to determine %O_2. JJ's %O_2 varied from 4.8% to 18.2%, increasing rapidly with increasing breath-hold duration (Figure B2.2).

Measurements of JJ's tidal lung volume were made with a large flowme-ter that fitted snugly over her blowhole to measure respiratory flow rates dur-ing each exhalation/in-halation cycle (Figure B2.3). Signals were recorded on a lap-top computer and were integrated over the time durations of each exha-lation to determine V_T (Figure B2.4). As JJ grew during her year of rehabil-itation at Sea World, her V_T increased substantially, and in addition to her increasing body size, the increase in V_T was a predictable consequence of the duration of each exhalation. To monitor her exhalation duration while she swam freely in her pool, JJ's exhalation sounds (her blows), were recorded, and the duration of each exhalation was measured from this audio signal. This technique is easily transferrable to noncaptive whales and has been used successfully in their natural habitat. Calculated V_T for JJ derived from exhalation duration (T_E) ranged from 2.9 to 3.4% of body mass (W). These values are typical for mammals, supporting the growing awareness that whales do not have larger lungs than do other mammals. These estimates indicate that meta-bolic rates of gray whale calves, when scaled for body mass, are similar to those of other young and growing mammals.

Additional Reading

Sumich JL. Oxygen extraction in free-swimming gray whale calves. *Marine Mammal Science* 1994;10:226–230.

Sumich JL. Direct and indirect mea-sures of oxygen extraction, tidal lung volumes and ventilation rates in a young captive gray whale, *Eschrichtius robustus. Aquatic Mammals* 2001;27:279–283.

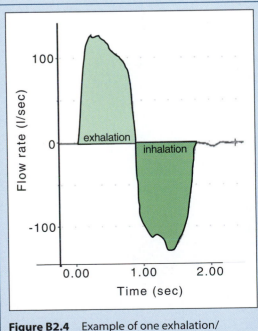

Figure B2.4 Example of one exhalation/inhalation cycle from JJ.

must depend on autotrophs for nourishment. These are **heterotrophs**, each having some specialization in terms of nutrition yet all ultimately dependent on the autotrophs for their energy. Animals that feed on autotrophs are **herbivores** and occupy the second trophic level; those that prey on other animals are **carnivores** and occupy the third and higher trophic levels. The decomposers, primarily bacteria and fungi, exist on **detritus**, the waste products and dead remains of organisms from all the other trophic levels. Whatever their specialized feeding role may be, all heterotrophs metabolize the organic compounds synthesized originally by primary producers to obtain usable energy.

Organic compounds produced by autotrophs become the vehicle for the transfer of usable energy to the other inhabitants of the ecosystem. A distinction must be made between the flow of essential nutrients and the flow of energy in an ecosystem. The movement of nutrient compounds and dissolved gases is cyclic in nature, going from autotrophs to consumers to decomposers, and then eventually back to the autotrophs (FIG. 2.10). Most ecosystems function as nearly closed systems, so materials move from one ecosystem component to another in **biogeochemical cycles**. These cycles link living communities of organisms with nonliving reservoirs of important nutrients within their respective ecosystems (FIG. 2.11).

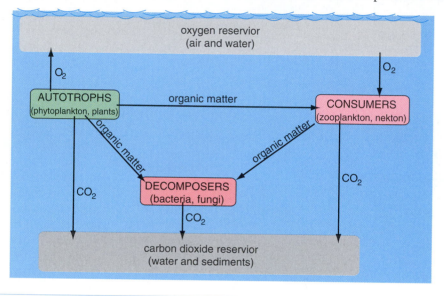

figure 2.10

Simplified paths of the flow of oxygen and carbon in an idealized marine ecosystem.

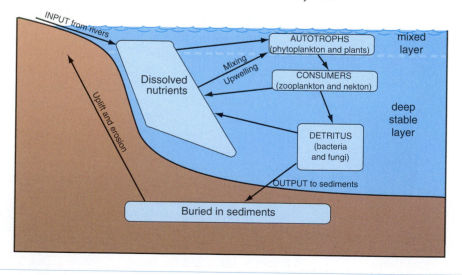

figure 2.11

Biogeochemical cycle of P or N nutrients, showing its major marine reservoirs. Few nutrients are available to autotrophs in the upper mixed reservoirs; much more is dissolved in deep stable waters. Similar cycles exist for biologically important chemicals such as carbon.

Food Chains and Food Webs

In contrast to the cyclic flow of materials in ecosystems, the flow of energy is only in one direction, from the sun through the autotrophs to the consumers and decomposers. Living organisms are not highly efficient in their use of energy. Less than 1% of the solar energy that makes it through the atmosphere to the sea surface is absorbed by autotrophs. Then a substantial portion of the energy captured in the photosynthetic process is used for cellular maintenance, growth, and reproduction. So only a small fraction of the energy from photosynthesis is available to the consumers.

A similar decrease in available energy occurs between the herbivores and carnivores (FIG. 2.12). Laboratory and field studies of marine organisms place the efficiency of energy transfer from one trophic level to the next at between 6% and 20%. In other words, only 6% to 20% of the energy available to any trophic level is usually passed on to the next level. A widely accepted average efficiency is 10%; however, recent studies of some benthic communities and fish populations provide examples of energy efficiencies substantially higher. These relationships can be illustrated as a food pyramid, arranged in a linear fashion to illustrate the decrease in available energy and material from lower to higher trophic levels. FIGURE 2.13 illustrates such a food pyramid, proceeding from phytoplankton to herring at the third trophic level.

However, marine communities seldom exist as simple straight-line food chains. Rather, a complexly interconnected food web provides a more realistic model of the paths that nutrients and energy follow through the living portion of ecosystems as its members feed on each other. They may be grazing food webs, commencing with autotrophs and progressing through a succession of grazers and predators, or they may be parallel detritus food webs, built on the waste materials and dead bodies from the grazing food webs.

With only a few near-shore exceptions, the first trophic level of marine food webs is occupied by widely dispersed microscopic phytoplankton. This microscopic character of most marine primary producers imposes a size restriction on many of the occupants of higher marine trophic levels. Because very few animals are adapted to feed on organisms much smaller than themselves, marine herbivores are usually quite small. Large marine animals are carnivores and usually occupy higher levels in the food web. In contrast, the plants of most terrestrial ecosystems are generally quite large. As a result, most large land animals are herbivores, and fourth-trophic-level animals are extremely rare.

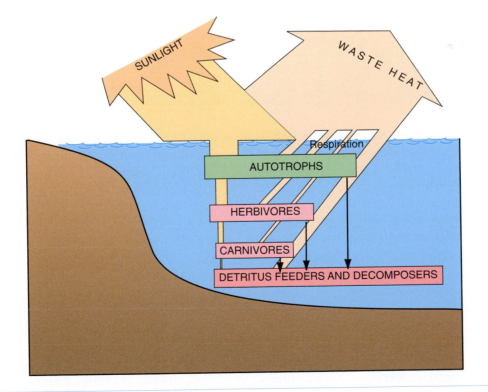

figure 2.12

Energy flow in a marine ecosystem. Sunlight first captured by autotrophs is eventually degraded by their own cellular respiration or that of their consumers and eventually is lost as waste heat.

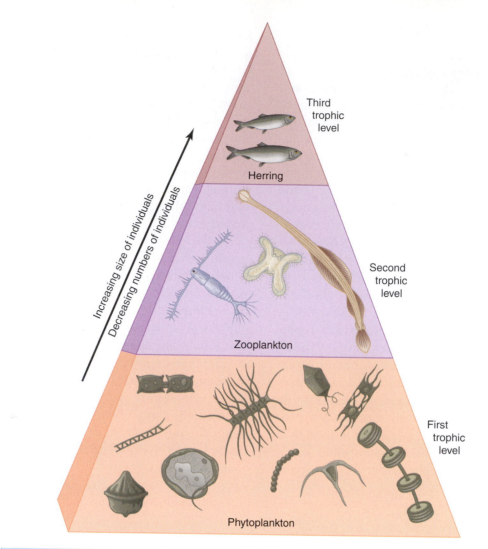

Third
trophic
level

Herring

Second
trophic
level

Zooplankton

First
trophic
level

Phytoplankton

Increasing size of individuals

Decreasing numbers of individuals

figure 2.13

Food pyramid that leads to an adult herring.

FIGURE 2.14 outlines the major trophic relationships of the members of a typical temperate-climate pelagic marine community. The herring, like many of the other organisms of this food web, is an opportunistic feeder and does not specialize on only one type of food organism. Because of the complex feeding relationships of the herring, it is very difficult to place it in a particular trophic level. The adult herring occupies the third level when feeding on *Calanus* copepods, the fourth level when feeding on sand eels, and either the fourth or fifth trophic level when feeding on the amphipod *Themisto*. Even the complex of feeding relationships outlined in Figure 2.14 is an oversimplification, because it ignores other marine animals that compete with the herring for the same food sources as well as the detritus food webs that develop in parallel with grazing food webs. FIGURE 2.15 outlines

some of these more common energy and nutrient pathways of a pelagic marine ecosystem.

Some marine organisms obtain their food by establishing highly specialized symbiotic relationships. The term **symbiosis** denotes an intimate and prolonged relationship between two (or more) species in which at least one species obtains some benefit from the relationship. Commonly, that benefit is food. Symbiotic relationships can be viewed as a spectrum of interactions (FIG. 2.16), ranging from **mutualism**, which provides an obvious benefit to the **symbiont** and to its **host**, to **parasitism**, which directly benefits the symbiont at the expense of the host. Common marine examples of mutualism include several different types of cleaning symbiosis, in which small fish or shrimp maintain obvious cleaning stations that are visited by larger host fish to have damaged tissues or parasites removed.

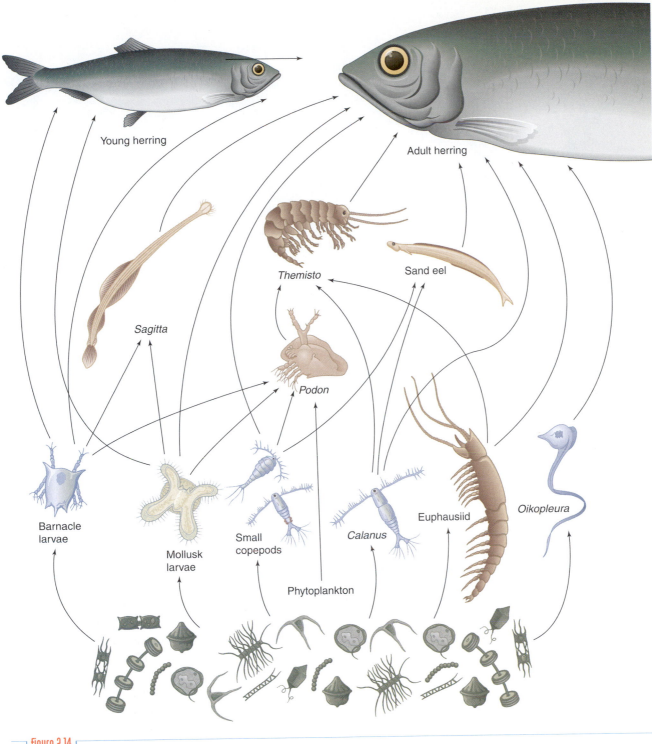

Young herring

Adult herring

Themisto

Sand eel

Sagitta

Podon

Euphausiid

Oikopleura

Barnacle
larvae

Mollusk
larvae

Small
copepods

Calanus

Phytoplankton

figure 2.14

A marine food web, illustrating the major trophic relationships that lead to an adult herring. (Adapted from Hardy 1924.)

Parasites live on or in their hosts and make their presence felt, not by killing their hosts but by reducing the host's food reserves, resistance to disease, and general vigor. The infected host then is more likely to become a casualty of infection, starvation, or predation but not of the parasite directly. Inter-mediate between mutualism and parasitism is a broad category of **commensal** interactions, in which the symbiont benefits but there is an insignificant, or at least poorly known, effect on its host. Later chapters examine some representative commensal and mutu-alistic symbiotic relationships.

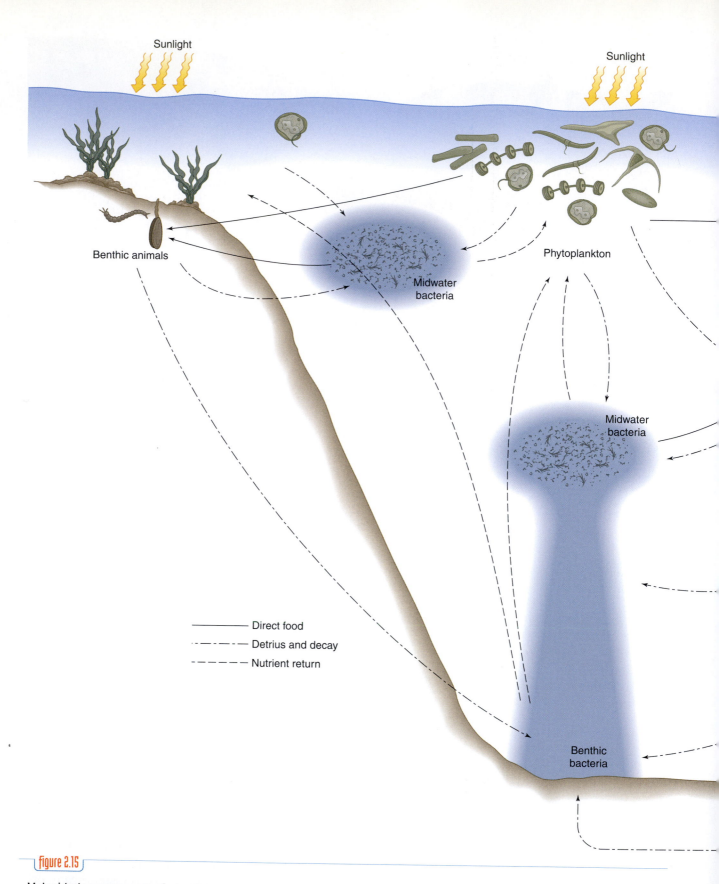

Sunlight

Sunlight

Benthic animals

Midwater
bacteria

Phytoplankton

Midwater
bacteria

——————— Direct food

—·—·—·— Detrius and decay

– – – – – Nutrient return

Benthic
bacteria

figure 2.15

Major biotic components of a marine ecosystem with their interconnecting paths of energy and nutrient exchange. (Adapted from Russell-Hunter 1970.)

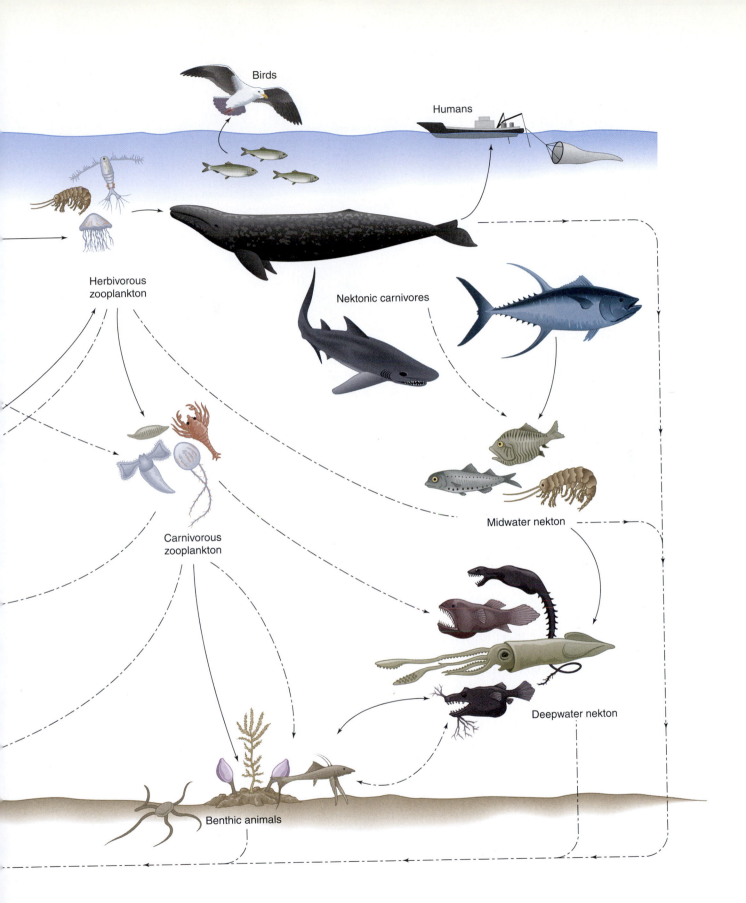

Birds

Humans

Herbivorous
zooplankton

Nektonic carnivores

Carnivorous
zooplankton

Midwater nekton

Deepwater nekton

Benthic animals

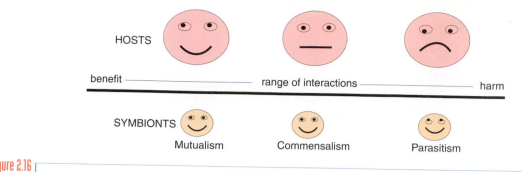

HOSTS

benefit ————————————— range of interactions ————————————— harm

SYMBIONTS

Mutualism   Commensalism   Parasitism

figure 2.16

The range of common symbiotic interactions between a host and its symbiont.

2.4 The General Nature of Marine Life

Although modern marine organisms share many basic structural and behavioral characteristics with their terrestrial relatives, marine life is unique in several important ways. Marine organisms exist within a dense, circulating, interconnected seawater medium. The movement of waves, tides, and currents stirs and mixes organisms, their food, and their waste products so that these organisms are never completely isolated from the effects of their neighbors. Populations of even the smallest unicellular planktonic organisms can become widely distributed by moving currents and water masses.

The biology of marine organisms is, to a large extent, the biology of the very small. It is the phytoplankton that initially establish much of the structural character of marine life. Even in very productive areas of the open ocean, the concentration of phytoplankton is thousands of times more dilute than a healthy meadow. The dispersed nature, extremely small size, and rapid reproductive rates of phytoplankton limit the size and abundance of other life in the sea. Most of the heterotrophs are congregated near the photic zone and its supply of food. At greater depths, the density of marine populations tends to decrease as the food supply diminishes. Below the photic zone, most marine life is dependent on the rain of detritus from above. There are few plant-dominated communities in the sea (a few notable exceptions are depicted in Chapters 4 and 7). Instead, living communities in the sea are organized around coral reefs, mussel beds, and other assemblages of large and dominant animal members.

Many of the substances produced by marine primary producers are not consumed directly by herbivores but are dissolved into seawater. These substances, including lipids and amino acids, are eventually absorbed by suspended bacteria at all depths. These bacteria in turn become food for consumers capable of harvesting them. These microscopic phytoplankton or even smaller bacteria are food for **suspension feeders**, small and large, that use filtering and trapping techniques and devices to collect these minute food particles suspended in seawater.

In a typical two-layered ocean system described in Figure 1.33, most marine organisms and all photosynthetic ones occupy the very shallow near-surface, nutrient-rich, sunlit photic zone. The photic zone is separated vertically by the pycnocline from the colder darker aphotic zone, below which is sparsely populated by animals, fungi, and bacteria, organisms dependent on the rain of food from the photic zone above.

Finally, the sea provides buoyancy and structural support to many strikingly beautiful organisms. But if these organisms are removed from the water, their delicate and fragile forms collapse into shapeless masses. Seawater also supports some extremely large animals. Deep-sea squids over 15 m long have been observed, and squids 20 or even 30 m in length are not improbable. Some blue whales approached weights of 200 tons before the largest members of their populations were removed by commercial whaling. But these animals are exceptional and stand out in sharp contrast to the generally diminutive nature of life in the sea.

Summary Points

Patterns of Associations

- Humans have difficulty conceptualizing vast amounts of complex information if it is not organized. Hence, we tend to classify and organize everything around us, from types of movies in a video rental store to items in grocery stores and even types of organisms on Earth.
- A classification scheme, from items in a grocery store to types of fishes in the sea, must present the information in an intuitive and generally accepted manner.
- No single classification scheme is inherently better than any other, and living organisms can easily be categorized in any number of organizational systems. This chapter presents three commonly used criteria for classifying marine organisms: their geographic distribution, their evolutionary interrelationships, and their trophic interactions.

Spatial Distribution

- Marine organisms are either pelagic (found in the water column) or benthic (living on, in, or near the seafloor). Benthic organisms often move on the sea bottom (epifaunal) or through the sediment beneath its surface (infaunal).
- Pelagic species found in the water column are either swimmers (nekton) or drifters (plankton). Planktonic drifters are termed phytoplankton if they are photosynthetic or zooplankton if they are heterotrophic consumers.

Evolutionary Relationships and Taxonomic Classification

- Living organisms are capable of both ecological adaptation during their lifetime and evolutionary adaptation during a population's existence.
- Natural selection, the mechanism that drives evolutionary adaptation, predicts that, on average, only those individuals that are best adapted to current local conditions will survive and reproduce, thus donating their genetically heritable traits to their offspring.
- All species that have ever existed on Earth are classified in a taxonomic system based on their phylogenetic interrelationships. These systems are often expressed as branching figures, or cladograms, that diagrammatically represent current hypotheses concerning evolutionary history.
- Today we follow a system of naming species that is a modification of nomenclature first formulated by Linneaus more than two centuries ago. This system recognizes five fundamentally different types of living creatures. Monera includes the unicellular procaryotic bacteria, Protista houses the unicellular eucaryotes that possess organelles, Fungi encompasses heterotrophic organisms with chitinous cell walls and no vascular tissue, Plantae is restricted to multicellular photosynthetic organisms, and Animalia includes all multicellular heterotrophs that lack cell walls.

Trophic Relationships

- All organisms must acquire energy and matter to survive, and two sources of each are known to be used by living creatures.
- Organisms capable of obtaining energy directly from sunlight are termed phototrophic; those that obtain their energy from ingesting molecular compounds are chemotrophic.
- Organisms capable of creating their own organic compounds from inorganic molecules are autotrophic; those species that must ingest preformed organics are termed heterotrophic.

- Both energy and matter flow through ecosystems, typically in extremely complex patterns in the sea that can only be properly termed food webs.

The General Nature of Marine Life

- Unlike terrestrial creatures, marine organisms exist in a dense, circulating, salty medium that constantly repositions them along with their food, waste products, and offspring.
- Most photosynthetic species in the sea are microscopic organisms that exist in concentrations that are thousands of times more dilute than that observed in a healthy lawn. Hence consumers that feed on particulate matter, such as detritus, suspension, and filter feeders, are much more common in the sea than on land.
- The buoyancy and structural support provided by the sea enable extremely fragile, gelatinous marine organisms to attain extraordinary sizes and beauty.

Topics for Discussion and Review

1. Consider the many ways in which one could organize and classify items commonly found in the grocery store. Is there a best criterion to use under all circumstances?
2. Label the following organisms (crab, squid, shark, jellyfish, sand worm, clam, dolphin, sea anemone) with one or more of the following adjectives (benthic, infaunal, epifaunal, pelagic, nektonic, planktonic). Consider whether or not these labels are consistent throughout the entire lifespan of each species.
3. List five differences between an amoeba and a bacterium, and five others between a plant and an animal.
4. Some mammals possess five fingers or digits (humans and cats), some possess three (tapirs and rhinos), some possess two (deer, pigs, hippos, and goats), and some possess only one (horses). Discuss how one could determine whether horses have lost digits over time or whether humans have gained digits during their evolutionary history, and present your conclusion in the form of a cladogram.

5. What is the taxonomic hierarchy for our species, *Homo sapiens*?
6. Compare and contrast the following three divisions of Monera: Bacteria, Cyanobacteria, and Archeobacteria.
7. Do plants perform photosynthesis, respiration, or both?
8. Organisms, including humans, perform anaerobic respiration at specific times or specific places. Generate a list of times and places that anaerobic respiration is common.
9. Using common names (such as dolphin or phytoplankton), provide an example of a marine organism that plays each of the following ecological roles: producer, consumer, decomposer, photosynthesizer, chemosynthesizer, heterotroph, herbivore, carnivore, detritus feeder, suspension feeder, filter feeder, predator, parasite, mutualistic symbiont, and commensal symbiont. (Hint: many of these terms are not mutually exclusive.)
10. Although all living organisms are fundamentally similar, generate a list of contrasting characteristics between marine and terrestrial photosynthesizers (such as a diatom and an oak tree).

Suggestions for Further Reading

Allen, T. F. H., and T. B. Starr. 1988. *Hierarchy: Perspectives for Ecological Complexity.* University of Chicago Press, Chicago.

Cambell, N. A., J. B. Reece, and L. G. Mitchell. 1999. *Biology.* Addison Wesley Longman, Inc., Menlo Park, California.

Caron, D. A. 1992. An introduction to biological oceanography. *Oceanus* 35:10–17.

Groves, D. L, J. S. R. Dunlop, and R. Buick. 1981. An early habitat of life. *Scientific American* 245:64–73.

Guttman, B. S. 1976. Is "levels of organization" a useful biological concept? *Bioscience* 26:112–13.

Lewin, R. A. 1982. Symbiosis and parasitism: definitions and evaluations. *Bioscience* 32:254.

Pomeroy, L. R. 1992. The microbial food web. *Oceanus* 35:28–35.

Vernberg, W. B. 1974. *Symbiosis in the Sea.* University of South Carolina Press, South Carolina.

The enormous capacity of the mouth of this blue whale is apparent as it feeds on krill at the sea surface.

CHAPTER
3

A micrograph of pelagic diatoms.

Phytoplankton

As indicated in Chapter 2, much of the special nature of life in the sea is due to the extremely small cell sizes of the most abundant marine primary producers, the phytoplankton. In this chapter, the general features of these important primary producers are introduced. The principal factors that affect the rates in which they produce material to fuel the rest of the marine ecosystem are then examined.

3.1 Phytoplankton Groups

Almost all marine phytoplankton belong to three divisions in the kingdoms Monera and Protista (TABLE 3.1 and FIG. 3.1). They are all single-celled microscopic organisms found dispersed throughout the photic zone of the oceans where they accomplish the major share of primary productivity in the marine

Table 3.1

Major Divisions of Marine Phytoplankton and Their General Characteristics

Division (common name)	Approx. no. of living species	Percent of species marine	General size and structure	Photosynthetic pigments	Storage products	Habit
Cyanobacteria (Blue-green algae)	200	~75	Unicellular, procaryotic, nonflagellated, microscopic	Chlorophyll *a* Carotenes Phycobilins	Starch	Mostly benthic
Chrysophyta (Golden-brown algae) (Coccolithophores) (Silicoflagellates) (Diatoms)	650 200 ? 6000–10,000	~20 96 Most 30–50	Unicellular, often flagellated, microscopic	Chlorophyll *a, c* Xanthophylls Carotenes	Chrysolaminarin Oils	Planktonic and benthic
Dinophyta (Dinophytes)	1100+	93	Unicellular or colonial, flagellated, microscopic	Chlorophyll *a, c* Xanthophylls Carotenes	Starch Fats Oils	Planktonic

Adapted from Seagel *et al.* 1980, Dawson 1981, and Kaufman *et al.* 1989.

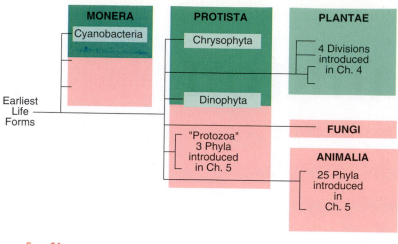

figure 3.1

The phylogenetic tree introduced in Figure 2.7, illustrating the evolutionary relationships of the major groups of marine organisms. The three divisions comprising most marine phytoplankton described in this chapter are listed.

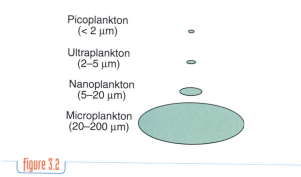

figure 3.2

Relative sizes of phytoplankton groups. All are enlarged 1000X. At the same magnification, a human hair will be as thick as this page is wide.

environment. FIGURE 3.2 compares the range of sizes of phytoplankton cells and the categories into which they are grouped. Only in the last few decades has it been possible to collect representative samples of the exceptionally small **picoplankton** and **ultraplankton.** As our knowledge of these very small phytoplankton groups improves, our understanding of their contribution to marine food webs is also increasing. Presently, it is thought that the most important primary producers in all marine environments, but especially in oceanic waters, are **nanoplankton** sized or smaller.

Cyanobacteria

Marine cyanobacteria have been the object of much recent study. Their small cell size (most are less than 5 μm) makes them very difficult to collect and study.

Their cell structure (FIG. 3.3) is typical of procaryotes, with only a few of the complex membrane-bound organelles so obvious in larger eucaryotic cells (see Fig. 2.9). Photosynthesis in cyanobacteria is similar to that in eucaryotic autotrophs, requiring chlorophyll *a* and producing oxygen. Cyanobacteria are not newcomers to marine environments. Fossil stromatolites made by cyanobacteria over 3 billion years ago are remarkably similar to modern ones found at the edges of tropical lagoons in Australia (FIG. 3.4) and the Bahamas. Marine cyanobacteria are especially abundant in intertidal and estuarine areas, with a smaller role in oceanic waters. Some species of cyanobacteria produce dense blooms in warm water regions. The red phycobilin pigment of *Oscillatoria* is responsible for the color and name of the Red Sea.

Benthic cyanobacteria can be found almost everywhere light and water are available. These organisms are individually microscopic and usually inconspicuous, but they may aggregate to produce macroscopic colonies. One abundant coral reef form, *Lyngbya,* develops long strands or hollow tubes of cells nearly a meter in length. Reproduction of cyanobacteria is usually accomplished by cell fission. Occasionally, a growing colony will fragment to disperse the cells. More complex modes of reproduction, involving motile or resistant stages, are also known.

On temperate seashores, some species of cyanobacteria develop interwoven strands (FIG. 3.5) that appear as tarlike patches or mats encrusting rocks in intertidal or splash zones. Other species can be found in abundance on mudflats of coastal marshes, estuaries, and in association with tropical coral reefs. One well-studied cyanobacterium, *Microcoleus,* is a major component of complexly laminated microbial mats that construct modern stromatolites (Fig. 3.4).

Several species of cyanobacteria are capable of **nitrogen fixation,** the conversion of nonreactive atmospheric nitrogen (N_2) to nitrates, ammonia, or other reactive forms of nitrogen to satisfy their metabolic needs. The process of nitrogen fixation is not well understood, but it is known to be limited to cyanobacteria and several types of bacteria. Nitrogen-fixing cyanobacteria are fairly common in near-shore regions and are often associated with large marine plants, but they are rare in oceanic waters. Apparently, even though growth of most oceanic

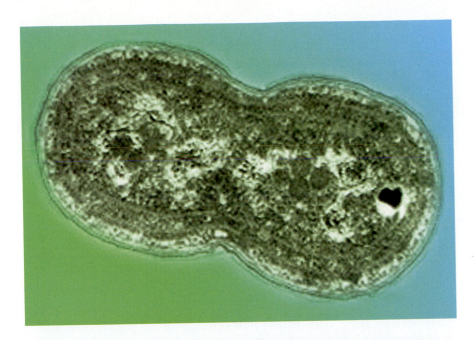

A transmission electron micrograph of a marine cyanobacterium, *Synechoccus*.

Stromatolites, resembling mushrooms 1 m high, grow on a shallow sandy bottom of Shark Bay, Australia.

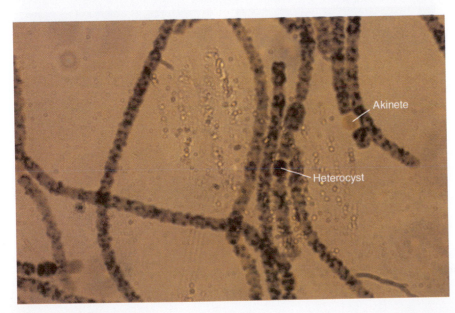

Micrograph of the cyanobacterium, *Anabaena,* showing spores (akinetes) and nitrogen-fixing heterocysts among chains of vegetative cells.

photosynthesizers is nitrogen limited, oceanic N_2-fixing cyanobacteria remain rare.

Cyanobacteria exhibit a strong tendency to form symbiotic associations with other organisms. Examples of symbiosis with animals are common, and two genera can even be found inhabiting marine planktonic diatoms, such as *Rhizosolenia*. Other cyanobacteria live as **epiphytes**, attached to larger plants. Some epiphytic cyanobacteria inhabit turtle grass beds along the Gulf Coast of the United States where they play an important role as nitrogen fixers in the overall fertility and productivity of these seagrass beds.

Chrysophyta

The division Chrysophyta consists of two classes: the Chrysophyceae and the Bacillariophyceae. Like all other eucaryotic autotrophs, their primary photosynthetic pigment is chlorophyll *a* contained in their chloroplasts. In addition, chrysophytes have accessory chlorophyll *c* and golden or yellow-brown xanthophyll pigments. Most chrysophytes have mineralized cell walls or internal skeletons made of silica or calcium carbonate. Some species possess flagella for motility but, like other planktonic organisms, can do very little to counter the drift caused by ocean currents.

Most species of Chrysophyceae are found in fresh water; however, two groups, the coccolithophores and silicoflagellates, are relatively abundant in some marine areas. Most marine coccolithophores and silicoflagellates are nanoplanktonic. Only in recent decades has the use of membrane filters, fine collection screens, and the wider application of scanning electron mi-croscopic techniques provided us with a much better look at these very small nanoplankton (FIG. 3.6).

Coccolithophores are unicellular, with numerous small calcareous plates, or **coccoliths**, embedded in their cell walls (Fig. 3.6). Although these plates are commonly observed in marine sediments, it was not until 1898 that the photosynthetic cells producing coccolith remains from seafloor sediments were directly observed. It has been suggested that coccoliths serve as a "sunscreen" to reflect some of the abundant light in clear tropical waters, permitting these organisms to thrive in areas of very high light intensity. Coccolithophores are found in all warm and temperate seas and may account for a substantial portion of the total primary productivity of tropical and subtropical oceans. In the Sargasso Sea, for instance, a single species, *Emiliania huxleyi* (Fig. 3.6a), seems to be responsible for most of the photosynthesis occurring there. However, the photosynthetic role of coccolithophores in global marine primary production is not yet well measured.

The silicoflagellates, like the coccolithophores, were first recognized and identified from fossil skeletons in marine sediments. Silicoflagellates have internal, and often ornate, silicate skeletons. They have one or two flagella and many small chloroplasts (FIG. 3.7). The significance of silicoflagellates as marine primary producers has not been evaluated, but their contribution is thought to be small. Reproduction in coccolithophores and silicoflagellates is mostly by cellular fission.

The most obvious and often the most abundant members of the phytoplankton and of the division Chrysophyta are the diatoms (class Bacillariophyceae).

(a)   (b)   (c)

figure 3.6

Scanning electron microscopy of three coccolithophore cells, each showing clearly their dense coverings of coccoliths. All are magnified approximately 4500×. (a) *Emiliania huxleyi*, (b) *Umbilicosphaera* sp., and (c) *Anthosphaera* sp.

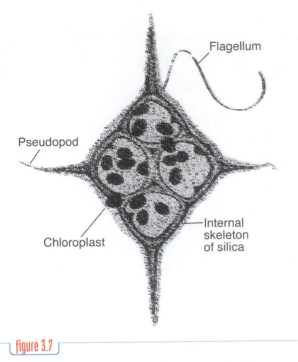

Flagellum

Pseudopod

Chloroplast

Internal skeleton of silica

Dictyocha, a common marine silicoflagellate slightly smaller than the coccolithophores shown in Figure 3.6.

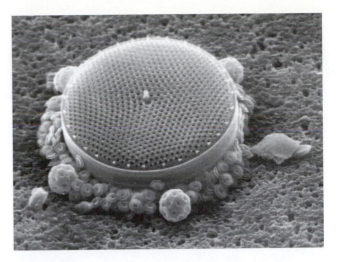

This scanning electron micrograph illustrates the size difference between a typical centric diatom, *Thalassiosira,* and four coccolithophore cells, *Crenalithus;* both the diatom and the coccolithophore cells adhere to a pad of coccoliths on the underside of the diatom cell. The association between these two species is thought to be symbiotic, but its precise nature is still unresolved.

Although diatoms are unicellular, some species occur in chains or other loose aggregates of cells. Cell sizes range from less than 15 μm to 1 mm (1000 μm). Most diatoms are between 50 and 500 μm in size and are typically much larger than coccolithophores or silicoflagellates (FIG. 3.8). Diatoms have a cell wall, or **frustule,** composed of pectin with large amounts (up to 95%) of silica. The frustule consists of two closely fitting halves, an **epitheca** and a larger **hypotheca,** which fits tightly inside the epitheca (FIG. 3.9). Planktonic diatoms usually have many small chloroplasts scattered throughout the cytoplasm, but in low light intensities the chloroplasts aggregate near the cell ends.

Diatoms exist in an immense variety of forms derived from two basic cell shapes. The frustules of most planktonic species appear radially symmetric from an end view. Circular, triangular, and modified square shapes are common. These are known as **centric** diatoms. Other diatoms, especially the benthic forms, tend to be elongated and display varying types of bilateral symmetry. These are termed **pennate** diatoms. Only pennate diatoms are capable of locomotion. The mechanism for locomotion is thought to involve a wavelike motion on the cytoplasmic surface that extends through a groove (the **raphe**) in the frustule. This flowing motion is accomplished only when the diatom is in contact with another surface. Diatoms capable of locomotion are generally

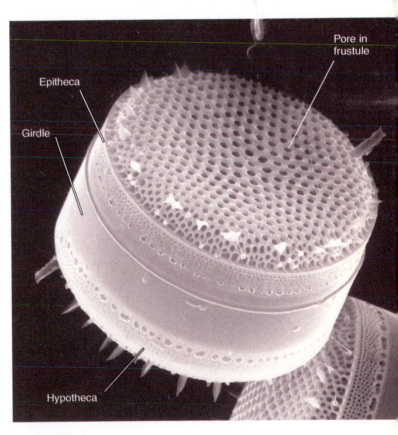

Epitheca

Girdle

Pore in frustule

Hypotheca

Another scanning electron micrograph view of *Thalassiosira,* a coastal diatom, clearly showing the epitheca, hypotheca, and a connecting girdle of cell wall material.

(a)   (b)

figure 3.10

Scanning electron micrographs of a centric diatom, *Asteromphalus heptacles.* (a) The entire cell (4700X); (b) a highly magnified portion of a similar cell showing detail of the perforations through the frustule.

restricted to shallow-water sediments or to the surfaces of larger plants and animals.

The silica frustules of diatoms exhibit sculptured pits arranged irregularly or in striking geometric patterns (FIG. 3.10A). Each pit penetrates a structural unit of the frustule, the **areolus**, which is usually hexagonal in shape (Fig. 3.10b). The outer pit connects with fine inner pores to facilitate exchange of water, nutrients, and waste between the diatom's cytoplasm and the external environment. The complex sculpturing of diatom frustules reduces the amount of silicate required for cell wall construction while dramatically increasing its mechanical strength to better resist the crushing effect of predators' jaws. The highly sculptured, yet very transparent, frustules of diatoms also act like fiberoptic light guides, possibly to direct available ambient light efficiently to the photosynthetic chloroplasts contained in the interior of the cell.

Diatoms and most other protists reproduce asexually by simple cell division. An individual parent cell divides in half to produce two daughter cells (FIG. 3.11). This method of reproduction can yield a large number of diatoms in a short period of time. When conditions for growth are favorable, a single diatom requires less than 3 weeks to produce 1 million daughter cells. Populations of diatoms and other rapidly dividing protists thus have the capacity to respond rapidly to take advantage of improved growth conditions. However, their enormous reproductive potential usually is limited by predation or availability

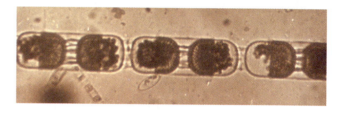

figure 3.11

Cells in a chain of *Stephanopyxis* just after synchronized division was completed. The darker half of each cell is the newly formed hypotheca, still connected by a girdle of silicate.

of light or nutrients. These limitations are addressed later in this chapter.

The sizes and shapes of diatoms imposed by their rigid frustules create a peculiar pattern of cellular reproduction (FIG. 3.12). During diatom cell division, two new frustule halves are formed inside the original frustule (Fig. 3.12b). One is the same size as the hypotheca of the parent cell (Fig. 3.12a) and is destined to become the hypotheca of the larger daughter cell (Fig. 3.12c). The other newly formed frustule half becomes the new hypotheca for the smaller daughter cell. Each daughter cell receives its epitheca from the original frustule of the parent cell. The daughter cells grow (Fig. 3.12c) and repeat the process (Fig. 3.12, d and e). This method of cell division efficiently recycles the old frustules, resulting in a slight decrease in the average cell size with each successive

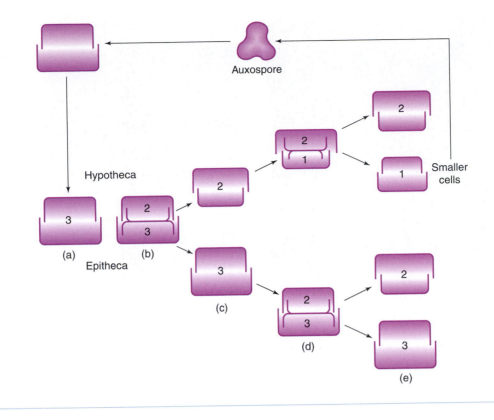

Hypotheca

Epitheca

(a) (b) (c) (d) (e)

Auxospore

Smaller cells

figure 3.12

Diatom cell division and subsequent size reduction. Numerals represent distinct cell sizes.

cell division. This size reduction, however, has not been observed in all natural diatom populations, suggesting that continual readjustment of cell diameter occurs in some species.

When cells reach a minimum of about 25% of the original cell size, these small diatoms shed their enclosing frustules (now very small) and the naked cell, known as an **auxospore**, flows out. The auxospore enlarges to the original cell size, forms a new frustule, and begins dividing again to repeat the entire sequence. Occasionally, diatoms in the auxospore stage fuse with others in a form of sexual reproduction.

The variety of planktonic diatom species existing in temperate waters is impressive. FIGURE 3.13 illustrates a few of the more common types. Benthic diatoms can be found on almost any solid substrate in shallow seawater: on mud surfaces, rocks, larger marine plants, human-made structures, and the hard shells of marine animals. One type, *Cocconeis*, even forms a thin film on the bellies of blue whales. Other benthic diatoms secrete a sticky mucilage pad to glue adjacent cells into complex chains and branching colonies a few centimeters long (FIG. 3.14).

When compared with planktonic diatoms, the geographic distribution of benthic diatoms is severely restricted because of their need for light and for solid substrates. Still, benthic diatoms make a significant contribution to the total amount of primary production in estuaries, bays, and other shallow-water areas. Some species of these diatoms are also key components in the ecologic succession of organisms that culminates in a rich growth of organisms on docks, boats, and other human-made structures. Studies have found that marine bacteria are usually the first organisms to settle and grow on new or freshly denuded underwater structures. Development of a diatom film a few cells thick quickly follows and is succeeded by more complex populations of larger algae and invertebrate animals.

Dinophyta

The division Dinophyta includes a few species that are not photosynthetic. Like other marine heterotrophs, they obtain energy from organic compounds dissolved in seawater or by ingesting particulate bits of food. However, most marine dinophytes are photosynthetic, and their share of the total marine plant production is significant. In warm seas, it often surpasses that of diatoms.

Dinophytes are typically unicellular, with a large nucleus, two flagella, and several small chloroplasts

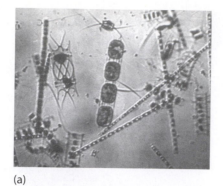

(a)

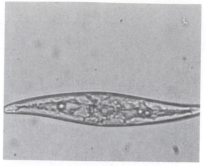

(b)

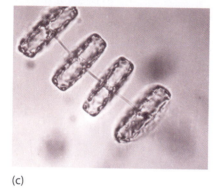

(c)

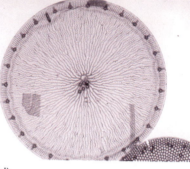

(d)

(e)

(f)

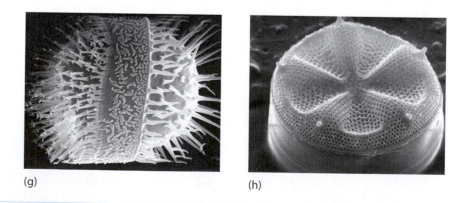

(g)

(h)

figure 3.13

Light (a–c) and scanning electron (d–h) micrographs of several common types of temperate-water planktonic diatoms: (a) mixed sample of spinous and chain-forming diatoms; (b) *Gyrosigma;* (c) *Thalassiosira;* (d) *Coscinodiscus;* (e) *Nitzchia;* (f) *Biddulphia;* (g) *Chaetoceros* resting spore; (h) *Actinoptychus.*

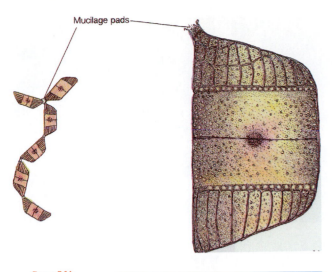

Mucilage pads

figure 3.14

A benthic diatom, *Isthmia*, forming long complex chains of cells. A closeup of a single cell is shown at right.

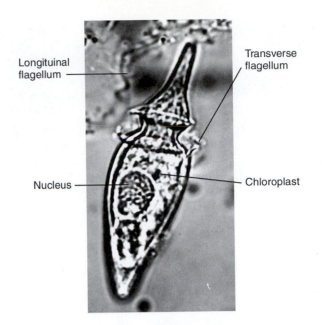

Longituinal flagellum

Transverse flagellum

Nucleus

Chloroplast

figure 3.15

This light micrograph of a dinophyte, *Oxytoxum*, illustrates the major cellular features.

containing photosynthetic pigments similar to those of diatoms (FIG. 3.15). One broad ribbonlike flagellum encircles the cell in a transverse groove and spins the cell on its axis. The other flagellum projects forward and pulls the cell, providing forward motion. Cell sizes range from 25 to 1000 µm. In armored forms, the cell wall consists of irregular cellulose plates arranged over the cell surface. The plates may be perforated by many pores. Spines, wings, horns, or other ornamentations also may decorate the cell wall. FIGURE 3.16 illustrates a few common marine dinophytes.

Dinophytes reproduce asexually by longitudinal cell division. Each new daughter cell retains part of the old cell wall and quickly rebuilds the missing part after cell division. Intermittent sexual reproduction has been reported in a few species; it is rapid and usually occurs in the dark, making it difficult to observe in natural conditions. In good growing conditions, the rate of cell division is extremely rapid and is similar to that of diatoms. Under optimal growth conditions, dense concentrations of dinophytes are produced quickly. Cell concentrations in these **blooms** are often so dense (up to 1,000,000 cells/liter) that they color the water red, brown, or green.

At night, dense blooms of luminescent forms (such as *Noctiluca* or *Ceratium*) become visible as a faint glow when disturbed by a ship's bow, a swimmer, or a wave breaking onshore. This luminescent glow is often highlighted by pinpoint flashes of light from larger crustaceans or ctenophores. This biologic production of light, or **bioluminescence**, occurs in several species of dinophytes, some marine bacteria,

and all major phyla of marine animals. Bioluminescence is produced when luciferin, a relatively simple protein, is oxidized in the presence of the enzyme luciferase. The light-producing reaction is a very energy-efficient process, producing light but almost no heat. In some species of the dinophyte, *Gonyaulax,* light production follows a daily rhythm, with maximal light output occurring just after midnight.

In the warm coastal waters of the East and Gulf coasts of the United States, the dinophyte, *Ptychodiscus,* produces toxins that in bloom conditions are known as toxic **red tides.** Toxic red tides can cause high mortality in fishes and other marine vertebrates. These dinophyte toxins either interfere with nerve functions, resulting in paralysis, or irritate lung tissues of air-breathing vertebrates, including humans. Widespread mortality of coastal fish and mammal populations sometimes occurs after particularly intense toxic red tides, fouling beaches and near-shore waters with their decomposing bodies.

Another form of indirect toxicity associated with dinophytes is created when animals (particularly shellfish) feed on dinophytes and accumulate toxins that render their flesh toxic. People who eat butter clams (*Saxodomus*) during the summer, for instance, occasionally experience paralytic shellfish poisoning from saxitoxin. However, this toxin is actually produced by the dinophyte *Alexandrium*, which is

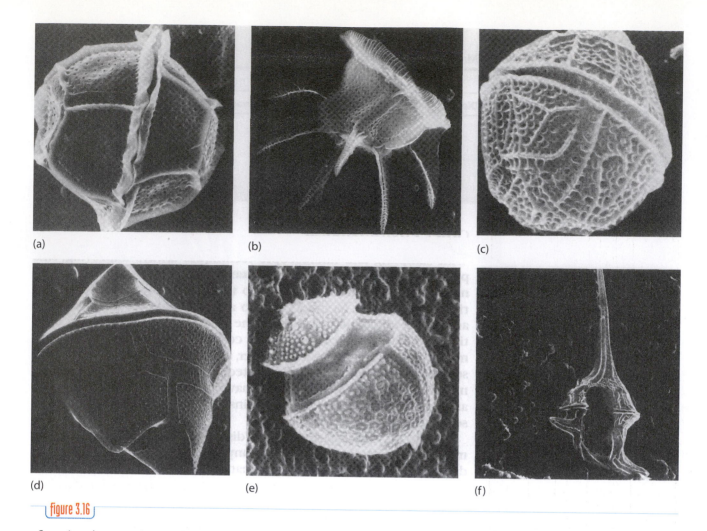

figure 3.16

Scanning electron micrographs of some common marine dinophytes: (a) *Heteroaulacus*, (b) *Ceratocoup*, (c) *Alexandrium*, (d) *Protoperidinium*, (e) *Oxytoxum*, and (f) *Ceratium*. All are 50–100 μm.

ingested and concentrated by *Saxodomus*. Other toxic conditions caused by dinophytes include ciguatera fish poisoning (from *Gambierdiscus*) and neurotoxic shellfish poisoning (from *Ptychodiscus*). A diatom, *Pseudonitzschia*, has been implicated in several recent domoic acid poisoning events affecting marine birds and mammals along the US west coast. Like saxitoxin, domoic acid is transferred to its ultimate victims through an intermediate victim, in this case, herbivorous krill.

A small group of specialized dinophytes known as zooxanthellae form symbiotic relationships with a wide range of animals, including corals, giant clams, anemones, sea urchins, and some flatworms. These symbiotic zooxanthellae account for substantial primary production in warm water marine communities, and their role in coral reef communities is described in Chapter 9.

Other Phytoplankton

With sampling and microscopic techniques continuing to improve, tiny phytoplankton from several other taxonomic groups are being recognized as important contributors to the trophic systems of many marine communities. Members of two additional classes of Chrysophyta and three unicellular classes of the division Chlorophyta are sometimes found in filtered samples of coastal seawater. Chlorophytes are much more common in fresh water, and that is where most of the research on this group is concentrated. Because most of the identified marine unicellular chlorophytes have been obtained from estuaries and coastal waters, freshwater origins for many of the species found in seawater samples is likely. TABLE 3.2 compares the distribution of cell sizes for the major groups of marine phytoplankton, based on the size terms used in Figure 3.2.

Table 3.2

Size Ranges of the Major Groups of Marine Phytoplankton

	Cyanobacteria	Chrysophyta			Dinophyta	Chlorophyta
		Diatoms	Silicoflagellates	Coccolithophores		
Picoplankton	+	+	+	+		+
Ultraplankton	+	+	+	+		+
Nanoplankton		+	+	+	+	+
Microplankton		+			++	

Adapted from Platt and Li 1986.

3.2 Special Adaptations for a Planktonic Existence

The evolutionary success of all phytoplankton hinges on their ability to obtain sufficient nutrients and light energy from the marine environment. Phytoplankton cells must be widely dispersed in their seawater medium to increase their ability to absorb dissolved nutrients, yet they must remain in the relatively restricted photic zone to absorb sufficient sunlight. These opposing conditions for successful planktonic existence have established some fundamental characteristics to which all phytoplankton and, indirectly, all other marine life have become adapted.

Phytoplankton have little or no ability of horizontal movement under their own power and must depend on the ocean's surface currents for dispersal. Adaptive features that prolong their residence time in the horizontally moving surface currents of the photic zone also serve to increase their geographic distribution.

Size

One of the most characteristic features of all phytoplankton is their small size. Almost without exception they are microscopic, which suggests that a strong selective advantage accompanies smallness in phytoplankton. Why? In contrast to land plants, phytoplankton are constantly bathed in seawater that not only provides nutrients and water but also carries away waste products. Exchange of these materials in a fluid medium is accomplished by diffusion directly across the cell membrane.

The quantity of materials required by the cell depends on factors such as the rate of photosynthesis and growth. But if these factors are held constant, the basic material requirements of the cell are proportional to the size or, more precisely, to the volume of the cell. However, the ability of the cell to satisfy its material requirements is not a function of its volume but rather the extent of cell surface across which the materials can diffuse. Thus, the ratio of cell surface area to cell volume is critical to these small cells. Smaller cells with higher surface area-to-volume ratios achieve an advantage in the competition to enhance diffusive exchange between their internal and external fluid environments (FIG. 3.17). In the same way, a nanoplankton-sized diatom 10 µm in diameter has

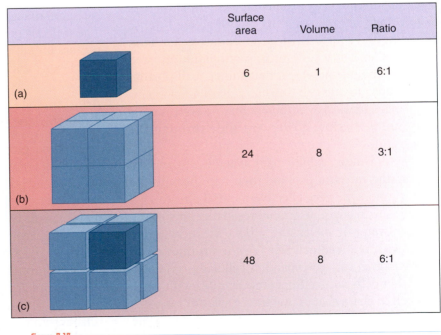

	Surface area	Volume	Ratio
(a)	6	1	6:1
(b)	24	8	3:1
(c)	48	8	6:1

figure 3.17

With increasing size (a and b) the ratio of surface area to volume decreases, unless the larger structure (c) remains subdivided so that the interior surfaces are exposed.

a 10-fold larger surface area-to-volume ratio than a microplankton-sized diatom with a cell diameter of 100 μm. And it presumably has an equally large advantage when competing for nutrients.

Reduction of cell size is an effective and widespread means of achieving high surface area-to-volume ratios, but there are other means. Many phytoplankton species have evolved complex shapes that increase the surface area while adding little or nothing to the volume. Cell shapes resembling ribbons, leaves, or long bars and cells with bristles or spines are all common mechanisms to increase the amount of surface area without adding much to their volume. Cell vacuoles filled with seawater are common in diatoms. These vacuoles make cells larger, but the actual volume of living protoplasm requiring nutrients is only a fraction of the total volume of the cell.

Long spines and horns may also make some phytoplankton less desirable to herbivorous grazers. There is some evidence to suggest that copepods, for instance, prefer nonspiny diatoms to spiny ones. Spines, cell chaining, and cell elongation all may be economic methods of increasing apparent cell size to discourage predators and reduce mortality.

Sinking

Phytoplankton cells are generally a bit more dense than seawater and tend to sink away from surface waters and sunlight. The problem for phytoplankton is to not float, because floating would create intense crowding and competition at the sea surface for light, nutrients, and space. Instead, phytoplankton need to sink, but sink slowly so that some small portion of any reproducing cell line has at least a few members carried upward by turbulent mixing even as most continue their slow downward slide through the photic zone.

Phytoplankton exhibit many adaptations that slow the sinking rate and prolong their trip through the photic zone. One very effective method already discussed is to increase frictional resistance to sinking through water by increasing surface area-to-volume ratios with reduced cell sizes or the production of spines or other surface-expanding cellular projections.

Other cells reduce their sinking rates with complex cell or chain shapes that trace zigzag or long spiral paths down through the water column. The asymmetrically pointed ends of individual *Rhizosolenia* cells create a "falling-leaf" pattern that prolongs its stay in the photic zone, whereas *Asterionella* forms long curved chains of cells that spiral slowly

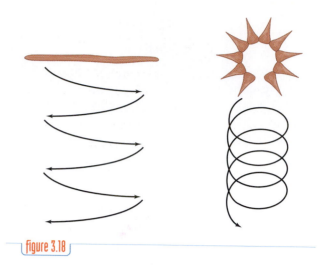

figure 3.18

Sinking patterns of the elongate diatom, *Rhizosolenia* (left), and the spiral chain-forming diatom, *Asterionella* (right).

through the water (FIG. 3.18). *Eucampia, Chaetoceros,* and many other diatoms form similar spiraled or coiled chains of cells.

Adaptations for reducing sinking rates of phytoplankton are not limited to variations in shape. Planktonic diatoms generally produce thinner and lighter frustules than do benthic diatoms. The diatom, *Ditylum,* excludes higher density ions (calcium, magnesium, and sulfate) from its cell fluids and replaces them with less dense ions. *Oscillatoria* and some other planktonic cyanobacteria have evolved relatively sophisticated internal gas-filled vesicles to provide buoyancy. The walls of these vesicles are constructed of small protein units that can withstand outside water pressures experienced anywhere within the photic zone.

Adjustments to Unfavorable Environmental Conditions

The optimal growth period for phytoplankton in nonupwelling temperate and polar seas is limited by reduced sunlight in winter and limited nutrient supplies in summer. Faced with the prospect of weeks or months with reduced photosynthesis, phytoplankton in these regions have limited options. Some move, some switch to other energy sources, and others simply persist until conditions improve. The first strategy does not generally apply to diatoms, but motility, limited as it is, is extremely important to flagellated cells. A swim of merely one or two cell lengths is often sufficient to place the cell away from its excreted wastes and into an improved nutrient supply. Toxins of dinophytes also serve to discourage predation by herbivores and sometimes inadvertently improve their own nutrient supply by causing extensive fish kills to hasten decomposition and more quickly renew limiting nutrients.

Strictly photosynthetic organisms must rely on stored lipids or carbohydrates for their short-term energy needs. When that source is depleted, some phytoplankton still have alternatives. Some species can improve their ability to harvest light by producing more chloroplasts that contain photosynthetic enzymes and pigments or by moving those chloroplasts closer to cell edges. Other species can absorb dilute but energy-rich dissolved organic material from surrounding seawater to tide them over. When these strategies have been exhausted, many diatoms produce dormant **cysts**, capsules that have reduced metabolic activity and increased resistance to environmental extremes (FIG. 3.19). Many near-shore

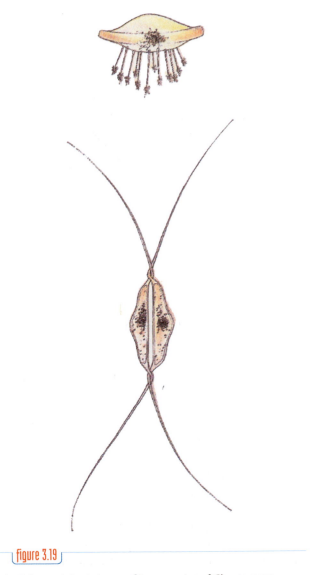

figure 3.19

Inactive resistant stages of two species of *Chaetoceros.*

species of dinophytes also produce dormant stages during periods of unfavorable growth conditions. With the return of improved growing conditions, these dormant cells germinate and commence photosynthesis and growth. At this point, the growing phytoplankton populations come under the regulatory influence of complex physical and biologic factors considered in the next sections.

3.3 Primary Production in the Sea

The two major categories of autotrophs in the sea, the pelagic phytoplankton and the attached benthic plants (described in Chapter 4), differ in much more than their physical appearance. These differences reflect adaptations to the very different physical and chemical terrains of the benthic and pelagic divisions of the marine environment. The narrow sunlit benthic fringe of the ocean is home to a variety of large, relatively long-lived, attached plants. Yet these plants account for only about 5 to 10% of the total amount of photosynthetically produced material in the ocean each year.

From our shore-based perspective, benthic plants gain immediate attention because of their high **standing crop** (the amount of plant material alive at any one time), but this is a poor indicator of their share of overall primary production. Most marine primary production is accomplished by the small dispersed pelagic phytoplankton. On an oceanic scale, the larger near-shore plants are minor players in the process of marine photosynthesis.

Primary production is a term more or less interchangeable with autotrophy or photosynthesis; it is the biologic process of creating high-energy organic material from carbon dioxide, water, and other nutrients. The organic material synthesized by the primary producers ultimately is transferred to other trophic levels of the ecosystem. Consider this example: A neatly trimmed lawn contains an easily measured amount of living plant material, its standing crop. If the lawn is maintained throughout a summer, it will be periodically mowed to maintain the same height or, in other words, the same standing crop. During that summer, the lawn clippings will total much more than the standing crop, but the lawn clippings are not part of the lawn. They represent the primary production that occurred during the summer. In an analogous sense, it is the rapidly consumed phytoplankton production, with typically very low standing crops, that fuels the metabolic processes of most of the consumers living in the sea.

Standing crop sizes at any given moment are governed by a balance between crop increases (cell growth and division) and crop decreases (sinking and grazing). Most of the primary production of a healthy actively growing phytoplankton population is not used in respiration but instead contributes to the existing standing crop. Old populations or healthy cells in poor growing conditions use a larger portion of their gross production in respiration, and net production declines.

Like our lawn analogy, the standing crop of a healthy phytoplankton population measured on successive days may demonstrate little or no increase, suggesting that no net production occurred from one day to the next. A more likely explanation is that significant net production did occur, but it replaced the portion of the crop lost to grazers (the "mowers") and to sinking. Thus, the relationship between standing crop and productivity depends to a large degree on the **turnover rate** of newly created cells.

The turnover rate of phytoplankton populations is extremely rapid. In good growing conditions, many species of large phytoplankton divide once each day, and several of the smaller species divide even faster. The coccolithophore shown in Figure 3.6a, for instance, undergoes almost two divisions per day. Its population can be completely replaced or turned over, twice each day, so comparatively few cells exist in the water at any one time. Even higher turnover rates are expected for the smaller picoplankton and ultraplankton and in benthic algae.

The total amount of organic material produced in the sea by photosynthesis represents the **gross primary production** of the marine ecosystem. Gross primary production is difficult to measure in nature; nonetheless, it is useful as a base of reference for understanding the production potentials of marine communities and ecosystems. A portion of the organic material produced by photosynthesis is used in cellular respiration to sustain the life processes of the photosynthesizers. Any excess production is used for growth and reproduction and is referred to as **net primary production.** Net marine primary production represents the amount of organic material available to support the consumers and decomposers of the sea. Different types of living material contain varying proportions of water, minerals, and energy-rich components. To avoid some of the problems encountered when comparing different types of primary producers, we commonly report standing crops in grams of organic carbon (g C). This unit represents approximately 50% of the dry weight and 10% of the live, or wet, weight of the standing crop. Primary production rates are listed in units of grams of organic carbon fixed by photosynthesis under a square meter of sea surface per day or per year (g C/m^2 per day or g C/m^2 per year).

The following discussion of the global aspects of marine primary production will necessarily be biased toward phytoplankton, but the general concepts discussed here also apply to the attached plants described in Chapter 4.

Measurement of Primary Production

Rates of primary production in the sea vary widely in time and in space, and animals exploiting the autotrophs must adapt to those patterns of variation. These production rates, and the ecologic factors that affect them, have become clearer with the development of techniques for measuring primary production in the sea. Theoretically, the net photosynthetic rate of a phytoplankton population can be estimated by measuring the rate of change of some chemical component of the photosynthetic reaction (p. 53), such as the rate of O_2 production or CO_2 consumption by phytoplankton.

The light bottle/dark bottle (LB/DB) technique was the classic approach used to study primary production in marine phytoplankton throughout much of the last century. With this method, measured changes in O_2 consumption and production were used to estimate phytoplankton respiration and photosynthetic rates. FIGURE 3.20 describes an idealized version of the LB/DB concept and is used here to introduce some of the basic concepts involved in directly measuring marine primary productivity.

The extraordinary amount of time and material resources that were needed to conduct a LB/DB study meant that locations actually sampled in a single study might be several hundred kilometers and many days apart. Environmental changes that occurred as the research vessel steamed from one station to the next could not be measured nor were the details between stations examined. It simply was assumed that the data collected at the sample stations could be averaged over the vast areas between stations and between sampling periods. The complexity and richness of small- to moderate-scale spatial variations in phytoplankton abundance were missed, as were the day-to-day variations occurring at any sampling station. Even when a procedure that used radioactive carbon (^{14}C) as a tracer of CO_2 in photosynthesis was introduced in the middle of the last century, most of the logistical limitations of the LB/DB approach remained.

In contrast to ship-based sampling, satellite sampling can provide a general, and instantaneous, overview of a large portion of ocean (FIG. 3.21). Satellites

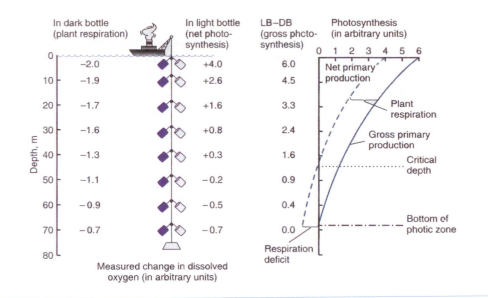

figure 3.20

The results of a hypothetical light and dark bottle experiment. Water samples from 10-m depth increments are replaced at original depths in paired light and dark bottles. After a period of time, the bottles are retrieved and changes in O_2 are determined. Dark bottle (DB) values indicate O_2 decreases at each depth due to respiration without photosynthesis. Light bottle (LB) values represent O_2 changes from photosynthesis and respiration (net primary production) in the light. The difference between the two values (LB − DB) is the gross primary production. With these values, net and gross primary production curves can be drawn to represent the variation in photosynthesis with depth.

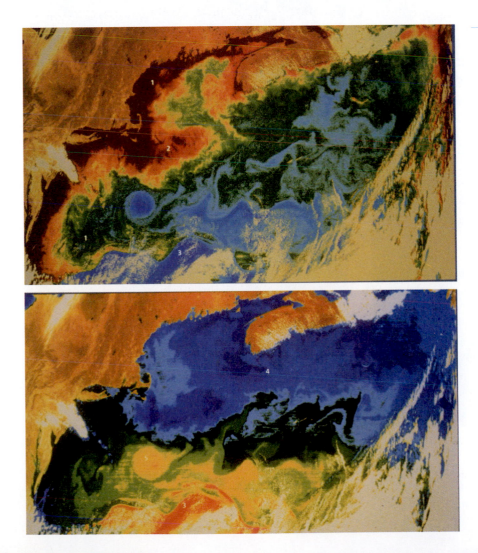

figure 3.21

Composite satellite views of the North Atlantic Ocean along the northeast coast of the United States. (Top) Phytoplankton concentrations, ranging from low (dark blue) to high (red). (Bottom) Corresponding sea surface temperature of the area shown, ranging from warm (red) to cold (dark blue). Generally, phytoplankton concentrations are highest where water is coldest.

cannot directly measure marine primary productivity. Instead, subtle changes in ocean surface color, which signify various types and quantities of marine phytoplankton, are measured. The more phytoplankton present, the greater the concentration of chlorophyll pigments and the greener the water. In the 1980s, a coastal zone color scanner (CZCS) aboard the NIMBUS 7 weather satellite was used to make the first global-scale measurements of ocean surface color that, with calibration from ship-based direct measurements, were used to estimate phytoplankton standing crops and growth rates and to extrapolate shipboard productivity measurements to large oceanic areas.

Remote sensing of ocean color by satellite was the first technique to measure marine primary productivity on a global scale with enough resolution to permit analyses of phytoplankton changes over time

scales of weeks or years. Since the CZCS was lofted into orbit, satellite imagery has revolutionized our view of primary productivity patterns in the ocean. As is apparent in Figure 3.21, distribution patterns of phytoplankton are complex and show some similarities with sea surface temperature distributions. Patches and eddies of phytoplankton are common. In upwelling areas, plumes of phytoplankton-rich water (known as squirts) extend as much as 200 km offshore. Before satellite observations, these dominant features of marine phytoplankton distribution were completely unknown (FIG. 3.22).

The sea-viewing wide field-of-view sensor (SeaWiFS) now operates as a follow-on sensor to the CZCS, which ceased operations in 1986. The SeaWiFS mission is a part of NASA's Earth Science Enterprise, which is designed to better understand our planet by examining it from space. The SeaWiFS

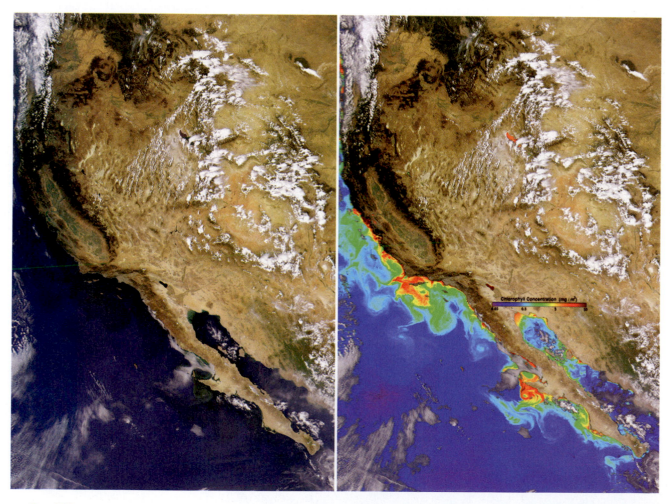

figure 3.22

This phytoplankton bloom along the California coast was imaged by SeaWiFS on 10–11 August 2003 for true color (left) and for chlorophyll *a* concentrations (right), which are proportional to the abundance of phytoplankton. Although the true color image is impressive, it does not provide the information about the exact quantity of phytoplankton that the image at right does.

Oceanography from Space

Observations of ocean conditions at sea have been the basis for amassing the extensive store of information that underlies much of our present understanding of oceanic structures and their functions. Typically, this information has been collected from ships, submersibles, and unmanned instrument buoys. These approaches were (and still are) appropriate for studying such conditions as water temperature, waves and currents, salinity, and the change in marine organisms from the sea surface downward; however, they were less effective in providing detailed views of how the same features varied through time or were distributed across oceanic distances. So, until the middle of the 20th century, long-time series of data collections were combined and averaged to provide a "typical" view of any of these features at large oceanic scales. It was not possible to obtain a synoptic, or instant snapshot, view of how any major component of the oceans was behaving at any particular time.

In 1959, the first of the TIROS series of weather satellites was placed in orbit. A remote eye-in-the-sky view of weather patterns over whole oceans has become standard (and expected) fare for television weather reports. The TIROS satellites have since been replaced with two newer series, NIMBUS and NOAA. Their capability to detect and track every hurricane on either side of North America has proved invaluable for saving lives and property.

Since TIROS I, numerous nondefense satellites capable of studying the oceans have been placed in orbit by the United States and many other nations. One of these, SEASAT, was the first specifically designed for ocean research. It lasted only 3 months before failing, but in that short time it provided an extensive global view of the variation in altitude of the sea surface. The radar beams directed from SEASAT could measure the distance between the satellite and the sea surface to within 5 to 7 cm. Useful information about the structure and composition of the seafloor was obtained by mapping the ocean surface from space. For example, the Gulf Stream varies about 100 cm in height across its width. Over seafloor trenches, the sea surface is as much as 60 m closer to the center of the earth, and a seamount causes the sea surface to bulge out about 5 m. SEASAT mapped these variations in sea surface topography, and the maps have been used to construct a composite views of how surface current patterns vary over time periods of a few weeks.

With the shutdown of SEASAT, oceanography from space has continued with instruments placed aboard satellites dedicated to other remote-sensing tasks. The radar altimeter measurements initiated on SEASAT were continued on the GEOS satellites and now on Ocean Topography Experiment (TOPEX)/Poseidon (Figure B3.1). This system provides global sea level measurements with even better precision than did SEASAT. The data are used to determine global ocean circulation and to understand how the oceans interact with the atmosphere, which improves our ability to predict global climate. JASON is the successor satellite mission to TOPEX/Poseidon and will monitor global ocean circulation to improve global climate predictions of events such as El Niño.

A multisatellite system currently provides continuous and precise three-dimensional global position fixing capabilities anywhere on Earth. Although none of these satellites was specifically designed for ocean observations, they have provided oceanographers with a new and valuable ability to pinpoint the location of an experiment or a sample site at sea. In the 1970s, the first of the LANDSAT series of satellites were

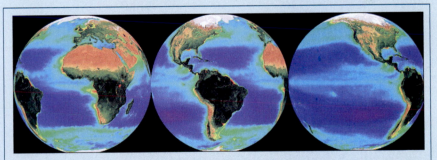

Figure B3.1 This TOPEX/Poseidon image is a global map of sea surface height, accurate to within 30 mm. The height of the water relates to the temperature of the water. Yellow and red areas indicate where the waters are relatively warmer and have expanded above sea level, green indicates near-normal sea level, and blue and purple areas show where the waters are relatively colder and the surface is lower than sea level.

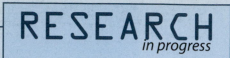

RESEARCH
in progress

Oceanography from Space

launched to assess land-based resources. The instruments on the LANDSAT satellites have recorded millions of images that are an important archival resource for global change research, agriculture, geology, and even national security.

The coastal zone color scanner (CZCS) on NIMBUS 7 monitored chlorophyll concentrations in surface waters and mapped the large-scale distribution and abundance of marine phytoplankton, with a capability to repeat each observation every six days. The Earth-orbiting ocean color sensor, SeaWiFS (Sea-

viewing *Wide Field-of-view Sensor*), was launched in 1997 to replace the CZCS system and has performed perfectly since. The purpose of Sea-WiFS is to provide quantitative data on global ocean color characteristics. Subtle changes in ocean color provide information on types and quantities of marine phytoplankton. SeaWiFS has continued the CZCS monitoring of short-term variability in marine phytoplankton productivity, contributing high-quality digital data sets to monitor the ocean's response to global warming and other global-scale changes.

Other satellites are equipped with radar, microwave, or infrared sensors and can measure sea-surface temperature, rain rate, wind speed, sea-surface roughness, and distribution of sea ice. The increased use of satellite-based remote sensing devices has opened exciting possibilities for acquiring previously unattainable synoptic views of large expanses of the world ocean and has ushered in a new era of global oceanography.

Presently, satellite remote sensing is the only method that can provide global-scale information about variations of sea-surface temperature, sea-surface altitude, and phytoplankton productivity. Many of the newer ocean-oriented remote-sensing satellite missions, ranging from sea-surface temperature mapping to tracking of individual whales, were initiated during the 1990s as part of NASA's Mission to Planet Earth.

Additional Reading
Hoffman R, Markman A, eds. *Interpreting Remote Sensing Imagery: Human Factors*. Boca Raton, FL: Lewis Publishers, 2001.

www.bioscience.jbpub.com/marinebiology
For images of the Earth's oceans from space, go to this book's website at http://www.bioscience.jbpub.com/marinebiology and look for the Oceanography from Space link in Chapter 3.

---

sensor provides 1-km resolution at the sea surface in eight color bands. Because this orbiting sensor can view every square kilometer of cloud-free ocean every 48 hours, these satellite-acquired ocean color data constitute a valuable tool for determining the abundance of ocean biota on a global scale and can be used to assess the dynamics of short-term changes in primary productivity patterns. Figure 3.22 and several of the images included at the end of Chapter 4 are from the SeaWiFS system.

Factors that Affect Primary Production

The continued synthesis of organic material by marine phytoplankton depends on a set of interacting biotic (biologic) and abiotic conditions. If nutrients, sunlight, space, and other conditions necessary for growth are unlimited, phytoplankton population sizes increase in an exponential fashion (FIG. 3.23).

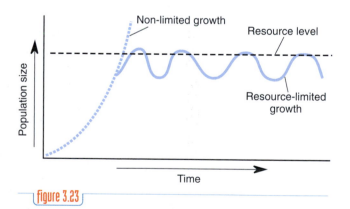
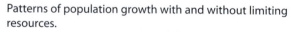

Patterns of population growth with and without limiting resources.

In nature, phytoplankton populations do not continue to grow unchecked as the unlimited growth curve in Figure 3.23 suggests. Rather, their sizes are controlled by their tolerance limits to critical environmental factors (including predators) or by the availability of substances for which they have a need. Any condition that exceeds the limits of tolerance or does not satisfy the basic material needs of an organism establishes a check on further population growth and is said to be a **limiting factor.**

Phytoplankton populations limited by one or a combination of these factors are forced to deviate from the exponential growth curve shown in Figure 3.23. Important limiting factors for phytoplankton are grazing by herbivores and the availability of light and nutrients. In the ocean, each major group of phytoplankton responds differently to combinations of these factors. In general, diatoms and silicoflagellates thrive in lower light intensities and colder water than do dinophytes and coccolithophores. Consequently, conditions that promote the growth of either group tend to exclude the other. These factors are examined, first alone and then in concert, in an attempt to convey some sense of the complex dynamic interactions that exist between photosynthesizers and their immediate surroundings. These same factors also influence or regulate the growth of large marine plants. These plants are introduced in Chapter 4, followed by an overview of patterns of total primary productivity in the sea.

Grazing

The trophic interrelationships of marine primary producers and their small herbivorous grazers (mostly zooplankton and small fishes) can be complex and are described in more detail in Chapter 10. Intensive grazing can decrease the standing crop and sometimes the productivity of a phytoplankton population. Ideally, grazing rates should adjust to the magnitude of primary productivity to establish a rough balance between producer and consumer populations. Photosynthetic rates do limit the average size of the animal populations that primary producers support, yet short-term fluctuations of both phytoplankton and grazer populations typically occur. The magnitude of these fluctuations tends to be moderated somewhat by stabilizing **feedback mechanisms** between all trophically related populations. An abundant food supply permits the grazers to reproduce and grow rapidly (FIG. 3.24). Eventually, however, they consume their prey more quickly than the prey can be replaced. Overgrazing reduces the phytoplankton population and its photosynthetic capacity, causing food shortages, starvation, and consequent reductions of the enlarged herbivore populations. When

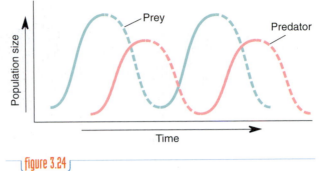

figure 3.24

Generalized population changes of a prey species and its predator, oscillating between unlimited (solid) and limited (dashed) phases of population growth.

grazing intensity is reduced after herbivore populations crash, the phytoplankton population may recover, increase in size, and again set the stage with an abundant food source to cause a repeat of the entire cycle. Similar fluctuations in population size may reverberate through many trophic levels of the food web.

In addition to the large-scale geographic variations in phytoplankton density observed from satellites (Figs. 3.21 and 3.22), marine phytoplankton also exhibit much smaller-scale localized patchiness. Dense patches of phytoplankton tend to alternate with concentrated patches of zooplankton. The inverse concentrations of phytoplankton and zooplankton densities stem in part from the effects of grazing and because of differences in their reproductive rates. Initially, a dense patch of phytoplankton provides favorable growth conditions for herbivores attracted from adjacent water into the phytoplankton patch. The grazing rate increases in the area of the patch and declines elsewhere. Production soon decreases in the original patch and increases in adjacent areas. Eventually, the original phytoplankton patch is eliminated by the increasing numbers of grazers. The adjacent areas become the new phytoplankton patches and attract herbivores from the recently overgrazed region to repeat the entire sequence. Similar patchy patterns of distribution may be established and maintained physically by **Langmuir cells** (named after Irving Langmuir, who first clarified their structure after he observed *Sargassum* in the North Atlantic floating in long rows parallel to the wind direction). This material is often evident at the surface as long parallel "slicks," foam lines, or rows of floating debris (FIG. 3.25). Although Langmuir cells extend only a few meters deep, they may create particle and nutrient traps

figure 3.25

Langmuir slicks.

under the convergences. Phytoplankton and particulate debris that accumulate under the convergences attract grazing zooplankton in concentrations often 100 times as dense as those in adjacent areas.

Light

The requirement for light imposes a fundamental limit on the distribution of all marine photosynthetic organisms. To live, these organisms must remain in the photic zone. The depth of the photic zone is determined by a variety of conditions, including the atmospheric absorption of light, the angle between the sun and the sea surface, and water transparency.

Water is not very transparent to light. This low transparency causes light intensity in seawater to diminish as it penetrates downward from the sea surface. At some depth, the light intensity is so faint that no photosynthesis will occur. This depth defines the bottom of the photic zone and varies from a few meters deep in coastal waters to more than 200 m in clear tropical seas. At a depth somewhat above the bottom of the photic zone, the rate of photosynthesis is balanced by photorespiration. This depth of no net primary production is the **critical depth** (Fig. 3.20). The critical depth is approximately equivalent to the depth at which the available light is reduced to about 1% of its summertime surface intensity (see Fig. 1.21). In clear tropical waters, the critical depth often extends below 100 m throughout the year. In higher latitudes, it may reach 30 to 50 m in midsummer, but it nearly disappears during winter. These are average critical depths for mixed phytoplankton communities composed of many different species; each species has its own peculiar critical depth.

In moderate and low light intensities, photosynthesis by phytoplankton exhibits a direct relationship to light intensity (FIG. 3.26). At higher light intensities, photosynthesis ceases to follow the light intensity curve; it may stabilize or even decrease nearer the sea surface because the photosynthetic machinery of phytoplankton cells is saturated with light. Higher light intensities nearer the sea surface fail to promote further increases in photosynthesis.

Phytoplankton from different environments exhibit some degree of photosynthetic adjustment to varying light intensities. Therefore, the saturation light intensity for any phytoplankton population changes with changing sets of environmental conditions. Variations in saturation light intensities are also found among major phytoplankton groups. Dinophytes seem to be better adapted than diatoms to intense light. As a result, their relative contribution to the total marine primary production is greater than that of diatoms in tropical and subtropical regions.

Photosynthetic Pigments

The photosynthetic apparatus of all marine primary producers except cyanobacteria is located in the chloroplasts of actively photosynthesizing cells. It is in the chloroplasts (or the whole cells of Cyanobacteria) that the pigment systems are located. Chloroplasts contain chlorophyll and varying amounts of other photosynthetic pigments (Tables 3.1 and 4.1). There, these pigments absorb light energy and convert it to forms of chemical energy that can be used

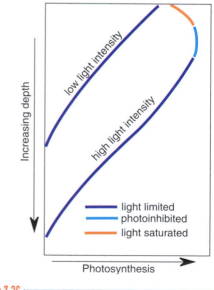

figure 3.26

Relationship between photosynthesis and depth at low and high light intensities.

by the photosynthesizer and by those intent on consuming phytoplankton.

Both cyanobacteria and eucaryotic autotrophs use an elaborate two-part photosynthetic process involving complex pigment systems and two distinct sets of chemical reactions. In the first set, the **light reaction** portion of photosynthesis (FIG. 3.27), photons of light are absorbed by chlorophyll molecules located in two separate pigment systems. The photons energize electrons and pump them through a series of other enzymes whose function is to manage some of that electron energy and transfer it to adenosine triphosphate (ATP) and another high-energy carrier molecule, NADPH. As the term implies, light is needed to drive the light reaction; without light, the reaction ceases.

The pigment systems and enzymes involved in the light reaction are housed within flattened sacs, which are stacked to form numerous **grana** within each chloroplast (FIG. 3.28). The **stroma** surrounds the grana and contains the enzymes needed for the next step of photosynthesis, the **dark reaction.** Light energy is not necessary to maintain the dark reaction, but the high-energy ATP and NADPH produced by the light reaction are. Energy from these substances is used in the dark reaction to synthesize carbohy-

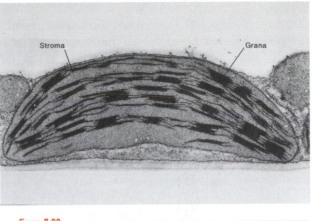

figure 3.28

A transmission electron micrograph of a chloroplast with stacked sets of parallel membranes (narrow dark bands).

drates, lipids, and the other organic compounds needed by the cell.

Chlorophyll appears green for the same reason coastal seawater appears green. Both absorb more of the available light energy from the violet and red ends of the spectrum, leaving the green light to be reflected back or to penetrate more deeply. Chlorophyll serves as the basic energy-absorbing pigment for land plants.

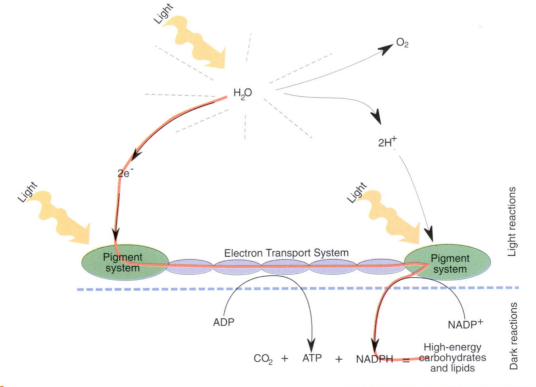

figure 3.27

Diagrammatic representation of the photosynthetic mechanism of eucaryotic autotrophs. The colored line traces the path of electrons initially activated by light energy.

However, a few meters of seawater absorbs much of the red and violet portions of the visible spectrum before it reaches the chloroplasts of most marine plants. Because chlorophyll best absorbs energy from red and violet light, its effectiveness as an absorber of available light energy is greatly reduced in seawater.

The evolutionary response of most marine primary producers has been to supplement the light-absorbing ability of chlorophyll with **accessory pigments** (FIG. 3.29). These pigments absorb light energy from spectral regions where chlorophyll cannot and then transfer the energy to chlorophyll for use in the light reaction. Figure 3.29 illustrates the complementary effect of chlorophyll *a* and accessory pigments such as fucoxanthin, which is found in diatoms, dinophytes, and brown algae. Fucoxanthin absorbs light primarily from the blue and green region of the spectrum, the region where chlorophyll absorbs light least effectively. In combination, chlorophyll and fucoxanthin are capable of absorbing energy from most of the visible light spectrum. Another group of accessory pigments, the phycobilins, are found in red algae and cyanobacteria. These pigments have absorption spectra much like that of fucoxanthin. These and other accessory pigments listed in Table 3.1 (see Table 4.1 for marine plants) have enabled various groups of marine plants to adapt to the limited conditions of light availability in seawater by absorbing light energy at almost any depth within the photic zone.

Nutrient Requirements

The nutrients required by all primary producers are a bit more complex than might be indicated by the general photosynthetic equation

$$6CO_2 + 12H_2O \rightarrow C_6H_{12}O_6 + 6H_2O + 6O_2$$

Proper growth and maintenance of cells depend on the availability of more than just water and carbon dioxide because plants are composed of compounds that cannot be assembled from C, H, and O alone.

These nutrient requirements can be best understood by determining the basic composition of the cell itself. Chemical analysis of a hypothetical "average" marine primary producer might yield the results shown in FIGURE 3.30. In general, marine primary producers experience no difficulty in securing an adequate supply of water. Most are continuously and completely bathed by seawater, and few cells of any marine plant are seriously isolated from the external water environment.

Coccolithophores and some seaweeds are equipped with cell walls or internal skeletons of calcium carbonate ($CaCO_3$). Carbon dioxide for carbonate formation and for photosynthesis exists in seawater as carbonic acid (H_2CO_3), bicarbonate (HCO_3^-), and carbonate (CO_3^{2-}) The abundance of these ions in seawater is influenced by photosynthesis, respiration, water depth, and pH balance. However, the concentration of total CO_2 present in seawater is not low enough to inhibit photosynthe-

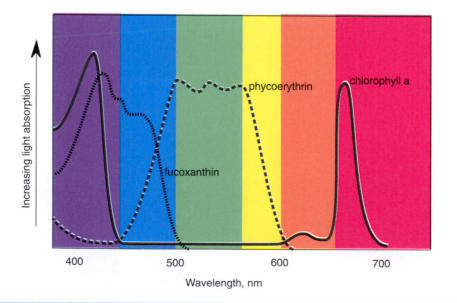

figure 3.29

Patterns of light absorption for three photosynthetic pigments: phycoerythrin (an accessory pigment found in Rhodophyta and Cyanobacteria), fucoxanthin (an accessory pigment found in Phaeophyta, Chrysophyta, and Dinophyta), and chlorophyll *a*. (Adapted from Saffo, 1987.)

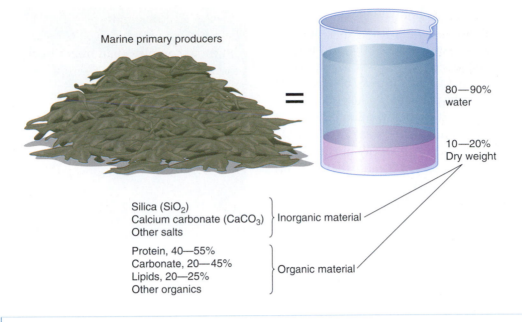

Marine primary producers

=

80—90% water

10—20% Dry weight

Silica (SiO$_2$)
Calcium carbonate (CaCO$_3$) } Inorganic material
Other salts

Protein, 40—55%
Carbonate, 20—45% } Organic material
Lipids, 20—25%
Other organics

figure 3.30

Chemical composition of typical marine autotrophs.

sis or the formation of CaCO$_3$. Calcium ions (Ca^{2+}) necessary for calcium carbonate formation are also very abundant in seawater at all depths (see Table 1.3). Silica (SiO$_2$) is required by silicoflagellates and diatoms, and concentrations of dissolved silica occasionally become so depleted that the growth and reproduction of these phytoplankton groups are inhibited.

Organic matter is a widely used term collectively applied to those biologically synthesized compounds that contain C, H, usually O, lesser amounts of reactive nitrogen (N) and phosphorus (P), and traces of vitamins and other elements necessary to maintain life. Proteins, carbohydrates, and lipids are the most abundant types of organic compounds in living systems. Each contains carbon, hydrogen, and oxygen in varying ratios. FIGURE 3.31 summarizes the generalized nutrient needs of photosynthetic cells.

How much of each of these elements do primary producers require? Chemical analyses of whole phytoplankton cells grown under various light conditions provide an average atomic ratio of approximately 110(C):230(H):75(O):16(N):1(P). Carbon, hydrogen, and oxygen are abundantly available from carbonate (CO$_3^{2-}$) or bicarbonate ions (HCO$_3^-$) and water (H$_2$O). Reactive nitrogen is much less plentiful but is present in seawater as nitrate (NO$_3^-$), with lesser amounts of nitrite (NO$_2^-$), and ammonium (NH$_4^+$). High concentrations of molecular nitrogen (N$_2$), which constitutes 78% of the Earth's atmosphere, are also

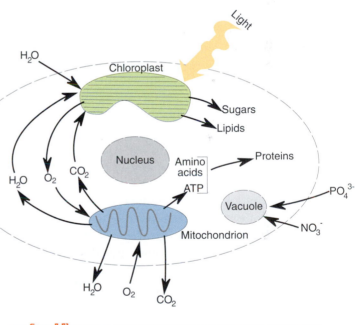

figure 3.31

A simplified photosynthetic cell, illustrating the chemical requirements and products of several components of the cell.

dissolved in seawater. However, most marine organisms are not metabolically equipped to use this nonreactive form of N. The cyanobacteria that can, however, contribute a substantial portion of the total N used by other phytoplankton in nutrient-depleted seas by converting nonreactive N$_2$ to more reactive NO$_3^-$ and NH$_4^+$. Phosphorus, present principally as phosphate (PO_4^{3-}), is less abundant in seawater than is nitrate. The biologic demands for phosphate are also less but just as critical (e.g., in the synthesis of ATP and cell membranes). The ratio of usable N and P in seawater is similar to the ratio of 16N:1P found in living cells of marine primary producers.

FIGURE 3.32 shows the vertical distribution patterns of silicate, nitrate, and phosphate in seawater. These nutrients are usually in short supply in the photic zone during the growing season because they are continually consumed by primary producers. In periods of rapid phytoplankton growth, needed quantities of one or more of these nutrients may not be available. In such circumstances, continued growth is limited by the rate of nutrient regeneration.

In addition to the major nutrient elements just described, marine autotrophs require several other elements in minute amounts. These **trace elements** include iron, manganese, cobalt, zinc, copper, and others. Depletion of iron in English Channel waters has been observed during spring diatom blooms, suggesting that iron availability may limit the size or composition of phytoplankton populations. More recently, the results of both laboratory and field attempts to enrich seawater artificially with iron demonstrate marked increases in phytoplankton growth. In natural systems such as the North Atlantic Ocean, massive inputs of iron from windborne dust carried from the Sahara Desert of Africa (FIG. 3.33) suggest that phytoplankton growth rates may be higher than in comparable areas isolated from similar inputs of iron-rich dust.

Vitamins too are crucial for the proper growth and reproduction of primary producers. Some species of diatoms, for example, require more vitamin B_{12} during auxospore formation than at other times. Some can synthesize their own vitamins; others must rely

figure 3.33

SeaWiFS image of airborne Saharan dust being carried westward over the Canary Islands and beyond into the North Atlantic Ocean.

on free-living bacteria to provide this and other critical vitamins that they cannot synthesize for themselves.

Nutrient Regeneration

Most of the biomass produced by marine photosynthesis is eventually consumed by herbivores and converted to more herbivore bodies or to be formed into fecal wastes. In either case, these compact particles quickly become colonized by bacteria and sink as "marine snow" to depths well below the photic zone. It is these deep cold portions of the world ocean that contain the large reserves of dissolved nutrients. Consequently, nutrient-rich waters are almost always cold waters, and water that is warm has likely been at the sea surface for some time during which its nutrient load has been depleted.

Regeneration of the nutrients initially used to produce phytoplankton cells or marine plants is dependent on respiration by consumers and on decomposition of organic material by bacteria and fungi living in the water column and on the sea bottom. Bacterial action decomposes organic material and returns phosphates, nitrates, and other nutrients to seawater in inorganic form for reuse by the primary producers. Bacteria also absorb dissolved organic compounds from seawater and convert them to living cells that become additional food sources for many benthic and small planktonic animals (FIG. 3.34). These **microbial loops** divert organic material from typical planktonic food chains (described in Fig. 2.13) to populations of bacteria that in turn feed a variety of microzooplankton. As much as half the total marine

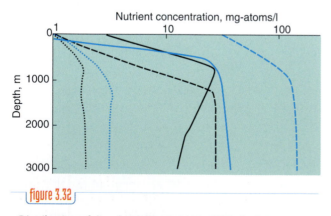

figure 3.32

Distribution of dissolved silicate (dotted lines), nitrate (dashed lines), and phosphate (solid lines) from the surface to 3000 m in the Atlantic (black) and Pacific (blue) Oceans. (Adapted from Sverdrup, Johnson, and Fleming 1942.)

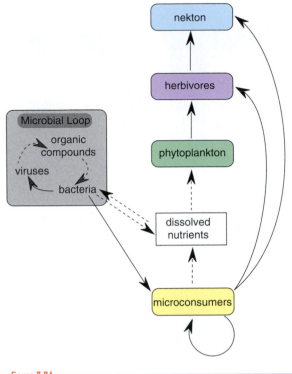

figure 3.34

Major pathways of cycling dissolved nutrients and food particles through a microbial loop (left) and a particulate food web (right). Dissolved organic materials and inorganic nutrients are indicated with dashed lines.

primary productivity may be directed into these microbial loops through planktonic bacteria.

Figure 3.32 indicates that major concentrations of limiting nutrients accumulate below the photic zone, where they cannot be accessed by photosynthesizers. These two fundamental needs, light from the sea surface and nutrients from below, impose severe restrictions on the rates of primary production. For much of the ocean, the sunlit photic zone is isolated from the nutrients of the deeper waters by a well-developed and permanent pycnocline. Here, very slow molecular diffusion is the only process to return nutrients to the photic zone. Marine primary producers really thrive only in those parts of the sea where dynamic processes move colder nutrient-laden waters upward into the photic zone. These large-scale mixing processes include small-scale turbulence and upwelling that rapidly transport nutrient-rich deep water upward.

Wind, waves, and tides create turbulence in near-surface waters and mix nutrients from deeper water upward. Turbulent mixing is most effective over continental shelves, where the shallow bottom prevents the escape of nutrients into deeper water. Tidal currents in the southern end of the North Sea and the

eastern side of the English Channel, for example, are sufficient to mix the water almost completely from top to bottom. As a result, summer phytoplankton productivity there remains high as long as sunlight is sufficient to maintain photosynthesis.

In tropical and subtropical latitudes of most oceans, the strong year-round thermocline and associated pycnocline near the base of the photic zone act as a strong barrier to inhibit upward mixing of deep nutrient-rich waters (FIG. 3.35 top). Consequently, these regions have very low rates of primary production, comparable with terrestrial deserts.

Pycnoclines also develop in temperate waters to restrict the return of deep-water nutrients, but only on a seasonal basis (Fig. 3.35, center). During winter, the surface water cools and sinks. The pycnocline disappears, and deeper nutrient-rich water is mixed with the surface water. As solar radiation increases in the spring, the surface water warms and the thermocline is reestablished. A well-developed summer pycnocline resembles the permanent pycnocline of tropical and subtropical waters and creates an effective barrier, blocking nutrient return to the photic zone. With shorter days and cooler weather in autumn, the pycnocline weakens and then disappears in winter. Without a pycnocline to interfere, **convective mixing** in temperate regions continues from late fall to early spring. But in higher latitudes, continuous heat loss from the sea to the atmosphere and low amounts of solar radiation produce year-round convective mixing (Fig. 3.35, bottom). Low light conditions rather than scarce nutrients usually limit the primary production in these polar regions.

Subsurface water rich in dissolved nutrients is carried up into the photic zone by several processes collectively termed **upwelling**. One type, coastal upwelling, is produced by winds blowing surface waters away from a coastline. The surface waters are replaced by deeper waters rising to the surface (FIG. 3.36). Near-shore currents, which veer away from the shoreline, produce the same result. Four major coastal upwelling areas occur in the California, Peru, Canary, and Benguela currents and lesser ones occur along the coasts of Somalia and western Australia. With the exception of the Somali Current, these currents are on eastern sides of subtropical current gyres (FIG. 3.37) and flow toward the equator. FIGURE 3.38 illustrates the influence of upwelling on nutrient availability in the photic zone. Note that the nitrate concentrations at a depth of about 50 m are 5 to 10 times higher in the upwelling systems than at similar depths in adjacent non-upwelled water.

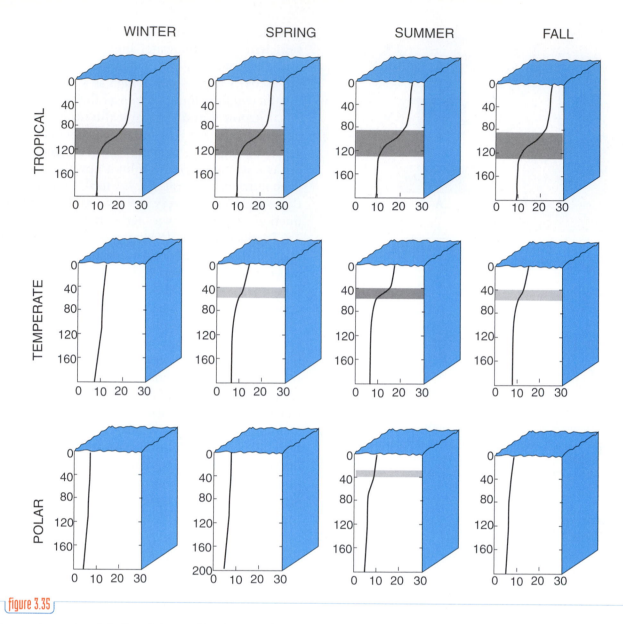

WINTER SPRING SUMMER FALL

TROPICAL

TEMPERATE

POLAR

figure 3.35

Seasonal growth and decline of thermoclines in tropical (top), temperate (center), and polar (bottom) ocean waters. Depth, in meters, is on vertical axes; temperature, in degrees Celsius, on horizontal axes.

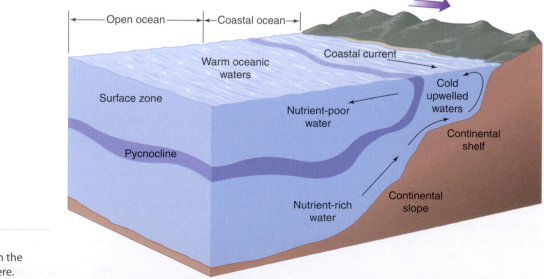

Winds

Open ocean — Coastal ocean

Warm oceanic waters

Coastal current

Surface zone

Cold upwelled waters

Nutrient-poor water

Continental shelf

Pycnocline

Nutrient-rich water

Continental slope

figure 3.36

Coastal upwelling in the Northern Hemisphere.

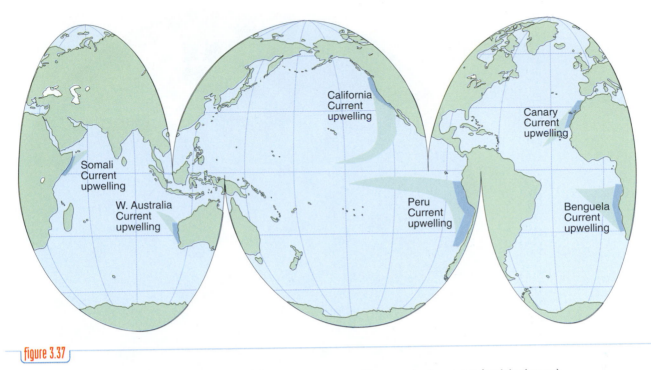

figure 3.37

Principal regions of coastal upwelling (blue) and down-current areas of increased primary productivity (green).

Another type of upwelling is more limited in extent and normally exists only in the central Pacific Ocean. The Pacific Equatorial Current flows westward, straddling the equator. The Coriolis effect causes a slight displacement to the right for the portion of the current in the Northern Hemisphere and to the left for the portion of the current in the Southern Hemisphere (see Fig. 1.36). The resultant divergence of water away from the equator creates an upwelling of deeper water to replace the water that has moved away (FIG. 3.39). The resulting increase in phytoplankton growth feeds a "downstream" population of zooplankton to the north and south of the Equator.

In large discontinuous patches around the Antarctic continent, another type of upwelling occurs. A consequence of thermohaline circulation patterns outlined in Figure 1.44, massive volumes of nutrient-rich North Atlantic Deep Water slowly, yet continuously, drift toward the ocean surface in the Antarctic Divergence Zone between 60 and 70° S latitude. For most of the year, phytoplankton production is inhibited by the absence of light (FIG. 3.40). In summer, however, this liquid conveyor belt flowing under much of the Atlantic Ocean delivers massive amounts of dissolved nutrients into the photic zone to support one of the richest communities in the marine environment.

Collectively, the dynamic interactions of all the processes described here that affect marine primary productivity create complex spatial and temporal

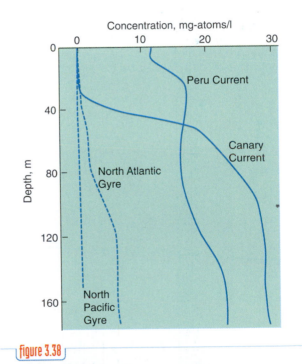

figure 3.38

A comparison of the vertical distribution of nitrate in upwelling areas (heavy curves) and in adjacent non-upwelling central ocean regions (light curves). (Redrawn from Walsh 1984.)

patterns of phytoplankton abundance. These are pulled together in Chapter 4 after the rest of the marine primary producers, the plants, have been introduced.

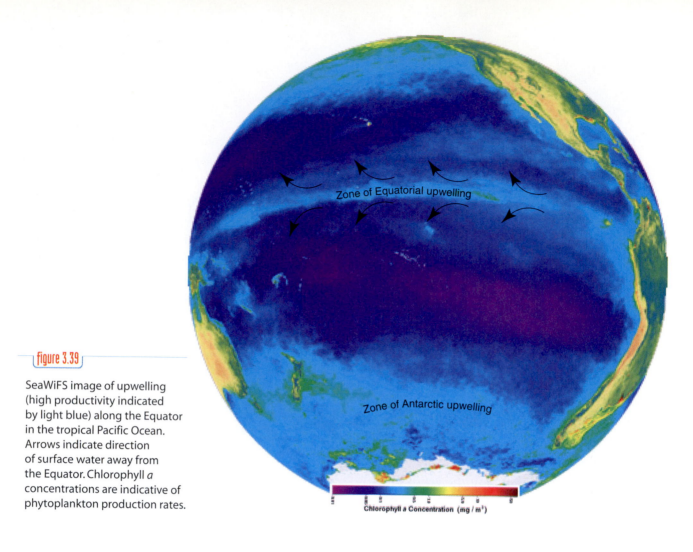

figure 3.39

SeaWiFS image of upwelling (high productivity indicated by light blue) along the Equator in the tropical Pacific Ocean. Arrows indicate direction of surface water away from the Equator. Chlorophyll *a* concentrations are indicative of phytoplankton production rates.

Zone of Equatorial upwelling

Zone of Antarctic upwelling

Chlorophyll a Concentration (mg / m³)

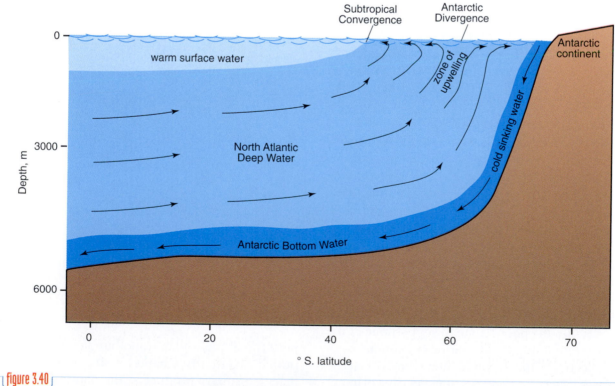

Subtropical Convergence

Antarctic Divergence

Antarctic continent

warm surface water

zone of upwelling

North Atlantic Deep Water

cold sinking water

Antarctic Bottom Water

Depth, m

0

3000

6000

0

20

40

60

70

° S. latitude

figure 3.40

Cross-section of the South Atlantic Ocean indicating the flow of water masses driving upwelling around the Antarctic continent. The area of upwelling is indicated in Figure 3.39

STUDY GUIDE

Summary Points

Phytoplankton

- Most primary production in the sea is accomplished by phytoplankton, unicellular, photosynthetic organisms. Hence, marine "plants" are fundamentally different from terrestrial producers. Unlike grasses, bushes, and trees, the phytoplankton have short lives, rapid turnover, low standing crops, severe mortality due to predation, and they do not directly contribute their nutrients to the next generation via decomposition immediately after they die.

Phytoplankton Groups

- Cyanobacteria have been producing oxygen in the sea for more than three billion years. Today they are everywhere, from intertidal rocks and estuaries to coral reefs and the open sea. Species capable of nitrogen fixation commonly form symbiotic associations with a variety of marine organisms.
- Division Chrysophyta is represented in the sea by nanoplanktonic coccolithophores and silicoflagellates and diatoms that can be macroscopic. Diatoms, often the most important members of cold-water phytoplankton communities, occur in a large variety of shapes, sizes, and locations, with some species forming multicellular colonies.
- Dinophytes dominate warm-water phytoplankton communities and are unique in their abilities to create light via bioluminescence and powerful toxins that become deadly via the phenomenon of biological magnification.

- With improving technology, our understanding of and appreciation for nano-, pico-, and ultraplanktonic species is increasing each year.

Adaptations for a Planktonic Existence

- Phytoplankton confront a persistent dilemma in that they must remain in the photic zone, yet nutrients occur in much greater concentrations near the seafloor. Most of their adaptations are responses to their need to linger near the surface while accumulating nutrients that are in extremely short supply.
- An extremely small cell diameter provides most phytoplankton with a relatively high surface area-to-volume ratio. This attribute increases frictional resistance to sinking and enables efficient uptake of very dilute nutrients.
- Many other antisinking and nutrient absorption adaptations have been documented, including variations in cell shape and architecture, the presence of gas-filled vesicles, and the phenomenon of ion exchange.

Primary Production in the Sea

- Estimation of gross and net primary production is necessary for understanding production potentials of and the quantity of organics available to marine communities. Such estimates have been made for nearly 100 years, starting with light-and-dark-bottle techniques and culminating with modern remote sensing via satellites.
- Although all photosynthetic organisms on Earth are influenced by many factors that affect rates of primary production, marine producers must overcome a number of unique limitations.
- Small herbivorous grazers routinely occur at such high concentrations that phytoplankton communities may be destroyed over a period of just a few weeks. Such devastating predation, the likes of which is rarely experienced on land, results in three well-known characteristics of phytoplankton communities: very low standing stocks, extremely patchy distributions, and the phenomenon of blooms.

- Light is of obvious necessity to all photosynthetic organisms, yet it occurs at sufficient intensity only in the relatively shallow photic zone. This fact, coupled with rapid loses in the spectral quality of light as it passes through seawater, results in accessory photosynthetic pigments (such as fucoxanthin) being more important in marine producers than terrestrial plants.
- Marine producers adapted to remaining at the surface in the photic zone must rely on a number of mechanisms of nutrient regeneration, such as turbulent mixing, convective mixing, and upwelling, to return inorganics to them from the seafloor.

Topics for Discussion and Review

1. Why are most marine photosynthesizers unicellular?
2. Describe the peculiar pattern of cellular reproduction observed in diatoms.
3. What is biological magnification? How does this phenomenon enable toxic unicellular dinophytes to harm humans?
4. Summarize all antisinking strategies used by marine phytoplankton.
5. Distinguish the terms "standing crop" and "primary production."
6. How do gross and net primary production differ?
7. How is the light bottle/dark bottle technique used to measure primary production in the sea?
8. Describe the population fluctuations commonly observed in phytoplankton and their herbivorous grazers.
9. What is the "critical depth?" Describe its significance to phytoplankton communities.
10. List and describe five mechanisms of nutrient regeneration in the sea.

Suggestions for Further Reading

Barber, R. T., and F. P. Chavez. 1991. Regulation of primary productivity rate in the equatorial Pacific. *Limnology and Oceanography* 36:1803–1815.

Capone, D. G. 2001. Marine nitrogen fixation: what's the fuss? *Current Opinion in Microbiology* 4:341–348.

Cerullo, M. M. 1999. *Sea Soup: Phytoplankton.* Tilbury House Publishers, Maine.

Chisholm, S. W., and K. Lawrence. 1991. What controls phytoplankton production in nutrient-rich areas of the open sea? *Limnology and Oceanography* 36:1507–1966.

Falkowski, P. G. 1980. *Primary Productivity in the Sea.* Plenum, New York.

Falkowski, P. G. 1992. *Primary Productivity and Biogeochemical Cycles in the Sea.* Plenum, New York.

Falkowski, P. G., R. T. Barber, and V. Smetacek. 1998. Biogeochemical controls and feedbacks on ocean primary production. *Science* 281:200–206.

Fernandez, D., and J. L. Acuna. 2003. Enhancement of marine phytoplankton blooms by appendicularian grazers. *Limnology and Oceanography* 48:587–593.

Field, C. B., M. J. Behrenfeld, J. T. Randerson, and P. Falkowski. 1998. Primary production of the biosphere: integrating terrestrial and oceanic components. *Science* 281:235–238.

Rost, B., U. Riebesell, and S. Burkhardt. 2003. Carbon acquisition of bloom-forming marine phytoplankton. *Limnology and Oceanography* 48:55–67.

Taylor, F. J. R. 1987. *The Biology of Dinoflagellates.* Blackwell, London.

Tomas, C. R. 1993. *Marine Phytoplankton: A Guide to Naked Flagellates and Coccolithophorids.* Academic Press, San Diego.

Tomas, C. R. 1997. *Identifying Marine Phytoplankton.* Academic Press, San Diego.

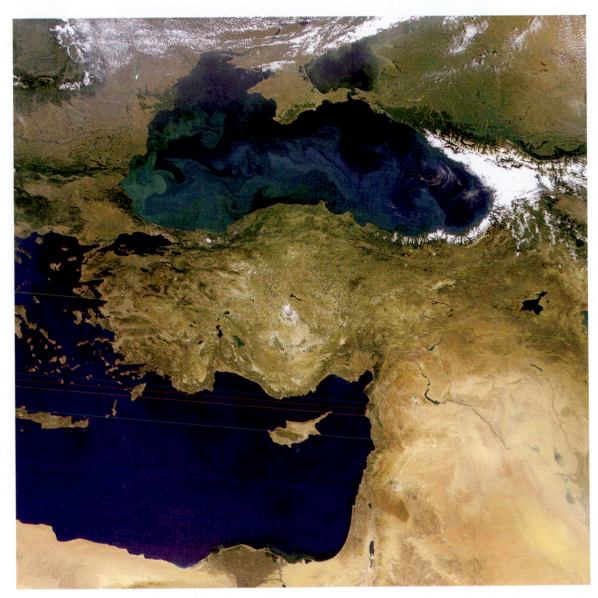

In this SEAWifs image, swirls of phytoplankton in the Black Sea (top) contrast strongly with the clear, blue waters of the eastern Mediterranean Sea (lower left).

CHAPTER
4

Giant kelp is just one of several marine plants that create unique habitats in the sea.

CHAPTER OUTLINE

Marine Plants

enthic marine plants are probably more familiar to even casual seashore observers than are phytoplankton because they are conspicuous, macroscopic, multicellular organisms typically large enough to pick up and examine. They all belong to a single Kingdom, the Plantae (FIG. 4.1). Like phytoplankton, these plants need sunlight for photosynthesis and are confined to the photic zone. But the additional need for a hard substrate on which to attach limits the distribution of benthic plants to that narrow fringe around the periphery of the oceans where the sea bottom is within the photic zone (the inner shelf of Fig. 1.46). Some benthic plants inhabit intertidal areas and must confront the many tide-induced stresses that affect their animal neighbors (discussed in Chapter 8). Their restricted near-shore distribution limits the global importance of benthic plants as primary producers in the marine environment. Yet within the near-shore communities in which they live, they play major roles as first-trophic-level organisms.

The abundant plant groups so familiar on land—ferns, mosses, and seed plants—are poorly represented or totally absent from the sea. Instead, most marine plants belong to two divisions, Phaeophyta and Rhodophyta, that are almost completely limited to the sea. Two other divisions, Chlorophyta and Anthophyta, are found most commonly in fresh water and on land, yet they are important members of some shallow coastal marine communities. The characteristics of these divisions are summarized in TABLE 4.1. We begin our examination of marine plants with a familiar group, the flowering plants, and then proceed to the seaweeds.

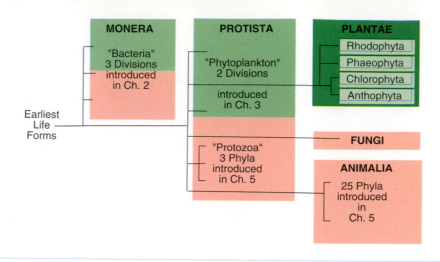

figure 4.1

The phylogenetic tree introduced in Figure 2.7, emphasizing the four Divisions of the kingdom Plantae described in this chapter.

Major Divisions of Marine Plants and Their General Characteristics

table 4.1

Division (common name)	Approx. no of living species	Percent of species marine	General size and structure	Photosynthetic pigments	Storage products	Habit
Phaeophyta (brown algae)	1500	99.7	Multicellular, macroscopic	Chlorophyll a, c Xanthophylls Carotenes	Laminarin and others	Mostly benthic
Rhodophyta (red algae)	4000	98	Unicellular and multicellular, mostly macroscopic	Chlorophyll a Carotenes Phycobilins	Starch and others	Benthic
Chlorophyta (green algae)	7000	13	Unicellular and multicellular, microscopic to macroscopic	Chlorophyll a, b Carotenes	Starch	Mostly benthic
Anthophyta (flowering plants)	250,000	0.018	Multicellular, macroscopic	Chlorophyll a, b Carotenes	Starch	Benthic

Adapted from Segal *et al.* 1980, Dawson 1981, and Kaufman *et al.* 1989.

4.1 Division Anthophyta

Marine flowering plants are abundant in localized areas along some seashores and in backwater bays and sloughs. Seagrasses are exposed to air only during very low tides, whereas salt marsh plants and mangroves are emergent and are seldom completely inundated by seawater. These plants represent a secondary adaptation to the marine environment by a few species of a predominantly terrestrial plant group, the flowering plants (division Anthophyta). Flowering plants are characterized by leaves, stems, and roots, with water- and nutrient-conducting structures running through all three of these basic structures.

Submerged Seagrasses

Twelve common genera of seagrasses, including about 50 species, are dispersed around coastal waters of the world. Half of these species are restricted to the tropics and subtropics and are seldom found deeper than 10 m. The four common genera found in the United States are *Thalassia*, *Zostera*, *Phyllospadix*, and *Halodule*. *Thalassia*, or turtle grass (FIG. 4.2A), is common in quiet waters along most of the Gulf Coast from Florida to Texas. *Zostera*, or eel grass (Fig. 4.2b), is widely distributed along both the Atlantic and Pacific coasts of North America. *Zostera* normally inhabits relatively quiet shallow waters but occasionally is found

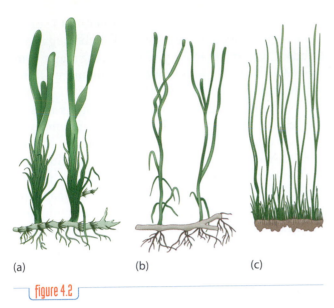

figure 4.2

Three common seagrasses from different marine climatic regions: (a) turtle grass, *Thalassia;* (b) eel grass, *Zostera;* and (c) surf grass, *Phyllospadix.*

as deep as 50 m in clear water. Surf grass, *Phyllospadix* (Fig. 4.2c), is found on both sides of the North Pacific and inhabits lower intertidal and shallow subtidal rocks that are subjected to considerable wave and surge action. *Halodule* prefers sandy areas with lower salinity.

Most seagrasses produce horizontal stems, or **rhizomes**, that attach the plants in soft sediments or to rocks (Fig. 4.2). From the buried rhizomes, many erect leaves develop to form thick green masses of vegetation. These plants are a staple food for nearshore marine animals and migratory birds. Densely matted rhizomes and roots also accumulate sediments and organic debris to alter further the living conditions of the area.

Seagrasses reproduce either vegetatively by sprouting additional vertical leaves from the lengthening horizontal rhizomes or from seeds produced in simple flowers. The purpose of most showy flowers on land plants is to attract insects or birds so that **pollen** grains are transferred from one flower to another and cross-fertilization occurs. Pollen grains contain the plant's sperm cells, but submerged seagrasses use water currents for pollen transport. In all seagrasses, pollination occurs underwater.

Some seagrasses, including *Zostera,* produce threadlike pollen grains about 3 mm long (about 500 times longer than their cargo, the microscopic chromosome-carrying sperm cells). After release, the pollen grains of *Zostera* become ensnared on the **stigma**, the pollen-receptive structure of the female flower, and fertilization occurs. *Thalassia* produces small round pollen grains released in a thread of sticky slime. When the slime thread lands on the appropriate stigma of another plant (also covered with a surface film of slime), the two slime layers combine to produce a firm bond between the pollen grain and the stigma, and fertilization follows. This two-component adhesive acts like epoxy glue to produce a strong bond after the separate components are mixed. It also provides a mechanism for selecting between compatible and foreign types of pollen grains. Only on contact with pollen of the same species will the stigma–pollen bond be formed. Foreign pollen grains do not adhere and are washed away, possibly to try again on another plant.

Mature seeds of each type of seagrass are adapted to their preferred habitat. Eel grass seeds drop into the mud and take root near the parent plant, whereas the fruits of *Thalassia* may float for long distances before releasing their seeds in the surf. The fruits surrounding individual seeds of *Phyllospadix* are equipped with bristly projections. When shed into the surf, these bristles snag branches of small seaweeds, and the seeds germinate in place.

Emergent Flowering Plants

Several other species of flowering plants often exist partially submerged on bottom muds of coastal salt marshes protected from strong ocean wave action. These plants are usually situated so that their roots are periodically, but not constantly, exposed to tidal flooding. They are terrestrial plants that have evolved various degrees of tolerance to excess salts from sea spray and seawater. Some even have special structural adaptations for their semimarine existence. The cord grass, *Spartina* (see Fig. 7.9a) for example, actively excretes excess salt through special two-celled salt glands on its leaves. Even so, several species of *Spartina* have higher experimental growth and survival rates in fresh water than in seawater. This difference strongly suggests that the salt marsh does not provide optimal growth conditions for *Spartina,* even though the salt marsh is its natural habitat. Competition with other land and freshwater plants may have forced *Spartina* and other salt-tolerant species into the more restricted areas of the salt marshes.

Salt marsh plants contribute heavily to detritus production in their protected environments as well as in nearby bays and estuaries. Some feature extensive stands containing several species of emergent grasses, especially various species of *Spartina*. At slightly higher elevations, these grasses give way to succulents (*Salicornia* and *Suaeda*), a variety of reeds and rushes, and the brush and smaller trees of the local woodland. These lush pastures are extremely productive and harbor a unique assemblage of organisms, including commercially important shellfish and finfishes. Yet as large urban centers develop near them, they have become popular sites for waste dumping, recreation, dredging and filling, and other detrimental uses. The degradation of salt marshes is a serious and worldwide problem that becomes more severe as human populations expand and place more pressure on these fragile habitats. The issues surrounding modification and degradation of salt marshes is discussed further in Chapter 7.

Several species of shrubby to treelike plants, the mangroves, create dense thickets of tidal woodlands known as **mangals** (FIG. 4.3). Mangals dominate large expanses of muddy shores in warmer climates and are excellent examples of emergent plant-based communities. Mangroves range in size from small shrubs to 10-m-tall trees whose roots are tolerant to seawater submergence and are capable of anchoring in soft muds. Collectively, mangrove plants, the major component of mangal communities, line about two-thirds of the tropical coastlines of the world (FIG. 4.4).

Members of these mangal communities are supported on their muddy substrate by numerous prop roots that grow down from branches above the water. The pattern of mangrove development illustrates well a series of adaptations needed to exist on muddy tropical shores (FIG. 4.5). Red mangroves (*Rhizophora*) produce seeds that germinate while still hanging from the branches of the parent tree. As the seedlings develop and grow longer, their bottom ends become heavier. When the seedlings eventually drop from the parent plant into the surrounding water, they float upright at the water's surface, are dispersed by winds or tides, and finally settle in shallow water on muddy shorelines. There, the seedlings promptly develop small roots to anchor themselves and continue to mature. The resulting tangle of growing roots traps additional sediments and increases the structural

figure 4.3

Dense mangal thicket lining a tidal channel.

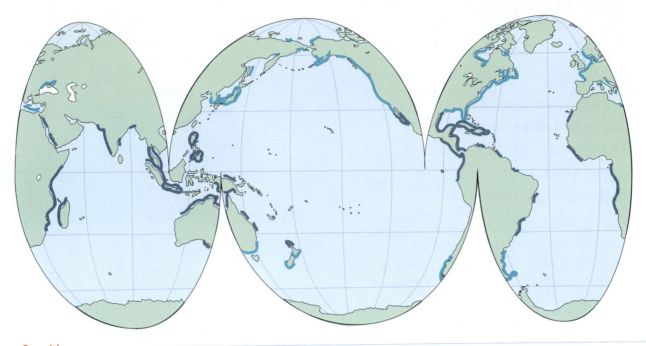

figure 4.4

Distribution of salt marshes (green) and mangals (blue).

figure 4.5

Germination cycle of a mangrove seedling.

complexity of mangal communities. Birds, insects, snails, and other terrestrial animals occupy the upper leafy canopy of the mangroves, and a variety of fishes, crustaceans, and mollusks lives on or among the root complex growing down into the mud. Because the leafy portions of these plants are above the water level, few marine animals graze directly on mangrove plants. Instead, leaves falling from these plants into the quiet waters surrounding their roots provide an important energy source for the detritus-based food webs of these communities.

In the United States, the distribution of mangals reflects their need for warm waters protected from wave action; they are found only along portions of the Gulf of Mexico and the Atlantic Coast of Florida. The south coast of Florida is dominated by extensive

Mangrove Trees: Connections Between Cattle Egrets and Yellow-Spotted Stingrays

Figure B4.1   Cattle egrets perched in mangroves.

Herons are a large family of birds that first appeared in the Lower Eocene epoch (about 55 million years ago). Most herons live in association with water and are adapted for preying on fishes, frogs, and other aquatic species. The cattle egret, a small entirely white heron with a dull orange bill and legs, apparently evolved on the plains of Africa (Figure B4.1). They soon spread through Asia and Europe and are known to have reached South America by 1880. Next, they invaded North America via Central America and various Caribbean islands during the 1940s. By 1962, they had reached California.

Cattle egrets are the most terrestrial heron. Throughout the Caribbean, they spend their days in pastures searching for food revealed by the activity of cows. Their prey primarily is insects, such as grasshoppers, crickets, flies, and moths, but routinely also includes spiders and frogs. Studies suggest that a cattle egret may obtain as much as 50% more food each day by associating with cows (while expending only two thirds as much energy to catch it).

As dusk approaches, the gregarious cattle egrets fly in flocks of 10 to 20 individuals to their roosting sites by following fixed routes, often along rivers. They roost in large colonies in mangroves or in trees near rivers. Interestingly, cattle egrets always seem to prefer one particular tree in a given area, and all birds in the vicinity roost in only that tree to the exclusion of all others. Their arrival in droves at dusk routinely transforms the spherical canopy of their selected tree into a large "snowball."

Yellow-spotted stingrays, *Urobatis jamaicensis,* are small Caribbean fishes with a disc width that rarely exceeds 20 cm (their maximum total length, including their tails, rarely exceeds 40 cm) (Figure B4.2). Their disc is yellowish on the upper surface with many brownish vermiculations and spots that interact to form a large variety of patterns. They are commonly found in shallow water, less than 25 m deep, along sandy beaches and especially in sandy areas adjacent to coral reefs. Yellow-spotted stingrays feed on shrimp, small fishes, clams, and worms. They are easily approached, perhaps because they are capable of inflicting painful wounds with their venomous caudal spine. Adult females give birth to three to four live young.

Finding cattle egrets is an easy task; one needs simply to look for an aggregation of pure white birds in a green meadow during the day or

interconnected shallow bays, waterways, and mangals. These mangals form a nearly continuous narrow band along the coast, with smaller fingers extending inland along creeks. Inland, toward the freshwater Everglades, the mangroves are not high, but tree height of all three species (red, black, and white mangroves) increases to as much as 10 m at the coast. It is these taller coastal members of mangal communities that are especially prone to hurricane damage. In 1992, Hurricane Andrew cut a swath of destruction across south Florida with sustained winds up to 242 km/hr. The accompanying storm surge lifted the sea surface over 5 m above normal levels. Some of the more exposed coastal mangal communities experienced over 80% mortality, due mostly to wind effects and lingering problems of coastal erosion.

4.2 The Seaweeds

By far, most large conspicuous forms of attached marine plants are seaweeds. The term *seaweed* is used here in a restricted sense, referring only to macroscopic members of the plant divisions Chlorophyta (green algae), Phaeophyta (brown algae), and scan the shoreline for a tree that gets transformed into a white ball at sunset. Finding yellow-spotted stingrays is more problematic; these nocturnal fishes spend their days resting on the sea floor, usually under a dusting of sediment that they have thrown on their backs via coordinated undulations of their expanded pectoral fins. Finding motionless buried rays while snorkeling through a shallow lagoon is difficult. Alternatively, one needs simply to snorkel around the periphery of an "egret tree," the red mangrove that is chosen every night as a roosting site by all cattle egrets for miles around. There,

among the mangroves prop and aerial roots, one will find as many as one resting stingray per square meter (rays are uncommon under mangrove trees that do not host cattle egrets each night). What is the connection between cattle egrets, yellow-spotted stingrays, and red mangrove trees? Students and researchers at the Hofstra University Marine Laboratory on the north coast of Jamaica are trying to determine just that.

They speculate that roosting cattle egrets produce a prodigious quantity of guano each night (indeed, most leaves of the egret tree often appear completely white from the guano coating that they accumulate during several days without rain). This guano, along with the many leaves that fall from the tree each day to decompose under its canopy, hypothetically results in sediment that is unusually rich in organic matter. Such nutritious sediment would be very attractive to deposit feeders, such as polychaete

Mangrove Trees: Connections Between Cattle Egrets and Yellow-Spotted Stingrays

worms and fishes, on which yellow-spotted stingrays are known to feed. Perhaps the rays preferentially select the shade of egret trees to exploit the unusually high abundance of deposit feeders living within the sediment between the tree's roots. And perhaps the deposit feeders that are prey for the rays are present only because the sediment itself is unusually rich in organic matter that is supplied directly by roosting cattle egrets. Only time and additional data will tell if this hypothetical scenario is accurate.

www.bioscience.jbpub.com/ marinebiology
For more information on this topic, go to this book's website at http://www.bioscience.jbpub.com/ marinebiology.

Figure B4.2   Yellow-spotted stingray.

Rhodophyta (red algae) (Table 4.1). These are plants that do not produce seeds or flowers, yet meet all the criteria for plants listed in Figure 2.7.

Seaweeds are abundant on hard substrates in intertidal zones and commonly extend to depths of 30 to 40 m. In clear tropical seas, some species of red algae thrive at depths as great as 200 m, and one species has been reported as deep as 268 m in the Bahamas. Many seaweeds tolerate or even require extreme surf action on exposed rocky intertidal outcrops, where they are securely fixed to the solid substrate. Where they are abundant, seaweeds can greatly influence

local environmental conditions for other types of shallow-water marine life by protecting from waves and providing food, shade, and sometimes a substrate on which to attach and grow.

Structural Features of Seaweeds

Seaweeds are not as complex as the flowering plants. Seaweeds lack roots, flowers, seeds, and true leaves. Yet, within these structural limitations, seaweeds exhibit an unbridled diversity of shapes, sizes, and structural complexity. Microscopic filaments of green

figure 4.6

The northern sea palm *Postelsia* (Phaeophyta) is equipped with a relatively large stipe and a massive holdfast.

and brown algae can be found growing side by side with encrusting forms of red algae and flat sheetlike members of all three divisions. The more obvious members of all three seaweed divisions typically develop into similar general forms, consisting of a **blade**, a **stipe**, and a **holdfast**, with many small fingerlike **haptera** (FIG. 4.6).

The Blade

The flattened, usually broad, leaflike structures of seaweeds are known as blades. Seaweed blades often exhibit a complex level of branching and cellular arrangement. Several larger species of brown algae produce distinctive blade shapes and blade arrangements (FIG. 4.7), yet each begins as a young plant with a single, unbranched, flat blade nearly identical to other young kelp plants.

The blades house photosynthetically active cells, but photosynthesis typically occurs in the stipes and holdfasts as well. In cross-section, seaweed blades (FIG. 4.8A) are structurally unlike the leaves of terrestrial plants (Fig. 4.8b). The cells nearer the surface of the blade are capable of absorbing more light and are photosynthetically more active than those cells near

the center of the blade. "Veins" of conductive tissue and distinctions between the upper and lower surfaces are lacking in the blades of seaweeds. Because the flexible blades usually droop in the water or float erect, there is no defined upper or lower surface. The two sides of the seaweed blade are usually exposed equally to sunlight, nutrients, and water and are therefore equally capable of carrying out photosynthesis. Unlike seaweeds, the flowering plants (including seagrasses) exhibit an obvious asymmetry of leaf structure, with a dense concentration of photosynthetically active cells crowded near the upper surface (Fig. 4.8b). Below the upper surface is a spongy layer of cells separated by large spaces to enhance the exchange of carbon dioxide, which is often 100 times less concentrated in air than in seawater.

Pneumatocysts

Several large kelp species have gas-filled floats, or **pneumatocysts**, to buoy the blades toward the sunlight at the surface. Pneumatocysts are filled with the gases most abundant in air, N_2, O_2, and CO_2, although some kelp pneumatocysts also contain a few percent of carbon monoxide, CO. Again, there is a large diversity in size and structure. The largest pneumatocysts belong to *Pelagophycus,* the elkhorn kelp (Fig. 4.7). Each *Pelagophycus* plant is equipped with a single pneumatocyst, sometimes as large as a basketball, to support six to eight immense drooping blades, each of which may be 1 to 2 m wide and 7 to 10 m long.

In strong contrast to *Pelagophycus, Sargassum* has numerous small pneumatocysts (FIG. 4.9). A few species of *Sargassum* lead a pelagic life afloat in the middle of the North Atlantic Ocean (the "Sargasso Sea"). In the Sargasso Sea, *Sargassum* creates large patches of floating plants that are the basis of a complex floating community of crabs, fishes, shrimp, and other animals uniquely adapted to living among the *Sargassum*. Large masses of this plant community sometimes float ashore on the US East and Gulf Coasts, creating odor problems for beachgoers as the dying plants decompose. In the Sea of Japan, other species of attached intertidal *Sargassum* break off and also become free-floating for extended periods of time.

The Stipe

A flexible stemlike stipe connects the wave-tossed upper parts of seaweeds to their securely anchored holdfasts at the bottom. An excellent example is *Postelsia*, the sea palm (Fig. 4.6), which grows attached to rocks only in the most exposed surf-swept por-

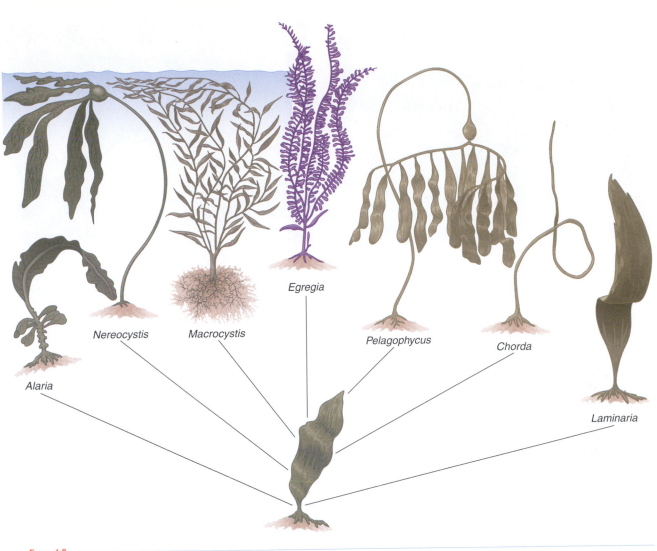

figure 4.7

Some large kelp plants of temperate coasts. Each mature plant develops from a young plant with a single flat blade.

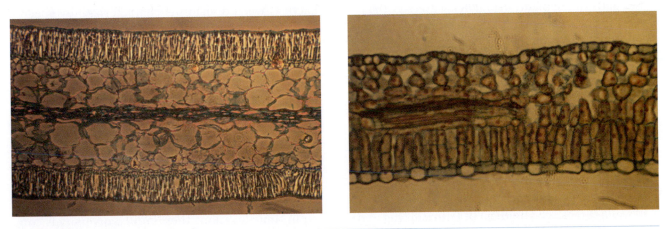

figure 4.8

Cross-sections of a blade of a typical marine alga, *Nereocystis* (left), and a typical flowering plant leaf (right). Note the contrasting symmetry patterns.

A portion of the floating brown alga, *Sargassum,* with numerous small pneumatocysts.

Complex interlocking mass of haptera that make up the holdfast of *Macrocystis.*

tions of the intertidal zone. Its hollow resilient stipe is remarkably well suited for yielding to the waves without breaking.

The blades of some seaweeds blend into the holdfast without forming a distinct stipe. In others, the stipe is very conspicuous and, occasionally, extremely long. The single long stipes of *Nereocystis, Chorda,* and *Pelagophycus* (Fig. 4.7) provide a kind of slack-line anchoring system and commonly exceed 30 m in length. The complex multiple stipes of *Macrocystis* are often even longer.

Special cells within the stipes of *Macrocystis* and a limited number of other brown and red algal species form conductive tissues strikingly similar in form to those present in stems of terrestrial plants. Radioactive tracer studies have shown that these cells transport the products of photosynthesis from the blades to other parts of the plant. In smaller seaweeds, the necessity for rapid efficient transport through the stipe is minimal, and such internal transport is lacking.

The Holdfast

Holdfasts of the larger seaweeds often superficially resemble root systems of terrestrial plants. However, the basic function of the holdfast is to attach the plant to the substrate. The holdfast seldom absorbs nutrients for the plant as do true roots. Holdfasts are adapted for getting a grip on the substrate and resisting violent wave shock and the steady tug of tidal currents and wave surges. The holdfast of *Postelsia* (Fig. 4.6), composed of many short, sturdy, rootlike haptera, illustrates one of several types found on solid rock.

Other holdfasts are better suited for loose substrates. The holdfast of *Macrocystis* has a large diffuse mass of haptera to penetrate muddy or sandy bottoms and stabilize a mass of sediment for anchorage (FIG. 4.10). Holdfasts of many smaller species do the same thing on a much smaller scale, with many fine filaments embedded in sand or mud on the sea bottom.

A variety of small red algae are epiphytes and demonstrate special adaptations for attaching themselves to other marine plants. FIGURE 4.11 illustrates two common red algal epiphytes attached to a strand of surf grass. Using other marine plants as substrates for attachment is a common habit of many smaller forms of red algae.

Photosynthetic Pigments

Each seaweed division is characterized by specific combinations of photosynthetic pigments that are reflected in their color and in the common name of each division (Table 4.1). The bright grass-green color of green algae is due to the predominance of chlorophylls over accessory pigments. Green algae vary in structure from simple filaments to flat sheets (FIG. 4.12) and diverse complex branching forms. They are usually less than half a meter long, but one species of *Codium* from the Gulf of California occasionally grows to 8 m in length. When compared with brown and red algae, the Chlorophyta have fewer marine species, yet in some locations their limited diversity is compensated with dense populations of individuals from one or two species.

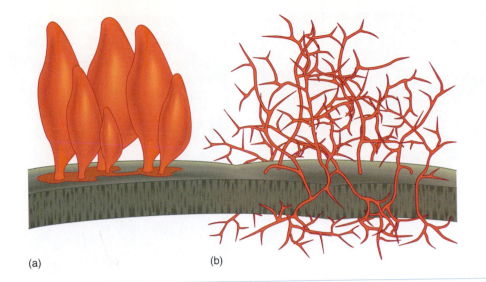

figure 4.11

Two red algal epiphytes: (a) *Smithora* and (b) *Chondria* attached to a leaf of *Phyllospadix*.

figure 4.12

Intertidal rocks covered with the green alga, *Ulva*.

The photosynthetic pigments of the Phaeophyta sometimes appear as a greenish hue. But more often the green of the chlorophyll is partially masked by the golden xanthophyll pigments, especially fucoxanthin, characteristic of this division. This blend of green and brown pigments usually results in a drab olive-green color (FIG. 4.13). Many of the larger and more familiar algae of temperate seas belong to this division. A number of species are quite large and are sometimes collectively referred to as **kelp** (Fig. 4.7). In temperate and high latitudes, these species usually dominate the marine benthic vegetation. Numerous smaller less obvious brown algae are also common in temperate and cold waters as well as in tropical areas.

figure 4.13

The brown alga, *Fucus,* from a rocky intertidal.

figure 4.14

Calcareous red algae, *Jania,* in a small tide pool.

Red algae, with red and blue phycobilin pigments as well as chlorophyll, exhibit a wide range of colors. Some are bright green, and others are sometimes confused with brown algae. However, most red algae living below low tide range in color from soft pinks to various shades of purple or red (FIG. 4.14). Red algae are as diverse in structure and habitat as they are in coloration, and they seldom exceed a meter in length.

The adaptive significance of accessory photosynthetic pigments for phytoplankton was described in Chapter 3. At first glance, it might appear that the green algae and seagrasses, with their preponderance of chlorophyll pigments, do not fare well at moderate depths because of their limited ability to absorb the deeper-penetrating green wavelengths of sunlight. But plants can adapt to low or limited wavelength light conditions in other ways. For example, because some green algae have dense concentrations of chlorophyll that appear almost black, they are able to absorb light at essentially all visible wavelengths. In addition, most green plants have chlorophyll *b* as well as chlorophyll *a*. Chlorophyll *b* has a strong light-absorbing peak in the blue region of the visible spectrum and can collect a good fraction of the deep-penetrating blue light available in tropical waters. Still, red and brown algae, with their abundant xanthophyll and phycobilin pigments working in concert with chlorophyll, generally have a slight competitive advantage in occupying the deeper portions of the photic zone in turbid coastal waters and function at no disadvantage in shallow waters or intertidal zones.

Reproduction and Growth

Reproduction in seaweeds, as well as in most other plants, can be either sexual, involving the fusion of sperm and eggs, or asexual, relying on vegetative growth of new individuals. Some seaweeds reproduce both ways, but a few are limited to vegetative reproduction only. The pelagic species of *Sargassum,* for instance, maintain their populations by an irregular vegetative growth followed by fragmentation into smaller clumps. The dispersed fragments of *Sargassum* are capable of continued growth and regeneration for decades. Sexual reproduction is lacking in the pelagic species of *Sargassum* but not in the attached benthic forms of the same genus.

Much of the structural variety observed in seaweeds is derived from complex patterns of sexual reproduction, patterns that define the life cycles of seaweeds. For our purposes, these complex life cycles can be simplified to three fundamental patterns. The sexual reproduction examples of the first two types described here are not meant to cover the entire spectrum of seaweed life cycles but are used to illustrate the basic patterns that underlie the com-

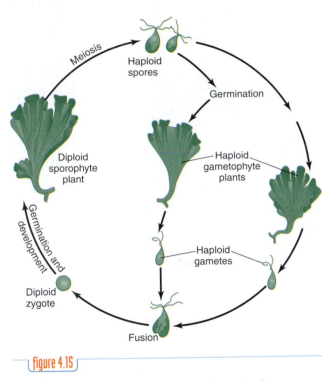

Meiosis
Haploid spores
Germination
Haploid gametophyte plants
Diploid sporophyte plant
Haploid gametes
Germination and development
Diploid zygote
Fusion

figure 4.15

The life cycle of the green alga, *Ulva*, alternating between diploid sporophyte and haploid gametophyte generations. (Adapted from Dawson 1966.)

plexity and variation involved in sexual reproduction of seaweeds.

In the life cycle of most of the larger seaweeds, an alternation of **sporophyte** and **gametophyte** generations occurs. The green alga *Ulva* represents one of the simplest patterns of alternating generations (FIG. 4.15). This basic life cycle is a hallmark of the kingdom Plantae. The cells of the macroscopic *Ulva* sporophyte are diploid; that is, each cell contains two of each type of chromosome characteristic of that species. Some cells of the *Ulva* sporophyte undergo meiosis to produce single-celled flagellated **spores**. As a result of meiosis, these spores contain only one chromosome of each pair present in the diploid sporophyte and are said to be haploid.

The spores of *Ulva* and other green algae have four flagella; each gamete has two flagella that are equal in length and project from one end of the cell. Spores produced by *Ulva* are capable of limited swimming and then settle to the bottom. There they immediately germinate by a series of mitotic cell divisions to produce a large multicellular, still haploid, gametophyte. Cells of the gametophyte in turn produce haploid gametes, each with two flagella, that are released into the water. When two gametes from dif-

ferent gametophytes meet, they fuse to produce a diploid single-celled **zygote**. By repeated mitotic divisions, the zygote germinates and completes the cycle by producing a large, multicellular, diploid sporophyte once again. In *Ulva*, the sporophyte and gametophyte generations are identical in appearance. The only structural difference between the two forms is the number of chromosomes in each cell; the diploid sporophyte has double the chromosome number of the haploid gametophyte cells.

The life cycles of numerous other seaweeds are characterized by a suppression of either the gametophyte or the sporophyte stage. In the green alga *Codium* and the brown alga *Fucus*, the multicellular haploid generation is completely absent. The only haploid stages are the gametes. In other large brown algae, the gametophyte stage is reduced. The life cycle of *Laminaria* is similar to that of most other large kelp plants and serves as an excellent generalized example of seaweeds with a massive sporophyte that alternates with a diminutive gametophyte (FIG. 4.16). Special cells (called **sporangia**) on the blades of the diploid sporophyte undergo meiosis to produce several flagellated microscopic spores. These haploid spores swim to the bottom and quickly attach themselves. They soon germinate into very small, yet multicellular, gametophytes. The female gametophyte produces large nonflagellated eggs. The egg cells are fertilized in place on the female gametophyte by flagellated male gametes, the sperm cells produced by the male gametophyte. After fusion of the gametes, the resulting zygote germinates to form another large sporophyte. The flagellated reproductive cells of brown algae always have two flagella of unequal lengths, and they insert on the sides of the cells rather than at the ends.

Red algae lack flagellated reproductive cells and are dependent on water currents to transport the male gametes to the female reproductive cells. The most common life cycle of red algae has three distinct generations, somewhat reminiscent of the reproductive cycle outlined for *Ulva* (Fig. 4.15). A diploid sporophyte produces haploid spores that germinate into haploid gametophytes. Instead of producing a new sporophyte, however, the gametes from the gametophytes fuse and develop into a third phase unique to red algae, the **carposporophyte**. The carposporophyte then produces **carpospores** that develop into sporophytes, and the cycle is completed.

The development of a large multicellular seaweed from a single microscopic cell (either a haploid spore

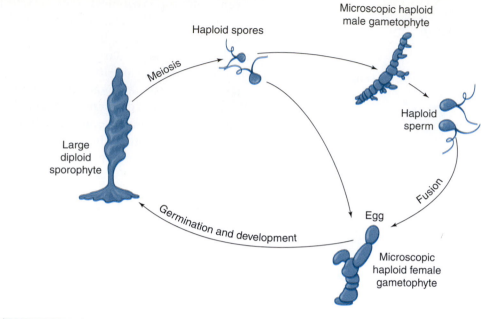

figure 4.16

The life cycle of *Laminaria* (similar to the cycles of other large kelps), alternates between large diploid sporophyte and microscopic haploid gametophyte generations.

or a diploid zygote) is essentially a process of repeated mitotic cell divisions. Subsequent growth and differentiation of these cells produce a complex plant with many types of cells, each specialized for particular functions. Once the plant is developed, additional cell division and growth occur to replace tissue lost to animal grazing or wave erosion. However, such cell division is commonly restricted to a few specific sites within the plant that contain **meristematic tissue** capable of further cell division. These meristems frequently occur at the upper growing tip of the plant. In kelp plants and some other seaweeds, additional meristems situated in the upper and lower portions of the stipe provide additional cells to elongate the stipe and blades. The meristematic activity of a cell layer near the outer stipe surface of some kelp species provides lateral growth to increase the thickness of the stipe. The stipes of a few perennial species of kelp, including *Pterygophora* and *Laminaria,* retain evidence of this secondary lateral growth as concentric rings that resemble the annual growth rings of trees.

In spring during periods of rapid growth, the rate of stipe elongation in large *Nereocystis, Pelagophycus,* and *Macrocystis* plants often exceeds 30 cm/day. Many kelp species produce kelp blades resembling moving belts of plant tissue (FIG. 4.17), growing at the base and eroding or being eaten away at the tips. At any one time, the visible plant itself (the standing crop) may represent as little as 10% of the total material it produced during a year.

4.3 Geographic Distribution

The interplay of a multitude of physical, chemical, and biologic variables influences and controls the distribution of marine plants on a local scale. For instance, on an exposed rock in the lower intertidal zone on the Oregon coast, *Postelsia* may thrive, but 10 m away the conditions of light, temperature, nutrients, tides, surf action, and substrate may be such that *Postelsia* cannot survive. Yet, on an ocean-wide scale, only a few factors seem to control the presence or absence of major groups of seaweeds. Significant among these are water and air temperature, tidal amplitude, and the quality and quantity of light. With these factors in mind, we can make a few generalizations concerning the geographic distribution of benthic plants.

In marked contrast to the impoverished seaweed flora of the Red Sea, the tropical western coast of Africa, and the western side of Central America, seaweeds thrive in profusion along the coasts of southern Australia and South Africa, on both sides of the North Pacific, and in the Mediterranean Sea. The US

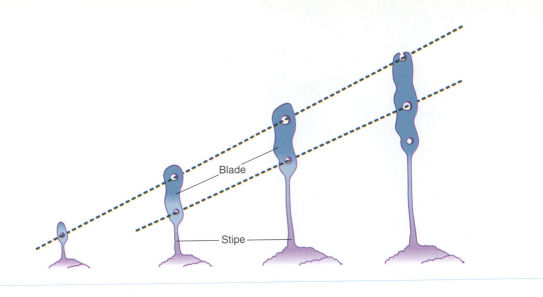

figure 4.17

Generalized growth pattern of a kelp. Punched holes and dotted lines indicate the pattern of blade elongation. (Adapted from Mann, 1973.)

West Coast is somewhat richer in seaweed diversity than is the East Coast. From Cape Cod northward, the East Coast is populated with subarctic seaweeds. South of Cape Cod, the effects of the warm Gulf Stream become more evident, until a completely tropical flora is encountered in southern Florida.

Red algae are not rare in cold-water regions but are more abundant and conspicuous in the tropics and subtropics. Calcareous forms of red algae (and some brown and green as well) are characterized by extensive deposits of calcium carbonate ($CaCO_3$) within their cell walls. The use of calcium carbonate as a skeletal component by warm-water marine algae is apparently related to the decreased solubility of $CaCO_3$ in water at higher temperatures. In the tropics, plants expend less energy to extract $CaCO_3$ from the water, and here coralline red algae contribute to the formation and maintenance of coral reefs. Encrusting coralline algae grow over coral rubble, cementing and binding it into larger masses that better resist the pounding of heavy surf. Some Indian Ocean "coral" reefs completely lack coral animals and are constructed and maintained entirely by coralline algae. The few calcareous forms of green algae that exist are also limited to tropical latitudes and play a large role in the production of $CaCO_3$ on some coral reefs.

The small green alga, *Halimeda*, is one of the few green algae to also secrete $CaCO_3$ skeleton, giving it a stony feel. *Halimeda* is a member of a remarkable group of Chlorophytes known as siphonous green algae. Although some siphonous green algae reach several meters in size, each plant consists of an enormously long tubular single cell containing millions of nuclei by uncoupling the process of nuclear division from that of cell division. Two other members of this group, *Caulerpa taxifolia* and *Codium fragile*, recently have become notorious for their explosively rapid invasions as introduced exotics, *Caulerpa* in the Mediterranean Sea (**FIG. 4.18**) and *Codium* in shallow

figure 4.18

A dense growth of *Caulerpa*.

figure 4.19

A kelp forest of the California coast, dominated by the brown alga, *Macrocystis*.

coastal waters of New Zealand and the US Northeast. These invasions have been enhanced by the ability of these plants to fragment in storms and quickly regrow from the wave-scattered pieces.

A few of the larger species of benthic marine plants flourish in such profusion that they dominate the general biologic character of their communities. Such community domination by plants is common on land but is exceptional in the sea. Away from the near-shore habitats occupied by benthic plants, the microscopic phytoplankton prevail as the major primary producers of the sea, and it is their larger animal consumers that define the visual character of their pelagic communities. In the near-shore fringe, however, mangals, salt marshes, seagrasses, and kelp beds thrive where the appropriate bottom conditions, light, and nutrients exist.

Kelp are temperate to cold-water species, with few tropical representatives. Large kelps are especially abundant in the North Pacific. Kelp beds abound with herbivores that graze directly on these plants and in turn become prey for higher trophic levels. The cool-water kelp plants form extensive layered forests of mixed species in both the Atlantic and Pacific Oceans. The blades of the larger *Macrocystis, Laminaria,* or *Nereocystis* form the upper canopy and the basic structure of these plant communities. Shorter members of other brown algal and red algal species provide secondary understory layers and create a complex three-dimensional habitat with a large variety of available niches (FIG. 4.19). The maximum depth of these kelp beds, usually 20 to 30 m, is limited by the light available for the young growing sporophyte. The larger kelp plants, with their broad blades streaming at the sea surface, create substantial drag against currents and swells and are susceptible to storm damage by waves and surge. Cast on the shore, these decaying plants are a major food source for beach scavengers.

4.4 Seasonal Patterns of Marine Primary Production

In Chapter 3, the influence of sunlight, nutrients, and grazers on marine primary productivity was considered. Here we put it all together to develop a dynamic coherent picture of how marine primary productivity patterns change over seasonal time scales and oceanic distances. The numbers needed from this summary are difficult to come by and are changing as new techniques for measuring marine primary productivity are developed. Since the first edition of this text was written, estimates of global marine primary productivity have approximately doubled, thanks in large part to satellite monitoring systems such as the sea-viewing wide field-of-view sensor (SeaWiFS) described in Chapter 3. Globally, the marine plants described in this chapter account for only about 2% of each year's total marine primary productivity; phytoplankton take care of the rest. So, again, the emphasis of the following discussion is on phytoplankton.

The spatial patchiness of marine primary production described in Chapter 3 is related on large scales to areas of nutrient abundance and on much smaller scales to the local influences of grazers, near-surface turbulence, and nutrient patches. Seasonal variations, or patchiness in time, occur in response to changes in light intensity, nutrient abundance, and

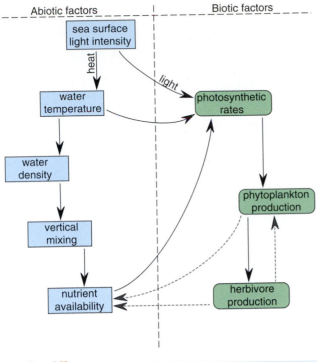

sea surface
light intensity

heat

light

water
temperature

photosynthetic
rates

water
density

phytoplankton
production

vertical
mixing

nutrient
availability

herbivore
production

figure 4.20

The seasonal variation of light intensity at the sea surface
sets in motion a cascading series of changes in the photic
zone. Eventually, these factors influence primary production,
either directly (solid arrows) or through feedback links
(dashed arrows).

grazing pressure. The underlying pulse for these time
changes is the predictable seasonal variation in the
intensity of sunlight reaching the sea surface. FIGURE 4.20
outlines the major links between factors involved in
defining the actual pattern of net primary production (NPP) through time. With these in mind, we can
develop some general pictures of the seasonal pattern of primary production for several different marine production systems.

Temperate Seas

FIGURE 4.21 depicts a somewhat idealized graphic summary
of major physical, chemical, and biologic events in
temperate oceanic areas well away from the influences of coastlines. These areas include broad swaths
of mid-latitude open ocean as well as locally defined
areas such as the Grand Banks of the North Atlantic
Ocean. These temperate oceanic systems are characterized by wintertime convective mixing between
surface and deep layers.

A prominent feature in the production cycle of
temperate seas is a spring diatom population explosion, or diatom bloom. Diatom blooms are the result

of combined seasonal variations of water temperature, light and nutrient availability, and grazing intensity. In early spring, water temperature and
available light increase, nutrients are abundant in
near-surface waters, and grazing pressure is diminished. As soon as the minimum threshold level of
sunlight needed for photosynthesis is achieved, conditions are ideal for rapid and abundant growth of
primary producers. If the bottom of the mixed layer
extends below the critical depth (determined by light
penetration, see Fig. 3.20), near-surface turbulence
will distribute the phytoplankton cells randomly
throughout the mixed layer, and primary production
will remain low. Cells in the deeper portions of the
mixed layer receive insufficient light, and no net production will occur. The spring bloom will commence
only after the thermocline thins the mixed layer to a
level above the critical depth. In general, bloom conditions in the open ocean occur as a broad band of
primary production sweeping poleward from mid-
latitudes in both hemispheres with the onset of spring.
The standing crop of diatoms increases quickly to
the largest of the year and begins to deplete nutrient
concentrations. The grazers respond to the additional
forage by increasing their numbers.

As spring warms into summer, sunlight becomes
more plentiful, but the now strongly developed seasonal thermocline effectively blocks nutrient return
from deeper water. The now warmer waters are "older"
surface waters, with most of their dissolved nutrients
depleted. Coupled with increased grazing, the diatom
population peaks and then declines and remains low
throughout the summer. With food more scarce, the
summer zooplankton population also drops off. Unlike diatoms, dinophyte populations increase slowly
during the spring, remain healthy throughout the
summer (although not as abundant as diatoms are in
spring), and decline in autumn because of diminished light intensity.

This replacement of diatoms by dinophytes is a
form of seasonal succession resulting from some basic ecologic differences between the two principal
groups of phytoplankton. Recall that diatoms lack
flagella, cannot swim, are more readily inhibited in
high-light intensities, perform better in low-light intensities, and have a nutrient need for silicate. These
features give diatoms a competitive advantage in less-
well-lit, colder, denser, nutrient-rich waters and dinophytes the advantage in warmer better-lit waters that
may be deficient in silicate. Some dinophytes deeper
in the photic zone may supplement their meager nutrient supply by migrating downward a few meters

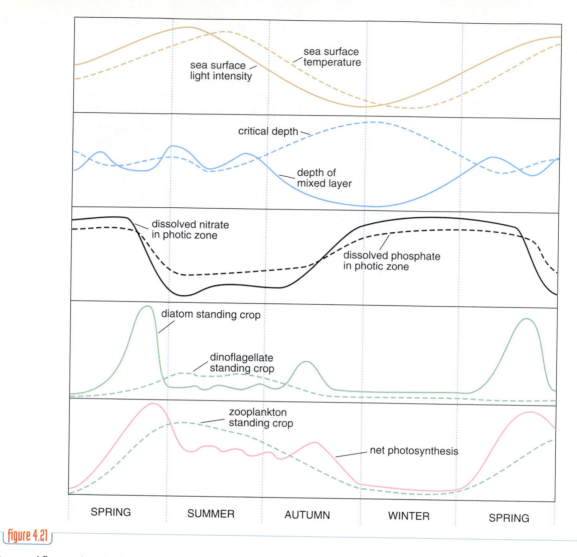

figure 4.21

Seasonal fluctuations in the major features of a temperate water marine primary production system.

during night hours to soak up additional nutrients from slightly deeper water.

Cooler autumn air temperatures begin to break down the summer thermocline and allow convection to renew nutrients to the photic zone. The phytoplankton respond with another bloom, which, although not as remarkable as the spring bloom, is often sufficient to initiate another upswing of the zooplankton population. As winter approaches, the autumn bloom is cut short by decreasing light and reduced temperatures. As production goes down, resistant overwintering stages of both phytoplankton and zooplankton become more abundant. Convective mixing continues to recharge the nutrient load of the surface waters in readiness for a repeat of the entire performance the following spring. It is now estimated that, on average, about 220 g C/m² per year

is produced in oceanic temperate and subpolar areas, with most of that total occurring during the spring diatom bloom.

Warm Seas

The production characteristics of tropical and subtropical oceanic waters closely resemble those of continuous summer in temperate regions as outlined in Figure 4.21. Sunlight is available in abundance, yet NPP is low (about 60 g C/m² per year) because a strong permanent pycnocline blocks vertical mixing of nutrients from below. The low rate of nutrient return is partially compensated for by a year-round growing season and a deep photic zone. Even so, NPP and standing crops are low (FIG. 4.22), and dinophytes are usually more abundant than diatoms.

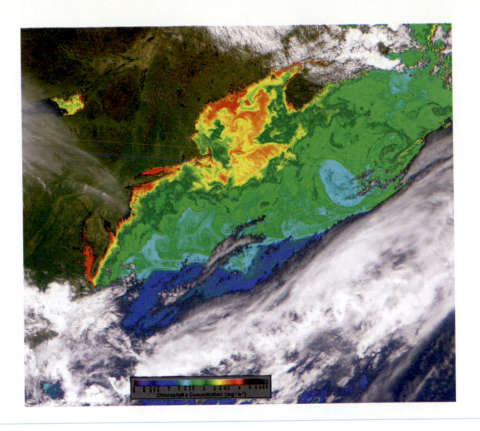

figure 4.22

SeaWiFS image of chlorophyll *a* concentrations in the area off the US East Coast, taken on 11 May 2002. Color bar is the same as those shown in other SeaWiFS images in Chapter 3.

Regions of upwelling in the equatorial Pacific and, to a lesser extent, the equatorial Atlantic are more productive than tropical open ocean areas, but they are very limited in geographic extent (see Fig. 3.39). So too are coral reef communities, with annual NPP rates up to 1000 g C/m^2 per year. Upwelling is described in the next section, and the reasons for exceptionally high productivity rates on coral reefs are discussed in Chapter 9.

Coastal Upwelling

Coastal upwelling in temperate seas alters the generalized picture presented in Figure 4.21 by replenishing nutrients during the summer, when they would otherwise be depleted. As long as light is sufficient and upwelling continues, high phytoplankton production occurs and is reflected in abundant local animal populations (FIG. 4.23). In some areas, the duration and intensity of coastal upwelling fluctuate with variations in atmospheric circulation. Along the Washington and Oregon coasts, the variability of spring and summer wind patterns produces sporadic up-

welling interspersed with short periods of no upwelling and lower NPP. Coastal upwelling zones have average NPP rates approaching 1000 g C/m^2 per year (FIG. 4.24). In the Peru Current, upwelling is massive year round and is interrupted only by major disturbances such as El Niño events.

El Niño is a phenomenon that represents a strong departure from the more typical current patterns in the central Pacific Ocean (see Fig. 1.38) that drive coastal and equatorial upwellings. El Niño is characterized by a prominent warming of the equatorial Pacific surface waters. El Niño occurs irregularly every few years, and each occurrence lasts from several months to well over a year. The El Niño phenomenon is associated with the Southern Oscillation, a trans-Pacific linkage of atmospheric pressure systems, and the climatic anomaly has come to be known collectively as El Niño/Southern Oscillation, or ENSO. Normally, the trade winds blow around the South Pacific high pressure center located near Easter Island and then blow westward to a large Indonesian low-pressure center. As these winds move water westward, the water is warmed and the thermocline is

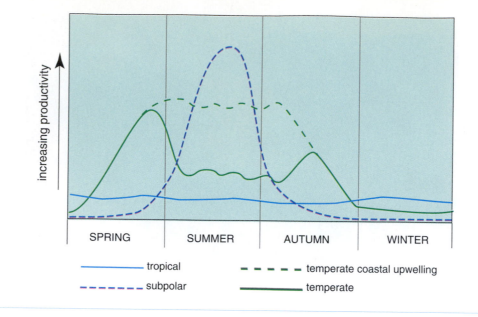

Comparison of the general patterns of seasonal variations in primary productivity for four different marine production systems.

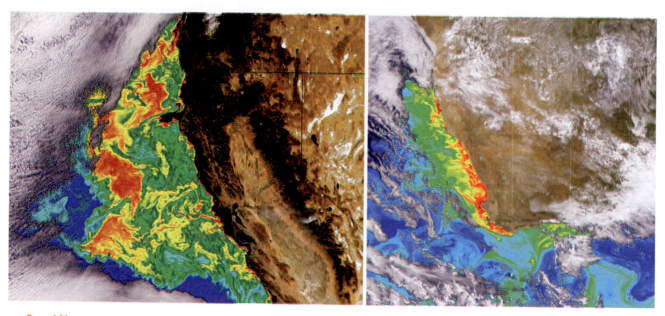

SeaWiFS image of chlorophyll *a* concentrations in the upwelling area of the California Current (left) and the Benguela Current off South Africa and Namibia. Note the pulse of high phytoplankton production off California being transported well offshore. Color bar is the same as those shown in other SeaWiFS images in this chapter and Chapter 3.

depressed from about 50 m below the surface on the east side of the Pacific to about 200 m deep on the west side. ENSOs occur, for reasons not yet understood, when this pressure difference across the tropical Pacific relaxes and both surface winds and ocean currents either cease to flow westward or actually reverse themselves. Although the effects of an ENSO event are somewhat variable, they are usually global in extent and occasionally severe in impact.

The 1982–1983 ENSO event, for example, was associated with heavy flooding on the West Coast of the United States, intensification of the drought in sub-Saharan Africa and Australia, and severe hurricane-force storms in Polynesia. Surface ocean water tem-

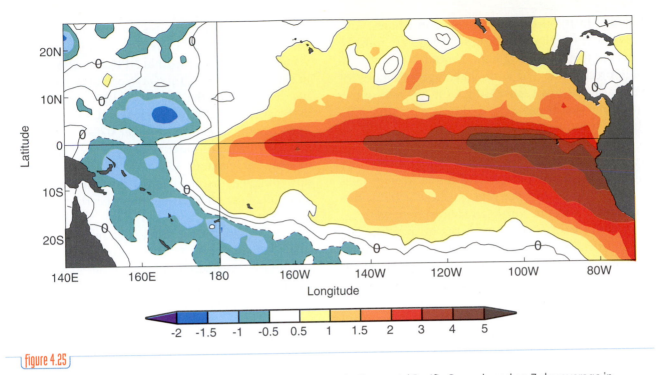

Observed sea surface temperature anomaly, in degrees Celsius, in the Equatorial Pacific Ocean based on 7-day average in mid-September 1997.

peratures from Peru to California soared to as much as 8°C above normal. The 1997–1998 ENSO event caused similar disruptions but was even more severe (FIG. 4.25). These strong El Niño events and the associated buildup of warm less-dense water blocks upwelling of nutrient-rich waters, and coastal marine populations decline. During severe El Niño years, some fish and fish-eating seabird populations almost completely disappear. Eventually, the area of warm tropical water dissipates, and El Niño conditions are replaced by cooler eastern tropical Pacific surface temperatures, low rainfall, and well-developed coastal upwelling along Peru and northern Chile.

Polar Seas

Seawater temperatures in polar regions are always low. Yet the north and south polar seas differ strongly from each other, and so do some of their productivity patterns. The Arctic is a frozen ocean surrounded by continents; the Antarctic is a frozen continent surrounded by ocean. The thermocline, if one exists at all, is poorly established and its associated pycnocline is not an effective barrier to nutrient return from deeper waters. Light, or more correctly the lack of it, is the major limiting factor for plant or phytoplankton growth in polar seas. Sufficient light to sustain

high phytoplankton growth rates lasts for only a few months during the summer. Even so, photosynthesis can continue around the clock during those few months to produce huge phytoplankton populations quickly. As the light intensity and day length decline, the short summer diatom bloom declines rapidly. Winter conditions closely resemble those of temperate regions, except that in polar seas the conditions endure much longer. There, the complete cycle of production consists of a single short period of phytoplankton growth, equivalent to a typical spring bloom immediately followed by an autumn bloom and decline that alternates with an extended winter of reduced net production.

In both the Arctic Ocean and around the Antarctic continent, the seasonal formation and melting of sea ice play a central role in shaping patterns of primary productivity. As ice melts in the spring, the low salinity meltwater forms a low-density layer near the sea surface. This increases vertical stability, which encourages phytoplankton to grow near the sunlit surface. The melting ice also releases temporarily frozen phytoplankton cells into the water to initiate the bloom. As the sea ice continues to melt toward the poles in early summer, the zone of high phytoplankton productivity follows, creating a very productive, seasonal, migrating ice edge community of diatoms, krill, birds, seals, fishes, and whales. Animals that do exploit this production system must be prepared to endure long winter months of little primary production. The annual average NPP rates for polar seas are low (about 30 g C/m^2/yr) because so much of the year passes in darkness with almost no phytoplankton growth.

Around the Antarctic continent, however, upwelling of deep nutrient-rich water supports very high summertime NPP rates and annual average rates of about 150 g C/m^2 per year. In these regions, water that sank in the Northern Hemisphere wells up to the surface between 60 and 70° S latitude (see Fig. 3.40), bringing with it a thousand-year accumulation of dissolved nutrients. The extraordinary fertility of the Antarctic seas stands in sharp contrast to the barrenness of the adjacent continent. Consequently, almost all Antarctic life, whether terrestrial or marine, depends on marine food webs supported by this massive upwelling.

4.5 Global Marine Primary Production

Mid- and high-latitude regions, shallow coastal areas, and zones of upwelling generally support large populations of marine primary producers, but most of this production is accomplished during the warm summer months when light is not a growth-limiting factor. Open-ocean regions, especially in the tropics and subtropics, where a strong thermocline and pycnocline are permanent features, and in polar seas, where light is limited through much of the year, have low rates of NPP (50 g C/m^2/yr).

TABLE 4.2 lists and compares the annual rates of marine NPP in several different regions (see FIG. 4.26 for a visual representation). Total NPP estimates included in syntheses such as this have been revised upward nearly 75% with the use of satellite-derived observations. About 55% of the total NPP occurs in the open ocean, spread thinly over 92% of the ocean's area. The more productive regions are very limited in geographic extent. Collectively, estuaries, coastal upwelling regions, and coral reefs produce only about 2 billion of the 45 billion tonnes of carbon produced each year.

The productivity numbers of Table 4.2 indicate that a total of nearly 45 billion tonnes of carbon are synthesized each year in the world ocean, and all but 1 billion tonnes (98%) are from phytoplankton. That number is equivalent to a bit more than 90 billion tonnes of photosynthetically produced dry biomass, or about 15 tonnes of phytoplankton dry biomass for each person on Earth.

When compared with land-based primary production systems, NPP on land is slightly higher (about 56×10^9 tonnes C/yr), even though oceans cover more than twice as much of the Earth's surface as does land. Although marine primary producers account for almost half of the total global NPP each year, at any one time phytoplankton represent only about 0.2% of the standing stock of primary producers because of their very rapid turnover rates. About 25% of ice-free land areas supports NPP rates over 500 g C/m^2/yr; in the ocean, this value is less than 2%.

The entire human population on Earth currently requires about 3 billion tonnes of food annually to

Table 4.2

Rates of Net Primary Production for Several Ocean Regions

Region	Area (×10^6 km^2)	Percent of Ocean	Average (g C/m^2/yr)	Total NPP (10^9 tonnes C)
Open ocean				
Tropics and subtropics	190	51	60	11.4
Temperate and subpolar (including Antarctic upwelling)	100	27	220	22.0
Polar	52	14	30	1.8
Continental shelf				
Nonupwelling	26.6	7.2	310	8.2
Coastal upwelling	0.4	0.1	970	0.4
Estuaries and salt marshes	1.8	0.05	1000	1.8
Coral reefs	0.1	—	1000	0.1
Seagrass beds	0.02	—	1000	0.02
				45.3

Data from Longhurst *et al.* 1995, Field *et al.* 1998, and Pauly and Christensen 1995.

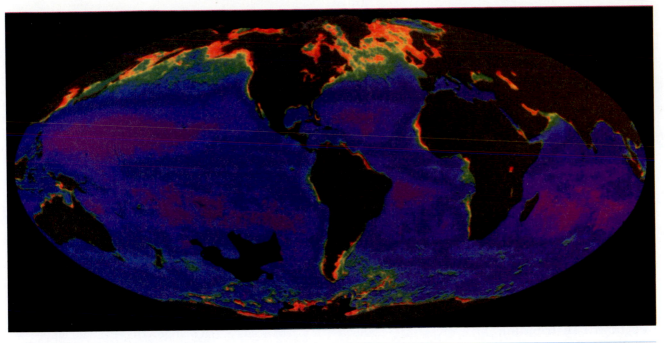

figure 4.26

The geographic variation of marine primary production, composed from over 3 years of observations by the satellite-borne coastal zone color scanner. Primary production is low (less than 50 g C/m²/yr) in the central gyres (magenta to deep blue), moderate (50–100 g C/m²/yr) in the light blue to green areas, and high (greater than 100 g C/m²/yr) in coastal zones and upwelling areas (yellow, orange, and red).

sustain itself, about 6% of the total annual marine NPP. Yet for several reasons to be discussed in Chapter 13, this abundant profusion of marine primary producers will probably never be used on a scale sufficient to alleviate the serious nutritional problems already rampant in much of our population. Instead, this vast amount of organic material will continue to do what it has always done: fuel the metabolic requirements of the consumers occupying higher marine trophic levels.

STUDY GUIDE

Marine Biology On Line

Connect to this book's web site: *http://www.bioscience.jbpub.com/marinebiology* *The site features eLearning, an online review area that provides quizzes, chapter outlines, and other tools to help you study for your class. You can also follow useful links for more in-depth information.*

Summary Points

Marine Plants

- Multicellular plants in the sea are dominated by brown and red algae, with green algae and some flowering plants also playing important roles.

Division Anthophyta

- About 50 species of seagrasses thrive throughout the world along subtidal soft-bottom shorelines. Most seagrasses reproduce vegetatively via horizontal rhizomes or sexually via underwater pollination of tiny flowers followed by fruit production.
- Additional flowering plants, such as marsh grasses and mangals, grow on soft bottoms in the intertidal zone. All types of marine flowering plants host a unique community of organisms within the habitat that they create.

The Seaweeds

- Most large conspicuous plants in the sea are macroalgae (seaweeds and kelps), growing from rocky substrates with their characteristic blades, stipes, holdfasts, and pneumatocysts (in some species).

- The common names of seaweeds often are motivated by their colors, which in turn reflect the various photosynthetic pigments that they contain. Just as in phytoplankton, there is adaptive significance for all accessory photopigments possessed by seaweeds.
- Reproduction in seaweeds can be either vegetative and asexual or complex and sexual. Sexual reproduction tends to follow three fundamental patterns, all variations of alternating sporophyte, gametophyte, and/or carposporophyte generations.

Geographic Distribution

- A complex interplay of a multitude of physical, chemical, geologic, and biologic factors determine the distribution of marine plants on both small and large scales. A knowledge of these variables helps one understand why opposite sides of an intertidal rock or an entire continent may host different species of plants.

Seasonal Patterns of Marine Primary Production

- Spatial and temporal variations in available sunlight, nutrients, and grazers cause significant seasonal and global differences in marine production. A diatom bloom during the spring, followed by the successional appearance of less-numerous dinophytes during the summer, is the prominent scenario of production in temperate seas.
- Tropical and subtropical waters exist in eternal stratification, with low rates of production and year-round plankton communities that resemble those of temperate regions during summer months.
- Rates of net primary production in areas of coastal upwelling are among the highest in the sea, but they are very limited in geographic extent and some of the most important ones are interrupted by El Niño events.
- The seasonal formation and melting of sea ice and tremendous variations in availability of sunlight greatly influences production in polar seas, yet photosynthesis can continue around

the clock during a few summer months to create dense populations of phytoplankton.

Global Marine Primary Production

- Spatial variations in primary production are common, with mid- and high-latitude regions, shallow coastal areas, and upwelling zones being the most prolific, but only during warm summer months when sufficient sunlight is available.

Topics for Discussion and Review

1. Terrestrial flowers are pollinated by a variety of insects, birds, and bats. How are the flowers of subtidal seagrasses pollinated?
2. The seeds of red mangroves germinate while their fruit still hangs from the parent tree. Summarize this unusual form of sexual reproduction.
3. *Sargassum* contains numerous small pneumatocysts to buoy the plant toward the sunlit surface and hosts a complex community of fishes and invertebrates that are uniquely adapted to living on this pelagic seaweed. Consider this paradox.
4. A life cycle consisting of alternating gametophyte and sporophyte generations is characteristic of almost all plants. How do the basic features of that life cycle differ among the different divisions of seaweeds?
5. What characteristics of green algae (Chlorophyta) support the hypothesis that they are ancestral to flowering plants (Anthophyta)?
6. How do local assemblages of kelp plants, seagrasses, and mangals influence and alter the physical characteristics of the shoreline on which they live?
7. Draw a generic graph of diatom concentrations in a temperate sea during the course of a single year, and then explain the factors that cause the observed peaks and valleys in this annual curve.
8. What is the El Niño phenomenon and how does it interrupt the massive upwelling of the Peru Current?
9. Describe how the formation and thawing of sea ice affects primary production in the Arctic and Antarctic.
10. Why are tropical waters usually as clear as gin, whereas the temperate ocean often seems as cloudy as soup?

STUDY GUIDE

Suggestions for Further Reading

Alongi, D. M. 2002. Present state and future of the world's mangrove forests. *Environmental Conservation* 29:331–349.

Bortone, S. A. 1999. *Seagrasses: Monitoring, Ecology, Physiology, and Management.* CRC Press, Boca Raton, FL.

Coimbra, C. S., and F. A. S. Berchez. 2000. Habitat heterogeneity on tropical rocky shores: a seaweed study in southern Brazil. *Journal of Phycology* 36:14–15.

Cole, K. M., and R. G. Sheath. 1990. *Biology of the Red Algae.* Cambridge University Press, New York.

Dawes, C. J. 1981. *Marine Botany.* John Wiley & Sons, New York.

De Lacerda, L. D. 2002. *Mangrove Ecosystems: Function and Management.* Springer-Verlag, New York.

Diaz, H. F., and V. Markgraf. 2000. *El Niño and the Southern Oscillation: Multiscale Variability and Global and Regional Impacts.* Cambridge University Press, New York.

Dring, M. J. 1982. *The Biology of Marine Plants.* University Park Press, Baltimore.

Druehl, L. 2003. *Pacific Seaweeds: A Guide to Common Seaweeds of the West Coast.* Harbour Publishing Company.

Duffy, J. E., and M. E. Hay. 1990. Seaweed adaptations to herbivory. *Bioscience* 40:368–375.

Falkowski, P. G., A. D. Woodhead, and K. Vivirito. 1992. *Primary Productivity and Biogeochemical Cycles in the Sea.* Plenum Press.

Fanshawe, S., G. R. Vanblaricom, and A. A. Shelly. 2003. Restored top carnivores as detriments to the performance of marine protected areas intended for fishery sustainability: a case study with red abalones and sea otters. *Conservation Biology* 17:273–283.

Foster, M. S. 1975. Algal succession in a *Macrocystis pyrifera* forest. *Marine Biology* 32:313–329.

Gaylord, B., M. W. Denny, and M. A. R. Koehl. 2003. Modulation of wave forces on kelp canopies by alongshore currents. *Limnology and Oceanography* 48:860–871.

Hemminga, M. A., and C. A. Duarte. 2001. *Seagrass Ecology.* Cambridge University Press, England.

Hogarth, P. J. 2000. *The Biology of Mangroves.* Oxford University Press, Oxford.

Jacobs, G. A. et al. 1994. Decade-scale trans-Pacific propagation and warming effects of an El Niño anomaly. *Nature* 370:60–63.

Joint, I., and S. B. Groom. 2000. Estimation of phytoplankton production from space: current status and future potential of satellite remote sensing. *Journal of Experimental Marine Biology and Ecology* 250:233–255.

Kaldy, J. E., and K. H. Dunton. 1999. Ontogenetic photosynthetic changes, dispersal and survival of *Thalassia testudinum* (turtle grass) seedlings in a sub-tropical lagoon. *Journal of Experimental Marine Biology and Ecology* 240:193–212.

Lobban, C. S., and P. J. Harrison. 1996. *Seaweed Ecology and Physiology.* Cambridge University Press, New York.

Mann, K. H., and P. A. Breen. 1972. The relation between lobster abundance, sea urchins, and kelp beds. *Journal of the Fisheries Research Board of Canada* 29:603–609.

Minchinton, T. E. 2002. Disturbance by wrack facilitates spread of *Phragmites australis* in a coastal marsh. *Journal of Experimental Marine Biology and Ecology* 281:89–107.

Palmer, J. R., and I. J. Totterdell. 2001. Production and export in a global ocean ecosystem model. *Deep Sea Research* 48:1169–1198.

Santelices, B. 1990. Patterns of reproduction, dispersal and recruitment in seaweeds. *Oceanography and Marine Biology* 28:177–276.

Teal, J., and M. Teal. 1975. *The Sargasso Sea.* Little, Brown, Boston.

Tiner, R. W., and A. Rorer. 1993. *Field Guide to Coastal Wetlands Plants of the Southeastern United States.* University of Massachusetts Press, Boston.

CHAPTER 5

Marine Protozoans and Invertebrates

In this chapter, the nonphotosynthetic members of the kingdom Protista (the "protozoans") are introduced, followed by the invertebrate phyla of the kingdom Animalia. Only the more abundant and obvious free-living phyla are described, and groups that are primarily or wholly parasitic are ignored. The purpose of this quick romp through the more common or familiar marine protistan and animal phyla is to provide an introduction to some of the major players making their appearance in subsequent chapters.

Of the many phyla introduced in this chapter, only a few—in particular, protozoans, cnidarians, nematodes, mollusks, annelids, echinoderms, and arthropods—clearly dominate the composition of most marine communities and monopolize the energy flow through those communities. These phyla are presented in an order that generally corresponds to a movement up the right, or heterotrophic, side of the phylogenetic tree in Figure 2.7, which is expanded in FIGURE 5.1 to include more detail. The structural features listed on the right side of Figure 5.1 (and others not included here) have long served as the basis for defining patterns of relationships between animal phyla. However, some of these traditionally accepted alliances are being challenged by new genetic comparison techniques. The recent and continuing accumulation of massive amounts of genetic data from these animal phyla suggest alternative phylogenies that are at odds with traditional ones, and also sometimes with each other. Genetic taxonomic methods are relatively new, and as more of this information is compiled, even newer arrangements of animal phyla can be expected.

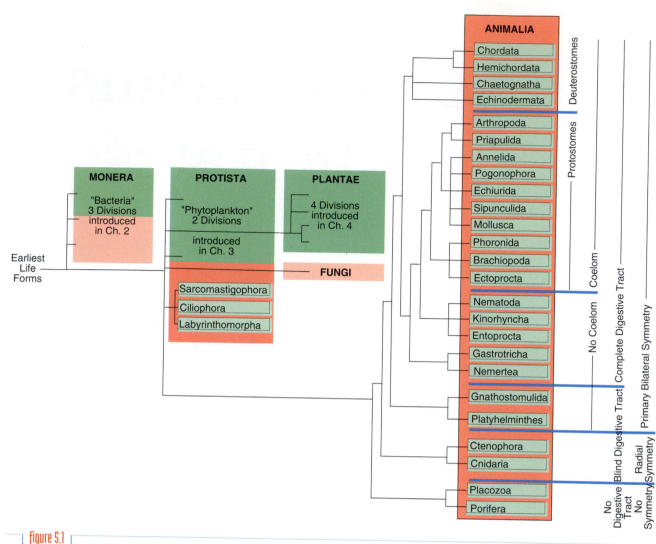

The phylogenetic tree in the figure includes the following labeled elements:

MONERA
"Bacteria"
3 Divisions
introduced
in Ch. 2

PROTISTA
"Phytoplankton"
2 Divisions
introduced
in Ch. 3

Sarcomastigophora
Ciliophora
Labyrinthomorpha

PLANTAE
4 Divisions
introduced
in Ch. 4

FUNGI

Earliest
Life
Forms

ANIMALIA
Chordata
Hemichordata
Chaetognatha
Echinodermata
Arthropoda
Priapulida
Annelida
Pogonophora
Echiurida
Sipunculida
Mollusca
Phoronida
Brachiopoda
Ectoprocta
Nematoda
Kinorhyncha
Entoprocta
Gastrotricha
Nemertea
Gnathostomulida
Platyhelminthes
Ctenophora
Cnidaria
Placozoa
Porifera

Deuterostomes
Protostomes
Coelom
No Coelom
Complete Digestive Tract
Blind Digestive Tract
No Digestive Tract
Primary Bilateral Symmetry
Radial Symmetry
No Symmetry

figure 5.1

The phylogenetic tree introduced in Figure 2.7, emphasizing the 3 phyla from the kingdom Protista and 25 phyla from the kingdom Animalia described in this chapter.

The sequence of phyla from the bottom to the top of Figure 5.1 includes significant trends and major milestones in the history of animal evolution. Increased complexity and specialization of body structures are evident, especially in the systems involved in gas exchange, excretion, feeding and digestion, circulation, and reproduction. Trends from radial to bilateral body symmetry and the evolution of body cavities are evident, as is the increased reliance on sexual reproduction. Improved sensory systems and increasingly complex brains able to integrate sensory information support expanding patterns of behavioral responses. In TABLE 5.1, the phyla introduced in this chapter along with some of their major taxonomic subgroups are listed with their English common names and approximate number of species.

5.1 Animal Beginnings: The Protozoans

The term "protozoa" encompasses a variety of nonphotosynthetic microscopic members of the kingdom Protista. (Marine photosynthetic protists were described in Chapter 3.) They are included in this chapter for convenience and because many biologists casually (and oxymoronically) consider them "single-celled animals." Individual protozoans consist of a single cell or loose aggregates of a few cells, and they share some other features of animals described later in this chapter, including ingestion of food particles for nutrition and an absence of cell walls and photosynthetic chloroplasts.

The number of protozoan phyla varies widely from one classification system to another and will

Table 5.1

A Partial (and Brief) Taxonomy of the Marine Animal and Non-photosynthetic Protist Groups Included in Figure 5.1[a]

Kingdom:	Protista
Phylum:	Sarcomastigophora (8700, all habitats)—unicellular animals; locomotion with flagella or pseudopodia
Phylum:	Ciliophora (9000, all habitats)—unicellular animals; locomotion with numerous cilia
Phylum:	Labyrinthomorpha—(40, all habitats) small colonies encrusting networks of cells
Kingdom:	Animalia
Phylum:	Porifera (10,000, mostly marine)—simple multicellular animals found attached to solid substrates in benthic habitats; reproduction is both asexual and sexual and results in free-swimming larval stages
Phylum:	Placozoa (1, marine)—small asymmetric plate of cells
Phylum:	Cnidaria (10,000, mostly marine)—radically symmetric animals with mouth, tentacles, cnidocytes, and simple sensory organs and nervous system; common in both benthic and pelagic habitats; reproduction is both sexual and asexual (by budding or fission)
Class:	Hydrozoa—solitary or colonial, with both polypoid and medusoid forms
Class:	Scyphoza—free-swimming medusoid forms (most jellyfishes)
Class:	Anthozoa—attached benthic polypoid forms (corals and anemones)
Phylum:	Ctenophora (100, marine)—biradially symmetric pelagic swimming animals with rows of cilia (ctenes)
Phylum:	Platyhelminthes (18,500, all habitats)—free-living and parasitic flatworms
Class:	Turbellaria—small free-living flatworms with incomplete digestive tracts and ciliated undersides; found in benthic habitats
Phylum:	Nemertea (900, mostly marine)—most are small, inconspicuous, wormlike benthic animals with complete digestive tracts
Phylum:	Gastrotricha (400, 50% are marine)—microscopic, with elongated bodies; in benthic habitats
Phylum:	Kinorhyncha (150, marine)—elongated, less than 1 mm in length; in benthic habitats
Phylum:	Gnathostomulida (80, marine)—small benthic worms
Phylum:	Priapulida (15, marine)—small, bethic worms
Phylum:	Nematoda (16,000, all habitats)—parasitic and free-living roundworms a few millimeters in length; mostly benthic
Phylum:	Entoprocta (150, mostly marine)—nearly microscopic benthic animals that form colonial encrustations on hard substrates
Phylum:	Bryozoa (4,000, mostly marine)—benthic, with the exception of one pelagic Antarctic species
Phylum:	Phoronida (12, marine)—tube-dwelling benthic worms
Phylum:	Brachiopoda (350, marine)—benthic animals; bodies covered with hinged shell
Phylum:	Mollusca (65,000, mostly marine)—unsegmented body usually covered with external shell of one, two, or eight pieces
Class:	Aplacophora—rare benthic mollusks without shells
Class:	Monoplacophora—rare, benthic
Class:	Amphineura—shallow-water benthic animals known as chitons; eight-piece shell
Class:	Gastropoda—mostly benthic; shell usually absent or of one piece; includes slugs, snails, and limpets
Class:	Scaphopoda—benthic; shell of one piece and elongated; known as tusk shells
Class:	Bivalvia—benthic; shell of two pieces; clams, oysters, and other bivalves
Class:	Cephalopoda—benthic and pelagic; shell usually absent, foot modified as tentacles with suckers; octopuses and squids
Phylum:	Sipuncula (350, marine)—benthic worms a few centimeters long; known as peanut worms
Phylum:	Echiurida (140, marine)—benthic; cylindrical worms
Phylum:	Pogonophora (120, marine)—deep-water benthic tube-dwelling worms; to 2 m in length
Phylum:	Hemichordata (85, marine)—elongated benthic worms; acorn worms
Phylum:	Chaetognatha (120, marine)—pelagic, active predators; to 15 cm long; known as arrow worms
Phylum:	Annelida (12,400, marine, fresh water, and terrestrial)—segmented worms, mostly small, but to 3 m in length
Class:	Polychaeta—mostly benthic, free living
Class:	Hirudinea—leeches, some parasitic
Phylum:	Arthropoda (1,000,000, all habitats)—segmented animals with bodies covered by exoskeleton of chitin; most a few centimeters or less in length; several classes not found in marine habitats
Class:	Merostomata—horseshoe crabs; benthic near-shore animals
Class:	Pycnogonida—sea spiders; benthic animals with four pairs of elongated legs
Class:	Crustacea—mostly marine; with two pairs of antennae; numerous pelagic and benthic species
Subclass:	Branchiopoda—brine shrimp
Subclass:	Ostracoda—seed shrimps; pelagic animals usually less than 1 cm
Subclass:	Copepoda—abundant animals in pelagic and benthic habitats; microscopic to about 1 cm
Subclass:	Cirripedia—barnacles; larger benthic, attached animals
Subclass:	Malacostraca

(continued on next page)

A Partial (and Brief) Taxonomy of the Marine Animal and Non-photosynthetic Protist Groups Included in Figure 5.1ᵃ—*Continued*

Table 5.1

Order:	Mysidacea—mysids; benthic and pelagic; size to a few centimeters	
Order:	Cumacea—burrow in mud and sand; size to a few centimeters	
Order:	Isopoda—benthic; body flattened dorsoventrally; size to a few centimeters	
Order:	Amphipoda—benthic and pelagic; body laterally flattened; size to a few centimeters	
Order:	Stomatopoda—mantis shrimps; benthic; size to 30 cm	
Order:	Euphausiacea—krill; pelagic; size to several centimeters	
Order:	Decapoda—crabs, shrimps, and lobsters; mostly benthic; several centimeters to 1 m in size	
Phylum:	Echinodermata (6500, marine)—five sided radial symmetry; most benthic	
Class:	Echinoidea—sea urchins, sand dollars	
Class:	Asteroidea—sea stars	
Class:	Ophiuroidea—brittle stars	
Class:	Crinoidea—feather stars, sea lillies	
Class:	Holothuroidea—sea cucumbers	
Class:	Concentricycloidea—sea daisies	
Phylum:	Chordata (62,000, all habitats)	
Subphylum:	Urochordata	
Class:	Ascidiacea—sea squirts; benthic; solitary or colonial	
Class:	Larvacea—pelagic; less than 1 cm	
Class:	Thaliacea—salps; pelagic; gelatinous	
Subphylum:	Cephalochordata—slender, laterally compressed; benthic	
Subphylum:	Vertebrata—fishes and tetrapods	
Class:	Agnatha—lampreys and hagfishes	
Class:	Chondrichthyes—sharks, skates, and rays	
Class:	Osteichthyes—bony fishes; includes about 30 orders with marine species	
Class:	Amphibia—frogs, toads, and salamanders	
Class:	Reptilia—marine turtles, iguanas, crocodiles, and sea snakes	
Order:	Testudinata—turtles	
Order:	Squamata—iguanas and snakes	
Order:	Crocodilia—caymens and crocodiles	
Class:	Aves—marine birds	
Order:	Sphenisciformes—penguins	
Order:	Procellariiformes—albatrosses, petrels, fulmars, shearwaters	
Order:	Pelecaniformes—pelicans, cormorants, gannets, boobies	
Order:	Charadriiformes—gulls, sandpipers, puffins	
Class:	Mammalia	
Order:	Carnivora—sea lions, seals, walruses, sea otters	
Order:	Cetacea—whales	
Order:	Sirenia—manatees and dugongs	

ᵃThe numbers in parentheses refer to the approximate numbers of described species in that group. Data mostly from Villee *et al.* 1984 and Hickman and Roberts, 1994.

remain confusing and unresolved because the kingdom Protista is an artificial catch-all category for many fundamentally different phyla and divisions clearly lacking an ancestral form common to all of them. Several protozoan phyla are mostly or completely parasitic; only three are described here that include marine nonparasitic species that thrive in benthic and planktonic communities. Asexual reproduction by cell division is common. Sexual reproduction, when it does occur, is often quite complex, with the process of meiosis separated from that of nuclear fusion by several cell generations. These three free-living protozoan phyla are most easily distinguished by their different methods of locomotion. When moving in water, these small cells encounter very high viscous forces between water molecules and, regardless of their mode of locomotion, do not swim so much as crawl through their watery environment.

Phylum Sarcomastigophora

A large and widespread phylum, the Sarcomastigophora use either whiplike flagella or extensions of their cellular protoplasm, pseudopodia (**FIG. 5.2**), or both, for locomotion. The foraminiferans and radiolarians are members of this phylum, as are a large variety of amoeboid and flagellated forms, including the familiar *Amoeba* from high school biology. Sometimes, the photosynthetic dinophytes (considered as phytoplankton in Chapter 3) are also included in this phylum under the label dinoflagellates.

About one half of all named protozoans are foraminiferans. Foraminiferans are shelled amoebae that are mostly marine. They are common in the plankton, yet more are benthic or live attached to plants and animal shells. Most foraminiferans are microscopic, although individuals of a few species grow to several millimeters in size. They have internal chambered shells, or tests, usually composed of either $CaCO_3$ or cemented sand grains. Penetrating this shell, or test, are numerous cytoplasmic filaments (Fig. 5.2a) that are used for locomotion, for attachment, and for collecting food. Some planktonic foraminiferans, such as *Globigerina* (Fig. 5.2b), are so widespread and abundant that their tests blanket large portions of the seafloor. After thousands of years of accumulation, this **globigerina ooze** may form deposits tens of meters thick. The famous chalk cliffs of Dover, England are composed mainly of foraminiferan tests and coccolith plates (described in Chapter 3) that accumulated on the seafloor and were subsequently lifted above sea level where they were eroded into vertical seaside cliffs. This fall of carbonate skeletal material from the sea surface has and continues to be a major process to remove CO_3 from the atmosphere and sequester it away for millions of years in marine sediments.

Radiolarians are entirely marine, and most members are planktonic. They are similar in size

(a)   (b)

figure 5.2

A planktonic foraminiferan, *Globigerina*. (a) Drawing of an intact animal emphasizing extended pseudopodia. (From Brady 1884.) (b) Photograph of *Globigerina* test.

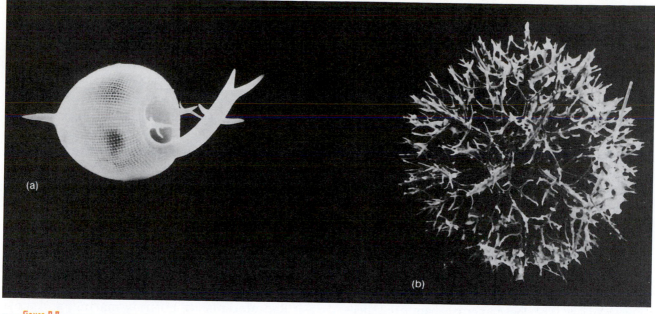

figure 5.3

Scanning electron micrograph of the silicate skeletons of two planktonic radiolarians: (a) *Euphysetta elegans*, 230×. (b) *Elatomma pinetum*, 110×.

to planktonic foraminiferans and occur in all latitudes and at all depths. An internal skeleton of silica forms the beautiful symmetry often associated with radiolarians (FIG. 5.3). Radiolarians also use cytoplasmic filaments to capture bacteria, detritus, and other protozoans for food.

Phylum Ciliophora

Members of the phylum Ciliophora are known as ciliates, because they possess **cilia** as their chief means of locomotion. Structurally, a cilium is much like a short flagellum; however, cilia are typically much more numerous than flagella, and they move in a coordinated manner (FIG. 5.4). Cilia move water parallel to the cell surface (flagella move water perpendicular to the surface of the cell). Tintinnids are probably the most abundant of the marine ciliophores. These planktonic cells live partially enclosed in a vase-shaped structure made of cemented particles or of a material secreted by the cell (FIG. 5.5). Ciliated tentacles at one end of the cell are used for collecting bacteria and other small protozoans. A large variety of other ciliates exists in marine planktonic and benthic communities, and many other marine ciliophorans exist as parasites.

Phylum Labyrinthomorpha

The small phylum Labyrinthomorpha includes a group of free-living organisms characterized by a net-work of slime through which colonies of small (about 10 μm long) spindle-shaped cells live and move. Most species are marine, forming their colonial networks on the surfaces of seagrasses and benthic algae. Their mechanism of gliding through their slime networks is not understood and is accomplished without pseudopodia, cilia, or flagella. Reproduction is both sexual and asexual in this phylum.

5.2 Defining Animals

The boundary separating the kingdoms Protista and Animalia is vague (see the criteria listed on pp. 52–53), and some colonial protozoans seem on the verge of crossing that boundary with their superficial appearance of multicellularity. Yet protozoans lack two of the hallmark features of animals: contractile muscles and signal conducting neurons. In addition, there is a greater dependence in the more complex animal phyla on sexual reproduction and less on asexual reproduction more commonly seen in the kingdom Protista. Sexual reproduction is not only more complex than asexual reproduction, it is also more costly. The high costs of sexual reproduction are related to the expenses of producing and maintaining males and to the expenses associated with meiosis. Regardless of the approach taken to accomplish reproduction, all sexually reproducing organisms include the same basic elements in the process

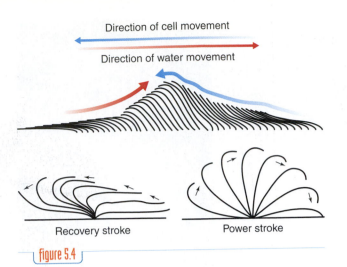

Direction of cell movement

Direction of water movement

Recovery stroke   Power stroke

figure 5.4

The pattern of ciliary movement appearing as waves of alternating recovery and power strokes sweeping over the cell surface.

figure 5.5

A marine tintinnid with a crown of cilia at one end.

(see Fig. 2.3). In **hermaphroditic** animals, such as most barnacles, many snails, and some fishes, all adults function in both female and male gender roles, some at the same time (simultaneous hermaphrodites); others function as one gender and then transform to the other (sequential hermaphrodites).

Most other animals retain the same gender role for their entire lives, with approximately equal numbers of males and females. Either way, they still express the basic attributes of sexual reproduction described in Chapter 2, namely meiosis followed by fertilization. The diploid zygote resulting from fertilization then develops by successive cell cleavages to form a hollow ball of a few hundred cells, the **blastula**, another developmental stage characteristic of animals. This process of sexual reproduction can create enormous genetic variety in populations within short time spans. This genetic diversity is the raw material on which natural selection processes act to produce adaptations in evolutionary time scales.

The first two animal phyla described, the Porifera and Placozoa, might be considered sidelines to the major trends of animal evolution. They are animals, yet with their individual cells functioning independently, they lack the coordinated patterns of cell functions seen in more specialized tissues or organs of other animal phyla.

Phylum Porifera

The Porifera is one of the few animal phyla with a widely accepted common name: the sponges. The sponges are among the simplest animals. Each sponge consists of several types of loosely aggregated cells organized into a multicellular organism with a distinctive and recognizable form. Despite their structural simplicity, sponges share several advantages with other multicellular animals. Cells within each individual sponge can divide repeatedly to achieve larger sizes and longer life spans than are possible for individual protists. In addition, the specialization of cells, although it is limited in sponges, does allow more efficient handling of food, protection, and other diverse chores of survival (**FIG. 5.6**).

The phylum name Porifera stems from the many pores, holes, and channels that perforate the bodies of sponges. Water is circulated through these openings into an internal cavity, the **spongocoel**, where food and oxygen are extracted by flagellated cells, the **choanocytes**, lining the spongocoel. The water then exits through a large excurrent pore, the osculum.

Sponges are mostly marine and are usually found attached to hard substrates such as rocks, pilings, or shells of other animals. Sometimes they are radially symmetric, but more commonly they conform to the shape of their substrate or to the sculpting influences of waves and tides. Some sponges are supported internally by a network of flexible **spongin** fibers. (Commercial bath sponges are actually the spongin skeleton with all living material removed.) Other sponges have skeletons composed of hard mineralized **spicules**. The spicules are either calcareous ($CaCO_3$) or siliceous (SiO_2). The silicate skeleton of the deep-water glass sponge *Euplectella* is one of the most complex and beautiful skeletons of all sponges (**FIG. 5.7**).

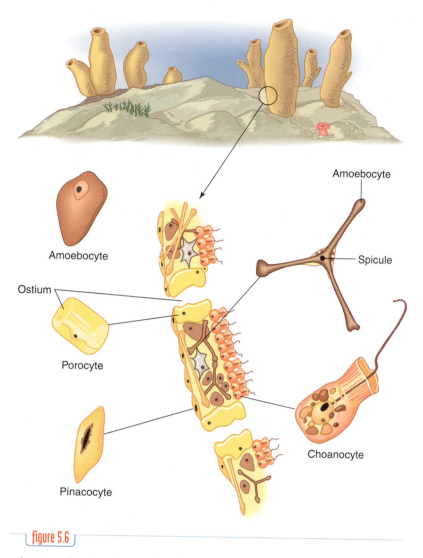

figure 5.6

A group of marine finger sponges and several of the specialized cell types that make up the sponge wall.

Phylum Placozoa

The phylum Placozoa is represented by only one known species, *Trichoplax adhaerens,* and until 1971 it was misidentified as a larva from another phylum. Each animal is 2–3 mm long and consists of a few thousand cells shaped like a flattened plate. Outside aquaria, it is found throughout the Pacific Ocean gliding on hard surfaces, although they seem to be pelagic for a part of their lives. Movement is accomplished via the flagella, which cover both sides of the animal. It has no obvious body symmetry. Both surfaces are covered with cilia for swimming, allowing it to move in any direction. Digestion of food is accomplished by secreting enzymes externally and then absorbing the digested molecules. Like sponges, these animals exhibit limited specialization and organization of cells and can regenerate a complete animal from a single cell. Although sexual reproduction is poorly known in placozoans, individuals have produced an oocyte or embryos in the laboratory.

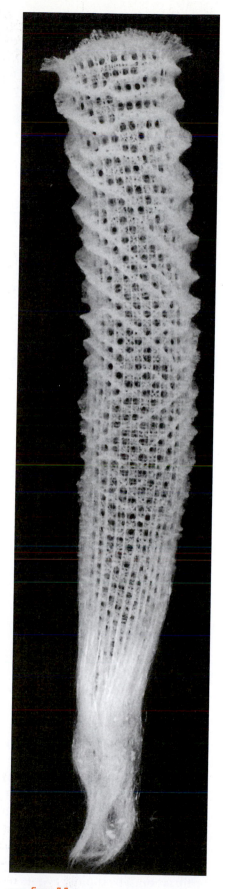

figure 5.7

Silicate skeleton of a glass sponge, *Euplectella.*

5.3 Radial Symmetry

Three animal phyla, the Cnidaria, Ctenophora, and Echinodermata, exhibit radially symmetric body plans. The circular shape of radially symmetric animals provides several different planes of symmetry to divide the animal into mirror image halves (FIG. 5.8). The mouth is located at the center of the body on the **oral** side; the opposite side is the **aboral** side. Radially symmetric animals possess a relatively simple and diffuse network of nerves that lack a central brain to process sensory information or to organize complex responses. Two of these phyla, the Cnidaria and Ctenophora, are described here because they represent an early evolutionary stage of animal body structures based on the presence of tissues and a primary, or fundamental, radial symmetry. The Echinodermata are introduced later in this chapter because they are evolutionarily only secondarily radially symmetric, having evolved from bilaterally symmetric ancestors.

Phylum Cnidaria

The phylum Cnidaria includes a large diverse group of relatively simple yet versatile marine animals, such as jellyfishes, sea anemones, corals, and hydroids. The inner and outer body walls of all cnidarians are separated by a gelatinous layer called the **mesoglea**.

A centrally located mouth leads to a baglike digestive tract, the **gastrovascular cavity**. The mouth is surrounded with tentacles capable of capturing a wide variety of marine animal prey. The tentacles and, to a lesser extent, other parts of the body are armed with batteries of microscopic structures, the **nematocysts**. Nematocysts are produced in special cells, the **cnidocytes**, and are a characteristic of this phylum. They are discharged when stimulated by contact with other organisms. Most nematocysts pierce the prey and inject a paralyzing toxin (FIG. 5.9), whereas others are adhesive and stick to the prey or become entangled in the prey's bristles or spines.

Cnidarians exist either as free-swimming **medusae** or as attached benthic **polyps**. The two forms have essentially the same body organization. The oral side of the medusa, bearing the mouth and tentacles, is usually oriented downward. The mesoglea of most medusae is well developed and is jellylike in consistency, thus earning them the descriptive, if inappropriate, name of jellyfish. In the polyp, the mouth and tentacles typically are directed upward. Many species of cnidarians have life cycles that alternate between a swimming medusoid generation and an attached benthic polypoid generation. In a generalized cnidarian

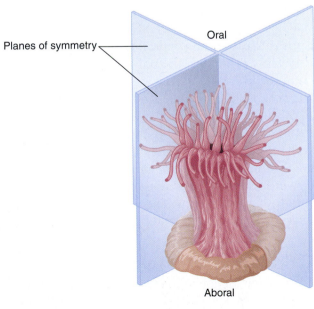

figure 5.8

Planes of symmetry in a radially symmetric animal.

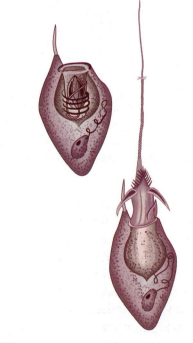

figure 5.9

(Top) Undischarged nematocyst coiled inside a cnidocyte. (Bottom) Discharged penetrant nematocyst.

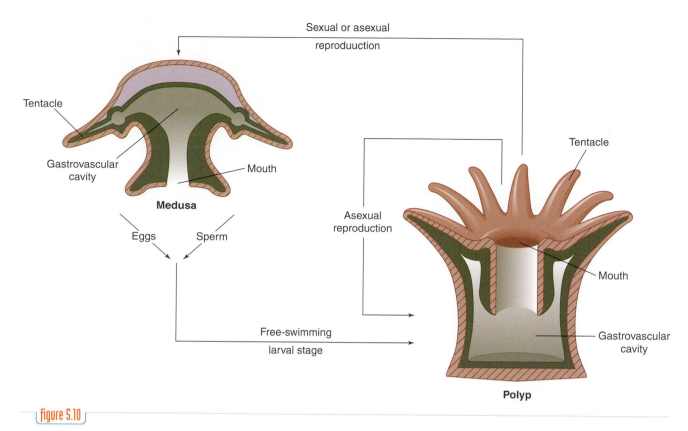

figure 5.10

Generalized cnidarian life cycle.

figure 5.11

A large jellyfish, *Pelagia*. (Photo by T. Phillipp.)

life cycle (FIG. 5.10), polyps can produce medusae or additional polyps by budding. The medusae in turn produce eggs and sperm that, after fertilization, develop into the polyps of the next generation.

The phylum Cnidaria consists of three classes, each characterized by its own variation of the basic cnidarian life cycle shown in Figure 5.10. The class Hydrozoa includes colonial hydroids and siphonophores, such as the Portuguese man-of-war, *Physalia*. Hydrozoans usually have well-developed medusoid and polypoid generations. Different individuals of the polyp colony are specialized for particular functions, such as feeding, reproduction, and defense. In the class Scyphozoa, the polyp stage is reduced or completely absent. This class includes most of the larger and better-known medusoid jellyfishes (FIG. 5.11). In the third class, the Anthozoa, the polyp form dominates and the medusoid generation is absent. Many anthozoans, including most corals (FIG. 5.12) and sea fans, are colonial, but some anemones exist as large solitary individuals. Unlike most cnidarians, corals and some other anthozoans (and a few hydrozoans) produce external, often massive, deposits of $CaCO_3$.

Phylum Ctenophora

The phylum Ctenophora consists of about 100 species. All are marine and most are planktonic, usually preying on small zooplankton. Most individuals are smaller than a few centimeters in size, but one tropical genus (*Cestum*) may grow to exceed 2 m in length.

Cnidarian polyps, with batteries of cnidocytes visible as beadlike structures on tentacles.

Ctenophores are closely related to cnidarians. Like cnidarians, ctenophores are radially symmetric, have a gelatinous medusa-like body, and, in some, colloblast cells that superficially resemble sticky cnidarian nematocysts. One species of ctenophore does possess true venomous nematocysts. Members of this phylum have external longitudinal bands of cilia, called **ctenes** (FIG. 5.13), that provide propulsion, whereas tentacles armed with colloblasts capture food.

5.4 Marine Acoelomates and Pseudocoelomates

With one exception (the echinoderms), the remainder of the animal phyla exhibit bilateral body symmetry. Bilateral symmetry refers to a basic animal body plan in which only one plane of symmetry exists to create mirror image left and right halves (FIG. 5.14). These animals exhibit definite head (**anterior**) and rear (**posterior**) ends, a top (**dorsal**) and bottom (**ventral**) surface, and right and left sides. Bilaterally symmetric animals typically possess specialized sensory organs and an anterior brain containing the complex of nerve cells needed to process the widening scope of information coming from the sense organs. This evolutionary trend is culminated in the

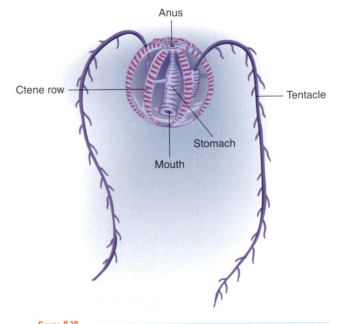

A ctenophore, *Pleurobrachia,* with tentacles and radial rows of visible ctenes.

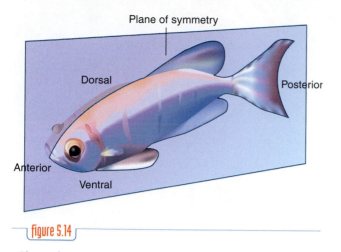

figure 5.14

Plane of symmetry in a bilaterally symmetric animal.

development of a defined head region containing specialized sensory receptors in close proximity to an anterior brain for vision and the detection of chemicals and sound vibrations.

The remaining animal phyla are divided into three different groups depending on their pattern of internal body cavity development. The simplest groups of bilaterally symmetric marine animals include small often overlooked inhabitants of soft mud and sand. These phyla lack an internal cavity between their body wall and digestive tract (acoelomates) or have a poorly developed one (pseudocoelomates). In these phyla, a circulatory system is either absent or is a vaguely defined open system relying on contractions of the body wall rather than a heart to circulate body fluids. Small body sizes also minimize the need for other sophisticated internal organ systems. These features contrast strongly with those of the coelomates with their well-defined fluid-filled body cavity, the **coelom**, described later in this chapter.

Phylum Platyhelminthes

Most Platyhelminthes, or flatworms, are parasitic (this group includes the well-known flukes and tapeworms). Only members of the class Turbellaria are free living. Turbellarians are primarily aquatic, and most are marine. There are a few planktonic species of flatworms, but most dwell in sand or mud or on hard substrates.

Marine turbellarians are usually less than 10 cm long, flat and thin, and are sometimes quite colorful (FIG. 5.15a). Cilia cover their outer surfaces and are best developed on the flatworms' ventral side. These cilia provide a gliding type of locomotion for moving over

solid surfaces. The mouth is usually centrally located on the ventral side and leads to a baglike digestive tract. Turbellarians are carnivorous, preying on other small invertebrates.

Phylum Gnathostomulida

Gnathostomulids, or jaw worms, are a small group of marine worms closely related to turbellarian flatworms. Like turbellarians, jaw worms have a mouth but no anus. Individuals seldom exceed 1 mm in length. They live in sea bottom deposits, where they scrape bacteria and algal film off sediment grains with their jaws and associated basal plate. Of the 80 or so species, most are hermaphroditic, and free-swimming larvae are absent.

Phylum Nemertea

Nemerteans are benthic animals, known as ribbon worms, that also are closely related to the flatworms but have a more elaborate body structure (Fig. 5.15b). They have a simple open circulatory system, a more complex nervous system, and a complete digestive tract with mouth and anus. Individuals of one species are over 2 m long, but most are much smaller. These shallow-water animals are equipped with a remarkable proboscis for defense and food gathering. The proboscis can be everted rapidly from the anterior part of the body to ensnare prey. The proboscis of some nemertean worms has a piercing stylet to stab prey and inject a toxin.

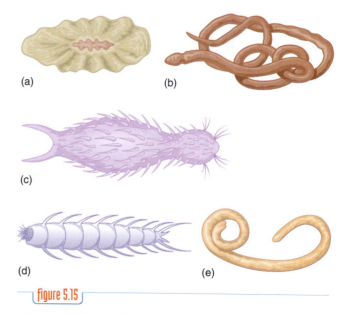

figure 5.15

Some simple wormlike animal phyla: (a) flatworm, (b) nemertean, (c) gastrotrich, (d) kinorhynch, and (e) nematode.

Phylum Gastrotricha

Gastrotrichs include a large variety of marine species, but most are so small (usually less than 1 mm) that they go unnoticed by most observers. They are cylindrical and elongated, with a mouth, feeding structures, and sensory organs at the anterior end (Fig. 5.15c). Marine gastrotrichs inhabit sand and mud deposits in shallow water and feed on detritus, diatoms, or other very small animals.

Phylum Kinorhyncha

Kinorhynchs are exclusively marine and resemble gastrotrichs. They also are cylindrical and elongated but are covered with a cuticle that is segmented (Fig. 5.15d). Marine kinorhynchs occur in sand and mud deposits, sometimes in densities of more than 2 million individuals/m², where they feed exclusively on bacteria.

Phylum Nematoda

Nematode worms (Fig. 5.15e) are among the most common and widespread multicellular animals. Some are parasitic, but many more are free living. Most marine nematodes live in bottom sediments and are found at virtually all water depths. In fact, nematodes are probably the most abundant multicellular animals in the marine benthic environment. Cylindrical in cross-section and greatly elongated, nematodes seldom exceed a few centimeters in length. Locomotion is not well developed; nematodes depend on quick bending movements of their small tubular bodies to wriggle through mud or water.

Phylum Entoprocta

Entoprocts are benthic, living on rocks, shells, sponges, and seaweeds. Most are colonial, secreting thin calcareous encrustations over rocks, seaweeds, and the hard shells of some other animals (FIG. 5.16). Superficially, they resemble small colonial hydroids. Their external appearance is also quite similar to that of members of another phylum, the Bryozoa.

Because entoprocts and bryozoans are superficially similar, they were formerly placed within one phylum. But recent studies of anatomy, embryonic cleavage, and their 18S rRNA genes have shown that two phyla are warranted. Individuals of both groups have U-shaped digestive tracts and a crown of tentacles projecting from the upper surface. The mouth and anus of entoprocts open within the ring of tentacles (hence the name Entoprocta, inner anus). They feed on a wide variety of small food particles.

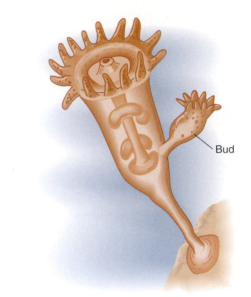

figure 5.16

A solitary entoproct, *Loxosomella*, with its crown of feeding tentacles.

5.5 Marine Coelomates

All remaining marine animal phyla described here are coelomates, characterized by a true internal body cavity, the coelom. The coelom was a major step in the evolutionary development of more complex animal phyla. This cavity originates during embryonic development and separates the digestive tract from the body wall. This separation allows fluids in the coelom to move and promote circulation of oxygen, wastes, and nutrients. Circulation in many coelomates is enhanced by a heart pumping blood through a closed circulatory system. With a coelom, the digestive tract has become more specialized and its efficiency improved. A more spacious coelom also allows larger gonads to increase the number of gametes for use in diverse reproductive strategies. Finally, body wall muscles function independently of the digestive tract and have a greater range of specialized actions.

Protostomes

Coelomates include most of the dominate and successful marine animal phyla as well as several smaller lesser known ones. Coelomates have evolved along two separate lineages, the protostomes and deuterostomes, reflecting features of their early embryonic development. This protostome–deuterostome split is characterized in modern animal groups by fundamental differences in the pattern of zygote division after fertilization. Protostomes express spiral patterns

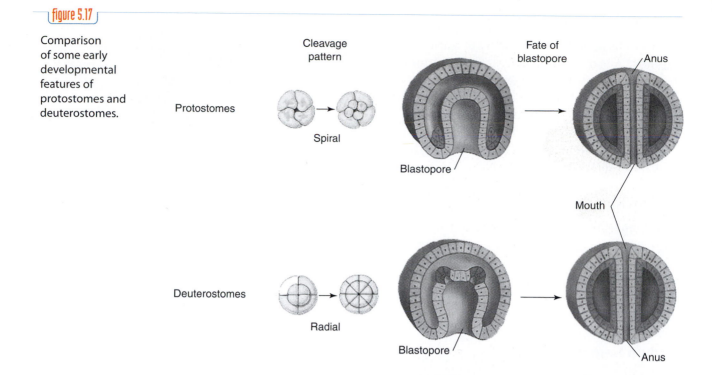

figure 5.17

Comparison of some early developmental features of protostomes and deuterostomes.

Cleavage pattern

Fate of blastopore

Protostomes

Spiral

Blastopore

Anus

Mouth

Deuterostomes

Radial

Blastopore

Anus

of egg cleavage and a determinate pattern of embryonic development, resulting in unequal-sized cells in the preblastula stage, each fated to develop into predetermined tissues or organs of the adult. As development proceeds, protostome embryos acquire an indentation, the blastopore, in the cells of the blastula that eventually becomes the mouth in the adult. Deuterostomes differ from protostomes in all these features. They exhibit radial egg cleavage resulting in equal-sized cells after fertilization, indeterminate development of preblastula cells without early embryonic determination of what adult tissues those cells will develop into, and a blastopore that develops into the anus of the eventual adult rather than the mouth. These fundamental differences are illustrated in FIGURE 5.17.

Three Phyla with Lophophores

A **lophophore** is a crown of ciliated feeding tentacles found in three structurally dissimilar phyla of marine animals: the Bryozoa, Phoronida, and Brachiopoda. The Bryozoa is a major animal phylum, with 4000 freshwater and marine species. They are primarily members of shallow-water benthic communities, occupying the same general habitats as entoprocts. The ectoproct mouth is located within the tentacles, but the anus is not (Bryozoa, outer anus). Like entoprocts, bryozoans are colonial and form encrusting or branching masses of small individuals (usually less than 1 mm in size; FIG. 5.18). Bryozoans and entoprocts provide ex-

figure 5.18

A magnified view of a branched colony of bryozoans with extended feathery lophophores.

figure 5.19

Brachiopods (*Neothyris,* at arrow) among scattered shell fragments at a depth of 40 m.

cellent examples of how the evolutionary pathways of quite different animal groups converge to similar adaptive forms and habits when exposed to similar environmental conditions. Such convergent evolution is a common theme in animal evolution.

The phylum Phoronida consists of about 12 species of elongated burrowing animals. All are marine and live in tubes in shallow water. When feeding, the lophophore projects out of the tube, but it can be rapidly retracted for protection. The phoronids seldom exceed 20 cm in length and have no appendages except for the lophophore.

The phylum Brachiopoda, the lamp shells, were a very successful group in the past, with more than 30,000 extinct species having been described. Fewer than 350 still survive. *Lingula,* for example, has an unbroken fossil history that extends back over the past one half billion years of the Earth's history. All brachiopods are benthic (**FIG. 5.19**) and live attached to the sea bottom by a muscular stalk. The outer calcareous shells superficially resemble those of bivalve mollusks (the next phylum to be described), but the symmetry of the shells is quite different. Bivalve shells are positioned to the left and right of the soft internal organs. In contrast, brachiopod shells are not symmetric and are located on the dorsal and ventral sides of the soft organs.

Living brachiopods occupy a wide variety of seafloor niches, from shallow-water rocky cliffs to deep muddy bottoms. As in the phoronids and bryozoans, the ciliated lophophore gathers minute suspended material for nutrition from seawater.

Phylum Mollusca

Members of the phylum Mollusca are among the most abundant and easily observable groups of marine animals because they have adapted to all the major marine habitats. It is difficult to characterize such a large and diverse group as the phylum Mollusca, but a few common traits are observable. Mollusks are unsegmented animals. Most mollusks have a hard external shell surrounding the internal organs and use a large muscular foot for locomotion, anchorage, and securing food. Most mollusks have an array of specialized sense organs in the anterior region of their body near the brain. (This pattern of body organization, known as **cephalization,** is most apparent in squids and octopuses.)

This phylum is commonly divided into seven classes. Representatives of five of these classes are

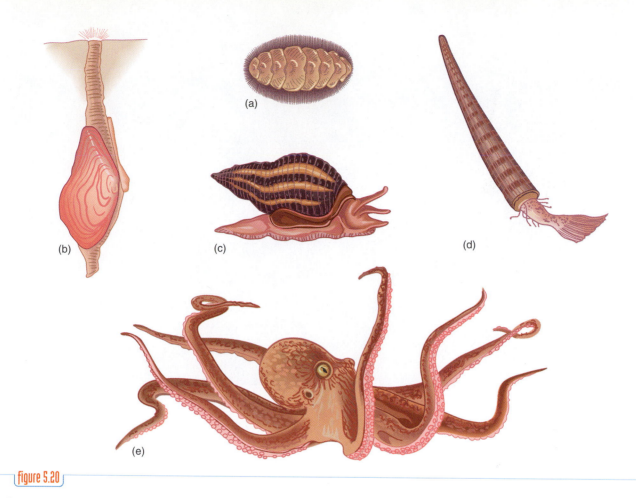

figure 5.20

Representatives of the common classes of mollusks: (a) Amphineura, (b) Bivalvia, (c) Gastropoda, (d) Scaphopoda, and (e) Cephalopoda.

quite common and are shown in FIGURE 5.20. In four of these classes, the early planktonic larval form is the **trochophore** larva (FIG. 5.21). This type of larva is also found in annelid worms. As the trochophore grows, a ciliated tissue called the velum develops, and the larva is then known as a **veliger.** The velum is used to collect food and to swim.

Chitons belong to the class Amphineura. They are characterized by eight calcareous plates embedded in their dorsal surfaces. These animals, found in rocky intertidal areas, use their large muscular foot to cling to protected depressions on wave-swept rocks. Chitons feed by grazing algae from rocks with a rasping tonguelike organ, the **radula** (FIG. 5.22). Radulas are also found in three other classes of mollusks: the gastropods, scaphopods, and cephalopods.

The class Gastropoda includes snails and slugs (marine, fresh water, and terrestrial), limpets, abalones, and nudibranchs. Although one-piece shells are characteristic of this class, several types of gastropods lack shells. Most gastropods are benthic; only

a few without shells, or with very light ones, have successfully adapted to a pelagic lifestyle. Like chitons, many gastropods graze on algae; others feed on detritus and organic-rich sediments. Numerous gastropods are also successful predators of other slow-moving animal species (FIG. 5.23)

The 350 species of tusk shells in the class Scaphopoda are found buried in sediments in a wide range of water depths. As the common name implies, the shells of these animals are elongated and tapered, somewhat like an elephant's tusk but open at both ends. The head and foot project from the opening at the larger end of the shell. Microscopic organisms from the sediment and water are captured by adhesive tentacle-like structures.

The class Bivalvia, which includes mussels, clams, oysters, and scallops, have hinged two-piece, or bivalve, shells. As adults, most are slow-moving benthic animals, and some, such as mussels and oysters, are permanently attached to hard substrates. This class has an extensive depth range, from intertidal

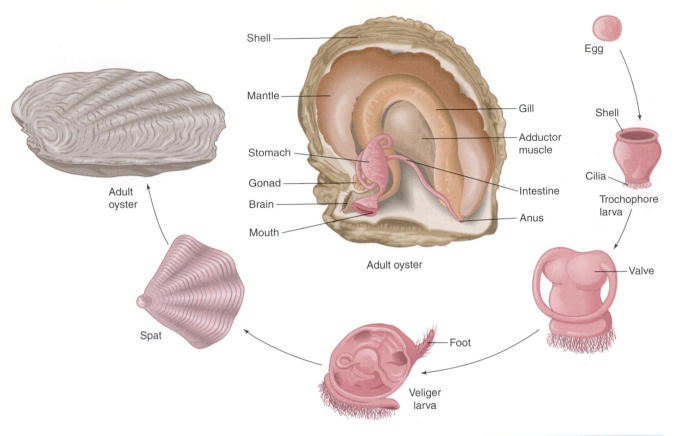

Shell

Mantle

Stomach

Gonad

Brain

Mouth

Gill

Adductor muscle

Intestine

Anus

Adult oyster

Egg

Shell

Cilia

Trochophore larva

Valve

Foot

Veliger larva

Spat

Adult oyster

figure 5.21

Oyster life cycle. About 2 weeks are needed to develop from egg to spat.

figure 5.22

Scanning electron micrograph view of the radula of a gastropod mollusk.

figure 5.23

The working end of a large (30 cm) predatory sea slug, *Navanax*.

areas to below 5000 m. All bivalves lack radulas and are unique among mollusks in having large and sometimes elaborate gills. These gills are covered with cilia (FIG. 5.24) that circulate water for gas exchange while sorting extremely small food particles entrapped on a thin mucous film secreted over the gill surfaces. Consequently, bivalves are specialized to feed on suspended bacteria, very small phytoplankton cells, and microscopic detrital particles found in sediment deposits.

Molluscan evolution has reached its zenith in the class Cephalopoda, which includes squids, octopuses, cuttlefish, and nautiluses. Members of this class are carnivorous predators with sucker-lined tentacles, well-developed sense organs and large brains, reduction or loss of the external shell typical of most other mollusks, and, in some species, very large body sizes. The eyes of squids and octopuses are remarkably similar to our own, with a retina, cornea, iris, and a lens focusing system. A unique propulsion system, using high-speed jets of water, provides excellent maneuverability and swimming speeds greater than those of any other marine invertebrate. Octopuses also use their eight arms to gracefully crawl over the seafloor. Octopuses sometimes have arm spans exceeding several meters. The giant squid, *Architeuthis*, which may reach 18 m in length and weigh over 1 tonne, is by far the largest living invertebrate species.

Rigid-walled gas containers used for buoyancy are found in only a few types of cephalopods. All cephalopods are believed to have evolved from an ancestral form that had an external shell. *Nautilus* is the only living cephalopod that has retained its external shell. The shells of other living cephalopods are either reduced to an internal chambered structure, as in the cuttlefish (*Sepia*) and *Spirula* (a deepwater squid), or are absent entirely, as in octopuses. In squids other than *Spirula*, a thin chitinous structure (the pen) extends the length of the mantle tissue and represents the last vestige of what was once an internal shell.

Nautilus, Spirula, and cuttlefishes all have numerous hard transverse septa, partitions that separate adjacent chambers of the shell. In *Nautilus*, only the last and largest chamber is occupied by the animal. As *Nautilus* grows, it moves forward in its shell and adds a new chamber by secreting another transverse septum across the area it just vacated. When a new chamber is formed, water is removed and is replaced by gases (mostly N_2) from tissue fluids. The gases diffuse inward, and the total pressure of the gases dissolved within the chambers never exceeds 1 atm.

These chambered cephalopods are confronted with the same depth-limiting factor that plagues submarines. Their depth ranges are limited by the resistance of their shells to increased water pressure. Each species has a critical implosion depth at which the external water pressure becomes too great for the design and strength of its shell and the shell collapses. The implosion depth of *Nautilus* shells, for example, is somewhat below 500 m, yet this animal does not normally live below 240 m.

figure 5.24

Lateral view (left) of the intact gill of a mussel, *Mytilus,* and a micrograph of the gill edge with cilia (right).

Some Wormlike Protostomes

Most wormlike marine invertebrates live in soft mud or sand deposits. Their elongated body forms permit effective burrowing movements despite a lack of rigid internal skeletons to support the muscles of locomotion. Muscles in the body wall work against the enclosed fluid contents of the body to allow burrowing actions and other body movements. The fluids cannot escape and are essentially incompressible. As such, they provide a **hydrostatic skeleton** for the muscles of the body wall. In the more effective burrowing worms, these muscles are arranged in two sets: circular muscle bands around the body and longitudinal muscles extending the length of the body. Like all other muscles, these muscles can work only by contracting. When the circular muscles of a worm's body contract, its diameter decreases, squeezing its hydrostatic skeleton and forcing the body to elongate, much as squeezing a long inflated balloon at one end causes it to expand at the other end. If the rear of the body is anchored, contracting the circular muscles pushes the anterior end forward. When the circular muscles relax, the longitudinal muscles can then contract to shorten the body and make it fatter or move it forward. These two types of muscles continue to work in opposition to each other to provide an effective sediment-burrowing motion for a large variety of marine worms.

Fewer than 20 species occur in the phylum Priapulida. Priapulid worms (FIG. 5.25) live buried in intertidal to abyssal sediments in warm or cold latitudes and seldom exceed 10 cm in size. Priapulid worms are detritus feeders or are predatory, feeding on soft-bodied invertebrates they capture with their eversible introvert. Priapulids are found in hypersaline ponds and anoxic muds, as well as more moderate benthic habitats.

Sipuncula, another phylum of wormlike animals, are found throughout the world ocean. The 350 species of sipunculids, or peanut worms, are entirely marine. Most species are found in the intertidal zone, but their distribution extends to abyssal depths.

Peanut worms are benthic. They live in burrows, crevices, or other protected niches, often in competition with other wormlike animals. Sipunculids range from 2 mm to over 50 cm in size and have a cylindrical body that is capped by a ring of ciliated tentacles surrounding the anterior mouth (FIG. 5.26).

The Echiurida is a small phylum of benthic marine worms that resemble peanut worms in size and general shape. Echiurids are common intertidally, but they are occasionally found at depths exceeding 6000 m. Most echiurids live in burrows in the mud. One remarkable feature of echiurids is their extensible proboscis, a feeding and sensory organ that projects from their anterior. The proboscis of some species is longer than the remainder of the body and is quite effective for gathering food by "mopping" the sediment while the worm remains in the protected confines of its burrow.

Urechis, an echiurid of the California coast known as the fat innkeeper, has a very short proboscis. The proboscis secretes a mucous net from the animal to the wall of its U-shaped burrow (FIG. 5.27). The burrow is also often inhabited by small crabs, shrimps, or other casual guests. Water is pumped through the burrow by repeated waves of contractions along the worm's body wall. As water passes through the mucous net, bacteria and other extremely small food particles are trapped. When the net is clogged with food, the worm consumes it and constructs another.

Members of the phylum Pogonophora are almost exclusively deep-water tube-dwelling marine worms. (Eighty percent of the known species live below 200 m.) This obscure phylum was not even

A marine priapulid worm.

figure 5.26

Sipunculid worms, *Sipunculus.*

Sediment surface

figure 5.27

The fat innkeeper, *Urechis,* in its burrow.

described until 1900. Since that time, about 120 species have been described. Adults are generally only a few centimeters in length. Pogonophorans are noted for their complete lack of an internal digestive tract and are thought to absorb dissolved organic material or to use symbiotic bacteria to provide their energy needs. The remarkably large tube worms discovered in deep-sea hot spring communities (see Fig. 11.13) were initially assigned to the phylum Pogonophora. However, ongoing studies suggest that they may be sufficiently distinct to warrant the erection of their own separate phylum, the Vestimentifera. The vestimentiferans of deep-sea hot spring communities are examined in Chapter 11.

Phylum Annelida

The phylum Annelida is usually represented by the familiar terrestrial earthworm. However, this phylum also contains a diverse and successful group of marine forms with more than 7800 species in the class Polychaeta. Polychaete worms, like other annelids, are segmented. Segmentation has evolved separately in both major lineages of coelomates, in annelids and arthropods in protostomes, and in chordates in the dueterostome line. The body cavity and internal organs contained within polychaete worms are subdivided into a linear series of structural units called **metameres.** The result of segmentation is a sequential compartmentalization of the worm's hydro-

static skeleton and surrounding muscles. This permits a greater degree of localized changes in body shape and a more controlled and efficient form of locomotion.

Some polychaetes ingest sediment to obtain nourishment, others are carnivorous, and many use a complex tentacle system (**FIG. 5.28**) to filter microscopic bits of food from the water. Suspension-feeding polychaetes often occupy partially buried tubes and are common in intertidal areas; however, they are also found in deeper water. The 40 species of *Tomopteris* are planktonic throughout their lives.

Phylum Arthropoda

Like annelids, arthropods are segmented along the length of their bodies. In addition to the advantages of segmentation, arthropods possess a distinctive hard **exoskeleton** made of a complex organic substance called **chitin**. This rigid outer skeleton serves not only as an impermeable barrier against fluid loss and bacterial infection but also as a system of levers and attachment sites for muscle attachment. Its structure resists deformations caused by contracting muscles, allowing faster responses and greater control of movements. Flexing of the body and appendages is limited to thin membranous joints located between the rigid exoskeletal plates. On the other hand, exoskeletons lack some of the shape-changing advantages of segmentation and hydrostatic skeletons found in soft-bodied annelid worms.

The exoskeleton also restricts continuous growth. Periodically, arthropods shed, or **molt**, their old exoskeleton, and it is replaced by a new larger one as the animal quickly expands to fill it (**FIG. 5.29**). Each molt is followed by an extended period of time with no growth. For some species of arthropods, the number of molts is fixed, with molting and growth ceasing at the adult stage. For others, growth and molting continues as long as they live, although the frequency of molting diminishes with increasing age and body size.

Members of the phylum Arthropoda account for almost three fourths of all animal species identified so far, now exceeding 1 million species. Most belong to the class Insecta, a group abundant on land and in freshwater habitats. Only a few species of insects, including pelagic water striders (see Fig. 1.15), several

figure 5.28

The filtering structures of a tube-dwelling polychaete worm.

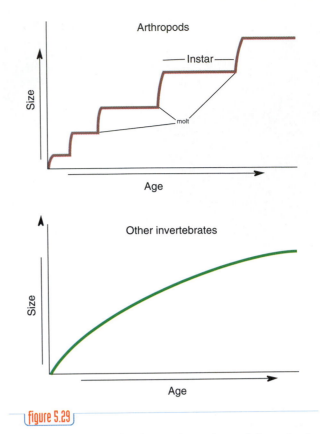

figure 5.29

Patterns of arthropod and nonarthropod growth. In contrast to the smooth curve of other animals, arthropods rapidly increase their body size in steps after each molt of the exoskeleton.

types of sand and kelp flies, and water beetles have evolved to thrive in seawater environments. However, three other classes of this phylum, the Crustacea, Merostomata, and Pycnogonida, are primarily or completely marine in distribution. Our view of their relationships to each other and to other arthropods is undergoing extensive revision. One widely accepted phylogeny, based on current genetic evidence, places Merostomata and Pycnogonida close to spiders (class Arachnida) and Crustacea as a sister group to insects.

Two classes of arthropods are completely marine, but their diversity is limited. The first class, Merostomata, has an extensive fossil history that includes extinct water scorpions, or eurypterids, up to 3 m long. Only three modern genera exist, including the horseshoe crab, *Limulus*, an inhabitant of the Atlantic and Gulf Coasts of North America (FIG. 5.30). The sea spiders of the class Pycnogonida are long-legged bottom dwellers with reduced bodies. Small pycnogonids only a few millimeters in size are quite common intertidally. They can be collected from hydroid or bryozoan colonies or from the blades of intertidal algae. Deep-sea pycnogonids are often much larger and may have leg spans of 60 cm (FIG. 5.31).

The third class of marine arthropods, the Crustacea, is an extremely abundant and successful group of marine invertebrates. Crustaceans are arthropods with two pairs of **antennae** and a larval stage known as a **nauplius.** Few other useful generalizations can be made concerning this class. Its members exhibit a tremendous diversity in body plans (FIG. 5.32) and modes of feeding. The range of habitats also varies greatly, from burrowing ghost shrimp to planktonic copepods and parasitic barnacles. Representatives of several of the crustacean subgroups listed in Table 5.1 are included in Figure 5.32. Obvious and well-known crustaceans include shrimps, crabs (Fig. 5.32f), lobsters (order Decapoda), and barnacles (order Cirripedia, Fig. 5.32g). These large mostly benthic crustaceans are not representative of the entire class, however, because most marine crustaceans are very small and are major components of the zooplankton.

Copepods (subclass Copepoda, Fig. 5.32d) are small crustaceans, mostly less than a few millimeters in length. Despite their small size, their efficient filter-feeding mechanisms (see pp. 301–303) and overwhelming numbers in pelagic communities dictate

figure 5.30

Two mating horseshoe crabs, *Limulus*.

figure 5.31

A deep-sea pycnogonid with leg spans of about 30 cm.

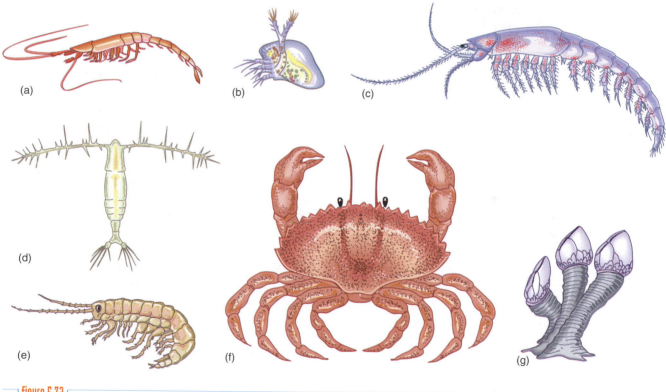

(a)

(b)

(c)

(d)

(e)

(f)

(g)

figure 5.32

A variety of marine crustaceans: (a) mysid, (b) cladoceran, (c) euphausiid, (d) copepod, (e) amphipod, (f) crab, and (g) barnacle.

that much of the energy available from the first trophic level of pelagic marine communities is channeled through copepods. They in turn are consumed by predators as diverse as minute fish larvae and huge right whales.

Euphausiids (Fig. 5.32c) are somewhat larger than copepods, but they fill similar niches in pelagic communities. These crustaceans have a global distribution (see Fig. 10.3). The largest species of this group, *Euphausia superba*, grows to 6–7 cm. Found

in the cold and productive waters around Antarctica, this species aggregates in large dense shoals sometimes tens of kilometers long. They are a favorite prey of fish, whales, seals, and penguins and are the basis for a potentially enormous single-species commercial fishery (see Chapter 13). Other common planktonic crustaceans include members of the orders Branchiopoda (brine shrimp and cladocerans, Fig. 5.32b), Ostracoda, and the early larval stages of most other marine crustaceans.

Two groups of smaller benthic crustaceans are isopods, similar to backyard sowbugs or pillbugs, and amphipods, including beach-hoppers. The bodies of isopods are somewhat flattened vertically, whereas amphipods are compressed laterally and exhibit more specialization of appendages, with some legs for swimming and others for jumping or digging (Fig. 5.32e). Both groups are found in a wide variety of marine habitats, with amphipods more commonly associated with soft bottom sediments.

Life cycles of crustaceans involve several definable stages: the nauplius, protozoea, zoea, and adult (FIG. 5.33). Each stage is punctuated by one or more molts of the exoskeleton and some accompanying structural metamorphosis. For example, the larvae of copepods enter the copepodite stage, a sexually immature form resembling adults after six molts. Five more molts lead to the adult stage; thereafter no more molts occur. The yearly growth and reproductive cycles for a planktonic copepod are shown in FIGURE 5.34.

Deuterostomes

Only four phyla of deuterostomes are found in the sea, yet they exhibit a wide range of body plans, from relatively simple worms to radially symmetric echinoderms to bilaterally symmetric and segmented chordates. The evolutionary affinities of these four phyla are not resolved, but it is clear that the developmental characteristics they share place them closer to each other than to any of the protostomes described to this point.

Two Deuterostome Wormlike Phyla

The phylum Hemichordata (acorn worms) is a small group of benthic marine worms resembling pogonophorans. They have an anterior proboscis and a soft flaccid body that is up to 50 cm long. These worms

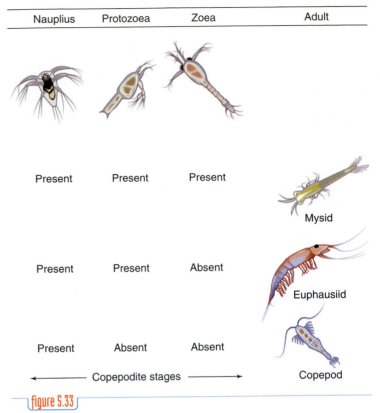

Nauplius	Protozoea	Zoea	Adult
Present	Present	Present	Mysid
Present	Present	Absent	Euphausiid
Present	Absent	Absent	Copepod

← Copepodite stages →

figure 5.33

Developmental stages of three groups of planktonic crustaceans.

are generally found in shallow water and live in protected areas under rocks or in tubes or burrows.

In contrast to the general benthic habitat of most marine wormlike animals, arrow worms (phylum Chaetognatha) are streamlined planktonic carnivores (FIG. 5.35). Although they seldom exceed 3 cm in length, they are voracious predators of other zooplankton, especially copepods. Arrow worms swim with rapid darting motions and capture prey with the bristles that surround their mouth. Only about 120 species of arrow worms exist, but they are frequently very abundant in the zooplankton. Certain species of arrow worms apparently respond to and associate with subtle chemical or physical characteristics of seawater. Consequently, they are useful biologic indicators of particular oceanic water types.

Radial Symmetry Revisited: Phylum Echinodermata

An exclusively marine phylum, the Echinodermata are widely distributed throughout the sea. They are common intertidally and are also abundant at great depths. Almost all forms are benthic as adults. Most are characterized by a calcareous skeleton, external spines or knobs, and a five-sided, or **pentamerous**, radial body symmetry (FIG. 5.36). Because echinoderms

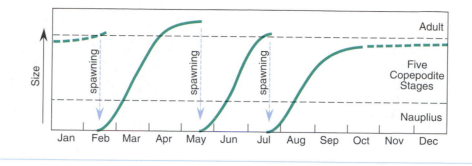

Growth and reproductive cycles of a North Atlantic copepod (*Calanus finmarchicus*). The adults of each brood produce eggs for the next brood (arrows). Dashed lines indicate overwintering of copepodites in deep water that experience little growth. (Adapted from Russell 1935.)

Sagitta, a chaetognath.

A sea star with obvious pentamerous body symmetry.

develop from bilaterally symmetric larval stages, radial body symmetry is a secondary condition in this phylum. This and other aspects of their evolutionary history separate them on the phylogenetic tree of animal evolution (Fig. 5.1) from Cnidaria and Ctenophora, the two phyla characterized by primary radial body symmetry. A unique internal **water-vascular system** functions as a simple circulatory system and hydraulically operates numerous tube feet. The tube feet extend through the skeleton to the outside and act as respiratory, excretory, sensory, and locomotor organs.

Six classes of echinoderms are usually described. Representatives of each are shown in FIGURE 5.37. The Echi-noidea are spiny herbivores or sediment ingesters variously known as sea urchins, heart urchins, and sand dollars (Fig. 5.37a). The Asteroidea, or sea stars (Fig. 5.37b), are usually five armed, but the number of arms may vary. Six-, 10- and 21-armed sea stars are known. Most sea stars are carnivorous, but a few use cilia and mucus to collect fine food particles. Feather stars and sea lilies (class Crinoidea, Fig. 5.37c) usually attach themselves to the sea bottom with their mouths oriented upward to trap plankton and detritus with their arms and with mucous secretions. Sea cucumbers of the class Holothuroidea are sausage shaped and have a mouth located at one end of their body (Fig. 5.37d). The body wall is muscular, with

The diversity of coral reefs, their importance as an ecosystem, and their economic potential are described in Chapters 9 and 13. Chapter 13 also details the fact that coral reefs worldwide are suffering mass mortalities due to pollution, nutrification, disease, overharvesting, and a number of other human-related causes. Nutrification from agricultural and sewage runoff, in particular, coupled with overexploitation of herbivorous fishes may be the cause of additional coral mortality via increasing macroalgae populations, which overgrow and kill corals.

The sea urchin *Diadema antillarum* is uniformly black, with long venomous spines that easily penetrate human skin if accidentally brushed against while swimming or diving. This sea urchin, a generalist grazer common to reefs and rocky areas throughout the Caribbean and southern Florida, often occurs in densities that are high enough to decrease the abundance of macroalgae on reefs due to its grazing activities. Thus, although a minor hazard and nuisance to divers and swimmers, this sea urchin can be an important moderator of competition for space on reefs between slow-growing corals and faster growing macroalgae.

As detailed in Chapter 9, *Diadema* experienced a catastrophic die-off throughout the Caribbean region in 1983. Immediately after the mass mortality of *Diadema,* the abundance of macroalgae on Caribbean reefs began to increase dramatically, especially in shallow areas less than 5 m deep where the sea urchins had been most common. This situation persisted for many years, as the density of *Diadema* remained extremely low for many years throughout the region. Then, about 15 years after the die-off, *Diadema* populations began to rebound and recover on several reefs along the north coast of Jamaica. Simultaneously, the abundance of reef-dwelling macroalgae decreased and the number of juvenile corals recruited to the reefs increased.

These observations persuaded Dr. Silvia Maciá of Barry University to hypothesize that transplantation of these long-spined sea urchins from areas to high density to areas of lower urchin density may promote coral recovery in additional parts of the reef. She suggested that we may be able to use sea urchins and their natural grazing abilities to mitigate the worldwide mortality of coral reefs, thus providing a practical conservation application that will improve the overall health and functioning of a human-disturbed ecosystem.

To test her hypothesis, Dr. Maciá and her students initiated an urchin-transplantation project on a Jamaican reef in the summer of 2003. Dr. Macia wrote:

If successful, this experiment will

Can Transplanted Sea Urchins Aid Reef Corals in Their Recovery?

provide a simple, inexpensive, and non-destructive method for improving the overall health of coral reefs. The decrease in macroalgal abundance resulting from the application of this method is likely to increase the abundance and growth rate of corals, the organisms on which the entire reef ecosystem is based. Such a shift from algae to corals will be especially important in areas where the *D. antillarum* die-off, overfishing, pollution, and hurricanes have acted in concert to degrade greatly the health of coral reefs. Many of these areas are in developing countries, where reef-associated activities such as fishing and tourism constitute a major portion of the local economies.

Additional Reading

Lessios, H. A. 1988. Mass mortality of *Diadema antillarum* in the Caribbean: what have we learned? *Annual Review of Ecology and Systematics* 19:371–393.

www.bioscience.jbpub.com/ marinebiology
For more information on this topic, go to this book's website at http://www.bioscience.jbpub.com/ marinebiology.

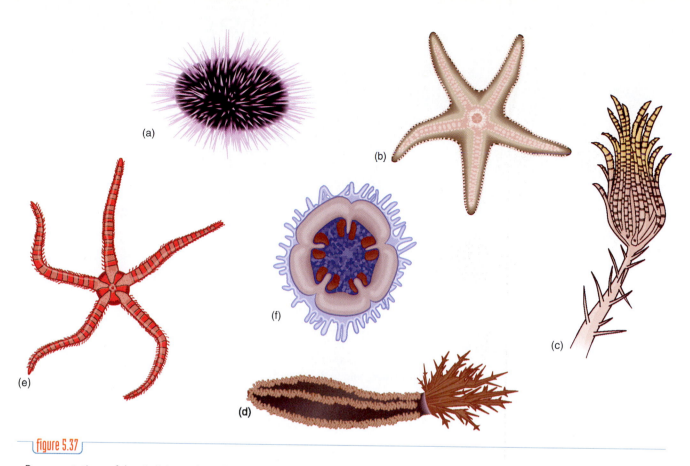

figure 5.37

Representatives of the six living echinoderm classes: (a) Echinoidea, (b) Asteroidea, (c) Crinoidea, (d) Holothuroidea, (e) Ophiuroidea, and (f) Concentricycloidea.

reduced skeletal plates and spines. A few sea cucumbers feed on plankton, but most ingest sediment and detritus. The brittle stars (class Ophiuroidea, Fig. 5.37e) are smaller than most other echinoderms but are very common animals in soft muds, rocky bottoms, and coral reefs. A recently described class, the Concentricycloidea, has been created for a few species of echinoderms known as sea daisies (Fig. 5.37f). These animals are essentially flattened armless sea stars and have been collected only at depths greater than 1000 m.

The Invertebrate Chordates: Phylum Chordata

Body segmentation evolved twice in the history of animal evolution, once in the protostome line, leading to annelids and arthropods, and again in the deuterostome line with the phylum Chordata. Like mollusks and arthropods, chordates represent a pinnacle in animal evolution. Chordates exhibit a remarkable variety of body forms, from small gelatinous zooplankton to large fishes and whales. Yet at some stage in their development, all members of this phylum possess these structural features at some time during their development: A supportive **notochord** made of cartilage extends along the midline of the body just below a hollow dorsal nerve cord. A postanal tail extends well beyond the posterior opening of the complete digestive tract. Finally, pharyngeal pouches or slits develop as openings on each side of the pharynx. These features are described in more detail in Chapter 6.

Two subphyla, the Urochordata and Cephalochordata, are relatively small and are completely marine in distribution. The urochordates are commonly called tunicates and are usually divided into three classes: benthic sea squirts (FIG. 5.38), gelatinous pelagic salps, and small planktonic larvaceans (see Fig. 10.18, *Oikopleura*). The other subphylum, Cephalochordata, contains about 25 species, including the lancelet, *Branchiostoma*. These animals are small and tadpole shaped and live partially buried tail first in near-shore sediments.

A unique feature of the third subphylum, the Vertebrata, is the vertebral column that replaces the notochord as the central skeletal structure of body support. Members of this subphylum include three classes of fishes and four classes of tetrapods and are the sole subject of Chapter 6.

figure 5.38

Nearly transparent sea squirts, each with a small incurrent and a large excurrent opening for circulating water through its body cavity.

STUDY GUIDE

Marine Biology On Line

Connect to this book's web site: **http://www.bioscience.jbpub.com/marinebiology** *The site features eLearning, an online review area that provides quizzes, chapter outlines, and other tools to help you study for your class. You can also follow useful links for more in-depth information.*

Summary Points

Marine Protozoans and Invertebrates

- The numbers of species of nonphotosynthetic protistans and the invertebrates in the world ocean are staggering. The more abundant and obvious free-living phyla that clearly dominate most marine communities and monopolize energy flow within them are presented in a manner that follows traditionally accepted interrelationships. The recent and continuing accumulation of tremendous quantities of genetic data are expected to suggest alternative phylogenies that contradict such traditional hypotheses.

Animal Beginnings: The Protozoans

- Kingdom Protista is an artificial catch-all category for many fundamentally different taxa clearly lacking a common ancestral form, which just happen to be unicellular and heterotrophic. Emphasis is placed on free-living marine groups that are distinguished by their different methods of locomotion.
- Sarcomastigophorans, which use either whip-like flagella or pseudopodia for movement, are represented by calcareous foraminiferans and siliceous radiolarians. Ciliates, which are represented in the sea by tintinnids, are so named because of their reliance on ciliary locomotion, and phylum Labyrinthomorpha includes an unusual group of small spindle-shaped cells that live and move through a network of slime via a poorly understood mechanism.

Defining Animals

- Members of kingdom Animalia are typically distinguished from the Protista by the presence of contractile muscles, signal-conducting neurons, and multicellular bodies (although a number of protistans flirt with forms of facultative multicellularity).

- The two most primitive animal phyla are sponges of Porifera and the placozoans, which both lack the coordinated patterns of intercellular interactions seen in the more specialized tissues and organs of more derived animal phyla.
- Sponges are loose federations of cells that interact sufficiently to allow more efficient handling of food, protection, and other sophisticated survival tasks. *Trichoplax,* the only living placozoan, is a poorly known, free-swimming, ciliated, amorphous mass of cells found throughout the Pacific Ocean.

Radial Symmetry

- Cnidarians are a large, diverse, and well-known assemblage of relatively primitive yet versatile marine invertebrates, including jellyfishes, sea anemones, corals, and hydroids. They are distinguished by their characteristic nematocyst-containing stinging cells (cnidocytes), some of which are painful and even deadly to humans.
- The 100 known species of ctenophores, or comb jellies, superficially resemble jellyfishes, yet they differ from jellyfishes and other cnidarians in that they capture food with sticky colloblast cells and move via highly coordinating contractions of external cilia located along eight longitudinal bands.

Marine Acoelomates and Pseudocoelomates

- All remaining animal phyla, except the echinoderms, are bilaterally symmetric and possess an anterior cephalization. The most primitive of these are small often-overlooked inhabitants of soft sediments that either lack an internal body cavity (the acoelomates) or have one that is very poorly developed (the pseudocoelomates).

Marine Coelomates

- Higher animals possess a true coelom, an embryonic development that separates the digestive tract from the body wall, and are separated into two fundamentally different lineages, the protostomes and the deuterostomes, which are distinguished by a number of developmental factors.

- The lophophorate phyla, Bryozoa, Phoronida, and Brachiopoda, all possess a crown of ciliated feeding tentacles, or lophophore.
- The snails, clams, mussels, oysters, scallops, octopus, and squid of phylum Mollusca are among the most abundant, easily observable, and best known marine invertebrates. The various groups of mollusks are distinguished by type, number, and location of shells that they possess (chitons have eight, octopods have zero).
- Many marine invertebrates are worm-like organisms that live in soft sediments. In fact, being worm-like (or vermiform), with a concomitant hydrostatic skeleton that facilitates movement through mud and sand, may be the most common and successful body plan among animals.
- Marine annelids are represented by about 7800 species of polychaetes, segmented relatives of the common earthworm. These seemingly simple organisms feed in an impressive variety of ways, including ingesting organic-rich sediments, preying on other animals, and using a complex system of tentacles to function as filter or suspension feeders that obtain food directly from the ambient water column.
- Most animals, and about two thirds of all known organisms, are segmented arthropods, which possess a characteristic exoskeleton of chitin that is molted periodically during growth. Many larval and adult arthropods are tiny and important members of the zooplankton, whereas others, such as lobsters and horseshoe crabs, can grow to a weight of several kilograms.
- Echinoderms, the familiar sea stars, sea urchins, sand dollars, and sea cucumbers, possess a body plan that is secondarily radial, a unique water-vascular system (which hydraulically operates numerous tube feet), and the ability to regenerate significant portions of their anatomy that have been lost to predators or injury.

- All chordates, including the vertebrate subphylum, possess a dorsal hollow nerve cord, a longitudinal stiffening notochord of cartilage, pharyngeal gill slits, and a postanal tail during at least some portion of their life. Primitive nonvertebrate chordates are small filter-feeding members of most shorelines and the open sea.

Topics for Discussion and Review

1. Why are protists placed in the same kingdom (in a five-kingdom classification system) as photosynthetic diatoms and dinophytes?
2. Placozoans and members of phylum Labyrinthomorpha are both seemingly multicellular, yet only the placozoans are classified as animals. Why is this?
3. What survival advantages and disadvantages might an animal such as a sea anemone with radial body symmetry have over an animal with bilateral symmetry?
4. Many common marine animals have worm-like body forms. Why might this body shape be advantageous for mud or sand dwellers?
5. Why do you think many critical sense organs are concentrated in the head region of "higher" animals rather than in other parts of their bodies?
6. Prepare a table of the following mollusks: snail, clam, chiton, abalone, nudibranch, scaphopod, squid, and octopus. Complete the table by listing the following characteristics of their shells: present or absent, shape, number present, internal or external.
7. Compare and contrast the methods of locomotion used by the above mollusks. (Hint: some mollusks are capable of more than one type of movement.)
8. List and discuss the advantages and disadvantages of the rigid arthropod exoskeleton in comparison with the fluid hydrostatic skeleton of annelid worms.
9. List the genus names of two common local intertidal animals that exhibit radial body symmetry, one that is primarily radial and one that is secondarily radial.
10. To be classified as chordates, humans must possess a dorsal hollow nerve cord, a longitudinal notochord, pharyngeal gill slit, and a postanal tail at some point during their life. When did you, or do you, possess each of these chordate characteristics?

Suggestions for Further Reading

Ayala, F. J., and A. Rzhetsky. 1998. Origin of the metazoan phyla: molecular clocks confirm palaeontological estimates. *Proceedings of the National Academy of Sciences* 95:606–611.

Buzas, M. A., and S. J. Culver. 1991. Species diversity and dispersal of benthic foraminifera. *Bioscience* 41:483–489.

Cairns, S. D. 1991. *Common and Scientific Names of Aquatic Invertebrates from the United States and Canada: Cnidaria and Ctenophora* (Special Publications 22). American Fisheries Society, Bethesda.

Capriulo, G. M. 1990. *Ecology of Marine Protozoa.* Oxford University Press, Oxford.

Coombs, G. H., K. Vickerman, M. A. Sleigh, and A. Warren. 1998. *Evolutionary Relationships Among Protozoa.* Chapman and Hall, London.

Debelius, H. 1999. *Crustacea Guide of the World: Shrimps, Crabs, Lobsters, Mantis Shrimps, Amphipods.* Hollywood Import and Export, Inc.

Debelius, H., and M. Norman. 2000. *Cephalopods: A World Guide.* Hollywood Import and Export, Inc.

Ellis, R. 1999. *The Search for the Giant Squid.* Penguin USA.

Erwin, D. et al. 1997. The origin of animal body plans. *American Scientist* 85:126–137.

Hadley, N. F. 1986. The arthropod cuticle. *Scientific American* 244:104–112.

Hanlon, R. T., and J. B. Messenger. 1998. *Cephalopod Behaviour.* Cambridge University Press, Cambridge.

Humann, P., and N. DeLoach. 2001. *Reef Creature Identification: Florida, Caribbean, Bahamas.* New World Publications.

Jahn, T. L., and F. F. Jahn. 1978 *How to Know the Protozoa.* McGraw-Hill Science.

Levinton, J. S. 1992. The big bang of animal evolution. *Scientific American* 252:84–91.

Margulis, L., and K. V. Schwartz. 1987. *Five kingdoms: An Illustrated Guide to the Phyla of Life on Earth*. W. H. Freeman, San Francisco.

Pechenik, J. A. 2000. *Biology of the Invertebrates*. McGraw-Hill, Dubuque.

Rouse, G. W., and F. Pleijel. 2001. *Polychaetes*. Oxford University Press, Oxford.

Simpson, T. L. 1984. *The Cell Biology of Sponges*. Springer-Verlag, New York.

Swearingen, D. C. III, and J. R. Pawlik. 1998. Variability in the chemical defense of the sponge *Chondrilla nucula* against predatory reef fishes. *Marine Biology* 131:619–923.

Turgeon, D. D. 1998. *Common and Scientific Names of Aquatic Invertebrates from the United States and Canada: Mollusks* (Special Publications 26). American Fisheries Society, Bethesda.

Wilbur, K. M., E. R. Trueman, and M. R. Clarke. 1988. *The Mollusca: Form and Function*. Academic Press.

Williams, A. B. 1989. *Common and Scientific Names of Aquatic Invertebrates from the United States and Canada: Decapod Crustaceans* (Special Publications 17). American Fisheries Society, Bethesda.

Willmer, P. 1990. *Invertebrate Relations*. Cambridge University Press, Cambridge.

A sally-light-foot crab, *Grapsus grapsus,* is a common intertidal crab of tropical and subtropical shores.

CHAPTER 6

Fish-eating birds are important vertebrate members of marine ecosystems.

Marine Vertebrates

After describing three protozoan and 25 invertebrate phyla in Chapter 5, including the invertebrate chordates, it may seem a strong bias to devote an entire chapter to a single subphylum of chordates, the Vertebrata. Yet the species and ecologic diversity of marine vertebrates rivals that of all other animal phyla except mollusks, arthropods, and perhaps the annelid polychaetes. Vertebrates occupy all major marine habitats and are especially dominant in the pelagic realm of the sea. As a group, adult vertebrates express an enormous range of body sizes, from 3-cm-long deep-sea lanternfishes to whales and sharks over 15 m long. Consequently, vertebrates exploit food sources of almost all sizes found in the sea and are important in almost all marine food webs.

6.1 Vertebrate Features

All chordates with a vertebral column are members of the subphylum Vertebrata. In addition to the basic chordate features listed at the end of Chapter 5, vertebrates also have specialized sensory organs and a brain consisting of a concentration of nerves positioned at the enlarged anterior end of the nerve cord. During development the notochord becomes segmented into a linear series of articulated skeletal units, the **vertebrae**. Skeletal muscles are segmented into **myomeres** that permit controlled and efficient movements. These minimum vertebrate structural features are illustrated in FIGURE 6.1.

Vertebrate circulatory systems are closed loops, each consisting of a multichambered heart, arteries, capillaries, and veins (FIGURE 6.2). The blood itself is unique in the animal kingdom, with all its oxygen-transporting hemoglobin completely contained in red blood cells that can circulate but cannot leave the blood vessels. The significance of these circulatory features are explained later in this chapter.

The details of the origin and early evolution of vertebrates have been blurred by the passage of nearly one-half billion years. Most biologists agree that early vertebrates evolved from filter-feeding chordate ancestors that had characteristics resembling the larval stages of invertebrate sea squirts or

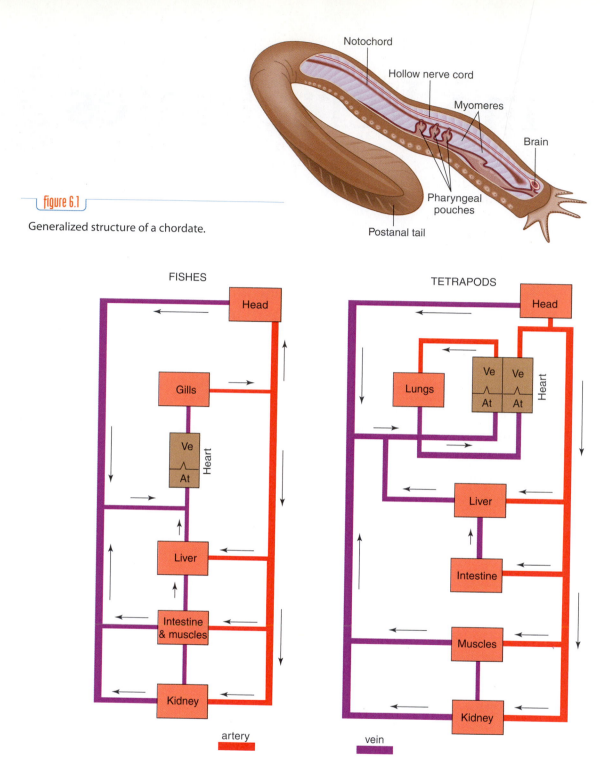

figure 6.1

Generalized structure of a chordate.

FISHES

TETRAPODS

artery

vein

figure 6.2

Comparison of circulatory patterns of typical fishes (left) and tetrapods (right). Heart chamber: Ve, ventricle; At, atrium.

lancelets. The earliest vertebrates may be conodont-bearing marine animals from the late Cambrian. Conodonts are common, tiny tooth-like fossils that belong to a 40-mm-long extinct animal that was discovered in 1983. These conodontophorans seem to possess

a notochord, myomeres, fin rays, and two eyes. Although a close relationship is not completely accepted, biologists agree that they should be included within vertebrate ancestry. Hagfishes, the most primitive surviving vertebrate (and therefore most likely to be representative of the primitive vertebrate osmoregulatory conditions), have body fluids isotonic to seawater, supporting arguments for a marine origin for early vertebrates. However, essentially all other vertebrate groups have body fluids with ionic concentrations less than half that found in seawater. When living in low salinity estuarine or freshwater conditions, reduced ion concentrations in the body fluids of vertebrates lessen the osmotic gradient and reduce the amount of metabolic energy expended on osmoregulation. However, normal nerve and muscle function requires that sodium ion concentrations of vertebrate body fluids are maintained at least 30–50% of its concentration in seawater.

One reasonable scenario proposes that early vertebrates moved between marine and freshwater habitats, possibly for spawning in fresh water as modern lampreys and salmon do now. Such migrations would place the young stages in nutrient-rich environments where they could feed and grow with perhaps fewer large active predators than would be encountered in the sea. In these brackish and freshwater environments,

the evolution of reduced-ion body fluids would be expected to conserve substantial amounts of energy that could be used for rapid growth. Only after gaining in body size would these fishes move into marine habitats to compete for larger prey items. This is precisely what salmon and lampreys do now. From this scenario for early vertebrates, it is suggested that each major group has evolved its own solution to the problem of maintaining osmoregulatory balance. These adaptations are described for each group in the next few pages.

6.2 Marine Fishes

Marine fishes include three of the seven classes of vertebrates (FIGURE 6.3); the other four classes are known collectively as tetrapods. Fishes are difficult to characterize precisely, but typically they live and grow in water, swim with fins, and use gills for oxygen and CO_2 exchange. Each class exhibits definite differences in body structure, in specialization of sense organs, in solutions to osmoregulatory changes, and ultimately in species diversity.

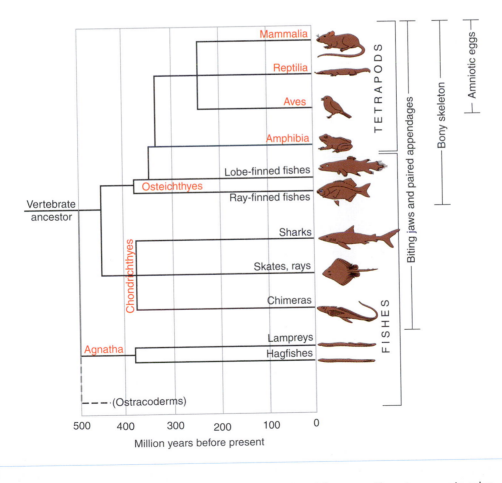

figure 6.3

Phylogenetic relationships of vertebrate classes, with emphasis on modern fish groups. Class names are in color.

Marine vertebrates are characteristically active animals. Their activity is fueled by the oxidation of lipids and other energy-rich foods. The oxygen used in the mitochondria of active cells necessarily either comes from the atmosphere (for tetrapods) or is dissolved in seawater (for fishes). For air-breathing tetrapods that swim below the sea surface, air is rich in O_2 (21%) but is available only when they are at the surface to breathe. Fishes, on the other hand, use their gills to extract constantly accessible O_2 from seawater, but water has a very low solubility for O_2 (just a few parts per million).

However they acquire their O_2, nearly all vertebrates use hemoglobin to store and transport it in the blood. Hemoglobin is found in other animal phyla (including several inhabitants of deep-sea hot vent communities; see Figure 11.13), but only vertebrates package their hemoglobin inside red blood cells. Hemoglobin is a pH-sensitive protein that has a high chemical affinity for O_2. Each hemoglobin molecule can bind with up to four O_2 molecules. In marine vertebrates, it matters little whether O_2 is obtained with gills or with lungs; the function of hemoglobin is essentially the same.

When oxygen is transported in vertebrate blood, little exists dissolved in the plasma of the blood, because a liter of blood plasma can only dissolve 2–3 ml of O_2. In chemical combination with hemoglobin, however, a liter of blood with red blood cells can hold about 200 ml of O_2. As an additional bonus, O_2 remains osmotically invisible when it is bound with hemoglobin, and the tendency for O_2 to continue diffusing into the red blood cells is enhanced.

All cartilaginous and bony fishes take water and dissolved gases into their mouths and pump them over their gills. Each **gill arch** supports a double row of blade-like gill filaments (FIGURE 6.4). Each flat filament bears numerous smaller secondary lamellae to further increase the gill surface available for gas exchange. Active fishes such as mackerel have up to 10 times as much gill surface as body surface. The O_2 requirements of sedentary bottom fishes are not as great, so their gill surfaces are not as extensive.

Microscopic capillaries circulate blood very near the inner surface of the secondary lamellae. As long as the O_2 concentration of the blood is less than that of the water passing over the gills, O_2 continues to diffuse across the very thin walls of the lamellae and

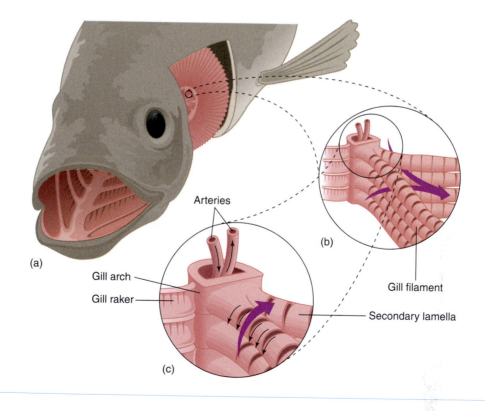

Arteries

(a)

Gill arch

Gill raker

(b)

Gill filament

Secondary lamella

(c)

figure 6.4

Cutaway drawing of a fish showing the position of the gills (a). Broad arrows in b and c indicate the flow of water over the gill filaments of a single gill arch. Small arrows in c indicate the direction of blood flow through the capillaries of the gill filament in a direction opposite that of incoming water.

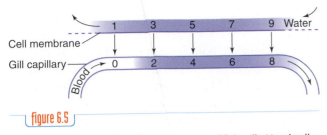

figure 6.5

A countercurrent gas exchange system of fish gills. Nearly all the oxygen from water flowing right to left diffuses across the gill membrane into the blood flowing in the opposite direction. Numbers represent arbitrary oxygen units.

into the bloodstream. In fish gills, the efficiency of O_2 absorption into the blood is enhanced by the direction of water flow over the gill lamellae (see Figure 6.4, arrows), a direction reverse that of the blood flow within the lamellae. Oxygen-rich water moving opposite to the flow of oxygen-depleted blood creates an effective countercurrent system for gas exchange (FIGURE 6.5). Blood returning from the body with a low concentration of O_2 enters the lamellae adjacent to water that has already given up much of its O_2 to blood in other parts of the lamellae. As the blood moves across the lamellae, it continually encounters water with greater O_2 concentrations, so a favorable diffusion gradient is maintained along the entire length of the capillary bed within the lamellae, and O_2 continues to diffuse from the water into the blood over most of the gill surface. With such a countercurrent O_2 exchanger, some fishes are capable of extracting up to 85% of the dissolved O_2 present in the water passing over the gills. In contrast, typical air-breathing vertebrates, such as humans, generally use less than 25% of the O_2 that enters their lungs.

Agnatha—The Jawless Fishes

Agnathans are vertebrates that are often characterized in negatives; they lack the paired fins, biting jaws, and skin scales so noticeable in most other fishes. Only two orders of these primitive fishes exist (TABLE 6.1), the hagfishes and the lampreys. Hagfishes are entirely marine benthic scavengers that secrete a thick viscous slime from skin glands to deter potential predators. Hagfishes periodically tie their flexible bodies into sliding knots to clean themselves of excessive slime buildup (FIGURE 6.6). This knot-tying behavior also provides stability and leverage while tearing apart large pieces of food.

Most lampreys are anadromous, hatching in fresh water and migrating to the sea to mature. In both the freshwater and marine portions of their life cycles, adults of some species of lampreys parasitize various species of bony fishes (FIGURE 6.7). Parasitic lampreys use numerous hook-shaped teeth within their oral disc to attach to their hosts. Then they rasp through the skin with a toothed tongue. Anticlotting salivary enzymes assist the continued flow of blood from the host in a manner reminiscent of vampire bats. The parasitic lifestyle of lampreys is likely a specialized form of feeding not representative of the way early vertebrates obtained their food. Neither hagfishes nor lampreys play major roles in present-day marine communities.

Chondrichthyes—Sharks, Rays, and Chimeras

The 14 orders included in the class of vertebrates called Chondrichthyes (Table 6.1) exhibit three basic body plans: streamlined sharks, dorsoventrally flattened rays, and the unusual chimeras (Figure 6.3). The evolution of paired fins and biting jaws armed with teeth (features found in all groups of marine fishes except agnathans) provide sharks, rays, and their allies with the structures needed for better maneuverability, faster swimming speeds, and ultimately for more effective predation.

Almost all species of Chondrichthyes are marine. Like most other vertebrate groups, the concentration of salts in their body fluids is much less (about 50%) than that of seawater. To achieve osmotic equilibrium with seawater, marine Chondrichthyes accumulate high concentrations of urea and trimethylamine oxide in their body fluids. Urea is a toxic nitrogen-containing waste product synthesized by the liver, and trimethylamine oxide is produced to protect proteins from its potentially damaging effects. The urea, together with trimethylamine oxide and NaCl, provides a total internal ion concentration equal to that of seawater outside the body. Ammonia, a breakdown product of urea, is responsible for the sharp smell associated with dead sharks.

Members of this class are often referred to as the cartilaginous fishes, because although the skeletons of the larger species may be strong, rigid, and mineralized, most smaller species use only cartilage for their skeletons, and the much stronger bone tissue characteristic of other vertebrates is absent in all members of this class. Chondrichthyes tend to be larger in body size than members of the other two classes of fishes. Adult body lengths range from less than 20 cm for some deep-sea sharks to over 15 m for whale sharks and basking sharks. With the exception of some whales, these are the largest living animals.

Classes and Orders of Marine Vertebrates

Class	Order	Common examples	No. of marine families
Agantha			**2**
	Myxiniformes	Hagfishes	1
	Petromyzontiformes	Lampreys	1
Chondrichthyes			**50**
	Hexanchiformes	Frill sharks	2
	Squaliformes	Dogfish sharks	4
	Heterodontiformes	Horn sharks	1
	Orectolobiformes	Whale sharks	5
	Lamniformes	Thresher and mackerel sharks	7
	Carcharhiniformes	Swell, tiger, and whitetip sharks	8
	Pristiophoriformes	Sawsharks	1
	Pristiformes	Sawfishes	1
	Squatiniformes	Angel sharks	1
	Rhinobatiformes	Guitarfishes, thornbacks	3
	Rajiformes	Skates	4
	Torpediniformes	Electric rays	4
	Myliobatiformes	Stingrays, eagle rays	8
	Chimaeriformes	Chimeras	1
Osteichthyes			**~209**
	Elopiformes	Tarpons, bonefishes	3
	Anguilliformes	Moray, conger, snake and snipe eels, gulpers	20
	Notacanthiformes	Spiny eels	3
	Salmoniformes	Salmon, smelts, hatchetfishes, dragonfishes	15
	Clupeiformes	Sardines, anchovies, herring, menhaden	5
	Myctophiformes	Laternfishes, lizardfishes, barracudinas	17
	Ophidiiformes	Cusk-eels, eelpouts	5
	Gadiformes	Cods, hakes, pollock, moras	12
	Batrachoidiformes	Toadfishes, midshipmen	1
	Gobiesociformes	Clingfishes	1
	Lophiiformes	Anglers, frogfishes	16
	Atheriniformes	Silversides, sauries, halfbeaks, flyingfishes	16
	Lampridiformes	Oarfishes, ribbonfishes, opahs	6
	Beryciformes	Deep-sea bigscales, whalefishes	7
	Gasterosteiformes	Sticklebacks, seahorses, pipefishes	10
	Scorpaeniformes	Rockfishes, scorpionfishes, sculpins, poachers	20
	Perciformes	Basses, groupers, remoras, jacks, dolphinfishes, pomfrets, snappers, croakers, surfperches, barracudas, wrasses, gobies, tunas, billfishes	~60
	Pleuronectiformes	Flatfishes	6
	Tetraodontiformes	Triggerfishes, boxfishes, puffers, molas	9

Most rays are lie-and-wait ambush specialists adapted to living on the sea bottom, whereas most sharks are streamlined fast-moving pelagic predators. When swimming, rays gracefully undulate the edges of their flattened pectoral fins or, in the cases of manta and eagle rays, flap their pectorals like large wings. Swimming sharks (and most pelagic bony fishes as well) develop thrust with their flattened caudal fin.

However, the asymmetric **heterocercal** tail (FIGURE 6.8), so characteristic of sharks, has a shape very different from that of most bony fishes (which are **homocercal**, or symmetric about the long axis of the body). When a typical shark's caudal fin is moved from side to side, a forward thrust develops, and because of the angle of the trailing edge of the tail, it produces some lift as well. The paired pectoral fins of sharks are flat and

Classes and Orders of Marine Vertebrates——Continued

Class	Order	Common examples	No. of marine families
Reptilia			**7**
	Squamata	Sea snakes, iguana	3
	Crocidilia	Estuarine, crocodiles, caymen	2
	Testudines	Sea turtles	2
Aves			**27**
	Podlcipediformes	Grabes	1
	Sp Sphenisciformes	Penguins	1
	Procellariiformes	Albatrosses, petrels, shearwaters, fulmars	4
	Pelacaniformes	Tropic birds, cormorants, boobies, gannets, pelicans, frigate birds	5
	Anseriformes	Ducks, geese	1
	Ciconiiformes	Herons	2
	Falconiformes	Osprey, eagles	2
	Gruiformes	Rails, coots	1
	Charadriiformes	Stilts, avocets, plovers, sandpipers, skuas, turnstones, phalaropes, jaegers, gulls, terns, skimmers, auks, murres, puffins	10
Mammalia			**17**
	Carnivora	Seals, sea lions, walruses, sea otters, polar bears	5
	Cetacea	Whales, dolphins, porpoises	10
	Sirenia	Manatees, dugongs	2

table 6.1

Examples and approximate number of marine families are listed as indications of group diversity.

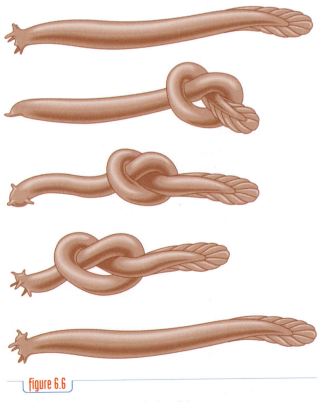

figure 6.6

Slime removal behavior of a hagfish.

large and extend horizontally from the body like stubby aircraft wings (FIGURE 6.9). The front part of the shark's underside is nearly flat and, with the flat extended pectoral fins, meets the water at an angle and produces lift for the front part of the body to balance the lift produced by the tail. Pelagic sharks and other cartilaginous fishes lack swim bladders and need this lift to maintain their position in the water column. This mechanism for achieving lift, however, does have its disadvantages. These fishes cannot stop or hover in midwater. To do so would cause them to settle to the bottom, a bottom that may be some distance away in the open ocean. Maneuverability is also reduced; the large and rigid paired fins that function as hydrofoils are not well suited for making fine position adjustments.

Cartilaginous fishes use a wide variety of reproductive strategies based on a general pattern of internal reproduction, leading to fairly small numbers of large eggs or offspring. Some rays and benthic sharks are **oviparous**, producing only a few large eggs (FIGURE 6.10) each reproductive cycle. The developing embryos are protected by a durable outer case and are nourished by the abundant yolk inside. A single egg of a whale shark can measure 30 cm in length. When

figure 6.7

Two sea lampreys, *Petromyzon,* attached to a host fish.

figure 6.8

Left side view of a typical shark.

the eggs hatch several months after laying, the young fishes are well developed and quite capable of surviving on their own.

Other sharks produce eggs that are maintained within the reproductive tract of the female until they hatch. This intermediate condition between viviparity (live birth) and oviparity (egg laying) is known as **ovoviviparity.** Ovoviviparity is an adaptation for incubating developing embryos inside the mother's reproductive tract where they obtain nourishment from the yolk of their own eggs (FIGURE 6.11). Several species of ovoviviparous fishes provide embryonic nutrition in addition to the nutrition contained in the yolk. Only a few examples, some of which are rather exotic, are described here. Several pelagic sharks and rays, for instance, have uteri and oviducts lined with numerous small projections called villi that secrete a nutritious uterine milk for the embryos. Other sharks, such as whitetips and hammerheads, absorb nutrients through a placenta-like connection between the yolk sac and the uterine wall. The embryos of the porbeagle shark, *Lamna,* for instance, lack structures with which to absorb nutrients from the reproductive tract of the female. Instead, when the oldest embryo within a female *Lamna* has used its own yolk, it simply turns on the other embryos within the oviduct and consumes them. With the nutrition gained from its potential siblings, the single embryo is developmentally much better prepared for a pelagic existence before leaving the protective confines of its

mother. A variation of that strategy is used by the sand tiger shark, *Carcharias.* After the two surviving embryos, one in each of the two uterine horns, have consumed their developing siblings, they remain in the uterus and consume thousands of additional pea-sized trophic eggs released by the female's single enormous ovary. This process may continue for a year, producing two shark pups each about 1 m long (a notable feat for a mother only 2.5 m in length).

Osteichthyes—The Bony Fishes

As the class name implies, a key characteristic of Osteichthyes is a skeleton of bone. Bone is stronger and lighter than cartilage and has permitted the evolution of smaller body sizes in this class, with most species growing to only a few centimeters in length. Of the two subclasses of Osteichthyes illustrated in Figure 6.3, the ray-finned fishes (Actinopterygii) reach their peak diversity in marine habitats, with more than 13,000 species. In contrast, the lobe-finned fishes (Sarcopterygii) have but one living marine member, the coelacanth of the Indian Ocean (the few other species of living Sarcopterygii are freshwater lungfishes). The coelacanth (FIGURE 6.12) shares some common characteristics with sharks and rays, including urea accumulation for osmoregulation. The coelacanth first appeared in the fossil record some 400 million

figure 6.9

A pelagic great white shark, *Carcharodon*. Lift is obtained from its herterocercal tail and the large pectorals extending from the flattened underside of the body.

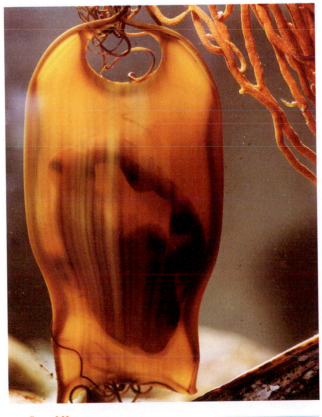

figure 6.10

Developing swell-shark embryo, *Cephaloscyllium,* enclosed in a tough protective egg case.

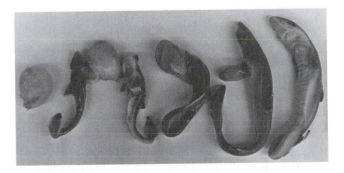

figure 6.11

A developmental series of the dogfish shark, *Squalus,* from an egg (left) to a completely formed embryo ready for birth (right). Note the twins (second from left) joined to a common yolk.

years ago and was assumed to have become extinct about 60 million years ago. Then, in 1938 a fresh (but dead) specimen was found in a South African fish market. Other individuals of this rare, but definitely not extinct, species have since been collected throughout the western Indian Ocean and Indonesia, where they live in rocky caves at depths below 100 m. Both the coelacanth and the freshwater lungfishes have been proposed as possible ancestors of the land-dwelling tetrapod vertebrates (Figure 6.3).

A coelacanth from the western Indian Ocean. The fish is about 1 m long.

The ray-finned fishes are divided into three major groups, of which the Teleostei is by far the largest. Of 39 orders of teleost fishes, 26 orders, hundreds of families, and thousands of species occupy marine habitats (Table 6.1). Nearly 60% of the species of living teleost fishes live in marine habitats and dominate the flow of energy in their communities. Several of the groups listed in Table 6.1 are familiar to most of us; others are not commonly seen except by professional biologists. A few representative species of seven of the common marine orders of teleost fishes are illustrated in FIGURE 6.13.

Regardless of where they live, teleosts share several fundamental features in common: All teleost fishes have skeletons of bone; thin and flexible skin scales; a gas-filled buoyancy organ, the **swim bladder**; and maintain body fluids hypoosmotic to seawater (see Figure 1.20 to review osmoregulatory processes). Buoyant swim bladders free the pectoral fins from the need to provide lift, as they do in sharks, to maintain or change their depth. Two features in particular, swim bladders and bony skeletons, complement the membranous fins supported by rays for fine control of swimming movements. Freed of some of the structural constraints seen in the remarkably uniform body plans of cartilaginous fishes, teleosts have adapted to enormously varied aquatic habitats and exhibit a diversity of body styles and of species unmatched in the phylum Chordata.

Swimming

The variety of ecological niches occupied by teleost fishes is reflected in their diversity of specialized approaches to swimming, with associated adaptations of body shape, fins, and muscle tissues. Most teleosts, such as the surfperch in FIGURE 6.14, are generalists. From this generalist body plan, other more specialized modes of swimming can be derived: sprinting (barracuda), fine maneuvering (butterflyfish), or nearly continuous high-speed cruising (tuna). Rapidly accelerating fishes, such as barracudas, tend to have thinner more elongated bodies, possibly to reduce their chances of being seen and recognized as they rush their prey. Butterflyfishes and other fine maneuverers are tall and elliptical in cross-section, with large fins extending even greater distances from the body. The increased amount of body surface, while adding to the overall drag, creates large control surfaces for making fine position adjustments. Tuna have large bulky bodies packed with swimming muscles and come closest to the ideal shape for a high-speed underwater swimmer (this topic is expanded in Chapter 10).

In fish species with any of the body shapes shown in Figure 6.14, the push, or thrust, needed for swimming is developed almost entirely by the backward component of the pressure of the animal's body and fins against the water. The bending motion of the anterior part of the body, initiated by the contraction of a few muscle segments (myomeres) on one side, throws the body into a curve (FIGURE 6.15). This curve, or wave, passes backward over the body by sequential contraction and relaxation of the myomeres (FIGURE 6.16). The contraction of each myomere in succession reinforces the wave form as it passes toward the tail. Immediately after one wave has passed, another starts near the head on the opposite side of the body, and the entire sequence is repeated in rapid succession.

The amount of forward thrust developed is magnified by the flared and flattened caudal fin at the posterior end of the body. Caudal fins typically flare

Order Clupeiformes
Anchovy

Order Atheriniformes
Flyingfish

Order Gasterosteiformes
Seahorse

Order Tetraodontiformes
Triggerfish

Yellowtail snapper

Striped bass

Order Perciformes

Damselfish

Grouper

Order Scorpaeniformes
Scorpionfish

Order Anguilliformes
Eel

Order Salmoniformes
Salmon

Order Pleuronectiformes
Flounder

figure 6.13

Some representative body types of nine common orders of marine fishes listed in Table 6.1.

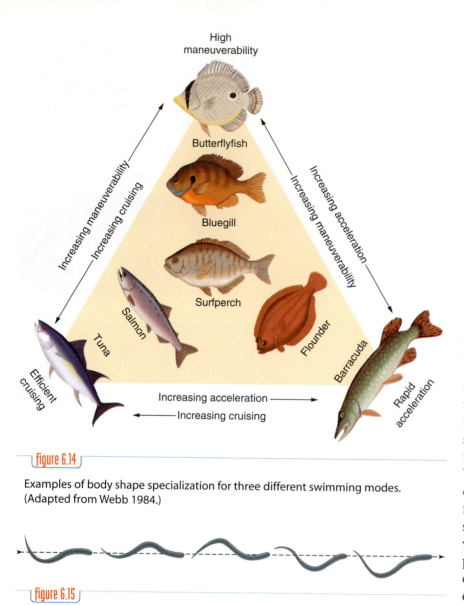

High maneuverability

Butterflyfish

Bluegill

Surfperch

Salmon

Flounder

Tuna

Barracuda

Increasing maneuverability
Increasing cruising

Increasing acceleration
Increasing maneuverability

Increasing acceleration →
← Increasing cruising

Efficient cruising

Rapid acceleration

figure 6.14

Examples of body shape specialization for three different swimming modes. (Adapted from Webb 1984.)

figure 6.15

Progression of a body wave as an eel swims from left to right.

dorsally and ventrally to provide additional surface area to develop thrust. One index of the propulsive efficiency of the caudal fin, based on its shape, is its **aspect ratio:**

$$\text{Aspect ratio} = (\text{fin height})^2/\text{fin area}$$

The caudal fins of fishes exhibit a range of profiles, illustrated with increasing aspect ratios in FIGURE 6.17: rounded, truncate, forked, and lunate. At the low end of the range of aspect ratios is the butterfly, with a rounded caudal fin that is soft and flexible. This flexibility permits the caudal fin to be used for accelerating and maneuvering. Truncate and forked fins have intermediate aspect ratios, produce less drag, and are generally found on faster fishes. These fins are also flexible for maneuverability. The lunate caudal fin

characteristic of the tuna, sailfish, marlin, and swordfish have high aspect ratios (up to 10 in swordfish) for reduced drag at high speeds. The shape closely resembles the swept-wing design of high-speed aircraft. These fishes are among the fastest swimming marine animals. The caudal fin is quite rigid for high propulsive efficiency but is poorly adapted for slow speeds and maneuvering. Fish with high aspect ratio caudal fins (especially forked and lunate types) are capable of long-distance continuous swimming.

Bony fishes equipped with swim bladders have their pectoral and pelvic fins free for other uses. In most bony fishes, the paired fins are used solely for turning, braking, balancing, or other fine maneuvers. When the fishes are swimming rapidly, these fins are folded back against their bodies. Yet several groups of bony fishes develop all their thrust without using their caudal fins. Wrasses and surgeonfishes, for example, swim with a jerky fanning motion of their pectorals and hold the remainder of their bodies straight. The greatly enlarged pectoral fins of the flyingfish in FIGURE 6.18 enable this animal to glide in air for long distances, apparently a means of escaping predators. Flyingfishes build up considerable speed while just under the sea surface and then leap upward with their pectorals extended. The length of the glide is dependent on wind conditions and the initial speed of the fish as it leaves the water, and glides up to 400 m have been reported.

Triggerfishes and ocean sunfishes swim by undulating their anal and dorsal fins only (FIGURE 6.19). For the triggerfishes, these fins extend along much of body. The large sunfish, which reaches lengths of nearly 3 m and attains weights up to a ton, is a sluggish fish and is often seen "sunning" at the surface. The little swimming it does is accomplished by its long dorsal and anal fins. Sea horses and the closely related pipefishes rapidly vibrate their dorsal and pectoral fins to achieve propulsion. Sea horses usually

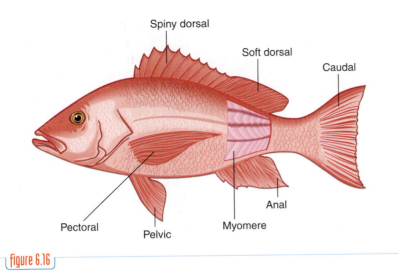

figure 6.16

Positions and names of fish fins. A portion of the musculature has been exposed to show the arrangement of myomeres.

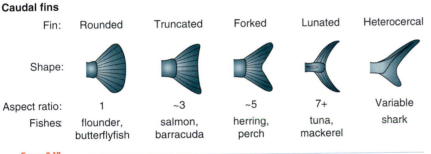

Caudal fins

Fin:	Rounded	Truncated	Forked	Lunated	Heterocercal
Shape:					
Aspect ratio:	1	~3	~5	7+	Variable
Fishes:	flounder, butterflyfish	salmon, barracuda	herring, perch	tuna, mackerel	shark

figure 6.17

Examples of shapes and aspect ratios for caudal fins.

figure 6.18

A flyingfish, *Exocoetus*, uses enlarged pectoral fins for gliding.

swim vertically with their heads at right angles to the rest of the body (FIGURE 6.20). The prehensile tail tapers to a point and is used to cling to coral branches and similar objects.

Gender in Bony Fishes

A common, although not exclusive, theme in sexually reproducing animals is to produce approximately equal numbers of female and male offspring. The maleness or femaleness of mammals, birds, and many kinds of invertebrates is determined by their complement of **sex chromosomes**. In these animals, the nucleus of each cell contains one pair of sex chromosomes and several other pairs (22 pairs in humans) of **autosomes** not directly involved in sex determination.

The influence of sex chromosomes on the gender of bony fishes is less straightforward and, in fact, is quite variable. Some guppies, for instance, reflect the mammalian pattern of sex chromosomes; XX is female and XY is male. Occasionally, these genders are reversed. The sex chromosomes of bony fishes lack the absolute control over gender determination found in birds and mammals, because some of the genes involved in gender determination, unlike those of mammals and birds, are also carried on the autosomes. In some fishes, these autosomal sex genes apparently influence or regulate the production of sex hormones, especially **androgen**, a male hormone, and **estrogen**, a female hormone. These hormones in turn influence the expression of several sexual characteristics and the determination of gender. In controlled experiments, a high percentage of salmon eggs treated with estrogen will hatch as females; if treated with androgen, most will hatch as males.

The fluid and unfixed nature of gender determination in bony fishes has been effectively exploited

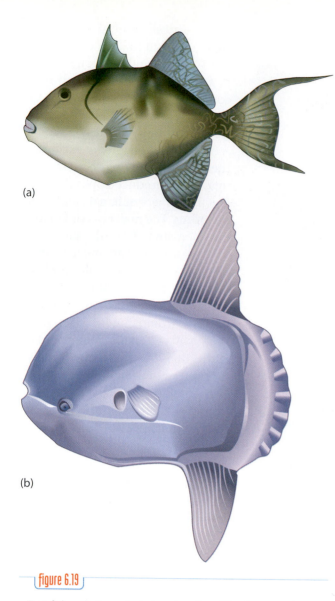

(a)

(b)

figure 6.19

Two fishes that use their dorsal and anal fins for propulsion: (a) triggerfish, *Balistes*; (b) ocean sunfish, *Mola*.

figure 6.20

The sea horse, *Hippocampus*, swims vertically, using its dorsal fin for propulsion.

through the evolution of a broad range of sex ratios and reproductive habits not common in other vertebrate groups. Part of this sexual diversity is due to the separation of sexes. Separate sexes housed in different individuals eliminates the possibility of self-fertilization and its accompanying reduction in genetic variation. Even in hermaphroditic fishes such as sea basses, *Serranus*, which function simultaneously as both males and females, specific behavioral interactions with mates ensure that cross-fertilization will occur.

Other species of bony fishes produce offspring that all clearly show the functional characteristics of one gender as they mature; then, at some point in their lives, some or all of them undergo a complete and functional transformation to the opposite gender. These fishes are hermaphroditic, but, unlike *Serranus*, they are sequential hermaphrodites. The entire gonad functions as one gender when the fish first matures and then changes to the other gender. California sheephead, for example, become sexually mature as females at about 4 years of age. Those that survive to 7 or 8 years of age undergo gender transformation, become functional males, and mate with the younger females. The actual ratio of females to males depends on the survival curve of the population (FIGURE 6.21) and on the age at which gender transformation occurs. For the sheephead, it is approximately five females to one male. In *Labrus* (belonging to the same family, Labridae, as the sheephead), gender transformation is size dependent. All *Labrus* mature as females and remain female until they reach a length of about 27 cm. Beyond that size about 50% of the surviving individuals change to males. A few species of sea basses do the reverse; they begin life as males and then change to females.

An extreme example of manipulating the gender ratio for increased reproductive fitness is found in the tropical cleaner fish *Labroides* (also in the family Labridae; see Figure 9.26). This inhabitant of the Great Barrier Reef of Australia occurs in small social groups of about 10 individuals. Each group consists of one dominant functional male and several females existing in a hierarchical social group. The single male accommodates the reproductive needs of all the

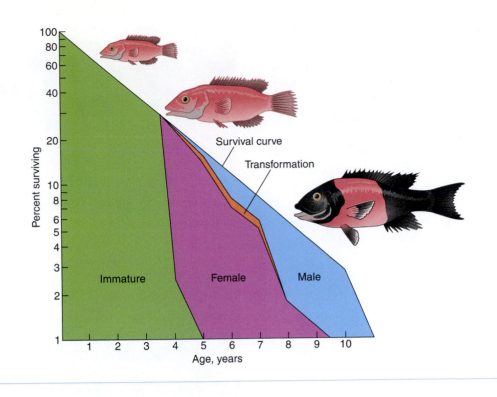

figure 6.21

Distribution of sexes according to age of the Catalina Island sheephead population. The survival curve is based on an assumed 30% annual mortality rate. The ages of sexual maturity and sexual transformation determine the relative numbers of females and males in the population at any one time. (Data from Warner 1973.)

females. This type of social and breeding organization is termed **polygyny**. In these polygynous groups, only the dominant most aggressive individual functions as the male and, by himself, contributes half the genetic information to be passed on to the next generation. In the event the dominant male of a *Labroides* social group dies or is removed, the most dominant of the remaining females immediately assumes the behavioral role of the male. Within 2 weeks, the dominant individual's color patterns change, the ovarian tissue is replaced with testicular tissue, and the population has a new male. In this manner, males are produced only as they are needed and then only from the most dominant of the remaining members of the population.

The physical characteristics associated with these gender changes seem to be controlled by the relative amounts of androgen and estrogen produced by the gonads as fishes grow and mature. Young female sheepheads, when artificially injected with the male hormone androgen, change to males at a younger age than normal. Injections of estrogen delay gender transformation and maintain the individual in a prolonged state of femaleness. Conditions of social stress imposed by the dominant male *Labroides* may induce estrogen production in the females and inhibit gender transformation. Removal of the male likely eliminates that imposed stress and permits the dominant female to transform.

6.3 Marine Tetrapods

Sharing the pelagic realm with the numerous fishes are three of the four living classes of air-breathing tetrapods (Figure 6.3): the reptiles, the birds, and the mammals. With the exception of a crab-eating frog that lives in the mangrove swamps of southeast Asia, amphibians are strictly nonmarine. Tetrapods are four-limbed air-breathing vertebrates with a terrestrial predecessor somewhere in their distant evolutionary past. Within each class, various groups have adapted to a marine existence independently of each other. Each of these groups depends on the sea for food and spends a good portion of time in the sea. Despite the obvious specializations of each of the three classes, these groups still share several important adaptations. They all use lungs to breathe air in an environment where air is only available at the sea surface. Many successfully prey on other active animals even though they cannot use their sense of smell underwater and have only limited vision. Like teleost fishes, tetrapods have body fluids that are hypoosmotic to seawater—they, too, lose water to the environment by osmosis, and with the exception of some marine reptiles and birds whose salt-secreting glands enable them to drink seawater, marine tetrapods must satisfy all their water needs from their food. Two of these tetrapod classes, the birds and the mammals, are homeothermic in an environment perpetually colder than their bodies. Despite these limitations imposed by their terrestrial ancestry, several groups of tetrapods have reinvaded the sea and have done so very successfully.

Marine Reptiles

A few species of snakes, turtles, and iguanas from three reptilian orders (Table 6.1) are succeeding quite well in the sea, and some caymens, alligators, and crocodiles occupy coastal estuarine habitats. The single species of marine iguana (FIGURE 6.22a) is restricted to rocky haulouts on the shores of the Galapagos Islands of the eastern tropical Pacific Ocean. There, they graze on seaweeds exposed at low tide or dive in shallow near-shore water for deeper marine plants.

Four families with more than 100 species of snakes (most of them venomous) live in the ocean or in brackish-water estuaries. The 61 species of true sea snakes (Figure 6.22b) are related to cobras and live in the warmer latitudes of the Indian and Pacific Oceans. The yellow-bellied sea snake, *Pelamis platuris,* is the world's most abundant reptile and, like many of its relatives, is highly venomous. Most sea snakes are ovoviviparous, retaining their eggs internally until they hatch and then giving birth to live young.

All seven species of sea turtles are primarily tropical and subtropical in distribution, although leatherback turtles have been tracked with radio transmitters into the cool southern waters of the Peru Current. Most species of sea turtles eat a variety of invertebrates, but the green turtle (Figure 6.22c) feeds exclusively on seagrasses in shallow waters. All sea turtles come ashore only to lay eggs in nests dug in sandy beaches above the high tide line.

Marine reptiles and birds have evolved a double-barreled solution to deal with the shortage of fresh water and with the extra salt loads associated with feeding at sea: special **nasal glands** and kidneys. Both reptiles and birds have complex salt-excreting glands (FIGURE 6.23), one above each eye, that concentrate NaCl to twice that of seawater. After concentration, this

(a)

(b)

(c)

figure 6.22

Some marine reptiles: (a) marine iguana, *Amblyrhynchus,* of the Galapagos Islands; (b) the sea snake; (c) green sea turtle, *Chelonia.*

salty solution drips down or is blown out the nasal passages. The kidneys of both birds and reptiles convert toxic nitrogen wastes from body metabolism to **uric acid** rather than urea, as do the kidneys of mammals and many fish. Uric acid is a nearly nontoxic

figure 6.23

A Lesser albatross with prominent nasal openings for salt excretion.

substance that requires only about 2 g of water to excrete each gram of uric acid. The white uric acid paste is mixed with feces for elimination, so urine production (and its associated water loss) as we know it in mammals does not occur in either birds or reptiles.

The ability to produce uric acid is related to another major evolutionary advancement of birds and reptiles, the shelled, or **amniotic** egg (FIGURE 6.24). Although most sea snakes are ovoviviparous, most reptiles and all birds are oviparous, laying large shelled eggs that require some period of incubation before hatching. During this period, the enclosed developing embryo floats in its own water-filled sac, the amnion, whereas its nitrogen wastes are stored as uric acid in the allantois because these wastes cannot be eliminated across the protective outer eggshell. Un-

til hatching occurs, gas exchange occurs across a third membrane, the chorion. When the egg hatches, its shell, inner membranes, and accumulated uric acid are abandoned (FIGURE 6.25).

Fertilization of shelled eggs must occur before the protective shell is in place, so all birds and reptiles fertilize their eggs internally. Shelled eggs, uric acid excretion, and even internal fertilization were early adaptations for life out of water. So, for birds, turtles, and iguanas that have returned to the sea, laying and incubating their eggs ashore is still the norm.

Marine Birds

Birds (class Aves) are tetrapods with feathers and front appendages adapted for flight. Several of the basic avian adaptations for flight, including streamlined and insulated bodies, are also useful for exploiting marine habitats as well. There is a greater diversity of birds (Table 6.1) than of either reptiles or mammals, and their role in marine communities is substantial. The term "marine" carries a variety of meanings for birds. A few birds, like penguins (FIGURE 6.26a), spend most of their lives at sea, going ashore only to breed and raise their young. At the other extreme are some ducks, geese, and coots that are common on inland ponds and lakes, but some species do move into coastal waters to feed. Herons, stilts, sandpipers, turnstones, and other shorebirds (Figure 6.26b) venture into shallow coastal waters only to feed on benthic animals. Others, including albatrosses, petrels, gannets, pelicans, gulls, terns, and murres, are more pelagic. These birds forage extensively at sea and often rest on the sea surface rather than returning to land to roost.

Birds from penguins to pelicans prey extensively on animals living in neritic waters. Their diving styles, patterns of pursuit, and even bill shapes differ greatly (FIGURE 6.27), depending on their prey preferences. Shorebirds also show a great deal of variation in bill shapes, again depending on their food preferences (FIGURE 6.28).

Birds and mammals are the only true homeotherms in the animal kingdom. Although some large fishes such as tunas and the leatherback sea turtle can maintain body temperatures somewhat higher than the waters in which they swim, these animals lack the integrated physiologic adaptations of true homeotherms. While at sea, birds and mammals live in direct contact with seawater much colder than their body temperatures. Most live in high-latitude food-rich waters where water temperatures always hover near the freezing point. But even in more temperate latitudes, the high heat capacity of water (about

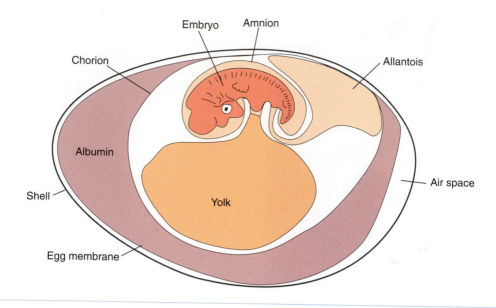

figure 6.24

Cross-section of a typical amniotic egg, showing the major internal membranes. The amnion, for which this type of egg is named, surrounds and protects the developing embryo.

figure 6.25

Marine turtle hatchling emerging from its egg.

25 times higher than air of the same temperature) is a major heat sink and makes serious inroads into the heat budgets of these homeotherms.

Marine birds and mammals exhibit several adaptations that reduce their body heat losses to tolerable levels. The streamlined bodies of marine birds and mammals adapts them to move through air and water with minimal resistance and may also reduce the extent of body surface in contact with seawater and the amount of heat transferred to the water. The major muscles of propulsion (which generate considerable heat) are located within the animal's trunk rather than on the exposed parts of the flippers, wings, or feet. For many species, a surface layer of dense feathers, fur, or blubber also insulates and also streamlines the body. Body heat losses are further limited by restricting the flow of warm blood from the core of the body to the cooler skin. Vascular countercurrent heat exchangers found in feet of birds and flukes (FIGURE 6.29), fins, and even the tongues of cetaceans also conserve heat. Arteries penetrating these appendages are surrounded by several veins carrying blood in the opposite direction. Heat from the warm blood of the central artery is absorbed by the cooler blood in the surrounding veins and carried back to the warm core of the body before much of it can be lost through the skin of the appendages. Collectively, these structures allow the relatively large-bodied birds and mammals to exploit the cold yet very productive parts of marine ecosystems.

(a)   (b)

figure 6.26

(a) Emperor penguins on the Ross Sea in Antarctica. (b) A mixed flock of shorebirds, including godwits, plovers, and an egret.

Pursuit diving with wings (penguins)   Pursuit diving with feet (cormorants)   Surface plunging (boobies)   Dipping (gulls)   Skimming (skimmers)

figure 6.27

Bill shapes and pursuit patterns of birds that feed at sea.

Marine Mammals

Of all the tetrapod classes, only the modern mammals (class Mammalia, subclass Eutheria) are characterized by **viviparity**, the internal nourishment and development of a fetus (not an egg, as in some cartilaginous fishes and sea snakes). Mammals also have body hair, milk-secreting mammary glands, specialized teeth, and an external opening for the reproductive tract separate from that of the digestive system. Thus, eutherian mammals lack the cloaca that is characteristic of all other nonmammalian vertebrates.

The three orders of marine mammals listed in Table 6.1 have experienced varying degrees of adaptation in their evolutionary transition from life on land to life in the sea. Seals, sea lions, walruses, sea otters, and polar bears (order Carnivora) are quite agile in the sea, yet all except the otter must leave the

Godwit

Ringed Plover

Curlew

Redshank

Turnstone

Oystercatcher

figure 6.28

Bill shapes of some common wading shore birds.

ocean to give birth. Sea otters (FIGURE 6.30) have retained a strong resemblance to their fish-eating relatives of freshwater lakes and streams. Sea otters prefer to eat benthic invertebrates they find along the shallow edges of the North Pacific. At some point in their evolutionary past, sea otters entered a tool-using stone age of their own. Using rocks carried to the surface with their food, they float on their backs and crack open the hard shells of sea urchins, crabs, abalones, and mussels to get at the soft tissues inside.

Recent studies comparing molecular structures and DNA from seals, walruses, and sea lions indicate that all pinnipeds share a common evolutionary ancestry and are placed in the suborder Pinnipedia. Pinnipeds evolved from terrestrial carnivores and, in the sea, have maintained their predatory habits. Only one species of walrus survives today in shallow Arctic waters where they feed on benthic mollusks. Seals and sea lions (FIGURE 6.31) are not as easily distinguished from each other as they are from walruses. In the open water, sea lions swim using a slow underwater "flying" motion of their front flippers (FIGURE 6.32). Seals propel themselves underwater with side-to-side movements of their rear flippers.

Manatees, dugongs, and sea cows (order Sirenia, FIGURE 6.33) are large ungainly creatures with paddle-like

figure 6.29

A cross-section of a small artery from the tail fluke of a bottlenose dolphin. The muscular artery in the center is surrounded by several thin-walled veins carrying blood in the opposite direction.

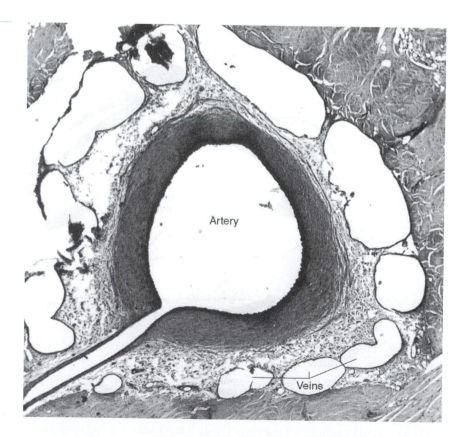

Artery

Veins

figure 6.30

A sea otter, *Enhydra*.

(a)

(b)

figure 6.31

Two types of pinnipeds: (a) harbor seals, *Phoca*; and (b) Steller sea lion, *Eumetopias*.

tails (manatees) or horizontal flukes (dugongs and sea cows) and no pelvic limbs. They are docile herbivorous animals and are now completely restricted to shallow tropical and subtropical coastal waters where they can secure an abundance of macroscopic marine and freshwater vegetation. They inhabit coastal regions along both sides of Africa, across southern Asia and the Indo-Pacific, and across the western Atlantic from South America to Florida and the Gulf of Mexico. At one time, the Steller's sea cow occupied parts of the Bering Sea and the Aleutian Islands. It took hunters and whalers less than 30 years, from the time the explorer Bering discovered these slow quiet animals in 1741, to exterminate the species.

The evolution of cetaceans from terrestrial ancestors has culminated in a remarkable assemblage of structural, physiologic, and behavioral adaptations to a totally marine existence. In contrast to typical mammals, cetaceans lack pelvic appendages and body hair, breathe through a single or a pair of dorsal blowholes, are streamlined, and propel themselves with broad horizontal tail flukes. Body lengths vary greatly from small dolphins a bit more than a meter long to blue whales exceeding 30 m in length and weighing over 100 tons (FIGURE 6.34). Convincing evidence of the tetrapod ancestry of whales can be seen during embryonic development. Limb buds develop (FIGURE 6.35) and then disappear before birth, leaving only a vestigial remnant of pelvic appendages.

The modern whales (order Cetacea) are of two distinct types. The filter-feeding baleen whales (suborder Mysticeti) lack teeth; in their place, rows of comb-like **baleen** project from the outer edges of their upper jaws (FIGURE 6.36). All other living whales (including porpoises and dolphins) are toothed whales (suborder Odontoceti). They are generally smaller than mysticetes and are most are equipped with numerous teeth to catch fish, squid, and other slippery morsels of food. The smaller odontocetes, especially, are very social and are thought by many to be highly intelligent animals. Ongoing studies evaluating cetacean intelligence and communication capabilities remain highly visible aspects of current marine

figure 6.32

California sea lion, *Zalophus*, swimming with typical "flying" motion of flippers.

figure 6.33

Manatee cow and calf (*Trichechus*).

mammal research. These studies are necessarily biased toward smaller species that are easily maintained in captivity and are complemented by information derived from examination of singly or mass stranded animals (FIGURE 6.37) and from research programs conducted in the animals' natural habitats.

Breath-Hold Diving in Marine Tetrapods

All marine tetrapods (reptiles, birds, and mammals) breathe air to obtain O_2. The length of time they can spend under water is controlled by an intricate balance of their capacity for storing O_2, their metabolic demands placed on that store of O_2, and their tolerance of low internal levels of O_2, or hypoxia. Marine

reptiles are poikilothermic, and their relatively low body temperatures drive correspondingly low metabolic rates. Homeothermic birds and mammals have much higher metabolic rates and activity levels that require more O_2 to support. The mechanisms for prolonged breath-hold diving in birds and mammals are similar, and only marine mammals are emphasized in the following discussion.

Aristotle recognized more than 20 centuries ago that dolphins were air-breathing mammals. Yet it was not until the classic studies conducted by Irving and Scholander nearly halfway into the 20th century that the physiologic basis for the deep and prolonged breath-holding dives by marine birds and mammals was defined. Their diving capabilities vary considerably. Some are little better than the Ama pearl divers of Japan who, without the aid of supplementary air supplies, repeatedly dive to 30 m and remain under water for 30 to 60 seconds. The maximal free-diving depth for humans is about 60 m; breath-holds lasting as long as 6 minutes have been independently achieved, although not while actively swimming. But even the best efforts of humans pale in comparison with the spectacular dives of some whales, pinnipeds, and penguins (TABLE 6.2). With dive times often exceeding 30 minutes, these exceptional divers are no longer closely tied to the surface by their need for air.

When diving, submerged marine tetrapods experience a triad of worsening physiologic conditions: Activity levels (and O_2 demands) are increasing at the very time their stored O_2 is diminishing, and CO_2 and lactic acid are accumulating in working tissues. Prolonged dives such as those listed in Table 6.2 are achieved with several respiratory adjustments in addition to just holding one's breath. As indicated in Table 6.2, breathing rates of marine mammals are decidedly lower than those of humans and other terrestrial mammals. The pattern of breathing is also quite different. In general, marine mammals exhale and inhale rapidly, even when resting at the sea surface, and then hold their breaths for prolonged periods before exhaling again. Even the largest baleen whales can empty their lungs of 1500 liters of air and refill them in as little as 2 seconds. In the larger species of whales, dives of several minutes' duration are commonly followed by several blows 20 to 30 seconds apart before another prolonged dive is attempted. This **apneustic breathing** pattern (FIGURE 6.38) is also common in diving penguins and pinnipeds.

Extensive elastic tissue in the lungs and diaphragms of these animals is stretched during inspiration and recoils during expiration to rapidly and

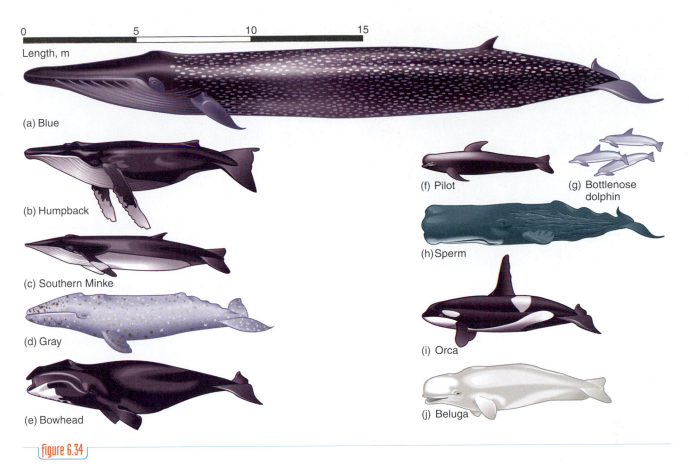

figure 6.34

A few species of cetaceans, showing the immense range of body sizes at maturity. Baleen whales are on the left, toothed whales on the right: (a) blue whale, *Balaenoptera*; (b) humpback whale, *Megaptera*; (c) southern minke whale, *Balaenoptera*; (d) gray whale, *Eschrichtius*; (e) bowhead whale, *Balaena*; (f) pilot whale, *Globicephala*; (g) bottlenose dolphin, *Tursiops*; (h) sperm whale, *Physeter*; (i) killer whale, *Orcinus*; (j) beluga whale, *Delphinapterus*.

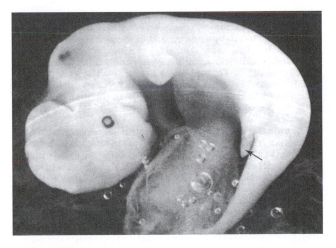

figure 6.35

A 70-day embryo of a gray whale. Note the definite rear limb buds (arrow).

figure 6.36

Left side baleen plates of a rehabilitating year-old gray whale, "JJ," with author.

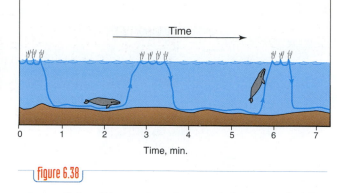

figure 6.38

Apneustic breathing pattern of a gray whale, observed while feeding. Blows at the surface represent individual breaths.

figure 6.37

Pilot whales, *Globicephala,* stranded on a New England beach.

Diving and Breath-Holding Capabilities of a Few Mammals		
Animal	Max. depth (m)	Max. duration of breath-hold (min)
Human (*Homo*)	67	6
Dolphin (*Tursiops*)	535	12
Sperm whale (*Physeter*)	3000	138
Fin whale (*Balaenoptera*)	500	30
Sea lion (*Zalophus*)	482	15
Weddell seal (*Leptonychotes*)	626	82
Elephant seal (*Mirounga*)		
Male	1530	89
Female	1273	120
Walrus (*Odobenus*)	100	13
Manatee (*Trichechus*)	600	6
Sea otter (*Enhydra*)	23	2.3

table 6.2

Data from Scheer and Kovacs (1997) and Berta and Sumich (1999).

nearly completely empty the lungs. Apneustic breathing provides time for the lungs to extract additional O_2 from the air held in the lungs. Dolphins can remove nearly 90% of the O_2 contained in each breath. Oxygen uptake within the **alveoli** (air sacs) of the lungs may be enhanced as lung air is moved into contact with the walls of the alveoli by the kneading action of small muscles scattered throughout the lungs. In some species, an extra capillary bed surrounds each alveolus and may also contribute to the exceptionally high uptake of O_2. Taken together, these features represent a style of breathing that permits marine

mammals increased freedom to explore and exploit their environment some distance from the sea surface. Still, apneustic breathing alone cannot explain how some seals and whales are capable of achieving extremely long dive times.

Cetaceans typically dive with full lungs, whereas pinnipeds often exhale before diving. Whether they dive with full or empty lungs is of little importance because the lungs and their protective rib cage smoothly collapse as the water pressure increases with increasing depth (FIGURE 6.39). For a dive from the sea surface to 10 m, the external pressure is doubled, causing the air volume of the lungs to be compressed by half and the air pressure within the lungs to double. Complete lung collapse for most diving tetrapods probably occurs in the upper 100 m; any air remaining in the lungs below that depth is squeezed by increasing water pressure out of the alveoli and into the larger air passages, the **bronchi** and **trachea**, of the lungs. Even the trachea is flexible and undergoes partial collapse during deep dives.

By tolerating complete lung collapse during dives, these animals sidestep the need for respiratory structures capable of resisting the extreme water pressures experienced during deep dives (over 300 atm for a sperm whale at 3000 m). And they receive an additional bonus: As the air is forced out of their collapsing alveoli during a dive, the compressed air still within the larger air passages is blocked from contact with the walls of the alveoli. Consequently, little of these compressed gases are absorbed by the blood, and marine mammals avoid the serious diving problems (decompression sickness and nitrogen narcosis) sometimes experienced by humans when they breathe compressed air at moderate depths while under water. After prolonged breathing of air under pressure (with hard hat, hooka, or scuba gear), large

figure 6.39

A self-portrait of Tuffy, a bottlenose dolphin, taken at a depth of 300 m. The water pressure at that depth caused the thoracic collapse apparent behind the left flipper.

quantities of compressed lung gases (particularly N_2) are absorbed by the blood and distributed to the body. As the external water pressure decreases during rapid ascents to the surface, these excess gases sometimes are not eliminated quickly enough by the lungs. Instead, they form bubbles in the body tissues and blood vessels, causing excruciating pain, paralysis, and occasionally even death. The excess N_2 absorbed from scuba gear also has a mildly narcotic effect on human divers, sometimes leading to inappropriate behaviors during the deepest parts of deep dives. Deep-diving marine mammals avoid both of these problems simply because the air within their lungs is forced away from the walls of the alveoli as the lungs collapse during a dive, thereby preventing excess N_2 from diffusing into the blood.

Because the collapsed lungs of deep-diving marine mammals are not effective stores for O_2, it must be stored elsewhere in the body or its use must be seriously curtailed during a prolonged dive (FIGURE 6.40). Both options are exercised by diving mammals. Additional stores of O_2 are maintained in chemical combination with hemoglobin of the blood or with myoglobin in muscle cells. There are more hemoglobin-containing red blood cells in deep-diving birds and mammals, and

the blood volume is also significantly higher than that of nondiving mammals. About 20% of the total body weight of elephant seals and sperm whales, for instance, is blood. Much of the additional blood volume is accommodated in an extensive network of capillaries (actually another rete mirablia) located along the dorsal side of the thoracic cavity (FIGURE 6.41). The **vena cava** (the major vein returning blood to the heart) in some species is bag-like and elastic. In elephant seals, it alone can accommodate 20% of the animal's total blood volume. These features all contribute to large reserves of stored O_2 for use during a dive.

The swimming muscles of marine mammals are highly tolerant to the accumulation of lactate, a metabolic product of hypoxic conditions during a breath-hold dive. Therefore, muscles and other non-critical organs (such as the kidneys and digestive tract) can be temporarily deprived of access to the reserve O_2 stored in the blood. To regulate the distribution of blood, muscles in the walls of arteries leading to these peripheral muscles and organs contract to reduce the flow of blood. The general term for this process is **vasoconstriction**. Most of the circulating blood is then shunted to other vital organs, primarily the heart and brain. Simultaneously, the heart rate slows dramatically to accommodate pressure changes in a much-reduced circulatory system comprising the heart, the brain, and connecting blood vessels. Other circulatory structures also help to smooth out and moderate fluctuations in the pressure of blood going to the brain. An elastic bulbous "natural aneurism" in the **aorta** (the large artery leaving the heart) and another rete in the smaller arteries at the base of the brain both help to dampen blood pressure surges each time the heart beats.

Bradycardia is a term used to describe the marked slowing of the heart rate that accompanies vasoconstriction and probably occurs in all vertebrates experiencing reduced access to their usual supply of O_2. These two responses, peripheral vasoconstriction and bradycardia, probably occur in all diving air-breathing vertebrates, including birds, reptiles, and mammals. The intensity of bradycardia varies widely

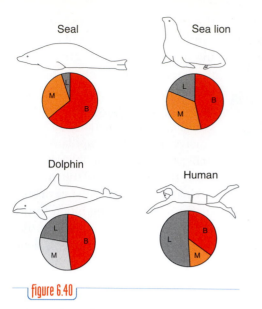

Seal   Sea lion

Dolphin   Human

Comparison of oxygen stores in blood (B), muscles (M), and lungs (L) for several different mammals.

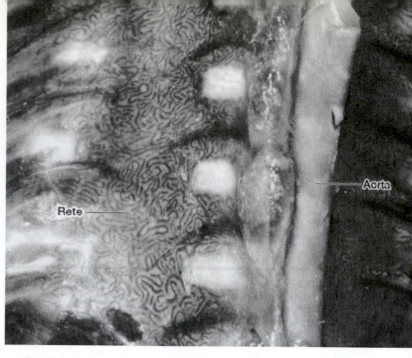

Rete   Aorta

The right thoracic rete mirabilia of a small porpoise, *Stenella*.

A Summary of the Mammalian Dive Response

Table 6.3

1. Cessation of breathing
2. Extreme bradycardia regardless of dive duration
3. Strong peripheral and some central vasoconstriction
4. Reduced aerobic metabolism in most organs
5. Rapid depletion of muscle O_2
6. Lactate accumulation in muscles
7. Variety of blood chemistry changes during and immediately after dive

between marine mammal groups. During experimental dives under laboratory conditions, heartbeat rates of restrained cetaceans are reduced to 20 to 50% of their predive rates. The combination of bradycardia, peripheral vasoconstriction, and other circulatory adjustments to diving has been commonly referred to as "the mammalian dive response." The major components of this response are summarized in TABLE 6.3.

Despite variable and even conflicting data for some cetaceans and semivoluntary diving pinnipeds, this set of characteristics was widely applied to explain long-duration breath-holds by marine mammals. Subsequent studies of free-diving Weddell seals in Antarctic waters support a substantially different picture of diving responses in unrestrained divers in their nat-

ural habitat (FIGURE 6.42). These researchers attached instrument packages to numerous free-diving seals to monitor and record dive time, depth, heart rate, core body temperature, and blood chemistry. Weddell seals were ideal subjects for this type of study because they breathe by surfacing at holes maintained in the fast sea ice. Each seal returning to its own hole to breathe after a dive offered convenient opportunities to attach and retrieve recording instrument packs.

Weddell seals perform breath-holding dives up to about 20 minutes in length without using the mammalian dive response presented in Table 6.3. This suggests that Weddell seals have sufficient stored oxygen at the beginning of a dive to last about 20 minutes. Only during dives lasting longer than about 20 minutes (their apparent **aerobic dive limit** [ADL]) are circulatory responses observed that resemble the expected dive response described in Table 6.3. An animal's ADL is defined as the longest dive that does not lead to an increase in blood lactate concentration during the dive. Therefore, if an animal dives within its ADL, there is no lactate accumulated to metabolize after the dive, and a subsequent dive can be made as soon as the depleted blood oxygen is replenished.

For dives longer than the ADL, the magnitude of physiologic responses is generally related to the length of the dive, yet the picture is not a simple one. Peripheral vasoconstriction and bradycardia do occur, with the magnitude of those responses generally proportional to the duration of the part of the dive that

figure 6.42

A Weddell seal (*Leptonychotes*) at a breathing hole.

Table 6.4

A Summary of Dive Responses in Weddel Seals (Compare with those listed in Table 6.3)

1. Cessation of breathing
2. Variable bradycardia depending on dive duration
3. Variable peripheral and central vasoconstriction depending on dive duration
4. Reduced aerobic metabolism in most organs
5. Rapid depletion of muscle O_2
6. Lactic acid accumulation in muscles beginning after 20 minutes
7. Variety of blood chemistry changes during and immediately after dive, depending on dive duration
8. Voluntary reduction of body core temperatures during very long dives

exceeded the animal's ADL. The more a dive exceeds the ADL, the greater the accumulation of lactate and the longer it takes for the level of lactate to return to resting levels. After very long dives (1 hour or more), Weddell seals are exhausted and sleep for several hours. Although Weddell seals are capable of remaining submerged for more than an hour, they seldom do, because about 85% of their dives are within their projected 20-minute ADL.

It now appears that Weddell seals anticipate before a dive begins how long it will last and then consciously make the appropriate circulatory adjustments before leaving the sea surface. This is why it should not be called a dive response, but rather an integrated set of related responses. During short dives (<20 minutes), no adjustments are necessary. For anticipated dives of long duration, both peripheral vasoconstriction and bradycardia occur maximally at the beginning of the dive and remain that way throughout the dive. On extended dives approaching 1 hour in duration, core body temperature can be voluntarily depressed to 35°C, kept depressed between dives, and then rapidly elevated after the last dive of a dive series. Together, these responses (TABLE 6.4) allow Weddell seals to accomplish some of the longest breath-holds known for mammals. They also indicate that, at least in this species, peripheral vasoconstriction, heart rate, and core body temperature are under conscious control and argue strongly against the earlier concept of a reflexive "dive response" for marine mammals.

Can these results be generalized to other species of marine mammals? The question is difficult to answer for more than a few species, because it is apparent that Kooyman's work with unrestrained Weddell seals, whose dive duration was under their own control, was a key to elucidating this species' physiologic response to breath-hold diving. Opportunities to attach instrument packages to monitor physiologic responses of a free-diving mammal, release the subject animal for a series of unrestrained dives, and then recapture the subject animal for instrument package recovery and blood sampling have been limited in the past. However, such opportunities are becoming more frequent as improvements in electronic monitoring and radio telemetry techniques expand our studies of animals diving without restraint.

Although not as well studied, elephant seals (see Figure 10.37) may surpass Weddell seals in their breath-holding ability. Elephant seals spend months at sea foraging for squids and fishes at depths between 300 and 1500 m (Table 6.2). Elephant seals exhibit diving patterns that suggest they also may play a role in the avoidance of predators: Diving deeply with no swimming at the surface, short surface intervals, and long duration dives may help elephant seals to minimize encounters with white sharks. Their feeding dives are typically 20 to 25 minutes long, with females usually going to depths of about 400 m and males to depths of 750 to 800 m. Both sexes dive night and day for weeks on end without sleeping and usually spend only 2 to 4 minutes at the surface between dives. These short surface times between long deep dives suggest that these are not unusual dives but are the norm for this species. Further studies may show that the dive

Serendipity in Science: The Discovery of Living Coelacanths

Coelacanths are primitive fishes with an extensive (about 90 extinct species are recognized) and well-studied fossil record that extends back about 400 million years (Figure B6.1). Their claim to fame is that they are a close relative of the lineage that presumably gave rise to terrestrial tetrapods, including humans. Before World War II, it was thought that all coelacanths had gone extinct along with the dinosaurs about 65 million years ago.

That all changed on December 22, 1938 when 31-year-old Marjorie Courtney-Latimer, the curator of a small South African museum, stopped by the docks while on her way home to pick up some fish for Christmas dinner. There she saw a very strange fish that had been taken in 70 m of water by the trawler *Nerine* just outside the mouth of the Chalumna River. It was a living coelacanth, and its arrival was as unlikely as the possibility of a living *Velociraptor* walking into your backyard tonight.

Courtney-Latimer did not know exactly what this strange fish was, but she was certain it was very unusual and worthy of collection. Unfortunately, its large size prevented traditional preservation in a bottle of alcohol or formaldehyde (Courtney-Latimer went so far as to ask the local mortuary for many gallons of embalming fluid—they were aghast at her request). Failing to obtain the necessary supplies on the eve of Christmas, she was forced to discard the rotting viscera and preserve only the head and skin of the fish.

She sent a note containing her sketch of this weird specimen to J. L. B. Smith, a chemistry professor at Rhodes University in Grahamstown who had a passion for ichthyology, the study of fishes. He recognized the fish in the drawing as a coelacanth but was understandably skeptical that a living representative had been captured. It took Smith 2 full months to arrange to travel the 100 miles to

Figure B6.1   Coelacanth.

visit Courtney-Latimer to see the specimen in person. He arrived on February 16, 1939, saw that the fish was indeed a living coelacanth, immediately understood the scientific importance of this finding, and proclaimed that "this fish will be on the lips of every scientist in the world."

Smith named the fish *Latimeria chalumnae* (in honor of its discoverer and the location of its capture) and immediately launched a campaign to obtain a second specimen, one that could be fully examined, dissected,

---

responses of Weddell seals, as outlined in Table 6.4, are essentially what all marine tetrapods do to varying degrees.

Deep-diving whales, such as sperm and beaked whales, are not accessible to researchers in the way Weddell and elephant seals are, yet they likely surpass either seal in both maximum dive depth and maximum dive duration (Table 6.2). They exhibit many of the features described here for deep and prolonged divers. We return to this topic in Chapter 10

to examine the potentially lethal effects of powerful human-introduced sound sources on the blood chemistry and diving ability of these supreme divers.

6.4 Vertebrate Sensory Capabilities

To participate successfully in migration, feeding, mating, or any other important life events, all animals must be able to evaluate their immediate surroundings and to update those evaluations more or

and properly preserved. He distributed leaflets all over southeastern Africa written in English, French, and Portuguese that offered £100, a great deal of money in 1939, for delivery of another coelacanth. Unbeknownst to Smith, the first specimen found by Courtney-Latimer in a South African fish market may have been a stray. The primary population center for living coelacanths seems to be the Comoro Islands, just north of Madagascar, so Smith wasted many years searching fruitlessly in the region from which the first coelacanth had been found.

It was nearly 14 years later, on December 18, 1952, when Smith received a telegram from Eric Hunt, an English sea trader, stating that he had obtained a second coelacanth from a Comoran fisherman named Ahamadi Abdallah. Smith was frantic; the Comoros were many miles distant, a journey there by boat would take many weeks, and this was just a few days before Christmas, so everyone was on holiday and formaldehyde was once again impossible to obtain. Smith contacted D. F. Malan, then Prime Minister of South Africa (and the creationist architect of apartheid), at his beach house and asked for a plane that could transport Smith to pick up the second coelacanth. Malan arranged for a military Dakota to carry Smith to this precious fish.

Unfortunately, France (who held the Comoros under colonial rule at the time) did not give permission for a South African military plane to land or to snatch away the fish. To make matters worse, the second specimen was missing its first dorsal fin, which fooled Smith into naming a new genus, *Malania* (in honor of his Prime Minister), for this second coelacanth. France immediately declared an international embargo that resulted in all subsequent coelacanths being sent exclusively to the Museum of Natural History in Paris for the next 23 years. (The embargo ended when the Federal Islamic Republic of the Comoros gained their independence in 1975.) Not surprisingly, this embargo resulted in the development of a black market for the capture and sale of coelacanths, which are valued for their scientific appeal and notochordal fluid (which is believed to prolong the life of a human just as it has prolonged the existence of this ancient fish).

Since then, nearly 200 coelacanths have been collected from all over the western Indian Ocean, including Madagascar, Mozambique, Kenya, and Tanzania. Then, in September 1997 Lady Luck would smile again. Mark Erdmann, a graduate student at Berkeley, and his new wife, Arnaz, were persuaded by their friends to visit a local fish market in North

Serendipity in Science: The Discovery of Living Coelacanths

Sulawesi, Indonesia on the last day of their honeymoon. There they saw and photographed a living coelacanth, 10,000 km from the site of capture of the first living coelacanth back in 1938. They returned in 1998 and captured a second Indonesian coelacanth, thus greatly extending the known range of coelacanths and providing hope that a vast population of these important animals may occur throughout the Indian Ocean.

Additional Reading

Musick, J. A., M. N. Bruton, and E. K. Balon. 1991. The biology of *Latimeria chalumnae* and evolution of coelacanths. *Environmental Biology of Fishes 32*: 1–435.

For more information on this topic, go to this book's website at http://www.bioscience.jbpub.com/marinebiology.

less continuously. These evaluations are accomplished with a variety of specialized sensory organs connecting their internal nervous systems with the myriad of chemical, mechanical, or electromagnetic stimuli coming from their external world. Sensory organs contain receptor cells specialized to convert these environmental stimuli into nerve impulses. These nerve impulses are conducted to the brain, where perception of the stimulus occurs and a response is initiated. This section focuses on the more obvious or important adaptations of vertebrate sensory abilities for underwater use.

Most of us have a general understanding of our five basic sensory capabilities: taste, smell, touch, vision, and hearing. However, some fishes exhibit electroreceptive and magnetoreceptive abilities that have no known counterparts in most terrestrial vertebrates, including humans. When animals are submerged in seawater, even comfortable human notions like the differences between taste and smell become confusing. Our ability to smell depends on tens of millions of ciliated sensory cells located in our nasal passages that detect and identify thousands of different chemicals carried to us dissolved in air. Taste, on the other hand, responds to a limited range of substances (sugars, acids, and salts) that must be dissolved in water and delivered to a few hundred taste buds on the tongue, mouth, and lips. These convenient distinctions between taste and smell become less clear when applied to fishes that live constantly under water and never breathe air.

Chemoreception

Both taste and smell are chemoreceptive senses. For marine fishes swimming in an aquatic medium of near-uniform salinity, with a buffered pH, and a nearly complete absence of sugar, an ability comparable with our sense of taste has little use. But **olfaction**, the detection with olfactory sensory cells of chemicals dissolved in water, is highly evolved in fishes. Salmon and some species of large predatory sharks respond to very low concentrations of odor molecules. Just a few parts per billion of chemicals in the water of their olfactory sacs are sufficient for recognition. With such olfactory capabilities, it is possible for predatory sharks to locate odor sources using very dilute chemical trails left by injured prey or for a migrating salmon to locate and identify the stream of its birth.

For air-breathing marine tetrapods, the olfactory situation is reversed. Regardless of how acute their olfactory sense is in air, when a reptile, bird, or mammal submerges, it closes its nostrils and leaves its olfactory sense at the sea surface. Cetaceans, because they spend most of their lives below the sea surface, have completely lost all traces of their ancestral nasal structures used for olfaction.

Electroreception and Magnetoreception

Humans are totally oblivious to the weak electrical and electromagnetic energy fields generated by contractions of muscles in swimming animals, by water currents moving past inanimate objects, and even by the Earth's own magnetic field. Yet organisms as small as bacteria and as large as sharks detect and respond to some of these signals. These specialized senses are known or suspected to exist in several classes of vertebrates, but the best studied examples are the cartilaginous fishes. Sharks and rays exhibit an extensive network of tiny pores or pits arranged on their snouts and pectoral fins. Each pit connects via a short jelly-

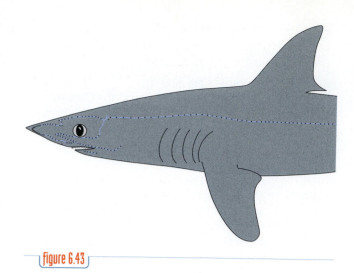

figure 6.43

The major branches of the left lateral line system (blue) of a shark. Pores scattered in clusters over the snout are the openings to the ampullae of Lorenzini.

filled canal to a flask-shaped ampulla of Lorenzini. These ampullae are associated with extensions of the lateral line system of cartilaginous fishes (FIGURE 6.43) and of at least one marine bony fish, the marine catfish. Electroreception is accomplished by sensory cells located at the bottom of each ampulla, possibly evolved from the basic lateral line sensory hair cell (described below). With this sensory system, some sharks and rays are able to detect (at distances of a meter or so) bioelectrical fields equivalent to those generated by the muscle contractions of typical prey species. Similar electrical fields are also produced by some metal objects in seawater. The seemingly erratic responses by some sharks to metal boat parts may be explained as the sharks' normal response to these artificial electric fields mimicking those produced by their usual prey.

The study of geomagnetic reception in animals is still in its infancy. It has been confirmed in cartilaginous fishes and is suspected in some bony fishes such as tuna and salmon, some birds, and possibly some whales. The ampullae of Lorenzini are thought to be the organs of detection in sharks and rays; in other vertebrate groups, the possible organs of detection have not yet been identified.

Vision

Most animals rely on ambient light from the sun, moon, or stars to illuminate what they see and also to provide the energy needed to stimulate their photoreceptor cells. (A few animals, especially mid- and deep-water fishes, have light-producing photophores to illuminate their own very small visual fields.) Although sunlight travels several kilometers through our atmosphere with little loss in intensity, an additional few hundred meters through the clearest ocean water so reduces the intensity that photosynthesis is impossible and vision is very limited. As light intensity is reduced, visual fields shrink to a few meters, and the range of colors available narrows to the green and blue portions of the visible spectrum (see Figures 1.21 and 1.23.)

All marine vertebrates except hagfishes obtain visual images of their surroundings with a remarkable organ, the camera eye (FIGURE 6.44). A similar type of eye has evolved independently in octopuses and other cephalopod mollusks. In vertebrate eyes, light is focused by a round lens through a light-tight and nearly spherical eye cavity to the light-sensitive receptor cells of the **retina** at the back of the eye. In contrast to eyes adapted for vision in air, fish and cetacean eyes must accommodate the higher refractive power of water. To do this, the eyes of fish and cetaceans are strongly flattened in front, with a round lens that focuses by moving nearer to or away from the retina rather than by changing shape as ours do.

The retina contains the light-sensitive rod and cone cells specialized for light detection. Typically, cones serve as high-intensity and color receptors, and rods serve as low-intensity receptors. Fishes living below the photic zone usually have fewer cones than rods, and deep-sea fishes often lack cones. For all fishes with only one type of cone (or no cone cells at all), vision is limited to detecting variations in light intensity; they see their world in varying shades, not varying colors. Many fish species nearer the surface and in better-illuminated marine environments such as coral reefs are capable of varying degrees of color vision. The retinas of these fishes contain either two or three different types of cone cells, each type sensitive to a different range of wavelengths. In bright light and clear water, fishes with three types of cones apparently possess a visual color acuity that may exceed that of humans.

Equilibrium

A critical aspect of orienting oneself in space is knowing which way is up or down. For human divers away from surface light or diving at night, the absence of a visual horizon or other clear notions of up and down can be very disorienting. It is also needed by actively moving animals to monitor their swimming speeds and changes in direction. With this information, body positions and orientations can be updated continuously. To do this, almost all vertebrates have two types of receptors within their organs of equilibrium: One responds to the pull of gravity, and the other detects acceleration forces.

Vertebrates use a basic **sensory hair cell** design (FIGURE 6.45) to serve a variety of mechanoreceptive purposes, including equilibrium. Their organ of equilibrium, the **labyrinth organ**, is located on each side of

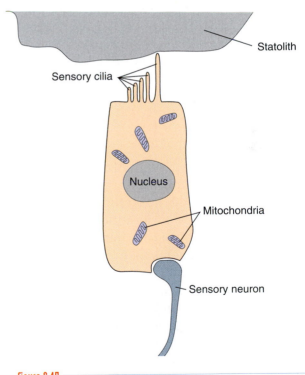

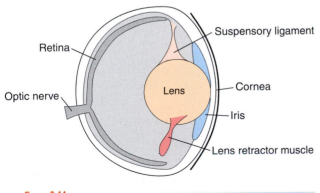

figure 6.44

Cross-section of a fish eye. Note the solid round lens that is focused by being moved toward or away from the retina by the retractor muscle.

figure 6.45

General structure of a mechanosensory hair cell. When the sensory cilia of the hair cell are bent, a nerve impulse is initiated and passed to the associated sensory neuron.

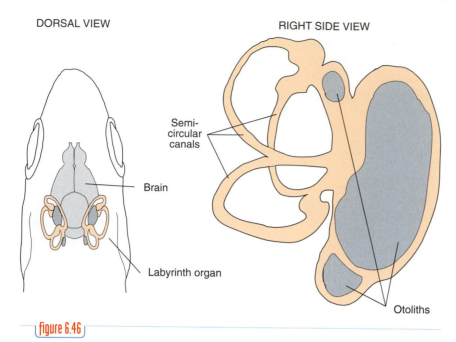

DORSAL VIEW

RIGHT SIDE VIEW

Semi-
circular
canals

Brain

Labyrinth organ

Otoliths

figure 6.46

Anatomic location (left) and general structure of a labyrinth/otolith organ of a bony fish.

the head (FIGURE 6.46). Each labyrinth organ consists of three semicircular canals and two smaller sac-shaped chambers. These sac-shaped chambers are the gravity detectors, with small stony secretions suspended by sensory hair cells known as **neuromasts**. The canals are the acceleration detectors. Each canal is filled with fluid and has an enlarged ampulla that is lined with sensory hair cells supporting a cupula. Acceleration in any direction moves the fluid against the cupula and stimulates the neuromasts of at least one canal.

Sound Reception

Sound is transmitted through air or water as spreading patterns of vibrational energy. This energy travels at 1500 m/s in water, about five times faster than in air. Although the sea may at times seem surprisingly silent to human divers, animals with sensory organs attuned to that medium must find it a noisy place. Swimming animals produce unintentional noises as they move, feed, and bend their bodies. Others make intentional grunts, groans, chirps, and other noises. Add those to surface wave noises, vocalizations from whales and seals, and the mechanical noises we introduce into the sea, and the ocean becomes a cacophony of sounds within, below, and well above the range of human hearing.

Fishes detect sounds with mechanoreceptive sensory hair cells nearly identical to those in their organs

of equilibrium. The lateral line system of fishes consists of canals contained in scales extending along each flank and in complex patterns over their heads (Figure 6.43). Within the canals are neuromasts. The cilia of the neuromasts are stimulated by water movement and by pressure differences at the fish's body surface. These are communicated to the lateral line canals through pores in the skin surface. In this way, the lateral line systems function to detect disturbances caused by prey or by predators, by swimming movements of nearby schooling companions, and by sound vibrations.

Another structure associated with sound detection in bony fishes is the otolith (Figure 6.46). Otoliths are small calcareous stones embedded in and associated with part of the labyrinth organ. Together they constitute the inner ear. On each side of the head, two or three otoliths are suspended in fluid-filled sacs where they contact neuromasts. Arriving sound waves move the fish very slightly, and the denser otoliths lag behind, bending the neuromast cilia and stimulating a nerve impulse. Hair cells with different orientations relative to the otoliths may provide some directional information about the sound source.

Tetrapods have very different hearing systems evolved for detecting sounds transmitted in air. Airborne sound vibrations are transmitted from the eardrum to the cochlea of the inner ear. The cochlea is a coiled tubular structure filled with fluid and lined with yet another type of sensory hair cells. In toothed whales, the sound-processing structures of the middle ear are enclosed in a bony case, **the tympanic bulla**. The bulla is supported only by a few wisps of connective tissue to isolate it from adjacent bones of the skull by air sinuses filled with an insulating emulsion of mucus, oil, and air.

The **external auditory canal** is the usual mammalian sound channel connecting the external and middle ears. However, the auditory canal of mysticetes is completely blocked by a plug of earwax; in toothed whales, the canal is further reduced to a tiny pore or is completely covered by skin. Mapping of acoustically sensitive areas of dolphins' heads has shown the external auditory canal to be about six times less sensitive to sound than the lower jaw. The unique sound reception system in toothed whales begins with the bones of the lower jaw, which flare toward the rear and are extremely thin. Within each half of the lower jaw is a fat body (or, in some cases, liquid oil) that directly connects with the wall of the bulla of the middle ear (FIGURE 6.47). These fat bodies act as a sound channel to transfer sounds from the flared portions of the lower jaw directly to the middle ear. An area on either side of the forehead is nearly as sensitive as the lower jaw, providing four very sensitive separate channels for sound reception. Sensitive underwater hearing is one of the two components necessary for marine mammals to echolocate. This topic is discussed in Chapter 10.

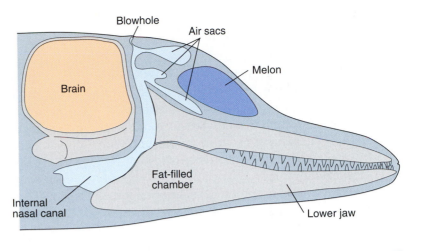

figure 6.47

Mid-section of a dolphin head, showing the bones of the head, the air passages, and the structures associated with sound production and reception.

STUDY GUIDE

Summary Points

Marine Vertebrates
- Although occupying a single subphylum, vertebrates rival nearly all other phyla in terms of diversity and ecology. They occupy all marine habitats and are important players in nearly all marine food webs.

Vertebrate Features
- All vertebrates possess the standard chordate features as well as segmental ossification of the notochord, a musculature that is segmented into myomeres, and a closed circulatory system with hemoglobin contained within red blood cells that cannot leave the blood vessels.
- Vertebrates are hypothesized to have evolved from filter-feeding chordates not unlike modern lancelets, and conodonts from the Late Cambrian period are the earliest vertebrates known.
- All living vertebrates, with the exception of hagfishes, osmoregulate in that they spend molecular energy to maintain a relatively constant concentration of solutes in their plasma and cytosol.

Marine Fishes
- Fishes, which include about 50% of all living vertebrates, are a diverse assemblage that is difficult to characterize but which usually live in water, swim with fins, possess scales, and use gills supplied with a countercurrent circulation for gas exchange.
- Agnathans are living jawless fishes that are not closely related but which are distinguished from other vertebrates in that they lack a jaw, paired fins, and scales. Hagfishes are marine benthic scavengers that possess unique slime glands, whereas lampreys are anadromous creatures that often parasitize teleosts as adults.

- The sharks, rays, and chimeras of class Chondrichthyes are mostly marine fishes that often grow to large sizes, retain metabolic waste products (urea and trimethylamine oxide [TMAO]) to achieve osmotic equilibrium with seawater, and possess a characteristic heterocercal caudal fin.
- Chondrichthyans reproduce via internal fertilization and possess internal embryos that develop inside egg cases that are deposited in the environment or are born live after completing their development inside their mothers, both with and without the benefit of a placental attachment to her blood supply.
- Many thousands of species of bony fishes are marine, and all but one are ray finned. The only exception is the lobe-finned coelacanth, first collected in the western Indian Ocean but now known from Indonesia as well.
- Most marine bony fishes are teleosts, advanced fishes with bony skeletons, thin and flexible scales, an adjustable swim bladder for buoyancy, and body fluids that are hypoosmotic to seawater.
- Fishes swim via thrust generated by highly coordinated movements of many different fins and fin combinations, including lateral movements of the caudal fin, flapping or undulating pectoral fins, fanning of the pectoral fins, or sculling movements of the dorsal and anal fins.

Marine Tetrapods

- Air-breathing tetrapods that have returned to a sometimes full-time life in the sea includes representatives of reptiles, birds, and mammals. All are hypoosmotic to seawater like the fishes, and two groups are homeothermic.
- About 100 species of sea snakes, seven species of sea turtles, and one representative each of iguana and crocodile are truly marine (other reptiles are estuarine). All marine reptiles, and the closely related sea birds, eliminate excess salts via their kidneys and specialized salt glands contained in their mouths, nostrils, or orbits.
- "Marine" birds range from those that are nearly full-time residents of the sea, such as penguins, to other species that simply visit the shoreline to feed, such as some ducks, geese, and coots. Be-

ing homeothermic and living in or near a cold dense medium necessitates the development of a wide variety of thermoregulatory adaptations, including streamlined bodies, primary propulsion muscles being located in their torso, a dense covering of feathers and blubber, and countercurrent vasculature in feet and wings.
- Three orders of mammals can be found in the sea, including carnivores (sea otters and pinnipeds), sirenian manatees and dugong, and cetacean whales, dolphins, and porpoises.
- All air-breathing tetrapods are capable of diving deeper and longer than humans, and some species, such as elephant seals, sperm whales, and emperor penguins, can dive to more than 1000 m for longer than an hour. To accomplish these impressive feats, marine tetrapods possess a great variety of anatomic and physiologic adaptations.

Vertebrate Sensory Capabilities

- Most marine vertebrates possess the standard five sensory systems of vision, touch, taste, hearing, and smell, although they favor different systems relative to their terrestrial relatives (vision is of little use under water) and even possess additional sensory systems that are only useful in water (such as electroreception).
- Chemoreception is very important to marine fishes, and they may possess some of the most sensitive noses of any animal, but tetrapods all close their nostrils while under water and therefore rely on their sense of smell very little.
- The ability to detect weak electric and electromagnetic fields has been demonstrated in bacteria, cartilaginous and bony fishes, some birds, and possibly some whales.
- Aquatic vision requires alterations to terrestrial visual systems to compensate for the quality and quantity of light available under water, such that marine vertebrates focus by moving the lens within the eye (rather than changing its shape) and detect light with altered ratios of photoreceptor cells and unique visual pigments.
- Most marine vertebrates can hear, and they use anatomic systems that are very similar to those seen in terrestrial species. Nevertheless, they need to compensate for the fact that sound

travels about five times more rapidly under water and often change the route by which sound travels on its way to the sensory hair cells in their cochlea.

Topics for Discussion and Review

1. List and describe the major evolutionary structural advances exhibited in bony fishes, cartilaginous fishes, and agnathans.
2. Compare the osmoregulatory adaptations of marine teleosts, sharks, reptiles, and mammals.
3. Summarize the great variety of developmental methods observed in living cartilaginous fishes.
4. Describe the adaptive significance of nasal salt glands and uric acid secretion for reptiles and birds feeding at sea.
5. Compare and contrast the thermoregulatory strategies of marine birds and mammals.
6. Describe how some marine tetrapods are able to dive to depths greater than 1000 m.
7. Describe how some marine tetrapods are able to hold their breath for more than 1 hour.
8. List two specific structural features that distinguish each of the following marine mammal groups from the others: baleen whales, toothed whales, sea lions, seals, manatees.
9. Why is olfaction more important to marine fishes as opposed to marine tetrapods?
10. Describe how sharks are able to detect weak electric fields in the sea.

Suggestions for Further Reading

Ancel, A., G. L. Kooyman, P. J. Ponganis, J. P. Gendner, J. Lignon, X. Mestre, N. Huin, P. H. Thorson, P. Robisson, and Y. Le Maho. 1993. Foraging behavior of emperor penguins as a resource detector in winter and summer. *Nature* 360:336–339.

Atema, J., R. R. Fay, A. N. Popper, and W. Travolga. 1988. *Sensory Biology of Aquatic Animals*. Springer-Verlag, New York.

Berta, A. 1994. What is a whale? *Science* 263:180–182.

Berta, A., and J. L. Sumich. 1999. *Marine Mammals: Evolutionary Biology*. Academic Press.

Castellini, M. A., P. M. Rivera, and J. M. Castellini. 2002. Biochemical aspects of pressure tolerance in marine mammals. *Comparative Biochemistry and Physiology—Part A: Molecular & Integrative Physiology* 133:893–899.

Croll, D. A., A. Acevedo-Gutierrez, B. R. Tershy, and J. Urban-Ramrez. 2001. The diving behavior of blue and fin whales: is dive duration shorter than expected based on oxygen stores? *Comparative Biochemistry and Physiology—Part A: Molecular & Integrative Physiology* 129:797–809.

Davoren, G. K., W. A. Montevecchi, and J. T. Anderson. 2003. Search strategies of a pursuit-diving marine bird and the persistence of prey patches. *Ecological Monographs* 73:463–481.

DeLong, R. L., and B. S. Stewart. 1991. Diving patterns of northern elephant seal bulls. *Marine Mammal Science* 7:369–384.

Diamond, A. W., and C. M. Devlin. 2003. Seabirds as indicators of changes in marine ecosystems: ecological monitoring on Machias Seal Island. *Environmental Monitoring and Assessment* 88:153–181.

Douglas, R. H., and M. B. A. Djamgoz. 1990. *The Visual System of Fish*. Chapman and Hall, London.

Drent J., W. D. Van Marken Lichtenbelt, and M. Wikelski. 1999. Effects of foraging mode and season on the energetics of the marine iguana, *Amblyrhynchus cristatus*. *Functional Ecology* 13:493–499.

Ellis, R. 2000. *Encyclopedia of Sea*. Knopf.

Enticott, J., and D. Tipling. 1997. *Seabirds of the World: The Complete Reference*. Stackpole Books.

Ernst, C. H., and R. W. Barbour. 1989. *Turtles of the World*. Smithsonian Institution Press, Washington, DC.

Folkens, P., R. R. Reeves, B. S. Stewart, P. J. Clapham, and J. A. Powell. 2002. *National Audubon Society Guide to Marine Mammals of the World*. Knopf, Boston.

Grigg G. C., F. Seebacher, L. A. Beard, and D. Morris. 1998. Thermal relations of large crocodiles, *Crocodylus porosus*, free-ranging in a naturalistic situation. *Proceedings: Biological Sciences* 265:1793-1799.

Hamlett, W. C. 1999. *Sharks, Skates, and Rays: The Biology of Elasmobranch Fishes*. Johns Hopkins University Press, Washington, DC.

Hara, T. J. *Fish Chemoreception*. Chapman and Hall, London.

Harvey, K. R., and G. J. E. Hill. 2003. Mapping the nesting habitats of saltwater crocodiles (*Crocodylus porosus*) in Melacca Swamp and the Adelaide River wetlands, Northern Territory: an approach using remote sensing and GIS. *Wildlife Research* 30:365–375.

Heatwole, H. 1987. *Sea Snakes*. New South Wales University Press, Kensington.

Helfman, G. S., B. B. Collette, and D. E. Facey. 1997. *The Diversity of Fishes*. Blackwell Science, Malden, MA.

Hindell, M. A., D. J. Slip, and H. R. Burton. 1991. The diving behavior of adult male and female southern elephant seals, *Mirounga leonina* (Pinnipedia: Phocidae). *Australian Journal of Zoology* 39:595–619.

Hoelzel, A. R. 2002. *Marine Mammal Biology: An Evolutionary Approach*. Blackwell Publishers.

Irving, L., and J. Krog. 1954. Body temperatures of Arctic and subarctic birds and mammals. *Journal of Applied Physiology* 6:667–680.

Kalmijn, A. J. 1982. Electric and magnetic field detection in elasmobranch fishes. *Science* 218:916–918.

Kooyman, G. L. 1988. Pressure and the diver. *Canadian Journal of Zoology* 66:84–88.

Kuchel, L. J., and C. E. Franklin. 1998. Kidney and cloaca function in the estuarine crocodile (*Crocodylus porosus*) at different salinities: evidence for solute-linked water uptake. *Comparative Biochemistry and Physiology—Part A: Molecular & Integrative Physiology* 119:825–831.

Kvadsheim, P. H., and L. P Folkow. 1997. Blubber and flipper heat transfer in harp seals. *Acta Physiologica Scandinavica* 161:385–395.

Love, J. A. 1990. *The Sea Otter*. Whittet Books, London.

Lutz, P. L., and J. A. Musick. 1996. *The Biology of Sea Turtles*. CRC Press, Boca Raton, FL.

Noren, S. R., and T. M. Williams. 2000. Body size and skeletal muscle myoglobin of cetaceans: adaptations for maximizing dive duration. *Comparative Biochemistry and Physiology—Part A: Molecular & Integrative Physiology* 126:181–191.

Owens, D. W. 1980. The comparative reproductive physiology of sea turtles. *American Zoologist* 20:549–563.

Peichl, L., G. Behrmann, and R. H. H. Kröger. 2001. For whales and seals the ocean is not blue: a visual pigment loss in marine mammals. *European Journal of Neuroscience* 13:1520–1528.

Perrin, W. F., B. Würsig, and J. G. M. Thewissen. 2002. *Encyclopedia of Marine Mammals*. Academic Press, New York.

Pough, F. H., C. M. Janis, and J. B. Heiser. 2002. *Vertebrate Life*. Prentice Hall, Upper Saddle River, NJ.

Riedman, M. 1990. *The Pinnipeds: Seals, Sea Lions and Walruses*. University of California Press, Berkeley.

Reynolds, J. E., and D. K. Odell. 1991. *Manatees and Dugongs*. Facts on File, New York.

Reynolds, J. E. III, and S. A. Rommel. 1999. *Biology of Marine Mammals*. Smithsonian Institution Press, Washington, DC.

Schmidt-Nielson, K. 1959. Salt glands. *Scientific American* 200:109–116.

Standora, E. A., J. R. Spotila, J. A. Keinath, and C. R. Shoop. 1984. Body temperatures, diving cycles, and movement of a subadult leatherback turtle, *Dermochelys coriacea*. *Herpetologica* 40:169–176.

Stirling, I., and D. Guravich. 1988. *Polar Bears*. University of Michigan Press, Ann Arbor.

Tavolga, W. N., A. N. Popper, and R. R. Fay. 1981. *Hearing and Sound Communication in Fishes*. Springer-Verlag, New York.

Thewissen, J. G. M., L. J. Roe, J. R. O'Neil, S. T. Hussain, A. Sahni, and S. Bajpai. 1996. Evolution of cetacean osmoregulation. *Nature* 381:379–380.

Walker, B. G., and P. D. Boersma. 2003. Diving behavior of Magellanic penguins (*Spheniscus magellanicus*) at Punta Tombo, Argentina. *Canadian Journal of Zoology* 81:1471–1483.

Waller, G. (Ed.). 1996. *SeaLife*. Smithsonian Institution Press, Washington, DC.

Watanuki Y., Y. Niizuma, G. W. Gabrielsen, K. Sato, and Y. Naito. 2003. Stroke and glide of wing-propelled divers: deep diving seabirds adjust surge frequency to buoyancy change with depth. *Proceedings: Biological Sciences* 270:483–488.

Whittow, G. C. 1987. Thermoregulatory adaptations in marine mammals: interacting effects of exercise and body mass. A review. *Marine Mammal Science* 3:220–241.

Wikelski, M. 1999. Influences of parasites and thermoregulation on grouping tendencies in marine iguanas. *Behavioral Ecology* 10:22–29.

Wilson, R. P., Y. Ropert-Coudert, and A. Kato. 2002. Rush and grab strategies in foraging marine endotherms: the case for haste in penguins. *Animal Behaviour* 63:85–95.

Winn, H. E., J. D. Goodyear, R. D. Kenney, and R. O. Petricig. 1995. Dive patterns of tagged right whales in the Great South Channel. *Continental Shelf Research* 15:593–611.

Withers, P. C., G. Morrison, and M. Guppy. 1994. Buoyancy role of urea and TMAO in an elasmobranch fish, the Port Jackson Shark, *Heterodontus portusjacksoni*. *Physiological Zoology* 67:693–705.

Withers, P. C., G. Morrison, G. T. Hefter, and T.-S. Pang. 1994. Role of urea and methylamines in buoyancy of elasmobranchs. *The Journal of Experimental Biology* 188:175–189.

Yancey, P. H., and G. N. Somero. 1980. Methylamine osmoregulation solutes of elasmobranch fishes counteract urea inhibition of enzymes. *The Journal of Experimental Zoology* 212:205–213.

CHAPTER
7

Coastal wetlands are critical habitats for birds.

Estuaries

Estuaries are semienclosed coastal embayments where freshwater rivers meet the sea. Here fresh water and seawater mix, creating unique and complex ecosystems. Familiar places, such as the Chesapeake Bay, San Francisco Bay, Great South Bay, Tampa Bay, Puget Sound, and the Mississippi River Delta, are among the 100 or so bodies of water officially designated as estuaries in the United States. Over one third of the US population lives within the drainage basins of our estuaries. More than any other ecosystem discussed in this book, these transitional coastal habitats encapsulate the problems and challenges created by human intervention into the workings and the very structures of marine ecosystems.

Estuaries are highly variable ecosystems that continually change in response to local physical, geologic, chemical, and biologic factors. The transition from fresh water to salt water in estuaries of large rivers like the Columbia River may extend over 100 km inland, whereas the estuaries of small streams may be only a few hundred meters in extent, and the water in them may be well-mixed, highly stratified, or any combination in between. The size and shape of an estuary is influenced by its depth and geologic history. Geologic movement of the Earth's crust has elevated and lowered coastal areas, and the formation and melting of Ice Age glaciers has alternately removed and returned water from the ocean basins. The resulting changes in sea level alter the size and shape of estuaries by altering the water depth and the extent of submerged coastal features.

Physical forces at work also influence the chemistry of an estuary. When fresh water draining from a coastal watershed mixes with the seawater pushing upstream during high tides, suspended river sediments settle to the bottom and become part of the accumulating sediment blanket of the estuary. Because many pollutants are transported downstream in the river water or are adsorbed onto sediment particles, individual estuarine conditions also influence the fate and availability of pollutants to the inhabitants of the ecosystem.

In their natural states, estuaries are among the most biologically productive ecosystems on Earth. Their rates of primary productivity rival and often exceed those of coral reefs, rain forests, and even intensively cultivated corn fields. These special habitats are created by the combination of turbulent mixing, daily fluctuating tidal cycles, and the downstream flow of fresh water that usually changes seasonally in velocity and volume. When these forces meet in an estuary, they exert considerable and complicated effects on the system, creating diverse aquatic habitats atypical of either the river or the sea. More than two thirds of the species of fish and shellfish harvested by commercial and sports fishers depend on estuaries for feeding or as nursery areas. Estuaries also provide critical habitat for terrestrial and freshwater organisms, including many threatened, endangered, and rare species.

Estuaries and their surrounding wetlands are recognized as fragile environments that have been heavily used for and disturbed by human activities. Dredging of navigation channels in estuarine ports, filling estuarine wetlands for development, disposing of wastewaters from coastal communities, diverting rivers for irrigation purposes, and allowing pesticide- and fertilizer-contaminated runoff to flow into coastal watersheds have changed the character of estuaries and threatened their ecologic integrity. Many estuaries that were once rich sources of fish, game, and shellfish have become stagnant and unproductive as a result of unregulated or poorly regulated economic exploitation and pollution. Restoration and enhancement efforts are underway to reverse some of the environmental degradation and biologic devastation of some of the world's major estuaries. Unfortunately, the double-edged sword of runaway human population growth coupled with a rapid increase in per capita human consumption make these efforts a truly daunting challenge.

7.1 Types of Estuaries

Most estuaries owe their present configuration to ancient patterns of river or glacial erosion that occurred during the last glacial maximum (LGM), when sea level worldwide was about 150 m lower than at present. These scoured river or glacial channels slowly assumed their present configuration as the great continental ice sheets melted and gradually flooded them. Some estuaries remain sensitive to slight changes in sea level; increases of only a few meters could dramatically increase the size of small estuaries (FIG. 7.1), and comparable decreases could shrink others or even cause them to disappear.

Estuaries are found in some form along most coastlines of the world, but most are evident in wetter climates of temperate and tropical latitudes. In such areas, drainage of inland watersheds provides the necessary freshwater input at the head of the estuary to keep salinities below those of adjacent open ocean waters. In North America, excellent examples of all major types of estuaries exist. Estuaries are classified by both their modes of formation and by their patterns of water circulations. **Coastal plain estuaries**, also known as drowned river valleys (FIG. 7.2a), lie along the north and central Atlantic Coast, the Canadian

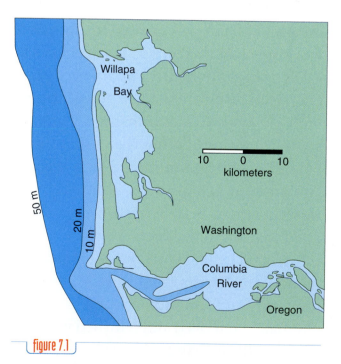

figure 7.1

Columbia River estuary and Willapa Bay with the present shoreline (green) and depth contours at 10, 20, and 50 m below.

The Environmental Protection Agency estimates that 24,000 acres of wetlands (marshes, bogs, swamps, and fens) are lost each year as a result of thousands of activities permitted through section 404 of the Clean Water Act. The U.S. Fish and Wildlife Service puts that estimate much higher (at 400,00–500,000 acres per year).

Wetlands are important to humans for flood control during times of heavy rainfall, soaking up and storing excess water. Moreover, as the heavy rainfall runs off the land, it transports sediments that normally settle in the wetland, thus preventing the clogging of streams, lakes, or rivers. In addition to soils and other sediments, runoff also contains pesticides, fertilizers, bacteria, road salts during winter, and other chemicals that wash from our roads and highways. Wetlands provide a filtrating process that improves the quality of the runoff water, an important source of water that can recharge ground water during dry periods. Finally, wetlands provide many pleasurable activities, such as fishing, hunting, wildlife observation and photography, and canoeing.

Wetlands may be most important as habitats for wildlife. In the United States alone, about 5000 plant species, 190 species of amphibians and reptiles, one third of our birds, and 35% of threatened and endangered species depend on wetlands for their survival. Many of these species live in wetlands throughout the year; others use them temporarily during essential portions of their life cycle (such as migration stops or nursery areas).

It might surprise you to learn that many species of sharks, often viewed as large open-oceanic predators, rely on shallow inshore areas, including wetlands and their associated tidal creeks, as their nursery grounds. For example, at least nine species of large sharks are known to use Bulls Bay, South Carolina, as a nursery area. Pregnant female blacktip, sandbar, dusky, smooth hammerhead, and spinner sharks, among others, enter the bay in the spring of each year to give live birth to their pups. Soon afterward, the large adults depart, but the juvenile sharks often remain in the shallow protected waters of the nursery for periods of up to 5 years.

Lemon sharks have been studied extensively in southern Florida, the Bahamian islands, and small islands off the coast of Brazil. Female lemon sharks arrive each May to drop 8–16 half-meter-long pups along the mangrove swamps that fringe most Caribbean shorelines. For the next 2–3 years, these young sharks remain in shallow tidal channels, swimming among thickets of mangrove roots while feeding on an assortment of fishes and spending 75% of their time in water no more than 50 cm deep.

Our understanding of the importance of wetlands to a great variety of marine species, including even large pelagic predators such as sharks, has been acquired in just the past 10 years and has resulted in a fundamental shift in our efforts to conserve many of these species. The Sustainable Fisheries Act of 1996 requires that regional fishery management councils and the Secretary of Commerce identify and describe Essential Fish Habitat (EFH) for species under federal Fishery Management Plans (FMP). EFH is defined in the Act as "those waters and substrate necessary to fish for spawning, breeding, feeding, or growth to maturity" (wherein "fish" includes finfishes, sharks, crabs, shrimp, lobsters, and mollusks). Thirty-nine species of sharks along the east coast of the United States have been managed since 1993 by an FMP. Therefore, EFH for these sharks must be identified and described by law. Imagine the surprise felt by many biologists and politicians when it was revealed that shallow coastal wetlands constituted an EFH for many species of large sharks. We now appreciate that loss of wetlands will result in the destruction of populations of large sharks living offshore. Many investigations into the utilization of wetlands by sharks as nursery areas are currently underway.

A Surprising Result of Wetland Loss

Additional Reading

Castro, J. I. 1993. The shark nursery of Bulls Bay, South Carolina, with a review of the shark nurseries of the southeastern coast of the United States. *Environmental Biology of Fishes* 38(1–3):37–48.

www.bioscience.jbpub.com/ marinebiology
For more information on this topic, go to this book's website at http://www.bioscience.jbpub.com/ marinebiology.

(a)

(c)

(b)

figure 7.2

Satellite images of three types of estuaries. (a) Chesapeake and Delaware Bays, two coastal plain estuaries on the US East Coast. (b) Several bar-built estuaries of the Texas coast. (c) Steep-sided fjords on the southwestern coast of Norway.

Maritime region, and many areas of the West Coast of North America; this type of estuary includes the Columbia River and Chesapeake and Delaware Bays. These estuaries are broad shallow embayments formed from deeper V-shaped channels as the sea level rose and flooded river mouths after the last episode of continental glaciation. The extent to which the sea has invaded these coastal river valleys since the LGM is determined by the steepness and size of the valley, its rate of river discharge, and the range and force of the tides of the adjacent sea. This type of estuary continues to be gradually modified as wave erosion cuts away some existing shorelines and creates others by building mudflats.

Bar-built estuaries are common along the south Atlantic Coast, the Gulf of Mexico (Fig. 7.2b) in North America, and along the coastal lowlands of northwestern Europe. These estuaries are formed as nearshore deposits of sand and mud are transported by coastal wave action to build an obstruction, or **barrier island**, in front of a coastal area fed by one or more coastal streams or rivers. Often, these small coastal rivers and streams have little freshwater flow so the estuary may be partially or completely blocked by sand deposited by ocean waves. During rainy seasons, however, the increased runoff often temporarily reopens the estuary mouth. Coastal **lagoons** (Fig. 7.2b) are similar to bar-built estuaries, but without a river or other source of fresh water input, coastal lagoons lack the strong salinity gradients and mixing patterns characteristic of estuaries.

Some estuaries have broad, poorly defined, fan-shaped mouths called **deltas.** The Mississippi and Colorado River deltas and other similar delta estuaries are created as heavy loads of sediments eroded from the upstream watersheds are deposited at the river mouth. Different still are **tectonic estuaries** such as San Francisco Bay, created when the underlying land sank due to crustal movements of the Earth. As coastal depressions created by these movements sank below sea level, they filled with water from the sea and also became natural land drainage channels, directing the flow of land runoff into the new estuary basin.

From the central West Coast northward, estuaries become more deeply incised into coastal landforms and gradually merge into the deep glacially carved **fjords** characteristic of British Columbia and southeastern Alaska (Fig. 7.2c). Fjords are also common on the coasts of Norway, southern New Zealand, and southern Chile. In cross-section, fjords resemble the Black Sea, with deeper regions at the upstream reaches of fjords. The shallow sills at their mouths partially block the inflow of seawater and lead to stagnant conditions near the bottoms of deeper fjords (see Fig. 1.45).

7.2 Estuarine Circulation

In addition to their structural differences, estuaries also exhibit differing patterns of freshwater and seawater mixing within the basin. The upstream-to-downstream variations in salinity, water temperature, turbidity, and current action are complex and change markedly during a single tidal cycle and in response to seasonal changes in the volume of freshwater stream discharge. Salinity values typically increase from the surface downward and from the estuary head downstream toward the mouth. As tides change sea level in a typical estuary, higher density seawater moves in and out along the estuary bottom and is gradually mixed upward into the outflowing low-salinity surface water (FIG. 7.3). Because of this mixing, there is an inward flow of nutrient-rich seawater along the bottom of the estuary and a net outward flow at the surface. On a localized scale, this upward mixing creates a process of estuarine upwelling that replenishes nutrients and promotes the growth of estuarine primary producers.

The shape of an estuary's basin is a major factor in determining its mixing pattern. A triangular estuary with a wide deep mouth allows seawater to move farther upstream. The currents may be strong and the water is well mixed, so salinity and water density are nearly the same from the surface downward at any location within the estuary (FIG. 7.4a). In narrow-mouthed estuaries, circulation is decreased, creating more pronounced vertical salinity gradients. These narrow-mouthed estuaries are often highly stratified. They have a well-defined seawater wedge under the less dense fresh water on the surface (Fig. 7.4b), with an abrupt salinity change, or halocline, where seawater and fresh water meet. By analyzing water samples taken at fixed depths throughout a tidal cycle and connecting points with the same salinity values, lines of equal salinity, called **isohalines,** can be plotted. The shape of the isohalines is useful in classifying types of estuaries and in understanding the distribution of estuarine organisms.

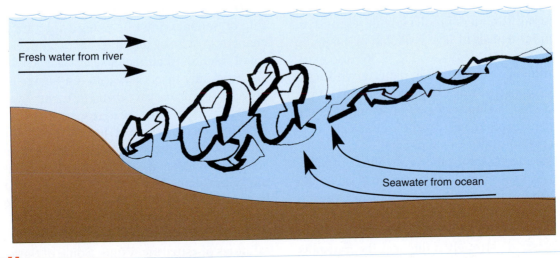

figure 7.3

The general pattern of freshwater and seawater mixing in an estuary.

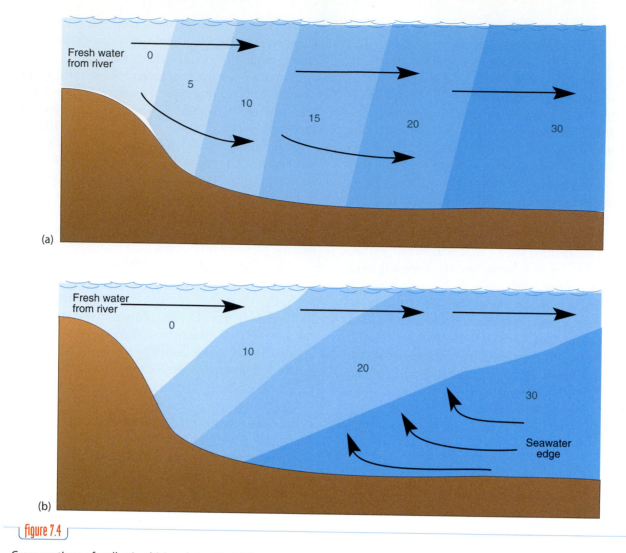

figure 7.4

Cross-sections of well-mixed (a) and stratified (b) estuaries, with the resulting salinities shown.

In addition to the more predictable effects of tides and river discharge, circulation in estuaries often changes rapidly and less predictably in response to short-term influences of heavy rainfall or changing winds. The Coriolis effect (see Fig. 1.36) also exercises its influence on circulation patterns within estuaries by forcing seawater farther upstream on the left sides (when facing seaward) of estuaries in the Northern Hemisphere and on the right sides of estuaries in the Southern Hemisphere.

The time necessary for the total volume of water in an estuary to be completely replaced is called the **flushing time.** Flushing times range from days to years depending on the estuary's combination of tides, river flow, wind, and salinity gradients. The flushing time in an estuary strongly influences the transport of nutrients and is also a crucial factor in determining the fates of pollutants in estuaries.

7.3 Salinity Adaptations

To survive in most estuarine conditions, benthic organisms must be able to tolerate frequent changes in salinity and internal osmotic stresses. A small fraction of animal species that live in estuaries, especially insect larvae, a few snails, and some polychaete worms, have their closest relatives in fresh water; however, most are derived from marine forms (FIG. 7.5) and include some of the same species found on nearby beaches. Some animal species have poorly developed osmoregulatory capabilities and avoid osmotic problems by not venturing too far into the low-salinity portions of estuaries. Others use a number of adaptive strategies to overcome the osmotic problems of recurring exposure to low and variable salinities of estuarine waters. Some of these adaptations are modifications of structural or physical systems already imperative for survival on exposed intertidal shorelines. Oysters and other bivalve mollusks, for instance, simply stop feeding and close their shells when subjected to the osmotic stresses of low-salinity water. Isolated within their shells, they switch to anaerobic respiration and await high tide, when water higher in salinity and O_2 returns. Other animal species retreat into mud burrows, where salinity fluctuations due to tidal cycles are usually much less severe (FIG. 7.6).

A few species of tunicates, anemones, and other soft-bodied estuarine epifauna are **osmotic conformers**. Osmotic conformers are unable to control the osmotic flooding of their tissues when subjected to low salinities, so their body fluids fluctuate to remain isotonic with the water around them (FIG. 7.7).

The most successful and abundant groups of estuarine animals have evolved mechanisms to stabilize the water and ion concentrations of their body fluids despite external variations. These mechanisms are as varied as the organisms themselves, yet all involve systems that acquire essential ions from the external medium and excrete excess water as it diffuses into their bodies. The body fluids of estuarine crabs remain nearly isotonic with their external medium when in seawater but become progressively hypertonic as the seawater becomes more dilute. When these partial osmotic regulators are subjected to reduced salinities, additional ions are actively absorbed by their gills to compensate for the ions lost in their urine (Fig. 7.7). Thus, these and most other estuarine crustaceans are osmotic conformers at or near normal seawater salinities and **osmoregulators** in dilute seawater.

Most estuarine animals are **stenohaline**; they tolerate exposure to limited salinity ranges and therefore occupy only a limited portion of the entire range of salinity regimes available within an estuary (FIG. 7.8).

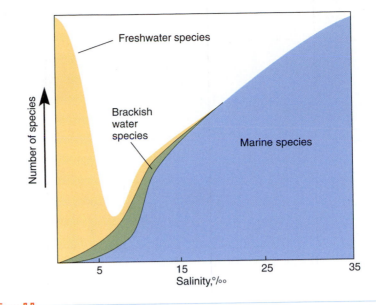

figure 7.5

Relative contributions of freshwater, brackish water, and marine species to estuarine fauna. (Redrawn from Remane 1934.)

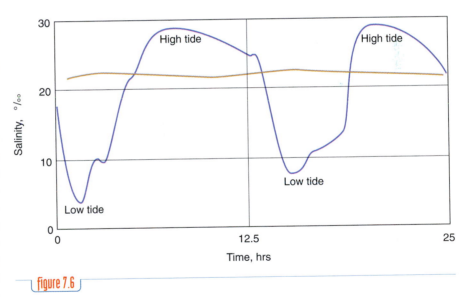

figure 7.6

Comparison of salinity variations through a typical tidal cycle of interstitial water (brown) with that of the overlying water (blue) in Pocasset Estuary, Massachusetts. (Adapted from Mangelsdorf 1967.)

A few opportunistic species of estuarine organisms are **euryhaline**, capable of withstanding a wide range of salinities. These species can be found throughout the range of estuarine salinities, with a limited number of euryhaline species also found in high salinity lagoons that fringe some of the world's arid coastlines. Lagoons such as those along the coast of Texas and both sides of northern Mexico have shallow bottoms, high summer temperatures, excessive evaporation, and high salinities. The osmotic problems experienced

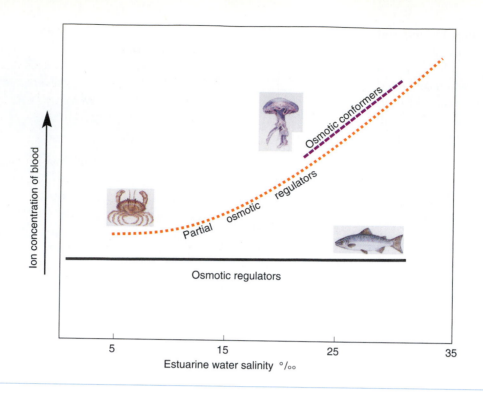

figure 7.7

Variations in ion concentrations of body fluids or blood with changing external water salinities for osmotic conformers, partial osmotic regulators, and osmotic regulators.

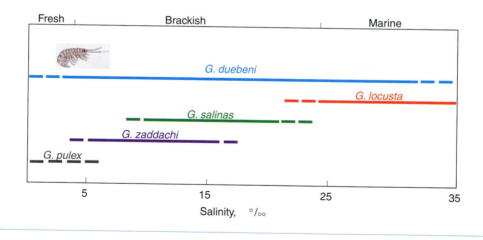

figure 7.8

Differing salinity tolerances of five species of amphipods (*Gammarus*). Of these, only *G. duebeni* is euryhaline. (Adapted from Nicol 1967.)

by animal species in these high-salinity lagoon populations are similar to those encountered by bony fish in seawater and are so severe that reproduction is seldom successful. Continued immigration of euryhaline species from nearby estuaries sustains these lagoon populations.

Species diversity and numbers of individuals usually decline considerably from a maximum near the ocean to a minimum near the headwaters of an estuary. The distributional patterns of estuarine animals are governed by salinity variations, patterns of food and sediment preferences, current action, water temperature variations, and competition between species. It is the collective interaction of all these factors that establishes and maintains the distribution limits of estuarine organisms.

7.4 Sediment Transport: Creating Habitats

Sediments are transported into estuaries from rivers that drain coastal watersheds and from coastal areas outside the estuary mouth. River sediment particles range in size from gravels and coarse sands to fine silts, clays, and organic detritus (see Fig. 8.2). They are derived from erosion of river banks denuded of their natural plant cover and from the scouring of meandering river channels. As fast-moving rivers widen and slow when they enter coastal flood plains, they begin to meander and their loads of suspended sediments settle to the bottom. Estuaries thus serve as effective catch basins for much of the fine suspended sediments washed off the land. The current speed necessary to keep the sediment load suspended diminishes in the protected and quiet waters of estuaries and the water slows to a point where only the finest silts and clays remain suspended in the water.

Storms and near-shore currents of the open ocean can also move coastal sand and detritus materials into the mouth of an estuary and add to the complex mix of estuarine sediments. Typically, these deposits show a characteristic distribution of different sediment types, with coarse particles deposited at the heads of estuaries and in shallow water, whereas finer particles settle nearer the mouth and in deeper water. These graded and sorted sediment deposits transported down from rivers and in from the sea provide a rich and varying substrate to support the estuarine communities.

7.5 Estuarine Habitats and Communities

Estuaries on both coasts of North America and in other areas of the world are ecologically critical areas that support a wide variety of biologic communities and serve as vital resting and feeding stops within the migratory flyways of ducks, geese, bald eagles, and many species of shorebirds. Salt marshes, in particular, are important natural filters to trap pollutants as resident bacteria convert some of them to less harmful substances. They also play a role in moderating flooding and sedimentation processes.

Estuarine communities include wetlands, mudflats, and channels. The areas of highest elevation are the wetlands; they are periodically covered by estuarine water at high tides and consist of dense plant communities that tolerate contact with seawater. Mudflats, or tideflats, are lower in elevation than the wetlands and are alternately submerged and exposed by changing tides. Channels, those areas that are under water even at the lowest tides, are prevented from filling with sediments by the scouring action of tides or river flow.

Salt Marshes

Salt marshes are essentially wet grasslands that grow along estuarine shores. The dominant members of these marshes are **halophytes** (FIG. 7.9), a few species of plants that require seawater or at least are tolerant to it. Salt marshes develop in the muddy deposits around the edges of temperate and subpolar estuaries, creating a transition zone between land and other estuarine communities. Salt marshes are inhabited by several plant species, each with its own specific set of sediment, water, and exposure requirements (FIG. 7.10). The lowest parts of a typical salt marsh, submerged for longer periods of time, are dominated by pickleweed, *Salicornia,* which stores excess salt in its fleshy leaves, and by salt grass, *Spartina,* which has special glands that excrete excess salt. In higher marsh zones, grasses, rushes, and sedges that cannot tolerate prolonged submersion by the tides dominate the landscape.

Salt marshes form an important part of the base of estuarine food webs. Some of the plants in the estuary are eaten directly by marsh herbivores, but most of the vegetation decays and enters estuarine food webs as detritus. The flooding and ebbing of the tides wash detritus from the marsh into the estuary and surrounding tidal creeks, where the detritus sinks and decomposes further. Each winter, salt marsh plants die back and tides (or, in cold climates, the shearing action of rising and falling tidal ice) harvest this grass and put it into the detrital food chain. There it becomes the target of decomposing bacteria. The microbial activities of the bacteria further break down the plant matter, especially the cellulose cell walls, which are undigestible by most estuarine animals, convert some of it to additional bacterial cells, and release dissolved organic materials and inorganic nutrients to be reused by other plants into the estuary.

In estuaries, bacteria and the phytoplankton of the overlying waters contribute heavily to the production of small, energy-rich, detrital food particles. These microorganisms then become a major source of food for large populations of estuarine particle consumers (FIG. 7.11). The particle consumers themselves produce still more food particles in the form of feces

(a)

(b)

figure 7.9

Two types of emergent salt marsh plants. (a) A stand of salt grass, *Spartina,* with taller mangroves behind. (b) Pickleweed, *Salicornia.*

and rejected food items. These particles are eventually recolonized by bacteria and recycled into the particle pool of the estuary. In this manner, estuarine bacteria and other microorganisms play a central role in transforming the productivity of estuarine margins into small detrital food particles available to numerous other species of estuarine animals.

Mudflats

Mudflats are estuarine expanses composed primarily of rich muds that are exposed at low tide. Only in these protected coastal environments can significant amounts of finer silt and clay particles settle out. Mudflats contain some sand, but the sand is mixed with varying amounts of finer silt and clay particles to produce mud. Where marine waters and rivers mix and salinity gradients are large, dissolved ions interact with the sediments and bind together to form larger particles and add to the accumulating richness of the bottom muds. These unstable soft mud deposits serve as the principal structural foundation of soft-bottom communities that thrive in estuaries.

Three groups of primary producers are found on mudflats: diatoms, larger algae, and seagrasses. Microscopic benthic diatoms coat the mud surfaces with a golden brown film. These photosynthesizers are a rich and important food source for benthic invertebrates. Green mats of macroscopic algae, such as sea lettuce, commonly cover rocks, shells, and pieces of wood debris on mudflats. These are important food sources for herbivores, especially certain worms, amphipods, and crabs.

Seagrasses are found at lower levels of mudflats. Eelgrass and a few other seagrasses comprise one of the few types of flowering plants that can survive completely submerged in salt water (FIG. 7.12). Eelgrass gets its name from the long (up to 2 m), thin, straplike leaves that weave back and forth in the currents. Eelgrass production contributes greatly to the pool of detrital particles within an estuary. In its detrital form, it is consumed by ducks and geese, invertebrates, fish, and larval stages of insects. Algae and diatoms grow on its leaves, as do many types of hydroids, clam larvae, tunicates, ectoprocts, and crustaceans. In addition to their roles in detritus food webs, seagrasses are usually the initial plants to stabilize shallow mudflats, with their roots and long leaves trapping even more fine particulate materials to add to this food-rich protected habitat. Finally, seagrasses also serve as nutrient pumps by taking nutrients from the sediment for growth and later releasing them to the water when they die at the end of the growing season.

In fine-grained muds, sediment particles pack together so tightly that little water can percolate through. Oxygen used by mud dwellers is not rapidly replenished, and their wastes are not quickly removed. Fine-grained muds with small interstitial spaces between sediment particles are effective traps for particles of organic debris. Much of the accumulated

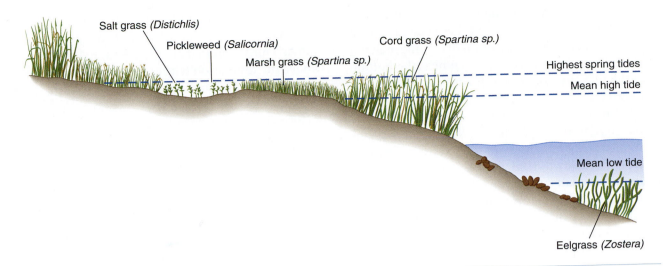

Salt grass (*Distichlis*)
Pickleweed (*Salicornia*)
Marsh grass (*Spartina sp.*)
Cord grass (*Spartina sp.*)
Highest spring tides
Mean high tide
Mean low tide
Eelgrass (*Zostera*)

figure 7.10

Plant-dominated salt marsh, mudflat, and channel habitats of Chesapeake Bay and other East Coast estuaries, with their vertical position relative to high tide indicated.

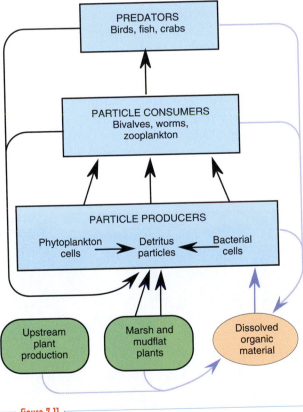

PREDATORS
Birds, fish, crabs

PARTICLE CONSUMERS
Bivalves, worms,
zooplankton

PARTICLE PRODUCERS
Phytoplankton cells → Detritus particles ← Bacterial cells

Upstream plant production

Marsh and mudflat plants

Dissolved organic material

figure 7.11

Food particle production and utilization in a typical estuary. (Adapted from Correll 1978.)

figure 7.12

A bed of eelgrass, *Zostera,* at low tide.

organic material is found in a thin, brownish, oxygenated surface layer about 1 cm thick. Aerobic decomposers dominate the surface oxygenated layer of mud, but their numbers decline rapidly with depth.

Beneath the oxygenated layer, anaerobes are active down to 40 to 60 cm, where their numbers also dwindle rapidly. The overwhelming abundance of both aerobic and anaerobic decomposers is responsible for most of the chemical changes that occur in estuarine sediments. The results of these chemical reactions include the decomposition of organic material, consumption of dissolved oxygen near the bottom, and the recycling of critical plant nutrients back to the water.

Below the thin oxygenated surface layer, the organic content of the muds usually decreases as animals and decomposing bacteria and fungi consume it. Respiration by the inhabitants of mudflats further reduces the available dissolved O_2 supply of the interstitial waters. The lower limit of O_2 penetration in organic-rich sediments is usually apparent as a color change, from a light color in the oxygenated surface layer to a dark or even black color in the anaerobic zone below. The anaerobic conditions of deeper muds inhibit, but do not halt, decomposition of the organic material.

Bacteria and fungi are the major groups of marine organisms capable of using the rich organic accumulations in the anaerobic portion of muddy sediments. Without O_2, these anaerobic decomposers must use other available elements for their respiratory processes. Sulfate, the third most abundant ion in seawater, is commonly reduced to hydrogen sulfide (H_2S), the gas responsible for the memorable rotten-egg odor and black color so characteristic of anaerobic estuarine muds.

Other types of sediment dwellers, such as clams and mud shrimp, burrow in the muds to seek protection from predators, to be sheltered from the drying effects of the sun at low tide, or to occupy an environment where the salinity of the mud is more constant than that of the overlying water (Figs. 7.6 and 7.13).

Animals living on or in mudflats have developed a variety of feeding habits. Many of the near-surface infauna are suspension feeders, gleaning small food particles from water currents above their burrows. Others, such as lug worms, are deposit feeders, digesting the bacterial and organic coatings of sediment particles passing through their digestive system. Other types of burrowing animals, such as the arrow goby, leave their burrows at high tide and forage for food over the mudflat.

The epifauna of mudflats are dominated by mobile species of gastropod mollusks, crustaceans, and polychaete worms. These organisms sometimes range over a wide area of the mudflat and demonstrate only blurred weakly established patterns of lateral zonation. Mud-dwelling organisms, however, do occupy vertically arranged zones in the sediment. A few cen-

timeters below the mud surface, the interstitial water is generally devoid of available oxygen, and the infauna must obtain their oxygen from the water just above the mud or do without. The numerous openings of tubes and burrows on the surfaces of estuarine mudflats (Fig. 7.13) attest to an unseen wealth of animal life underneath. Bivalve mollusks extend tubular siphons through the anaerobic mud to the oxygenated water above. The depth to which these animals can seek protection in the mud is limited largely by the lengths of their siphons and thus, indirectly, by their ages. Other infauna use the sticky consistency of fine-grained organically rich muds to construct permanent burrows with connections to the surface.

When the tide is out, the infauna of muddy shores must also cope with an absence of available O_2 and with changing air temperatures. Some switch from aerobic to anaerobic respiration. In doing so, many encounter a dilemma in trying to match the relative inefficiency of energy-yielding anaerobic respiration with the increased energy demands forced by increasing tissue temperatures and higher metabolic rates. Larger infauna exist anaerobically only temporarily and revert to aerobic respiration as soon as they are covered by the tides. But for many of their smaller burrowing neighbors with no direct access to the O_2-laden waters above, anaerobic respiration is a permanent feature of their infaunal existence in intertidal muds.

figure 7.13

Barren surface of a mudflat, with tubes, openings, burrows, and other evidence of abundant animal life beneath the surface.

At high tide, submerged mudflats are visited by shore crabs, shrimps, fishes, and other transients from deeper water. Some come to forage for food; others find the protected waters ideal for spawning. Their forays are only temporary, however, because they leave with the ebbing tide and surrender the mudflats to shorebirds and a few species of foraging coastal mammals. During low water, the long-billed curlews and whimbrels probe for the deeper infauna, whereas sandpipers and other short-billed shorebirds concentrate on the shallow infauna and small epifauna (see Fig. 6.28).

Channels

A channel is the part of an estuary that is filled with water under all tidal conditions (extreme right side of Fig. 7.10). A channel may be as broad as the entire estuary or it may be restricted to a narrow creek-like feature between mudflats. Numerous species of planktonic organisms inhabit channels, relying on the action of currents to move them around. Crabs, oysters, starry flounders, sculpins, anchovies, and killifish are also abundant in channels during certain times of the year.

Channel areas are also used as spawning and nursery areas for many animals. Herring and sole are ocean fish that move into the protected areas of the estuary to spawn and to allow the juveniles to feed on food available there. Crabs also use the estuaries as nursery areas. Anadromous fish such as salmon and shad may linger for a time in estuarine channels to feed when migrating between the ocean and their spawning areas in freshwater streams.

7.6 Environmental Pollutants

Throughout this book, the assumption is made that pristine ocean waters are essential for maintaining healthy marine communities. This assumption is being put to a global test as our growing human population generates an enormous and increasing burden of domestic and industrial wastes. Initially, these wastes may be dumped down sewers or up smokestacks but ultimately many make their way into estuaries on their way to the ocean (FIG. 7.14). The world ocean has a large but finite capacity to assimilate these waste materials without apparent degradation of water quality. However, that capacity is often

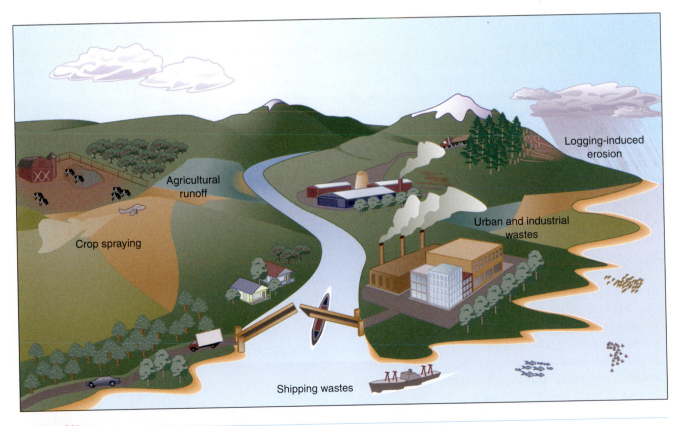

figure 7.14

Common sources of pollutants entering estuaries.

Table 7.1

A Summary of the Sources and Effects of Some Marine Pollutants

Pollutant	Sources	Effects
Particulate material	Dredged material, sewage, erosion	Smothers benthic organisms, clogs gills and filters, reduces underwater light
Dissolved nutrients	Sewage agricultural runoff	Increases phytoplankton blooms, decreases dissolved oxygen
Toxins	Pesticides, industrial wastes, oil spills, antifouling paint	Increases incidence of disease, contaminates seafood, suppresses immune systems, contributes to reproductive failure
Oil	Tankers, drill sites, urban and industrial waste	Smothers organisms, clogs gills, mats fur or feathers, causes anatomic and physiologic abnormalities
Marine debris	Garbage, ship wastes, fishing gear	Causes physical injuries and mutilations, increases mortality

exceeded in the semienclosed conditions of estuaries where mixing processes are not sufficient to dilute or disperse wastes, creating localized water quality problems and subsequent biologic disturbances.

Waste materials discharged into estuarine waters are considered pollutants if they have measurable adverse effects on natural populations. Persistent contaminants such as heavy metals, pesticides, radioactive wastes, and petroleum products head the list of substances (TABLE 7.1) that, even at low concentrations, can adversely affect the health of marine organisms and the integrity of their natural ecologic relationships because they accumulate in marine organisms, and their concentrations can be magnified as they are transferred up food chains.

Municipal and industrial wastewater discharges (including storm drain and sewer overflows) are considered **point sources** of pollutants. Urban runoff and land-based agriculture and forest harvest activities contribute to **nonpoint sources** of estuarine pollutants. It is paradoxical that some of the worst pollutants in estuaries are the very pesticide and fertilizer products initially used to increase food production on land and that some of the most obvious disturbances appear in marine species harvested for human consumption.

Oxygen-depleting Pollutants

Excessive amounts of organic materials or fertilizers from agricultural runoff and from sewage outfalls contain large quantities of nitrogen compounds, a common limiting nutrient for phytoplankton. Discharged into semienclosed estuaries, these nutrients promote increased phytoplankton production and blooms of toxic dinophytes such as *Pfiesteria*. These high concentrations of phytoplankton eventually die and sink to the bottom. Decomposition of the excess

phytoplankton biomass by bacteria create a high **biochemical oxygen demand** (BOD) and can reduce an estuary's reservoir of life-supporting oxygen.

When an estuary's BOD is high and multicellular organisms cannot survive, seasonal or permanent "dead zones" occur. Dead zones now occur in Long Island Sound, Chesapeake Bay, and nearly 50 other estuarine or very near-shore locations around the world, mostly in the Northern Hemisphere. One of the largest dead zones is located just outside the mouth of the Mississippi River (FIG. 7.15). The extent of this dead zone approximately doubled during the decade of the 1990s and reaches its greatest extent in spring and summer months. From a few meters below the surface to a depth of about 60 m, fish and mobile invertebrates move away from the anoxic water, whereas attached and burrowing species are killed. As this dead zone continues to grow in size, it imposes an increasing threat to the commercially valuable fish and shrimp populations living there.

Toxic Pollutants

We do not yet know how to determine the extent or fate of many toxic substances in the marine environment or precisely how to evaluate their effects on marine life. Some of the better known trace metals and toxic chemicals include mercury, copper, lead, and chlorinated hydrocarbons. The most common chlorinated hydrocarbons are synthetic chlorine-containing compounds created for use as pesticides or are byproducts of the manufacture of plastics. They are among the most persistent and harmful of all toxic substances and include well-known products such as chlordane, lindane, heptachlor, dichloro-diphenyl-trichloroethane (DDT), dioxins, and polychlorinated biphenyls (PCBs).

figure 7.15

Sea-viewing wide field-of-view sensor (SeaWIFs) satellite view of the US Gulf Coast, with the dead zone at the mouth of the Mississippi River indicated by black.

DDT

By virtue of its long and widespread use, DDT and its effects on marine life can serve as a model for the behavior of other persistent toxic substances in estuarine and marine systems. DDT was the first of a new class of synthetic chlorinated hydrocarbons. It became available for public use in 1945 and quickly gained international acceptance as an effective killer of most serious insect threats: house flies, lice, mosquitoes, and several crop pests.

DDT is a persistent pesticide; it does not break down or lose its toxicity rapidly. Once in seawater, DDT is rapidly adsorbed by suspended particles. Because it is nearly insoluble in water, measurable levels of DDT in seawater practically never occur. Yet DDT contamination from land and air has been so pervasive that it can be found in nearly all parts of the world ocean. Antarctic penguins, arctic seals, Bermuda petrels, and fishes everywhere have measurable accumulations of DDT in their fatty tissues.

Although DDT is insoluble in water, it is quite soluble in lipids. Fatty tissues and oil droplets concentrate the DDT adsorbed on suspended particles in seawater. Phytoplankton and, to a much lesser extent, zooplankton are the initial steps in DDT's entry into marine food webs. Fishes, birds, and other predators eventually consume this plankton and its load of DDT, concentrating the toxin in their fatty tissues. At each step in the food web, further concentration, or **bioaccumulation**, occurs (FIG. 7.16), eventually reaching the top carnivores.

Marine birds suffered the most devastating effects of DDT poisoning. As fish-eating predators, bald eagles, ospreys, pelicans, and other species are sometimes four or five trophic levels removed from the phytoplankton that initially absorb the DDT. The bioaccumulation of DDT that occurred at each trophic level assured these predatory birds of high DDT loads in their food. DDT and its residues block normal nerve functions in vertebrates. DDT also interferes with calcium deposition during the formation of eggshells. The eggshells of birds with high DDT loads were very thin and fragile. They frequently broke when laid or failed to support the weight of adult birds during incubation. The broken eggs lay in abandoned nests, mute testimony to the insidious effects of DDT.

The United States banned the general use of DDT in 1972, yet DDT is still being used in other parts of the world, each year adding to the load already existing in estuarine and marine organisms. Animals from walruses in the Arctic to penguins in the Antarctic show elevated concentrations of DDT, particularly in their blubber and other fatty tissues. By itself, DDT is a threat to the continued existence of several species of marine animals. But it is only symptomatic of the greater danger posed by the many other persistent toxins of which we know even less.

Dioxins

Dioxins are another group of chlorinated compounds gaining international notoriety. The most potent dioxin, 2,3,7,8-tetrachlorodibenzo-p-dioxin (TCDD) is toxic to birds and aquatic life at concentrations as low as a few parts per quadrillion and is considered

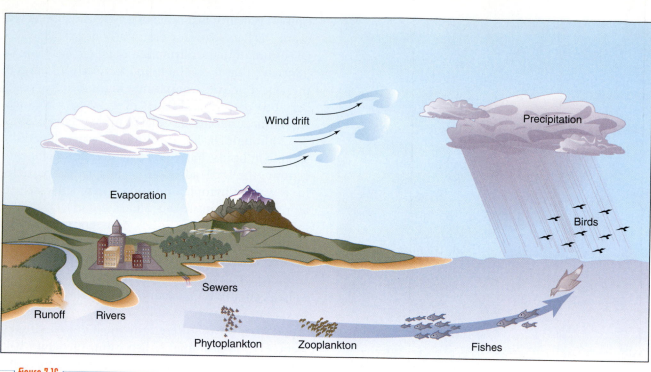

figure 7.16

Transfer of DDT to and within marine food webs (arrows). DDT is absorbed by phytoplankton and then concentrated at each step in the food web. (Adapted from Epel and Lee 1970.)

by the World Health Organization to be a class I, or known, human carcinogen. Like DDT, dioxins are fat soluble and stable, they bioaccumulate as readily as DDT, and they are suspected of causing developmental malformations and immune system and reproductive difficulties.

Dioxins enter estuaries from many sources, but most arrive in the effluent from pulp and paper manufacturing plants that chlorinate wood pulp to produce bleached paper. Trace levels of dioxins have been found in tissues of fishes collected near pulp and paper mills. Although the concentrations do not affect fishes, the bioaccumulation of dioxin may occur in the consumers of these fish, including humans. European paper-producing countries are working to reduce the amounts of dioxin discharged into the Baltic Sea by finding alternative bleaching techniques that do not use chlorine and by producing unbleached paper products. More studies to determine the extent and effect of dioxin in marine waters are currently under way.

PCBs

PCBs are a class of industrial chemicals now banned in the United States, but they are still present at toxic levels in many estuarine and coastal environments. Like dioxins, PCBs are carcinogenic and immunotoxic to many animals (including humans). Studies of stranded beluga whales from a population isolated in the St. Lawrence River estuary of Canada found high tissue levels of PCBs as well as DDT, heavy metals, Mirex, and other pesticides. These animals exhibited high rates of bacterial and protozoan infections, suggesting compromised immune system responses.

The North Sea, surrounded as it is by many of Europe's most industrialized nations, experiences some of the highest levels of marine pollution anywhere. High on the list of these pollutants are dioxins and PCBs. During the summer of 1988, more than 17,000 common and grey seals died of a viral infection that swept the Baltic, Wadden, and North Seas. By the time it ran its course, this epidemic killed as many as 80% of some North Atlantic populations of common and grey seals. Stressed seals exhibited symptoms including lesions, encephalitis, peritonitis, osteomyelitis, and premature abortions.

The source of this epidemic has not been established with any certainty, nor has an absolute link between this viral outbreak and any specific pollutant been demonstrated. However, PCBs in the seals' food is strongly suspected because strong experimental evidence indicates that low concentrations of PCBs (and possibly dioxins as well) lead to suppression of the immune systems of harbor seals. PCBs also interfere with embryo implantation in seals, leading to fewer births and lower birth weights. Seal pups born of PCB-contaminated mothers may still be confronted with higher mortality rates and suppressed immune system problems as they grow and mature.

To examine the extent of the pollution in the Baltic Sea, the signatory nations to the Helsinki Convention have developed pollution monitoring and control strategies. In the Pacific Rim nations, similar efforts are underway through the Pacific Basin Consortium on Hazardous Waste Research to identify and study pollutants present in the Pacific Ocean and to develop joint international partnerships to reduce existing and manage future marine pollutants.

Metal-based Antifouling Paints

For centuries, boat hulls have been sheathed in copper or other metals to attempt to retard the attachment and growth of barnacles and other hull-fouling organisms. More recently, antifouling paints using biocides made from these metals have been applied. The latest entrant in this race to find a substance to block fouling organisms is tributyltin (TBT). Paints with TBT are more effective and last longer than older copper-based antifouling paints. TBT leaches from boat hulls into harbor waters but in amounts difficult to detect with current analytical techniques. Yet even in these very low concentrations (a few parts per trillion), TBT harms nontarget organisms such as oysters and clams. TBT is known to deform oyster shells and to cause chronic reproductive failure in a variety of shellfish species. In 1987, the US Congress passed the Antifouling Paint Control Act that classified TBT as a restricted pesticide and severely limited its use and the amount that could be used in paints. Because TBT eventually degrades to less toxic forms in the marine environment, these new restrictions have already led to decreased concentrations of tin in estuaries and bays.

Even so, since 1987 large numbers of dolphins, seals, and sea turtles have been killed by disease in the Atlantic Ocean, Gulf of Mexico, North Sea, and Mediterranean. Bottlenose dolphins found dead on Atlantic and Gulf coast beaches in Florida between 1989 and 1994 had elevated levels of TBT, presumably derived from boats in coastal marinas. Concentrations of tin compounds (most likely degradation products of TBT) found in the Florida stranded bottlenose dolphins were much higher than concentrations of tin compounds found in offshore whales, presumably because bottlenose dolphins spend their lives close to shore, where antifouling paint from boats and ships has contaminated bottom sediments and local food chains. Accumulated tin compounds, combined with PCBs and DDT (which were also found at high levels), may have damaged their immune systems and left them vulnerable to the bacterial and viral infections that were their eventual cause of death.

7.7 The Chesapeake Bay System

Estuarine shorelands are used for a wide variety of commercial and recreational industries: agricultural and forest production; residential, commercial, industrial, and shipping facilities; and disposal sites for dredged materials. Most of the world's major seaports are situated in estuaries, so they become centers for industrialization and extensive population growth. Commercial and industrial activities lead to modification of estuaries. To facilitate shipping, estuary channels are often dredged to accommodate large ships. Docks, pilings, piers, and jetties alter the natural flushing and circulation patterns of the estuary. To facilitate development, many wetlands have been dredged and filled to create additional flat acreage. This disturbs the bottom, reduces the number of species that live in the estuary, and affects the primary productivity of the estuarine ecosystem. Over 75% of the estuarine wetlands in the United States have been lost due to draining and diking to create agricultural lands or areas for commercial or military development.

Although multiple shoreland uses are an integral part of all coastal economies, they contribute contaminants to the coastal river and estuary systems that can cause habitat loss and a change in the ecologic integrity that cumulatively affects public health, fish and wildlife habitat, and recreational resources. Despite past regulatory and management efforts, the water quality and habitat within most estuaries has been seriously degraded in most estuaries around the world, and it continues to decline.

To better understand some of the conflicts that surround the use of estuarine resources, we examine in more detail one major and well-known estuary, the Chesapeake Bay system on the East Coast of the United States. The Chesapeake Bay system exemplifies the physical, chemical, and biologic features of estuaries and the substantial social and political problems generated by conflicts between natural estuarine processes and the many additional uses imposed on these coastal habitats by humans.

The Chesapeake Bay system is a cascading series of five major and numerous smaller estuaries (FIG. 7.17). These estuaries were linked together when the ancestral Susquehanna River valley flooded after the LGM. The bed of the ancient Susquehanna River is now a deep channel, but most of the Bay is sufficiently shallow to allow sunlight to penetrate to the bottom.

The Chesapeake Bay drains a very large (166,000 km²), heavily populated, and agriculturally rich land area of the US central Atlantic coastal plain. Human activities have imposed some serious stresses on the Bay in the form of increasing loads of heavy metals, fertilizers, pesticides, and incompletely treated sewage.

Increased BOD in Bay waters is only one of several complications resulting from this input. Yet the Chesapeake Bay continues to be used as a protein factory. Millions of blue crabs, oysters, striped bass, and other finfish are harvested from these Chesapeake Bay waters each year, contributing several hundred million dollars annually to the economies of Maryland and Virginia. When the recreational, military, shipping, and other uses of the Chesapeake Bay are added to the mix of conflicting uses, we have in microcosm a picture of some of the same problems confronting other estuarine systems and the larger world ocean.

In the Chesapeake Bay, the existing salinity gradient creates an upper Bay low salinity zone, a mid-Bay brackish zone, and a lower Bay marine zone (Fig. 7.17). Although the tides in the Chesapeake Bay have an average vertical range of only 1–2 m, they are the major mixing influence in the Bay. On longer time

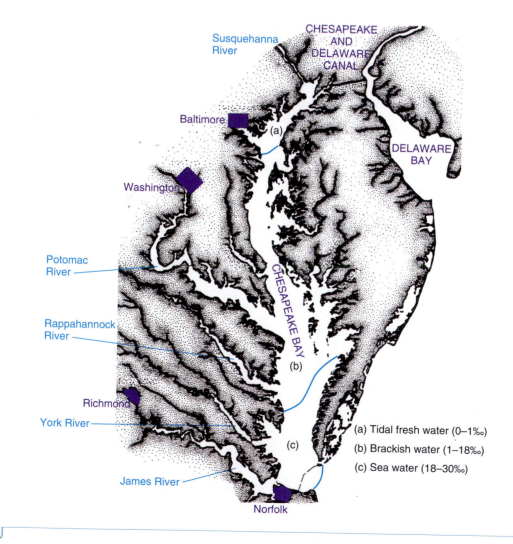

(a) Tidal fresh water (0–1‰)

(b) Brackish water (1–18‰)

(c) Sea water (18–30‰)

figure 7.17

Chesapeake Bay and its numerous smaller side estuaries, showing mean surface salinity zones.

scales, seasonal flooding and storms have additional effects on the salinity distribution patterns in the Bay.

For animals capable of tolerating the dynamic fluctuations of the Chesapeake Bay and other estuaries, these habitats offer nearly ideal nursery conditions for their young. Estuaries provide some protection against the physical stresses of nearby open coasts, and there is an abundance of food available in a large range of particle sizes. Most fish species commercially exploited along the Atlantic and Gulf Coasts of the United States use estuaries such as the Chesapeake Bay as spawning or juvenile feeding areas. Some species occupy estuaries throughout their lives; others occupy estuaries for only a particularly crucial stage of their development. FIGURE 7.18 illustrates this range of utilization patterns for a few commercially important or otherwise notable Chesapeake Bay species.

At one extreme are oysters, which typically spawn, mature, and die within the confines of the Bay (although some larvae occasionally may drift to other nearby estuaries). Oysters are broadcast spawners, with fertilization and a 2-week larval development period occurring in the moving water above the benthic habitat of the adults. The double-layered circulation pattern of the Bay is used by oyster larvae to avoid being washed out of the estuary. During ebb tides, when most of the tidal outflow is in the surface layers, the larvae remain in the deeper inflowing seawater. During slack or incoming tides, the larvae venture into the shallower portions of the larger

Bay or its smaller side estuaries, where high levels of larval settling and retention occur.

Horseshoe crabs and menhaden are marine species that use the Chesapeake Bay for early life stages only. Adult horseshoe crabs move into the Bay only to spawn. During high tides in the spring, these animals crawl into salt marshes at the water's edge. There the female digs a depression to deposit her eggs. The smaller male, who hitches a ride on the female's back (see Fig. 5.30), sheds sperm to fertilize the newly deposited eggs. The eggs are then covered to await hatching 2 weeks later, when they are again flooded by the next series of spring tides. After hatching, the larvae swim and feed near the surface while currents carry them out to sea where development continues through as many as 13 successive larval stages.

Menhaden is a commercially valuable fish species of the middle Atlantic Coast. Although the adults live and spawn in coastal waters, menhaden larvae drift into the Chesapeake Bay and other estuaries to continue their development. Juvenile sea turtles (loggerheads and olive ridleys are the most common) frequent the Bay, with green sea turtles grazing on its slowly shrinking eelgrass beds. The 200 to 300 olive ridleys seen each year in the Chesapeake Bay may represent nearly all the existing juveniles of this seriously endangered species.

For blue crabs, the pattern of Bay utilization is different still. Adults live in estuaries along most of the US Atlantic and Gulf Coasts. After mating, females seek higher salinities in the open sea before

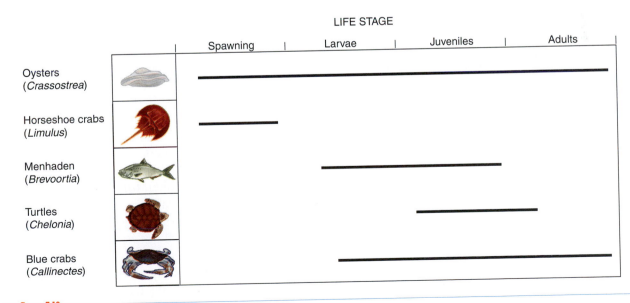

figure 7.18

Utilization of estuaries by differing life stages of five common inhabitants of Chesapeake Bay.

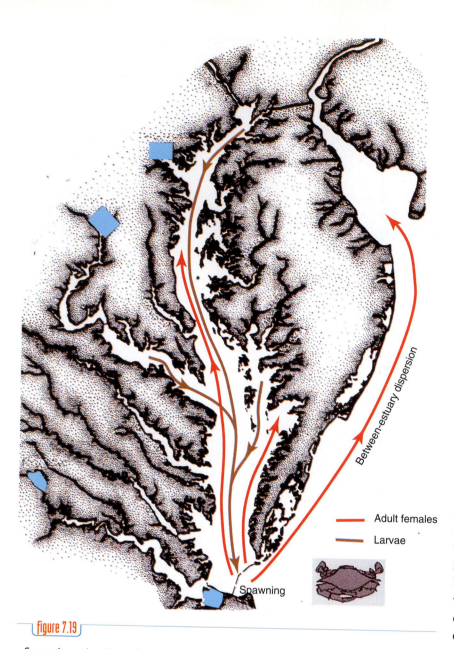

Adult females

Larvae

Spawning

figure 7.19

Spawning migration of adult female blue crabs and return routes of planktonic larval stages.

Between-estuary dispersion

they grow sufficiently to exploit other estuarine habitats.

Through their roles as nursery areas and feeding grounds, the value of estuaries such as the Chesapeake Bay extends far beyond the bounds of the estuary itself. Yet the very health of the Chesapeake Bay, as well as that of most of Earth's other major estuaries, has gradually but seriously deteriorated in the past few decades. Since 1960, submerged vegetation in the Bay, especially eelgrass beds, has declined in size and abundance. Most of the decline has been in the upper and western parts of the Bay, but the problem is moving down the Bay as well. Dissolved nutrient loads draining into the Bay have increased, causing changes in the species composition of the water. In the upper reaches of the Chesapeake Bay, concentrations of cyanobacteria and dinophytes have increased 250-fold since 1950 at the expense of diatom species. During the same time period, submerged eelgrass and marsh grass beds have declined dramatically. Presently, most of the Bay water deeper than 13 m (downstream from near the mouth to the Rappahannock River) is a dead zone, with little or no dissolved O_2. As a consequence of these biologic changes, recent harvests of alewives, shad, striped bass, and oysters also have dropped substantially.

Serious and sustained efforts on the part of state regulatory agencies of the six states with rivers that empty into the Chesapeake (Maryland, Delaware, Virginia, West Virginia, New York, and Pennsylvania), the District of Columbia, and several cooperating federal agencies are required to stem these adverse changes and reduce the nutrient and toxic substance loads presently carried into the Bay. It will not be easy, yet only when these efforts are successful can we ensure the continuing health of this hardy and beautiful estuary system that is the Chesapeake Bay.

releasing their larvae (FIG. 7.19). Larval development continues in coastal waters outside the Bay, where winds and coastal currents combine to keep blue crab larvae close to shore until they return to the Bay as young crabs. It is at this time that exchange of individuals between neighboring estuaries sometimes occurs, preventing genetic isolation of the crabs that occupy any one estuary. Within the estuary, young blue crabs seek eelgrass beds and salt marshes as winter nursery areas for food and protection until

Summary Points

Estuaries

- Estuaries are semienclosed coastal embayments where freshwater rivers meet and mix with the sea, creating unique and complex ecosystems. More than one third of Americans live within the drainage basins of our 100 or so estuaries; thus, these transitional coastal habitats encapsulate the opportunities and management challenges that humans face more than any other ecosystem discussed in this book.

- Because of variations in rainfall, riverine input, and tidal movements, estuaries are highly variable habitats that, by necessity, house characteristic species that are extremely tolerant to change.

- The combination of chronic runoff and tidal influx establishes a unique pattern of water circulation within estuaries that may function as a trap. This phenomenon is a double-edged sword in that beneficial items become trapped, such as nutrients that make estuarine systems among the most productive on Earth, along with harmful pollutants.

Types of Estuaries

- Estuaries are classified both by their modes of formation and by their patterns of water circulation. Coastal plain estuaries are broad embayments formed from flooded river mouths, bar-built estuaries are formed behind barrier islands created from sediments transported by adjacent streams, deltas are created as heavy loads of sediments are deposited in a fan-shaped pile at the mouth of a large river, and tectonic estuaries form due to crustal movements of the Earth.

Estuarine Circulation

- Although the upstream-to-downstream values of salinity, temperature, turbidity, and current speed change markedly during a single tidal cycle, salinity typically increases from the surface downward and from the estuary head downstream to its mouth.

- Because of the higher density of seawater relative to fresh water, incoming tides move along the floor of the estuary and riverine inputs exit along its surface. As such, detritus and other organic nutrients that settle out of the river water are transported back up into the estuary with each tidal cycle, thus creating a unique upwelling process that greatly augments estuarine production.

Salinity Adaptations

- Daily variations in tidal flow and seasonal variations in rainfall and freshwater input result in constantly fluctuating salinities within all estuaries; resident organisms must be able to tolerate these changes and concomitant osmotic stresses.

- Most slow-moving invertebrates have poorly developed osmoregulatory capabilities and avoid salinity fluctuations by remaining near the high-salinity mouth of the estuary or traveling in and out with the tides such that they always remain within water of a relatively constant concentration. Such species are osmoconformers.

- All estuarine vertebrates, and most fast-swimming invertebrates (such as blue-claw crabs) have evolved a great variety of mechanisms used to regulate and stabilize the water and solute concentrations within their body fluids in spite of ambient conditions. Such organisms are osmoregulators.

Sediment Transport: Creating Habitats

- Sediments are commonly transported into estuaries by rivers that drain coastal watersheds. These sediments not only help to create some types of estuaries, such as deltas and bar-built estuaries, but they also settle out such that large-grained course sediments dominate estuarine heads where current velocity is high and muddy sediments predominate the floor at the mouth of estuaries. Hence, sediments not only create the estuary itself, but also promote the development of various microhabitats within the estuary from its head to its mouth.

Estuarine Habitats and Communities

- Estuarine communities include salt marshes, high-elevation plant communities that tolerate periodic inundation during high tides, lower elevation mudflats that are alternately exposed and covered by changing tides, and subtidal channels that are always filled with flowing water.

- Salt marshes, or wetlands, are intertidal grasslands that grow along estuarine shores. Zonation of these halophytic, or salt-loving, plants is common as each species grows in a particular band given its preference for sediment type, frequency of tidal inundation, and exposure. Wetlands form an important component of estuarine food webs. Although some grasses are eaten directly by herbivorous residents and visitors, most grass production enters the estuary as decaying detritus. Hence, bacteria and other important microbes play a fundamental role in the transformation of marsh grass productivity into detrital food that is available to countless other estuarine species.

- Mudflats, or tideflats, are broad expanses of nutrient-rich fine-grained muds that are exposed during low tide and act as both home and food source for a wide variety of infauna and epifauna. They host three types of producers: diatoms that coat the mud surface with a photosynthetic film, macroalgae that attach to all manner of debris (such as rocks, shells, wood, and trash), and seagrasses that prefer complete submergence during periods of high tide.

- Channels are packed with all types of organisms that cannot tolerate intertidal life in wetlands or on mudflats. Some of these, such as crabs, oysters, and killifishes, live in estuarine channels throughout the year, whereas others simply visit the estuary to spawn (such as herring, soles, and sandbar sharks).

Environmental Pollutants

- Because estuaries are usually supplied with fresh water by countless rivers and streams that collectively drain vast watershed areas, many pollutants commonly get carried to estuaries where they are trapped by the unique type of water circulation therein.

- Dangerous and problematic pollutants include seemingly innocuous chemicals, such as agricultural fertilizers that farmers spread on our food, to obviously harmful compounds such as DDT (dichloro-diphenyl-trichloroethane), dioxins, polychlorinated biphenyls, and antifouling paints. Agricultural fertilizers lead to unnatural blooms of helpful and harmful phytoplankton, all of which ultimately die and decompose on the floor of the estuary, thus robbing it of essential oxygen. Heavy metals and chlorinated hydrocarbons are unquestionably toxic. Yet our ability to determine their fate in the environment, as well as their effects on the health of marine life and humans, is in its infancy.

The Chesapeake Bay System

- The Chesapeake Bay system, a cascading series of five major and numerous minor estuaries, exemplifies the physical, chemical, geologic, and biologic features of a typical estuary. It is also a microcosm for the substantial social, economic, and political problems generated by conflicts between natural estuarine processes and the many ways in which we choose to use these unique coastal habitats.

Topics for Discussion and Review

1. Compare and contrast the various types of estuaries, being sure to discuss their origin, size, shape, and likely location on Earth.
2. Describe the pattern of water circulation in a typical estuary during periods of high and low river input.
3. Describe the distribution of salinity values in a typical estuary, from surface to sediment and from head to mouth.
4. Compare and contrast the type of sediment found at the head of an estuary as opposed to its mouth. Describe the factors that result in this differential deposition.
5. What types of plants dominate low frequently submerged portions of wetlands? And higher marsh zones that do not experience prolonged submergence?

6. If you were to walk out onto a mudflat at low tide and dig a hole that is 2 feet deep, what types of organisms would you expect to find while digging this hole?

7. Why are estuaries important to nonestuarine species, such as migrating geese and anadromous shad?

8. Summarize the series of events leading from runoff of agricultural fertilizers to dangerously low depletion of dissolved oxygen within a nearby estuary.

9. What is the source of DDT? What is its residence time in the environment? What effect does it have on wildlife? Is it being produced or used in America today?

10. Draw a food web representative of the Chesapeake Bay system.

Suggestions for Further Reading

Bertness, M. D. 1991a. Interspecific interactions among high marsh perennials in a New England salt marsh. *Ecology* 72:125–137.

Bertness, M. D. 1991b. Zonation of *Spartina patens* and *Spartina alterniflora* in a New England salt marsh. *Ecology* 72:138–148.

Bertness, M. D. 1992. The ecology of a New England salt marsh. *American Scientist* 80:260–268.

Cloem, J., and F. Nichols. 1985. *Temporal Dynamics of an Estuary*. Kluwer Academic, Boston.

Cooke, S. S. 2003. *A Field Guide to the Common Wetland Plants of Western Washington & Northwestern Oregon*. Seattle Audubon Society, Seattle.

Cronk, J. K., and M. S. Fennessy. 2001. *Wetland Plants: Biology and Ecology*. Lewis Publishers, Inc., Boca Raton, Florida.

Dyer, K. R. 1997. *Estuaries: A Physical Introduction*. Wiley, Chichester.

Dennison, W. C. et al. 1993. Assessing water quality with submerged aquatic vegetation. *Bioscience* 43:86–93.

Flindt, M. R., M. A. Pardal, A. I. Lillebo, I. Martins, and J. C. Marques. 1999. Nutrient cycling and plant dynamics in estuaries: a brief review. *Acta Oecologica* 20:237–248.

Hart, J., and D. Sanger. 2003. *San Francisco Bay: Portrait of an Estuary*. University of California Press, Berkeley.

Hobbie, J. E. 2000. *Estuarine Science: A Synthetic Approach to Research and Practice*. Island Press, Washington, DC.

Horton, T. 2003. *Turning the Tide: Saving the Chesapeake Bay*. Island Press, Baltimore.

Keddy, P. A. 2000. *Wetland Ecology: Principles and Conservation*. Cambridge University Press, Cambridge.

Kennish, M. J. 1999. *Estuary Restoration and Maintenance: The National Estuary Program*. CRC Press, Boca Raton, FL.

Kusler, J. A., W. J. Mitsch, and J. S. Larson. 1994. Wetlands. *Scientific American* 270:64–70.

Leonard, G. H., J. M. Levine, P. R. Schmidt, and M. D. Bertness. 1998. Flow-driven variation in intertidal community structure in a marine estuary. *Ecology* 79:1395–1411.

Little, C. 2000. *The Biology of Soft Shores and Estuaries*. Oxford University Press, Oxford.

Margulis, L., D. Chase, and R. Guerrero. 1986. Microbial communities. *Bioscience* 36:160–170.

McLusky, D. S. 1989. *The Estuarine Ecosystem*. Blackie, Glasgow and London.

Milliman, J. 1989. Sea levels: past, present and future. *Oceanus* 32:40–43.

Mitsch, W. J., and J. G. Gosselink. 2000. *Wetlands*. John Wiley & Sons, New York.

Noble, P. A., R. G. Tymowski, and M. Fletcher. 2003. Contrasting patterns of phytoplankton community pigment composition in two salt marsh estuaries in southeastern United States. *Applied and Environmental Microbiology* 69:4129–4143.

Peterson, C. H., and R. Black. 1987. Resource depletion by active suspension feeders on tidal flats: influence of local density and tidal elevation. *Limnology and Oceanography* 32:143–166.

Pettitt, J., S. Ducker, and B. Knox. 1981. Submarine pollination. *Scientific American* 244:134–143.

Silberhorn, G. M., M. Warinner, and K. Forrest. 1999. *Common Plants of the Mid-Atlantic Coast: A Field Guide*. Johns Hopkins University Press, Baltimore.

Teal, J., and M. Teal. 1983. *Life and Death of the Salt Marsh*. Little, Brown, Boston.

Tiner, R. W., Jr., and A. Rorer. 1987. *A Field Guide to Coastal Wetlands Plants of the Northeastern United States*. University of Massachusetts Press, Boston.

CHAPTER

8

Coastal marine ecosystems include
sandy beaches and rocky shores
like these on the Oregon coast.

CHAPTER OUTLINE

8.1 Seafloor Characteristics

8.2 Animal–Sediment Relationships

8.3 Larval Dispersal

8.4 Intertidal Communities
Rocky Shores
Sandy Beaches
Oiled Beaches

8.5 Shallow Subtidal Communities

Temperate Coastal Seas

Over 90% of the animal species found in the ocean and nearly all the larger marine plants live in close association with the sea bottom. Collectively, these organisms are the benthos. Benthic primary producers, introduced in Chapter 4, exist only in the shallow near-shore fringe where the sea bottom coincides with the photic zone. Benthic animals range much more widely from high intertidal zones to cold perpetually dark trenches more than 10,000 m deep.

Benthic organisms live at the interface between the sea bottom and the overlying water; and the environmental conditions they experience are defined by the characteristics of the bottom materials and the overlying water, the exchange of substances between the sediments and the overlying water, and conditions established by the other members of their communities. In this chapter we examine the general conditions of life on temperate-climate coastal regions from intertidal shorelines to continental shelf areas just below the low tide line.

8.1 Seafloor Characteristics

The sea bottom supports the weight of many organisms considerably more dense than seawater. Some animals excavate burrows or construct tubes of soft sediments. On hard rock bottoms, animals and plants secure a firm attachment so they can resist the tug of waves and currents. Benthic organisms are adapted for life on or in particular bottom types, and the character of life there, to a large extent, is dependent on the properties of the bottom material, which varies from solid rock surfaces to very soft loose deposits.

The sea bottom also accumulates plankton, waste material, and other detritus sinking from the sunlit waters above. In some regions, fallout of organic detritus from the photic zone is the only source of food for the inhabitants on the bottom. A variety of worms, mollusks, echin-oderms, and crustaceans obtain their nourishment by ingesting accumulated detritus and digesting its organic material.

The composition of the sea bottom is determined by its constituent materials and, in shallow water, by the amount of energy available in the wind-driven waves and currents at the sea surface. Chapter 1 explained that the energy of the waves (and their ability to move particles) decreases from the sea surface downward (Figure 1.35) and disappears at depths equal to about one half the wavelength of surface waves. The coastline shape also strongly influences the amount of energy expended on the shore when waves break. Wave fronts approaching shore start to slow down just as they begin interacting with shallow reefs and bars. As a result, the wave fronts lag behind the rest of the wave as it approaches shore and changes shape to approximately match the curvature of an irregular coastline (**FIGURE 8.1**, and compare with Figure 8.14). This process redistributes the energy in the breaking waves, concentrating wave energy on headlands while spreading out and diminishing the energy of waves entering bays, coves, and other coastal indentations.

Taken together, the overall behavior of wind waves causes large amounts of energy to be expended on headlands in shallow waters (shallower than one half the wavelength of the waves) and lesser amounts of energy to be expended in coastal indentations and in deeper waters. On exposed headlands, erosion is the principal result of breaking waves. These high-energy environments are continuously swept clean of fine sediment particles, detritus, and anything else

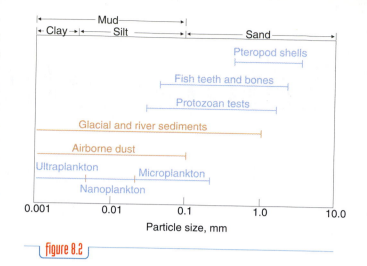

figure 8.2

Particle size ranges for some common sources of marine sediments. Biogenic particles are shown in blue, and terrigenous particles in tan.

not securely attached to the bottom. This debris is washed offshore into deeper water or along shore into bays or other calm-water coastal features, where it settles to the bottom and adds to the accumulating deposits already there.

Marine sediments near shore and on the continental shelves are largely the products of erosion on land and subsequent transport by rivers (and, to a lesser extent, winds) to the sea (**FIGURE 8.2**, and see Figure 3.33). Once in the ocean, suspended sediment particles are carried and sorted by current and wave action according to their size and density. Large dense sand grains quickly settle to the bottom near shore. Very fine clay particles are often carried several hundred kilometers out to sea before settling.

On an oceanic scale, rocky outcrops occur in association with deep-sea ridges, rises, and volcanoes, but these outcrops are comparatively small and geographically isolated. Other rocky substrates are scattered around the very edges of ocean basins where wave-driven erosional processes dominate. Although relative rare, these rocky shorelines have been more intensively studied than any other part of the ocean because of their easy access to land-based researchers and their students.

8.2 Animal–Sediment Relationships

Benthic animals that crawl about on the surface of the sea bottom or sit firmly attached to it are referred to as the epifauna. Epifauna are associated with rocky outcrops or the surface of firm sediment deposits. Other benthic animals, the infauna, find

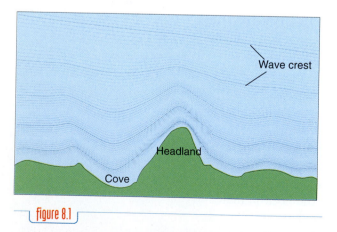

figure 8.1

In coves and bays, refraction of advancing ocean waves spreads out the wave crests (and their energies) and concentrates them on headlands and other projecting coastal features.

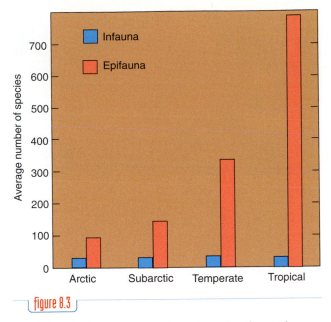

Variations in the average number of species of several bottom invertebrate groups from equal-sized coastal areas in different latitudes. (Adapted from Thorson 1957.)

Sandstone erosion pits created by the rasping actions of small chitons.

food or protection within the substrate forming the bottom. FIGURE 8.3 shows the relative abundance of epifauna and infauna at different marine climatic zones. Note that epifauna seem to be more sensitive than infauna to global-scale climatic differences.

Infaunal clams, worms, and crabs are macroscopic and are familiar to anyone who has spent a few moments digging in a sandy beach or mudflat. These **macrofauna** either swallow or displace the sediment particles around them as they move. Less obvious, but no less important, are **microfauna**, microscopic infauna less than 100 µm in size. Intermediate in size between the macrofauna and microfauna are the **meiofauna**, a very interesting and abundant group of animals. The meiofauna are also referred to as **interstitial animals,** because they occupy the spaces (the interstices) between sediment particles.

Benthic animals mix and sort the sediments through their burrowing and feeding activities. Oxygen and water from the sediment surface circulate down into the sediment through their burrows. Sedimentary characteristics are further modified when particles are cemented together to form tubes and when sediments are compacted to form fecal pellets and castings. On rocky bottoms, the grazing activities of chitons, gastropods, and sea urchins aid the erosive processes of waves by scraping away rock particles as well as food (FIGURE 8.4). A few benthic animals, such as boring clams, are especially adapted for boring into solid rock.

The distributional patterns of benthic plants and animals are strongly influenced by the firmness, texture, and stability of their substrate. These features govern the effectiveness of locomotion or, for nonmotile species, the persistence of their attachment to the bottom. Epifauna are most frequently associated with firm or solid bottom material.

The particle size and organic content of the bottom material limit the versatility and distribution of specialized feeding habits. Suspension feeders depend on small plankton or detritus for nutrition. Filtering devices (FIGURE 8.5) or sticky mucous nets or sheets collect minute suspended food particles from the water. Suspension feeders generally require clean water to avoid clogging their filters with indigestible particles. Therefore, they are usually found on rocks or are associated with coarse sediments.

Benthic environments abound with predators and scavengers who feed on the residents of the bottom or on their remains. Most bottom predators and scavengers are permanent members of the benthos and are eventually eaten by other benthic consumers. Fishes, however, often make serious inroads into intertidal animal populations during high tides, and shorebirds replace them as predators at low tide. Sea stars are also usually carnivorous (FIGURE 8.6), but a few species are quite opportunistic. When feeding on bottom sediments, the bat star extrudes its stomach outside its body and then digests and absorbs organic matter from the sediments.

figure 8.5

Barnacles, *Balanus,* attached to a small snail shell, with their feathery filtering appendages extended.

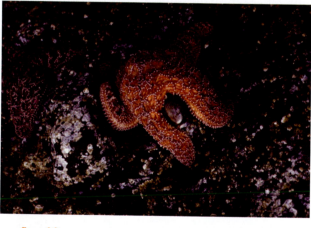

figure 8.6

A sea star, *Pisaster,* feeding on a mussel.

Regardless of the feeding habit used by benthic animals, the ultimate source of food is the primary producers of the photic zone. Intertidal and shallow-water benthic plants provide direct sources of nutrition for the abundant herbivorous **algal grazers.** Some algal grazers nibble away bits of the larger seaweeds (FIGURE 8.7). Most, however, rasp filmy growths of diatoms, cyanobacteria, and small encrusting plants from rocky substrates. Sea urchins use their five-toothed Aristotle's lantern to remove algal growths. Herbivorous gastropods and chitons accomplish similar results with their file-like radula (see Figure 5.22).

figure 8.7

A large snail, *Norrisia,* grazing on a kelp stipe.

8.3 Larval Dispersal

A sluggish benthic life-style does not need to limit slow-moving or sedentary benthic animals to narrow geographic ranges. One fifth of the common shallow-water animal species found at San Diego, California, for example, can also be found along the entire West Coast of the United States and in British Columbia. Other benthic species are even more widely dispersed, often in similar ecologic conditions on opposite sides of the same ocean basin. *Mytilus edulis,* variously known as the bay mussel, blue mussel, or edible mussel, is common to temperate coasts of both sides of the Pacific and Atlantic Oceans.

A few animals, including a small percentage of barnacles, hitch rides on floating debris, on the hulls of ships, or in the ballast water of ships to travel transoceanic distances. An Australian barnacle, *Elminius modestus,* was apparently introduced to England by supply ships during World War II. It has since colonized most of the north and west coasts of continental Europe. In many sheltered reaches of these coastlines, *E. modestus* is competing with and replacing native barnacle populations.

A far more common adaptation for extending the geographic range of temperate- and warm-water benthic species involves the production of temporary planktonic larval stages that account for most temporary plankton forms, the **meroplankton.** These small feeble swimmers, bearing little resemblance to their parents (FIGURE 8.8), drift with the ocean's surface currents for some time before they metamorphose and assume their benthic life-styles.

About 75% of shallow-water benthic invertebrate species produce larvae that remain planktonic for 2 to 4 weeks. Over 5% of the species examined in one study had planktonic larval stages exceeding 3 months, with a few as long as 6 months in duration

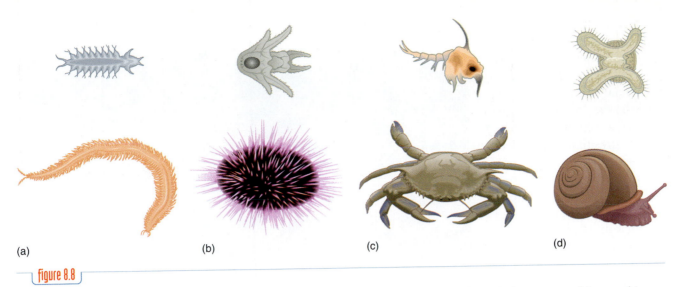

(a)   (b)   (c)   (d)

Planktonic larval forms (top) and adult forms (bottom) of some common benthic animals: (a) polychaete worm, (b) sea urchin, (c) crab, and (d) snail.

(FIGURE 8.9). Our understanding of ocean currents suggests that none but the most prolonged larval stages can make direct transoceanic trips before settling to the bottom. For each extra day the larvae remain in the plankton, they are exposed to additional threats of predation, increased pressures of finding food, and greater possibilities of being carried by the currents to areas where survival is unlikely. Even so, a temporary planktonic larval existence can provide several advantages to offset the enormous mortality experienced by these larvae.

Even in very slow ocean currents, drifting planktonic forms may spread far beyond the geographic limits of their adult population. Many are swept into unfavorable areas and perish. But survivors may expand their parents' original range or settle into and mix with other populations and reduce their genetic isolation. During their planktonic existence, many types of larvae react positively to sunlight and remain near the sea surface and their food supply, the phytoplankton. As their planktonic life draws to a close and they seek their permanent homes on the bottom, some larvae remain near the sea surface and ride into intertidal shores on waves and tides. Other larvae shun the light and swim near the bottom. Most enter a swimming–crawling phase and settle to the bottom and investigate it, and if it is not suitable, they swim up to be carried elsewhere.

Just how larvae know when a suitable substrate is encountered is an important, but as yet unanswered, question. Chemical attractants, current speeds, types and textures of bottom material, and the effects of light are only partial answers to the question. Spe-

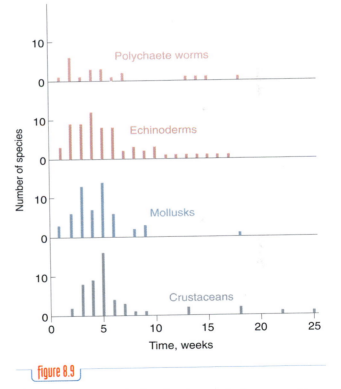

figure 8.9

Typical duration of planktonic existence for four common groups of marine benthic invertebrates. (Adapted from Thorson 1961.)

cific bottom types, such as sand or hard rock, do not attract larvae from a distance. Only after the appropriate bottom type is actually encountered may larvae be induced to remain and quickly metamorphose into a bottom-living stage. Alternatively, chemical substances diffusing from established populations of

Larval Dispersal 223

some attached animals, including oysters and barnacles, attract larvae of their own species. This attraction may be beneficial for oyster and barnacle larvae, for the mere presence of adults in the settling site ensures that physical conditions have been appropriate for survival. Also, the larvae's eventual reproductive success may be enhanced if they are in the vicinity of other members of their species. For many other larvae, however, settling among their adults can be disastrous. Older established individuals generally have relatively lower demands for food and oxygen, have more stored energy, and,

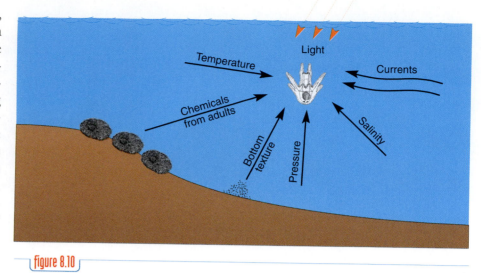

figure 8.10

Several major environmental factors that influence the selection of suitable bottom types by planktonic larvae.

in general, can compete more effectively for resources with newly settled young. In times of shortages, younger or smaller individuals are usually the first to suffer. FIGURE 8.10 illustrates the more obvious environmental features that may guide and influence larval settling.

Until they settle to the bottom and undergo metamorphosis to their juvenile benthic form, planktonic larvae are not in competition for food or space with the adults of their species. Even so, competition for food among plankton is often rigorous. A few species with long planktonic larval phases produce large yolky eggs that provide larvae with most or all their nutritional supply. However, most species hatch from small eggs with little stored food and must begin feeding and competing with each other almost immediately.

It is becoming apparent that for some species, the quality of larval feeding experiences strongly influence their juvenile growth and survival rates after metamorphosis. In laboratory studies, short-term food shortages of only a day or two during the larval stages of snails and barnacles led to delay of metamorphosis, increased larval mortality, and reduced growth or survival of juveniles after metamorphosis (FIGURE 8.11). It would be interesting, but is not yet possible, to follow the fates of individuals to determine whether the detrimental effects of food shortages in their early larval history persisted to adulthood and influenced their individual fitness by reducing fecundity or by delaying the age of sexual maturation.

Some benthic species, especially those in the tropics and the deep sea, spawn all year long; others have short and well-defined spawning seasons. In the latter group, the timing of reproduction or spawning is geared to produce young at times most advantageous to their survival. The spawning periods of many species of benthic animals are timed to place their larvae in the plankton community when phytoplankton are abundant and readily accessible during spring and fall diatom blooms (see Chapter 4). In shallow waters, temperature and day length provide two of the more obvious cues for timing reproduction. The gonads of spring and summer spawners develop in response to rising water temperatures and lengthening days. Oysters, for instance, refrain from spawning until a particular water temperature is reached.

Regardless of how many eggs are produced, the reproductive success or fitness of an individual requires that its **fecundity** (the production of eggs or offspring) exceed its offspring **mortality** (the rate at which individuals are lost). Any population whose mortality consistently exceeds its fecundity will shrink and eventually disappear. Adaptations that increase fecundity or reduce mortality improve the chances for successful reproduction. The fecundity of some shallow-water benthic animals is truly amazing, with females of many species producing several million eggs annually. A sea hare, *Aplysia,* weighing a few kilograms produced an estimated 478 million eggs during 5 months of laboratory observations. Such excessive reproduction enthusiasm would quickly place any shoreline knee deep in sea hares if all the spawning efforts of only a few adults survived.

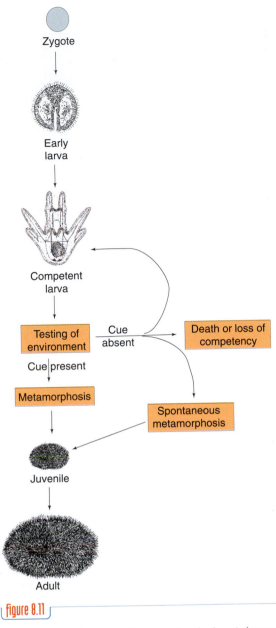

Zygote

Early
larva

Competent
larva

Testing of environment → Cue absent → Death or loss of competency

Cue present

Metamorphosis

Spontaneous metamorphosis

Juvenile

Adult

figure 8.11

Generalized developmental pattern for planktonic larvae, illustrating available options in response to food, substrate, or other environmental cues.

Obviously, the egg and larval mortality of these species is extremely high. Of the millions of potential offspring produced, very few attain sexual maturity. The eggs of some benthic animals are fertilized internally before they are released into the water. Some species of snails, crabs, sea stars, and other invertebrates retain their larvae internally or in special brood pouches until the larvae are at reasonably advanced stages of development. For some of these invertebrates, brooding may be an adaptive consequence of small adult size. Smaller species of benthic invertebrates, with correspondingly smaller gonads, are

less likely to produce sufficient planktonic larvae to equal their larger competitors. Consequently, they may opt for internal fertilization and larval incubation to reduce offspring mortality to some extent.

Many of the more abundant and familiar seashore animals are **broadcast spawners**; they spew great quantities of eggs and sperm into the surrounding water where fertilization occurs. These numerous eggs are necessarily small and hatch quickly into planktonic larval forms. Each larva, then, is a relatively low-cost genetic insurance policy for spawning adults, there to ensure that the genes of the parents survive another generation.

Fertilization in these broadcast spawners is neither a casual nor haphazard process. Chemical substances (known as **pheromones**) are present in the egg or sperm secretions of sea urchins, oysters, and other broadcast spawners. When shed into seawater, these pheromones induce other nearby members of the same population to spawn, and they in turn stimulate still others to spawn until much of the population is spawning simultaneously. As you might guess, spawning pheromones are typically species specific; pheromones of one species of broadcast spawner will induce only other members of the same species to spawn without influencing the spawning of any other species.

Once spawning has occurred, sperm must still make contact with the correct type of egg, possibly in an ocean of other eggs, for fertilization to occur. A structural protein, contained in the heads of sperm cells of broadcast spawners, binds only with other proteins on the surface coat of eggs from the same species (FIGURE 8.12). Together with pheromones, these substances regulate the timing of spawning, the specificity of sperm for eggs of the same species, and ultimately the overall prospects for successful fertilization by broadcast spawners.

8.4 Intertidal Communities

The coastal strip where land meets the sea is home to some of the richest and best-studied marine communities found anywhere. Although this coastal strip is narrow, its influence is enhanced by the wealth of marine organisms present. Typically, the total biomass in a square meter at the low tide line is at least 10 times as high as that of a comparable area on the bottom at 200 m and is several thousand times higher than that found in most abyssal areas.

The periodic rise and fall of the tides (see Chapter 1) has a dramatic effect on a portion of the coastal

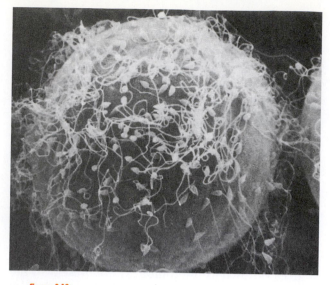

A scanning electron micrograph of a sea urchin egg with numerous sperm cells.

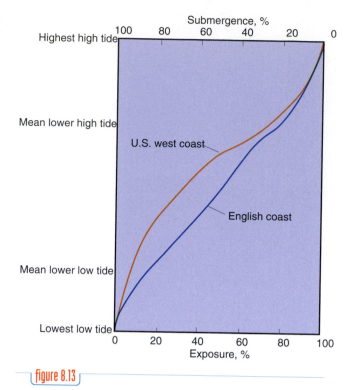

figure 8.13

Exposure curves for the Pacific Coast of the United States and the Atlantic Coast of England. (Adapted from Ricketts and Calvin 1968 and from Lewis 1964.)

zone known as the intertidal, or **littoral**, zone. In the littoral zone, the sea, the land, and the air all play important roles in establishing the complex physical and chemical conditions to which all intertidal plants and animals must adapt. Tidal fluctuations of sea level often expose intertidal plants and animals to severe environmental extremes, alternating between complete submergence in seawater and nearly dry terrestrial conditions. Local characteristics of the tides, including their vertical range and frequency, determine the amount of time intertidal plants and animals are out of water and exposed to air. Still, regardless of their locations, most intertidal regions have exposure curves that resemble FIGURE 8.13.

Most intertidal plants and animals are marine in origin and prefer to remain in seawater, so exposure at low tide is a time of serious physiologic stress for them. When the tide is out, exposed organisms are subjected to wide variations of atmospheric conditions. The air may dry and overheat their tissues in hot weather or freeze them during cold weather. Rainfall and freshwater runoff create osmotic problems as well. Predatory land animals, such as birds, rats, and raccoons, also make their presence felt in the intertidal zone at low tide. Only at high tide are truly marine conditions restored to the intertidal zone. The returning waters moderate the temperature and salinity fluctuations brought on by the previous low tide. Needed food, nutrients, and dissolved oxygen are replenished, and accumulated wastes are washed away.

Accompanying the beneficial effects of seawater is the physical assault of waves and surf. The influence of wave shock on the distribution of intertidal plants and animals is apparent on all exposed coastlines of the world. Surf, storm waves, and surface ocean currents shift and sort sediments, transport suspended food, and disperse reproductive products. Much of the wave energy expended on the shore and the organisms living there eventually serves to shape and alter the essential character of the shoreline itself. Continually modified by the power of ocean waves, shorelines assume a variety of forms (FIGURE 8.14). Rocky shorelines are constantly swept clean of finer sediments by heavy surf or strong currents. Waves on beaches remove fine silt and clay particles but leave well-sorted sand grains behind. The finer materials are washed out to sea or are deposited in the quiet protected waters of bays and lagoons.

The variety of tidal conditions, bottom types, and wave intensities along the shore creates a boundless assortment of living conditions for coastal plants and animals. It is difficult to characterize the prevailing conditions on long stretches of shoreline without risking overgeneralization. The west coast of North

The southern tip of Florida is the only shoreline on the continental United States to experience tropical conditions. Tropical shorelines are typically marked by large coral reefs or by extensive swampy woodlands of mangroves (see Chapter 9). The coral reef system off the Florida Keys, fairly typical for the Caribbean, is unlike most other parts of the continental United States. Mangroves grow in profusion along Florida's southern coast and recently have become established along the extreme southern Texas coast. Included among these mangroves are several types of shrubby and tree-like plants that grow together to form impenetrable thickets (FIGURE 8.15). Their branching prop roots trap sediments and detritus to extend existing shorelines or to form new low-lying islands.

The vertical distribution of intertidal plants and animals within the intertidal is governed by a complex set of environmental conditions that vary along gradients above and below the sea surface. Temperature, wave shock, light intensity, and wetness are some of the more important physical factors that vary

figure 8.14

An infrared aerial photograph of a portion of the Oregon coast with protected coves, exposed headlands, sandy beaches, and offshore rocky reefs. Note the complex refraction of surface waves around the offshore reefs.

(a)

(b)

figure 8.15

(a) A dense mangrove thicket in Southern Florida. (b) The network of prop roots traps sediment and detritus.

America, for instance, has many rugged rocky cliffs exposed to the full force of wave action. Yet interspersed between these cliffs and headlands are numerous sandy beaches and quiet mud-bottom bays and estuaries. On the east coast, conditions vary from the spectacular rugged coastline of northern New England and Canada's Maritime Provinces to extensive sandy beaches in the mid-Atlantic states. From Chesapeake Bay south to Florida, numerous coastal marshes are protected from extensive wave action by long low barrier islands that parallel the mainland. Similar conditions with smaller tidal ranges exist along much of the Gulf Coast (see Chapter 7).

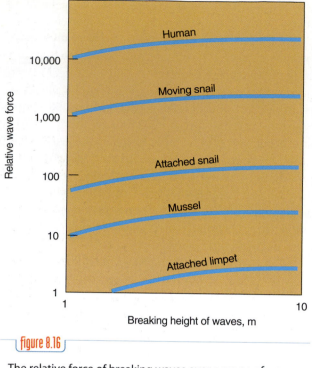

figure 8.16

The relative force of breaking waves over a range of wave heights for several different intertidal animals. (Adapted from Denny 1985.)

along such gradients. Breaking waves can impose large forces on intertidal organisms (FIGURE 8.16). These organisms in turn demonstrate a remarkable variety of adaptations to deal with the forces associated with large waves. Several biologic factors, including predation and competition for food and space, are superimposed on the physical gradients to delineate further the life zones of the shoreline. The combination of physical and biologic factors frames and limits the range of an organism's existence. It, as well as the biologic role the organism plays in its habitat, defines the organism's **niche**. The complex interplay of physical and biologic conditions and the variety of shore life itself create an abundance of niches for intertidal organisms.

Within a particular type of coastal environment, the interrelated influences of tidal exposure, bottom type, and intensity of wave shock produce an infinitely varied set of vertically arranged habitats. The vertical distribution of coastal plants and animals reflects the vertical changes in the shoreline's environmental conditions. Different species tend to occupy different levels or zones within the intertidal shoreline. Quite often, each zone is sharply demarcated from adjacent zones immediately above or below by the color, texture, and general appearance of the

species living there. The result is a well-defined vertical series of horizontal life zones, sometimes extending for substantial distances along a coastline.

As the tides advance and recede over the intertidal portions of the shoreline, so too do the vertically graded changes in temperature, light intensity, degree of predation, and other environmental factors. Ocean tides are not the sole cause of vertical zonation within the intertidal shoreline. They do, however, modify and compress the pattern of zonation to make it more pronounced. (Vertically arranged sequences of plant and animal populations also can be found on mountainsides, in the sea below low tide, and in tideless lakes and ponds.) These unifying and quite visible patterns of vertical zonation on rocky and sandy shorelines provide an appropriate framework within which to compare and describe the marine life of a few selected shores.

Rocky Shores

Chapter 2 emphasized that ecosystems have two complementary pathways for energy transfer: a grazing food web and a detritus food web. The grazing food web routes energy and nutrient material from the primary producers through the grazers and predators. Detritus feeders use the bits and pieces of dead and decaying matter available from all trophic levels within the ecosystem. Within the coastal zone, neither rocky shores nor sandy beaches appear to be complete ecosystems by themselves. The trophic relations within rocky shore communities exhibit well-developed and complicated grazing food webs but very little in the way of detritus food webs. The erosional nature of rocky shores simply prohibits the accumulation of detritus and the existence of those animals dependent on it for food.

In rocky intertidal communities, patterns of distribution and abundance, as well as trophic relationships, are complex and sometimes change dramatically over short distances and from season to season. A shaded northern exposure may harbor several species absent from nearby sunny slopes. Tide pools contain an assemblage of plants and animals quite different from well-drained platforms a meter away. The variety of life on one side of a boulder may differ markedly from life on the other side. And if you look under the boulder, still other species may be found.

With such a bewildering array of niches available on a small stretch of shoreline, it might seem improbable to find recurring themes of vertical zonation on widely separated shorelines. Yet similar patterns of zonation do exist on temperate rocky shores, whether in New England, Australia, British Columbia, or South Africa. Vertical zonation is such a compelling feature of life on rocky shores that considerable effort has been expended in devising schemes to identify and describe distinct intertidal subzones and their inhabitants. For those of you who desire a detailed account of life on intertidal rocky shores, several works with regional emphasis are listed at the end of this chapter.

The following sections examines some of the more conspicuous intertidal plants and animals and the adaptations that permit them to remain conspicuous. Three rather ill-defined zones, the upper, middle, and lower intertidal zones, are used for general reference. Because the boundaries separating these zones are artificial, they are frequently violated by their residents. You are likely to see species that dominate one zone scattered throughout other zones as well. This intertidal overview begins at the top, the upper intertidal zone, and proceeds downward to the shallow subtidal portions of temperate coastal shorelines.

The Upper Intertidal

In the upper intertidal, living conditions are sometimes nearly as terrestrial as they are marine. The area is wetted only infrequently by extremely high tides and splash from breaking waves and is sparsely inhabited by marine organisms. Scattered dark mats of the cyanobacterium *Calothrix* or the lichen *Verrucaria* frequently form bands or series of tar-like patches to mark the uppermost part of the rocky intertidal. Small tufts of *Ulothrix,* a filamentous green alga, may also extend into the highest parts of the intertidal. These plants are tolerant to large temperature changes and are adapted to resist desiccation. The small tangled filaments of *Calothrix* are embedded in a gelatinous mass to maintain their store of water and to reduce evaporation. Lichens such as *Verrucaria* are symbiotic associations of a fungus and a unicellular alga (FIGURE 8.17). In the case of *Verrucaria,* the fungal part absorbs and holds several times its weight of water, water used by the fungus as well as by the photosynthetic algal cells that produce food for the entire lichen complex.

Only a few species of snails, limpets, and occasional crustaceans graze on the sparse and scattered vegetation of the upper intertidal (FIGURE 8.18). Unlike its other close marine relatives, the small snail *Littorina* is an air breather. *Littorina* uses a highly vascularized mantle cavity in much the same manner as land snails do for gas exchange. Some species of littorine snails

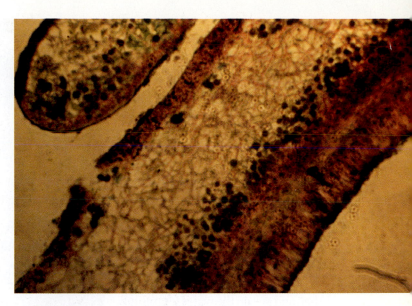

figure 8.17

Magnified cross-section of a lichen with algae cells (dark spots) embedded in fungal filaments (light stands).

figure 8.18

Two members of a sparsely populated rocky upper intertidal: the shore crab, *Pachygrapsus,* and the periwinkle snail, *Littorina.*

are so well adapted to an air-breathing existence that they drown if forced to remain underwater. Like littorines, limpets of the upper intertidal (especially *Acmaea*) are amazingly tolerant of temperature changes. Both littorines and limpets can seal the edges of their shell openings against rock surfaces to anchor themselves and to retain moisture. They are algal grazers and use their file-like radulae to scrape the small algae and lichens from the rocks.

A conspicuous zone of small barnacles (FIGURE 8.19) frequently appears just below the lichens and cyanobacteria. Barnacles are filter feeders, but in the high intertidal they are able to feed only when wetted by high spring tides a few hours each month. While submerged, their feathery feeding appendages extend from their volcano-shaped shell and sweep the water for minute plankton (Figure 8.5). Between high tides, a set of hinged calcareous plates blocks the entrance to the shell and seals in the remainder of the animal.

Most barnacles are hermaphroditic; they contain gonads of both sexes. Yet most generally refrain from fertilizing their own eggs. During mating, a long tubular penis is extended into a neighboring barnacle, and the sperm are transferred to fertilize the neighbor's eggs. The eggs develop and hatch within the barnacle's shell and are released as microscopic free-swimming planktonic organisms known as **nauplii** (FIGURE 8.20a). After several molts of its exoskeleton, the nauplius develops into a **cypris** larva (Figure 8.20b). The cypris eventually settles to the bottom, selects a permanent settling site, and then cements itself to the bottom with a secretion from its antennae. Cypris larvae are attracted by the presence of other barnacles, thus ensuring settlement in areas suitable for barnacle survival and for obtaining future mates. Soon after settling, the cypris turns over, loses its larval appearance, and begins to surround itself with a wall of calcareous plates (Figure 8.20c) characteristic of the adult it will become.

Two species of intertidal barnacles provide a clear example of how the interplay of physical and biologic factors influences the eventual vertical distribution of adult barnacles. In England, the larvae of one barnacle, *Chthamalus*, settle principally in the upper half of the intertidal, whereas the larvae of the other barnacle,

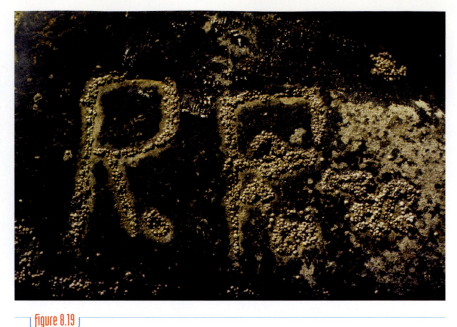

figure 8.19

Stunted acorn barnacles, *Chthamalus,* survive in the shallow depression of carved letters.

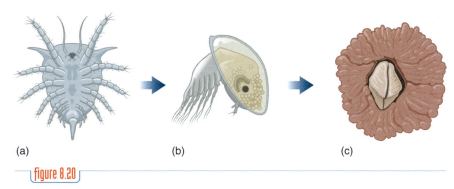

(a)   (b)   (c)

figure 8.20

Planktonic and early benthic stages of the barnacle *Balanus:* (a) nauplius stage, (b) cypris stage, and (c) early benthic stage.

Balanus, settle throughout the entire intertidal range (FIGURE 8.21). Desiccation in the upper extremes of the intertidal rapidly eliminates a good number of the settled *Balanus* but has little effect on *Chthamalus* at the same levels. Below the level of significant desiccation effects, however, *Balanus* is clearly the better competitor for space, overgrowing and undercutting *Chthamalus* wherever the two species overlap. The resulting distribution of adult forms of both species provides a useful generalization applicable to other attached animals living in space-limited intertidal situations: The upper limit of species distribution is restricted by the species' ability to cope with environmental stresses and other physical factors, such as temperature or desiccation, whereas a species' lower vertical range is limited by biologic factors, especially competition with other species.

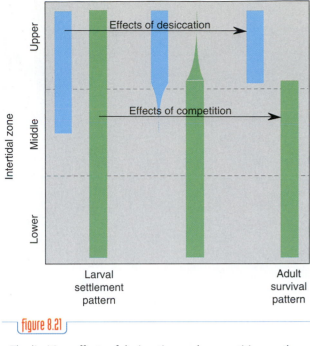

figure 8.21

The limiting effects of desiccation and competition on the vertical distribution of two species of intertidal barnacles, *Chthamalus* (blue bars) and *Balanus* (green bars). (Adapted from Connell 1961.)

figure 8.22

The aggregate sea anemone, *Anthopleura elegantissima*. Individuals exposed during low tide have retracted their tentacles and covered themselves with light-reflecting shell fragments.

The Middle Intertidal

The middle intertidal is occupied by greater numbers of individuals and species than is the upper intertidal zone. This zone, sufficiently inundated by tides and waves, provides an abundance of plant nutrients, oxygen, and plankton food for filter-feeding animals. The lush growths of green, red, and brown algae also furnish a bountiful supply of locally produced food for grazers.

Occasional small, water-filled, tide pools protect hermit crabs, snails, nudibranchs, anemones, and a few small fish species from exposure and the physical assault of the surf. Aggregate anemones of the Pacific Coast (*Anthopleura elegantissima*) are known to withstand internal temperatures as great as 13°C above the surrounding air temperature. Yet serious water loss will destroy them. These anemones combat extreme desiccation and temperature fluctuations by retracting their tentacles and attaching bits of light-colored stone and shell to themselves, presumably to reflect light and heat (**FIGURE 8.22**).

The clumped mats characteristic of the aggregate anemone are the result of a peculiar mode of asexual reproduction. To divide, these anemones pull themselves apart by simultaneously creeping in opposite directions. Each half quickly regenerates its missing portion, producing two new individuals to replace the original. All the members of a clump resulting from this asexual fission are **clones**, or genetically identical individuals, with the same sex and color patterns. The clonal clumps of anemones are uniformly spaced and are separated from adjacent clones by bare zones about the width of a single anemone. These anemones also have separate sexes and can reproduce sexually by releasing eggs and sperm into the water.

The bare zones between anemone clumps result from a subtle type of competition between dissimilar individuals from adjacent clones. They are armed with special tentacles, or **acrorhagi**, that can inflict serious damage to anemones of opposing clones but that have no effect on individuals of the same clump. In these border wars, anemone clumps rely on a mechanism of self-recognition so that the aggressive response is directed only to members of genetically dissimilar clones.

The dominant and conspicuous members of the middle intertidal zone (**FIGURE 8.23**) are mussels (*Mytilus*), barnacles (usually *Balanus*), some chitons and limpets, and several species of brown algae (especially *Fucus* or *Pelvetia*). The animals securely anchor themselves to the substrate and generally present low rounded profiles to minimize resistance to breaking waves

figure 8.23

Close-up view of mussels, *Mytilus*, and acorn barnacles, *Balanus*, in the middle intertidal.

figure 8.24

Fucus, a brown alga, thrives on a Maine intertidal rock.

figure 8.25

A bed of acorn barnacles (*Balanus*) severely deformed because of crowding. Because the barnacles were inhibited by their neighbors from expanding laterally, they instead grew upward.

(Figure 8.16). Plants living here are secured by strong holdfasts and usually have sturdy but flexible stipes to absorb much of the wave shock. Both *Fucus* and *Pelvetia* have thickened cell walls to resist water loss during low tide (FIGURE 8.24).

In the densely populated middle intertidal zone, mussels, barnacles, brown algae, and other sessile creatures are limited by two commonly shared resources: the solid substrate on which they live and the water that provides their dissolved nutrients and suspended food. Mussels, barnacles, and algae compete for these critical resources in different ways. In regions where physical factors permit each to survive, these three competing groups interact by dominating the available attachment space or by overgrowing their competitors and monopolizing the resources available from the water (FIGURE 8.25).

Available space on the rock surfaces of the middle intertidal is crucial for survival, yet it is seldom fully used. A number of interacting biologic and physical processes occasionally create patches of open space. Sea stars and predatory snails continually remove patches of barnacles and mussels. Seasonal die-offs of algae, and even battering by heavy surf and drifting logs, also clear patches for future settlement and competition.

Chance events and their relationship to seasonal patterns of reproduction wield appreciable influence in settlement patterns. Most of the middle intertidal animals have planktonic larval stages capable of set-

tling almost anywhere within the intertidal zone. Algal spores and barnacle larvae simultaneously settling on bare rock eventually grow and compete for available space. Because of the space limitations that may exist on exposed coasts, the algae are usually squeezed out as the barnacles increase in diameter and dislodge or overgrow them. On some sheltered rocky coasts, recruitment of barnacle larvae is prevented by sweeping action of wave-tossed algal blades. Only in this manner can *Fucus* achieve and maintain spatial dominance over barnacles. If the barnacles are not removed before they are securely cemented into place, they escape the adverse effects of algal blades because of growth and may eventually force *Fucus* off the rocks.

Barnacles are not necessarily safe once they have outgrown or overgrown their algal competitor. They are consumed in prodigious numbers by sea stars, carnivorous snails, and certain fishes. Even herbivorous limpets have a detrimental, and sometimes severe, impact on barnacle populations. Young barnacles are eaten or dislodged by limpets bulldozing their way through their grazing activities. Limpets seem to have less effect on the small crack-inhabiting *Chthamalus* than on the larger more exposed *Balanus*. Thus, in the presence of limpet disturbance, *Chthamalus* gains a slight competitive advantage over the otherwise dominant *Balanus*.

The larval stages of mussels do not require bare rock exposures; they will settle on algae and barnacles and among aggregates of adult mussels. After settling, the young mussels crawl over the bottom, seeking better locations before they permanently attach themselves to the substrate with several strong elastic **byssal threads** (see Figure 5.24). Byssal threads are formed from a fluid secreted by an internal byssal gland. The fluid flows down a groove in the small tongue-shaped foot and onto the substrate. On contact with seawater, the fluid quickly toughens to form an attachment plate and a thread. Then the foot is moved slightly and additional threads and plates are formed.

If left undisturbed, mussels eventually overgrow barnacles and algae. Seldom, however, do rocky intertidal conditions remain undisturbed for long. Mussels are extensively preyed on by sea stars (such as *Pisaster* on the Pacific Coast and *Asterias* on the Atlantic Coast). These sea stars are quite sensitive to desiccation and are limited to sites that remain submerged most of the time (FIGURE 8.26). Consequently, their impact on mussel populations is much more severe in the lower portions of the mussels' intertidal range. Young mussels also are devoured by *Nucella* and other predatory snails. Some mussels survive these predatory onslaughts by numerically swamping an area with more individuals than the local predators can consume. In time, the mussels may also escape, through growth, becoming so large that sea stars cannot open them and predatory snails are incapable of drilling through their shells to consume the soft flesh within.

As patches of mussels are cleared out by predators or broken off by waves, they are temporarily replaced by algae or barnacles, but gradually the mussels regain their ascendancy. In this way, diversity of species is maintained. It is ultimately the dynamic interplay resulting from competition between these dominant organisms and the patterns of disturbance that affect their survival that shapes the biologic character of the middle intertidal. These organisms, in turn, influence the distribution and abundance of other plant and animal species.

Living on the mussel shells or among the thick masses of byssal threads beneath is a complex community of more fragile and often unseen animals. This submussel habitat protects several common species of clams, worms, shrimps, crabs, hydroids, and many types of algae. Many of these species use mussel shells as available solid substrate for attachment. Others exist because they are unable to survive in the same area without the protection afforded by the canopy of mussels overhead.

This complex association of organisms is wholly dependent on the existence of thick masses of well-anchored mussels. When natural disturbances remove the mussels and disrupt the stability of the community, they are quickly replaced by a predictable succession of nonmussel populations (FIGURE 8.27). Eventually, this process of biologic succession may reach the climax stage: in this case, the mussel bed–gooseneck barnacle–sea star community. However, the structure

figure 8.26

Sea stars, *Pisaster*, aggregating near the low tide line to avoid desiccation.

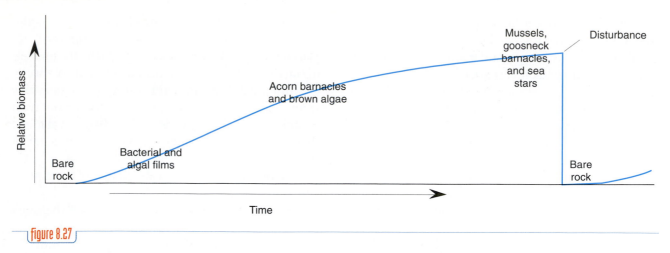

figure 8.27

General pattern of succession through time on temperate rocky shores. Blue curve indicates relative biomass.

of these communities is seldom stable for long; rather, they achieve a state of dynamic equilibrium between the stabilizing effect of succession and the many disruptive factors that reduce that stability.

The Lower Intertidal

The biologic character of lower intertidal rocky coasts differs markedly from the zones above it. It is difficult for some species and impossible for others to tolerate the more exposed conditions found in the upper and middle intertidal zones. The few species that do are often present in vast numbers. In the lower intertidal, the emphasis changes to a community with a high diversity of species, often without the conspicuous dominant types so characteristic of the middle and upper intertidal.

The lower intertidal of rocky shores abounds with seaweeds. Brown, red, and even a few species of green algae of moderate size spread a protective canopy of wet blades over much of the zone. In other places, extensive beds of seagrasses achieve a similar effect (FIGURE 8.28). Tufts of small filamentous brown and red algae carpet many of the rocks. Calcareous red algae become especially prolific at these levels. The pinkish hue of *Lithothamnion* encrusting rocks and lining the sides of tide pools is a common sight (see Figure 4.14).

The animals of the lower intertidal include species from several animal phyla. It is here that the diversity, complexity, and sheer beauty of intertidal marine life abound. On the east coast of the United States, a large white anemone, *Metridium,* occurs in tide pools and on exposed portions of the lower intertidal. *Anthopleura,* a beautiful green anemone (FIGURE 8.29), occupies a similar habitat on the west coast. Securely anchored

figure 8.28

Surf grass interspersed with brown algae in the lower intertidal.

by discs at their bases, anemones are predators of planktonic animals and small fishes. They capture prey by discharging many microscopic nematocysts from special cells in their tentacles. When touched with a finger, nematocysts of sea anemones produce a slight tingling sticky sensation.

The batteries of nematocysts found on anemone tentacles effectively discourage the hostile intentions of most predators, but they do not guarantee complete immunity against predation. A few snails and sea spiders penetrate the sides of anemones and feed on the unprotected tissues. Eolid nudibranchs also commonly graze on anemones and the closely related

figure 8.29

The green anemone, *Anthopleura xanthogrammica*.

figure 8.30

A nudibranch, *Hermissenda,* with long finger-like cerata projecting from its dorsal surface.

hydroids. These nudibranchs possess mechanisms, not yet completely understood, that block the discharge of the toxic nematocysts. During digestion, the undischarged nematocyst-containing cnidocytes are preserved and passed to special storage sacs in the rows of finger-like **cerata** along the back of the nudibranch (FIGURE 8.30). There the cnidocytes serve as second-hand defensive stinging cells to be used against potential predators.

The echinoderms are another familiar group of animals in the lower intertidal. Sea stars, sea urchins, brittle stars, and sea cucumbers are all quite sensitive to desiccation and salinity changes and are seldom seen in abundance above the lower intertidal zone. Although slow moving, sea stars are voracious predators of mussels, barnacles, snails, an occasional anemone, and even other echinoderms (Figure 8.6).

Sea stars continuously emit chemical substances that initiate alarm reactions in their prey; actual contact usually leads to even more vigorous escape movements. Prey species apparently recognize and identify their sea star predators by the substances they exude. They react violently to the touch or presence of sea stars that usually prey on them, but they seldom respond to those not encountered in their normal habitat. When approached by some species of sea stars, many normally sessile species execute remarkable escape responses. Scallops swim jerkily away, clams and cockles leap clear of the sea star, and sea urchins and limpets crawl away relatively rapidly. Some sea anemones detach themselves and somersault or roll aside when touched by certain sea stars (FIGURE 8.31).

figure 8.31

A swimming anemone, *Actinostola,* evading a predatory sea star. The anemone has detached itself from the bottom and is somersaulting away.

In summary, rocky intertidal zones of temperate shores have a dynamic pattern of organization dominated by physical forces at the upper extreme that diminish in influence downward and are gradually replaced by competition, predation, and other biologic interactions. FIGURE 8.32 shows the general pattern of vertical zonation of intertidal plants and animals on temperate rocky shores.

figure 8.32

Vertical zonation patterns on a 3-m-high rock on the coast of Oregon.

Sandy Beaches

Beaches and mudflats, the ecologic complement to rocky shores, are depositional features of the coastal zone and are best developed along sinking (or subsiding) coastlines. Beaches and mudflats are unstable and tend to shift and conform to conditions imposed by waves and currents. Large plants find the shifting nature of soft sediments difficult to cope with, and few exist there. The few plants that have managed to adapt support even fewer grazers. Detritus food webs dominate on these depositional shores. Bits of organic material washed off adjacent rocky shores and the surrounding land or drifted in from kelp beds farther offshore sustain the detritus feeders of sandy and muddy shores.

Beaches are made of whatever loose material is available. Quartz grains, black volcanic sand, or pulverized carbonate plant and animal skeletons are most common. Beaches occur where waves are sufficiently gentle to allow sand to accumulate but still strong enough to wash the finer silts and clays away. A good portion of the sand on many beaches is eroded away by large winter waves and deposited as underwater sandbars offshore. Smaller waves the following summer move the sand back on shore. Longshore currents also slowly move beach sands parallel to the shore. In response, populations of beach inhabitants may fluctuate widely from season to season and from one year to the next.

Several properties of marine sediments are defined by the size and shape of sediment particles. The size of spaces between sediment particles (the interstitial spaces) decreases with finer sediments. Interstitial-space size, in turn, regulates the porosity and permeability of sediments to water. In coarse sands, water flows freely between sand grains, recharging the supply of dissolved oxygen and flushing away wastes. Sand beaches also drain and dry out quickly during low tide.

When compared with the teeming populations of the rocky intertidal, beaches appear to be biologically desolate. Macroscopic algae and large obvious epifauna are rare. Shifting unstable sands are unsuitable platforms for surface anchorage, and nearly all the permanent residents of beach communities dwell below the sand surface. Patterns of zonation are more difficult to demonstrate, yet under the sand there are distinguishable life zones comparable with those on rocky shores (FIGURE 8.33).

The upper portions of sandy beaches along temperate coasts are occupied by a few species of amphipods, particularly *Talorchestia*. The common name of "beach hopper" reflects the unusual bounding mode of locomotion these small crustaceans use. Beach hoppers prefer to burrow a few centimeters into the sand during the day but are most active at night. Occasionally, they make excursions down the beach face as the tide recedes.

The middle beach is frequently populated by a variety of other amphipods, lugworms (*Arenicola*), dense concentrations of isopod crustaceans, and the sand crab, *Emerita*. *Arenicola* occupies a U-shaped burrow, with its head usually buried just below a sand-filled surface depression. The burrows are more or less permanent, because waves stir up and move sediment and detritus to the head region where they are consumed. Mounds of coiled castings indicate the location of the other end of this sediment ingester (FIGURE 8.34).

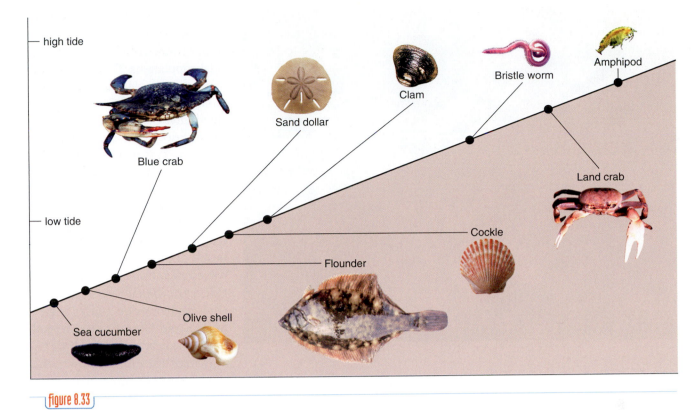

figure 8.33

Sandy beach zonation along the east coast of the United States. The species change rapidly from the portion permanently under water at left to the dry part of the beach above high tide at right.

figure 8.34

Coiled fecal casting of the lugworm, *Arenicola*.

The sand crab illustrates another feeding mode common to many beach macrofauna. When feeding, *Emerita* burrows tail first into the sand and faces down the beach. Only its eyes and a pair of large feathery antennae protrude above the sand (FIGURE 8.35). When a wave breaks over the crab and begins to recede, the antennae are extended against the rush of water. En-

trapped phytoplankton (and possibly even large bacteria) are swept into the filtering antennae and then moved to the mouth by other feeding appendages.

Small isopods, usually less than 1 cm long, actively prey on even smaller interstitial animals that inhabit the interstitial spaces between sand grains. Many animal phyla are represented in the interstitial fauna of beaches, and a few groups such as harpacticoid copepods and gastrotrichs are practically confined to the interstices of beach sands. Despite their divergent backgrounds, most interstitial animals exhibit a few basic adaptations needed for life between sand grains. They tend to be elongated, small (no more than a few millimeters), and move with a sliding motion between sand grains without displacing them. Examples of interstitial animals from different phyla are shown in FIGURE 8.36. Some interstitial animals are carnivorous; others feed on detritus deposits and material in suspension. A specialized feeding habit, unique to interstitial animals, is sand licking. Individual sand grains are manipulated by the animals' mouthparts to remove minute bacterial growths and thin films of diatoms.

figure 8.35

Sand crabs, *Emerita,* with feeding antennae extended.

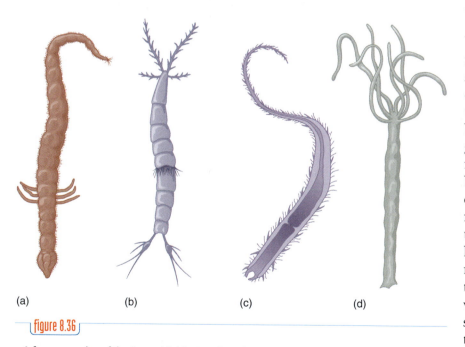

(a)   (b)   (c)   (d)

figure 8.36

A few examples of the interstitial fauna of sandy beaches. Each is of a different phylum, yet all exhibit the small size and worm-shaped body characteristic of interstitial fauna: (a) a polychaete, *Psammodrilus,* (b) a copepod, *Cylindropsyllis,* (c) a gastrotrich, *Urodasys,* and (d) a hydra, *Halammohydra.* (Redrawn in part from Eltringham 1972.)

In the lower portion of intertidal beaches, the diversity of life increases. Polychaete worms, still other amphipods, and an assortment of clams and cockles appear. Many of these lower beach inhabitants, such as soft-shelled clams and cockles of the Atlantic Coast, represent the upper fringes of much larger subtidal populations. The small wedge-shaped bean clam, *Donax,* of the Atlantic and Gulf Coasts (but not the Pacific Coast species) migrates up and down the beach with the tides, yet it is usually considered an inhabitant of the lower beach. *Donax* responds to the agitation of incoming waves of rising tides by emerging from the sand to feed. After the wave carries the clam up the beach, it digs in to await the next wave and another ride. During ebb tides, the behavior is reversed. *Donax* emerges only after a wave breaks and begins to wash back down the beach. So, with little energy expenditure of its own, this small clam capitalizes on the abundance of available wave energy to carry it up and down the beach face.

Donax is one of many sandy beach inhabitants to exhibit a rhythmic behavior that corresponds to the tidal cycle. Fiddler crabs quietly sit submerged at high tides and then emerge from their burrows at low tide to feed or engage in social activities. This tide-related cycle of activity exhibited by fiddler crabs is known as a **circalunadian rhythm**; it is synchronized to moon-related tidal cycles that repeat every lunar day (24.8 hours). When fiddler crabs are removed to the laboratory, their activity rhythms remain in concert with the changing tidal cycle for some time despite the absence of tidal cues. The same species of fiddler crabs are light colored at night but darken during the day; this cycle of color changes depends on **circadian rhythms** based on a solar day length of precisely 24 hours.

Grunion (*Leuresthes,* FIGURE 8.37) are small fish inhabiting coastal waters of southern California. They too exhibit a very tightly timed spawning behavior related to ocean tide cycles. On the second, third, and fourth nights after each full or new moon of the spring and summer spawning season, grunion move up on the beach by the thousands to deposit their eggs in the sand and away from water. Their precise timing is remarkable; they spawn only during the first 3 hours immediately after the highest part of the highest spring tides. During the spring and summer, these tides occur only at night.

As the highest spring tides occur at the time of full and new moons, the grunion spawn immediately after high tides, but they spawn on successively lower tides each night (FIGURE 8.38). Thus, the eggs are buried by sand tossed up on the beach by the succeeding lower tides, and they are not washed out of the sand until the next series of spring tides. Nine or 10 days after the last spawning, tides of increasing height reach the area where the lowest eggs were buried (Figure 8.38). Wave action erodes the sand away and bathes the eggs with seawater. Almost immediately after being agitated and wetted by the waves, the eggs hatch and the young grunion swim away. They remain in shallow coastal waters to feed and grow, reaching sexual maturity and their first spawning about 1 year later.

Like fiddler crabs and grunion, most, and probably even all, organisms have an innate time sense, a "biological clock." Rhythmic cycles of body temperatures, activity levels, oxygen consumption, and a host of other physiologic variations occur independently of changes in or signals from the external environment.

figure 8.37

Grunion, *Leuresthes,* spawning in the sands of a southern California beach. Several males are coiled around a female who has dug herself into the sand to deposit her eggs.

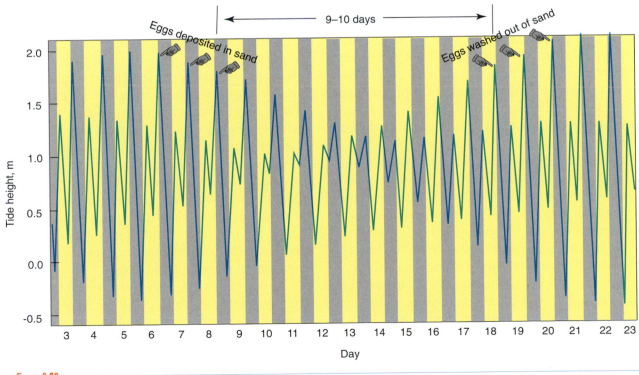

figure 8.38

Predicted tide heights for a 3-week period at San Diego, California. Spring tides appropriate for grunion spawning occur on days 6, 7, and 8 (pointers at left). Nine to 10 days later, the next set of spring tides (pointers at right) wash the eggs from the sand and they hatch. Shaded portions indicate night hours.

The internal "mechanism" of the clock is not known, but its existence has been demonstrated in a wide variety of organisms ranging from diatoms to humans.

Oiled Beaches

Oil naturally floats on water. This seemingly simple observation means that when oil is released (intentionally or not) into the ocean, most of it will eventually come to rest where the sea surface meets land: the intertidal. Problems of oil contamination resulting from tanker spills, well seepages or blowouts, or intentional dumping of waste oil occur throughout the world ocean, and temperate coasts have experienced a large portion of these.

figure 8.39

Oil-soaked murre after the *Exxon Valdez* oil spill.

Despite the spectacular nature of major spills from tankers or offshore drilling and production platforms, more oil actually enters the marine environment during an average year as runoff from urban streets and parking lots, from leaking underground storage tanks and improperly dumped waste oil, and in bilge water from nearly every freighter, fishing vessel, and military ship afloat. The numerous sources and more mundane aspects of these oil sources make them more difficult to manage than a single dramatic spill of the same volume of oil from a wrecked tanker. Yet it is tanker spills that engender the public attention, scientific scrutiny, and political activism needed to tackle the problem of oil contamination in the marine environment.

Amoco Cadiz (1978, France), *Exxon Valdez* (1989, Alaska), *Brear* (1993, England), *Prestige* (2002, Spain)—these are some of the more infamous catastrophic oil tanker spills that have occurred in the past quarter century. Because spilled oil floats on seawater, it provides a constant visible (and olfactory) reminder of its presence until it is washed ashore, it sinks, or it evaporates. The impact of oil on seashore life is immediate and can be long lasting, suffocating benthic organisms by clogging their gills and filtering structures or fouling their digestive tracts. Marine birds and mammals are especially vulnerable if their feathers or fur become oil-soaked and matted (FIGURE 8.39), causing them to lose insulation and buoyancy.

In March 1989, the supertanker *Exxon Valdez* ran aground on Bligh Reef in Alaska's Prince William Sound. The damage to the tanker resulted in the largest oil spill to date in US waters (242,000 barrels, or nearly 40,000,000 liters). The spill occurred in an area noted for its rich assemblages of seabirds, marine mammals, fishes, and other wildlife. It was one of the most pristine stretches of coastal waters in the United States, with specially designated natural preserves such as the Kenai Fjords and Katmai National Park.

The area of the spill, because of its gravel and cobble beaches, was particularly sensitive to oil. In places, the thick tarry crude oil penetrated more than a meter below the beach surface (FIGURE 8.40). High winds, waves, and currents in the days after the accident quickly spread oil over 26,000 km². The toll on wildlife was devastating. More than 33,000 dead birds were recovered, and hundreds of sea otters, seals, sea lions, and other marine mammals, as well as many thousands of commercial and noncommercial fishes, were killed. Traditional fishing and cultural activities of the native communities in Prince William Sound were halted for more than a year, creating enormous social and economic costs.

In November 2002, the even larger supertanker *Prestige* split apart about 200 km off the northwest coast of Spain and lost much of its cargo of oil. The seas at the site of the *Prestige* spill were too rough, so little could be done but watch the oil wash ashore, eventually to form a 1-m deep foamy mixture of oil and sea water known as "chocolate mousse." The volume of oil released from the *Prestige* was greater than that from the *Exxon Valdez* and, because of higher

A heavily oiled beach in Prince William Sound after the *Exxon Valdez* oil spill.

water temperatures along the Spanish coast, was also more toxic. The slick threatened one of Europe's most picturesque and wildlife-rich coasts, with more than a thousand Spanish and French beaches contaminated with spilled *Prestige* oil. Local fisheries have been destroyed, and current studies indicate that the environmental damage and the clean-up and remediation costs will be substantially greater than those incurred in the *Exxon Valdez* spill.

Once oil washes ashore, the issue becomes how to preserve affected plants and animals while minimizing damage to the surrounding ecosystem. One solution is to leave it alone and let natural processes take their course. Eventually, even the oiliest beaches will recover, but without human intervention it can take a long time. Fifteen years after the *Exxon Valdez* spill, patches of hardened oil similar to asphalt remain scattered throughout the area affected by the spill.

Cleaning oiled beaches is best done manually, with shovels and absorbent bags wielded by people careful to remove the oil while preserving the clean areas in between. Recovering oiled animals and removing oil from a beach is a difficult and messy job (FIGURE 8.41). These and other intensive clean-up options have their drawbacks. When high-pressure hot water was used to scrub parts of the Alaskan shoreline

oiled by the *Exxon Valdez* spill, plant and animal populations inhabiting these beaches recovered more slowly than those that were left alone.

Another reasonably gentle option is bioremediation, a process that encourages the growth of natural populations of oil-digesting bacteria. Bioremediation is not viable for all spills. The light crude oil from the *Exxon Valdez* was biodegradable and easily broken down; the oil from the *Prestige* was heavier and thicker and is not expected to respond as well to bioremediation.

Using a combination of manual labor and heavy machinery, the clean-up of the *Prestige* spill continues as this is being written. As the coastline gradually becomes cleaner, and clean-up efforts cost more for progressively less reward and more environmental damage, the difficult decision becomes when to stop. Once the bulk of the oil is recovered, the remaining traces of oil still may be toxic, and oil that has seeped into sediments may continue to harm organisms for decades. The legal dispute over the damage caused by the *Exxon Valdez* continues more than 10 years after that spill, as investigators monitor ongoing damages and detect new ones from the remaining oil. Similar biologic, legal, and political difficulties are sure to arise in the aftermath of the *Prestige* oil spill.

figure 8.41

Rescue workers collecting harbor seal pup covered with oil from the *Exxon Valdez*.

8.5 Shallow Subtidal Communities

At the low tide line, the lower intertidal merges with the uppermost part of the inner shelf zone. Where rocky bottoms extend below the low tide line and are not covered by sediments, the transition from intertidal to subtidal is gradual. Many plant and animal species common to the lower intertidal also are abundant in neighboring shallow subtidal regions. However, rocky substrates eventually give way to soft sediments. In protected stretches of coastlines or below the sea surface, wave action diminishes and loose sediments and detritus begin to accumulate. Organisms characteristic of the rocky shore are replaced by those typical of sand or mud bottoms.

Extensive studies of the life in shallow-water soft sea bottoms were initiated by the Danish biologist, C. G. J. Petersen, in the early part of the 20th century. His intent was to evaluate the quantity of food available for flounders and other commercially useful bottom fish. After sorting and analyzing thousands of bottom samples from Danish seas, Petersen concluded that large areas of the level sea bottom are inhabited by recurring associations, or communities, of infaunal species (FIGURE 8.42). Each community has a few very conspicuous or abundant macrofauna as well as several less obvious forms. On other bottom types, different distinct communities of other species can be found. When exposed to similar combinations

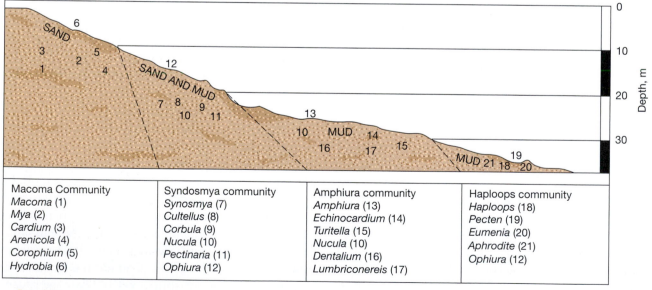

Macoma Community	Syndosmya community	Amphiura community	Haploops community
Macoma (1)	*Synosmya* (7)	*Amphiura* (13)	*Haploops* (18)
Mya (2)	*Cultellus* (8)	*Echinocardium* (14)	*Pecten* (19)
Cardium (3)	*Corbula* (9)	*Turitella* (15)	*Eumenia* (20)
Arenicola (4)	*Nucula* (10)	*Nucula* (10)	*Aphrodite* (21)
Corophium (5)	*Pectinaria* (11)	*Dentalium* (16)	*Ophiura* (12)
Hydrobia (6)	*Ophiura* (12)	*Lumbriconereis* (17)	

figure 8.42

A series of soft-bottom benthic communities found at different depths in Danish seas, including bivalve mollusks (1, 2, 3, 7, 8, 9, 10, 19), polychaete worms (4, 5, 17, 20, 21), gastropod mollusks (6, 15), scaphopod mollusks (16), ophiuroid echinoderms (12, 13), echinoid echinoderms (14), and arthropod crustaceans (18).

As described in this chapter, the *Exxon Valdez* accident released 42 million liters of oil into Prince William Sound, the worst oil spill in U.S. history. Yet this oil spill ranks an unimpressive 53rd in terms of largest oil spills in world history according to the *Oil Spill Intelligence Report*. Much larger spills include the 260 million liters released by the *Amoco Cadiz* off Portsall in 1978 and the 162 million liters released by the collision between the *Atlantic Express* and *Aegean Captain* off Trinidad and Tobago in 1979, with an additional 158 million liters being lost while the ruptured ships were being towed to port (some experts estimate that as many as 420 million liters were spilled by this collision).

As impressive and devastating as these tanker accidents were, they rank only third on the list of sources of the largest oil spills in world history. The second largest source of oil spilled into the sea is blowouts of wells during drilling operations (caused by equipment failure, human error, and natural events, such as seismic activity, ice movements, and hurricanes). The largest blowout to date was the spill off the Mexican coast in 1979–1980. This blowout caused an oil spill that continued for about 9 months, eventually releasing about 530 million liters of oil into the ocean.

The source of the largest oil spills in world history is not tanker accidents or well blowouts; it is war. During the Iran-Iraq war of 1980–1988, both sides attacked each other's oil installations in an effort to cause economic loss. Some burning platforms spilled oil for several years before being capped. In total, as much as 570 million liters of oil may have been released.

The largest oil spill in world history was also the work of Saddam Hussein and his military forces during Desert Storm. Beginning in late January 1991, just days after Coalition forces began their assault against Iraq for invading Kuwait, Iraqi troops sabotaged oil terminals, tankers, and other installations, thus initiating the largest oil spill in world history (Figure B8.1). For months, Iraqi forces dumped an estimated 3000 barrels (over 495,000 liters) of oil per day into the Persian Gulf—about one *Exxon Valdez* oil spill every 12 weeks. Oil continued to leak into the Persian Gulf until at least late May 1991, spilling an estimated 910 million liters of oil. Oil contaminated nearly 780 km of shoreline and 1.3 million cubic meters of sediment.

War: The Forgotten Source of Oil Spills

Salt marshes and mangrove swamps were smothered, with 1000 dead crabs littering a single meter of the shore. The 1992 shrimp season in Saudi Arabia fell to 33% of the normal harvest. Hundreds of endangered sea turtles died in oily inshore waters. As many as 30,000 pelagic seabirds were killed, and the feeding grounds for 100,000 or more shorebirds were destroyed. Although the absence of baseline data is a problem, a survey of humpback whales, bottlenose and common dolphins, and finless porpoises revealed that the number of observed calves was small relative to the number of observed calves in other marine environments of the world.

As this chapter is being written, another war continues in the oil-rich Middle East. We can only hope that the environmental disasters caused by two previous wars in that region are not repeated.

Additional Reading

Earle, S. 1992. Persian Gulf pollution: assessing the damage one year later. *National Geographic* 181:122–134.

www.bioscience.jbpub.com/marinebiology

For more information on this topic, go to this book's website at http://www.bioscience.jbpub.com/marinebiology.

Figure B8.1 Oil deliberately spilled by the Iraqis during the Gulf War in 1991.

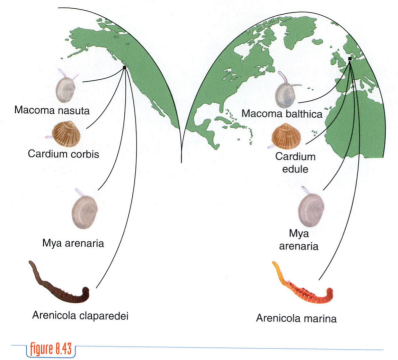

Macoma nasuta

Cardium corbis

Mya arenaria

Arenicola claparedei

Macoma balthica

Cardium edule

Mya arenaria

Arenicola marina

figure 8.43

Diagram showing the close similarity in composition of soft-bottom communities in the northeast Pacific and the northeast Atlantic. (Adapted from Thorson 1957.)

of environmental conditions, widely separated shallow-bottom communities in temperate waters closely resemble each other in structure and species composition (FIGURE 8.43).

Later benthic ecologists have found parallel shallow-water communities in much of the cold and temperate regions of the world ocean. This parallel community concept has been extended beyond obvious animal associations to include bottom type, depth, and water temperature as additional key factors in shaping benthic community structures. These infaunal communities exist only in soft muddy marine sediments that dominate the continental shelves of the world.

Where rocky outcrops exist on the shallow ocean floor in temperate areas and infaunal communities can not establish themselves, communities dominated by assemblages of brown algae develop instead. The large size (up to 30 m) of some kelp plants adds an important three-dimensional structure to kelp "forests" analogous to the canopy structure of terrestrial forests. Consequently, in the more complex ecologic terrain of kelp forests, more niches exist than do on nearby soft sediments, and these niches are occupied by a rich diversity of invertebrates on the sea floor and numerous fish species in the kelp canopy (FIGURE 8.44).

Most kelp plants are perennial. Although they may be battered down to their holdfasts by winter waves, their stipes will regrow from the holdfast for several successive seasons. Thus, the extent of the kelp canopy and the overall three-dimensional structure of the kelp forest is quite variable over annual cycles. Occasionally, herbivore grazing or the pull of strong waves frees the holdfast and causes the plant to wash ashore. More commonly, small fragments of blades and stipes are continually eroded away to decompose into food for detritus feeders.

Along most of the North American west coast, subtidal rocky outcrops are cloaked with massive growths of several species of brown algae, dominated by either *Macrocystis* or *Nereocystis* (FIGURE 8.45). West coast kelp forests occur as an offshore band paralleling the coastline because wave action tears these plants out nearer to shore and light does not penetrate to the sea floor farther offshore. In the dimmer light below the canopy of these large kelps exists a shorter understory of mixed brown and red algae. Together, these large and small kelp plants accomplish very high rates of primary production and support a complex community of grazers, suspension feeders, scavengers, and predators (FIGURE 8.46). From Central California northward, kelp abundance varies seasonally, and the fishes are dominated by several species of rockfishes in the genus *Sebastes*. In contrast, southern California kelp forest abundance varies irregularly and is especially vulnerable to the influences of El Niño–Southern Oscillation events. Here, the dominant fishes are not rockfishes; instead, perches, damselfishes, and wrasses abound, reflecting the more tropical affinities of these fishes.

Compared with the richness of species observed in western North American kelp forests, the kelp beds of the northwestern Atlantic Coast exhibit low diversity in most taxonomic groups. Unlike the US west coast, the rocky intertidal and subtidal shores of New England states and neighboring Canadian Maritime Provinces were scoured to bare rock (in places to several hundred meters below sea level) by several episodes of continental glaciation. Only since the retreat of the most recent glacial episode 8000 to 10,000 years ago have these shores been recolonized, and that recolonization is not yet complete.

The lower species diversity of northwestern Atlantic kelp beds leads to somewhat simpler trophic interactions than those occurring in US west coast kelp forests; still, similar species occupy the same major trophic roles (FIGURE 8.47). The macroscopic primary producers are dominated by the kelp, *Laminaria*, with an understory of mixed red and brown foliose algae. In clear patches below about 10 m, encrusting coralline red algae cover rock surfaces with a bright pink pavement of $CaCO_3$. These coralline crusts are maintained indirectly by the constant grazing actions of sea urchins on the larger kelp plants. On this coast, the lower limit of growth for *Laminaria* and other large kelps is controlled, not by low light intensity as it is on the west coast but by the presence of grazing urchins.

On the US west coast, too, kelp beds exist in a delicate balance with their major grazers, sea urchins. Since World War II, California beds on both coasts have been devastated by dense aggregations of sea urchins grazing on the holdfasts, causing the remainder of the plant to break free and wash onto the shore. These large urchin populations, capable of completely eliminating local kelp beds, seem free of the usual population regulatory mechanisms—predation and starvation. A major predator of West Coast kelp bed urchins is the sea otter (*Enhydra*, see Figure 6.30). East Coast sea urchins are similarly preyed upon by the lobster (*Homarus*). However, both of these predators have been subjected to intensive commercial harvesting and have experienced major population reductions in the past two centuries.

figure 8.44

Several species of kelp-community fishes sheltering under a large bladder kelp, *Macrocystis*.

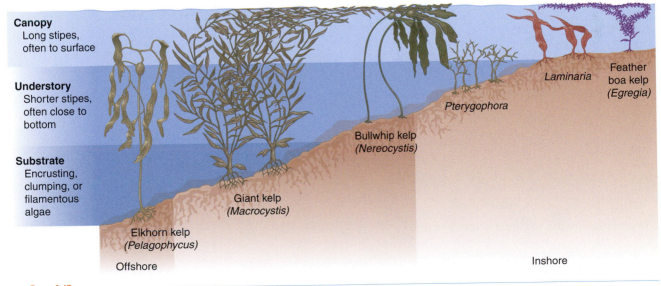

Canopy
Long stipes, often to surface

Understory
Shorter stipes, often close to bottom

Substrate
Encrusting, clumping, or filamentous algae

Elkhorn kelp
(*Pelagophycus*)

Offshore

Giant kelp
(*Macrocystis*)

Bullwhip kelp
(*Nereocystis*)

Pterygophora

Laminaria

Feather boa kelp
(*Egregia*)

Inshore

figure 8.45

General structure of a US west coast kelp forest, with a complex understory of plants beneath the dominant *Macrocystis* or *Nereocystis*.

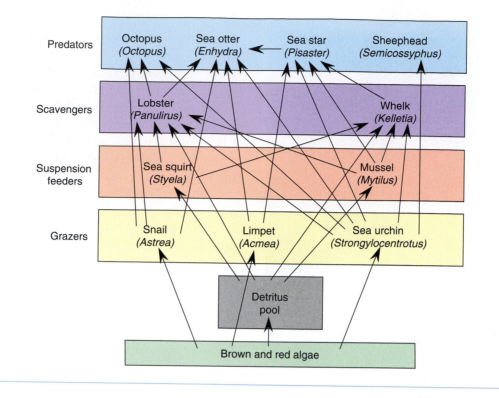

figure 8.46

Trophic relationships of some dominant members of a southern California kelp community.

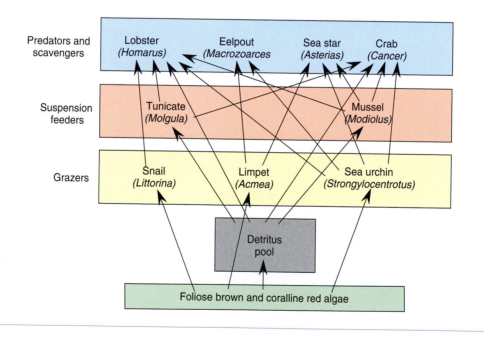

figure 8.47

Trophic relationships of the common members of a New England kelp community.

Available evidence indicates that the effects of this reduced predation have been magnified by increased concentrations of dissolved and suspended organic materials in coastal waters (mostly from urban sewage outfalls). The US Office of Technology Assessment has identified over 1300 major industries and 600 municipal wastewater treatment plants that discharge into the coastal waters of the United States. Standard secondary treatment of sewage is intended to separate solids and to reduce the amount of organic matter (which contributes to biochemical oxygen demand), nutrients, pathogenic bacteria, toxic pollutants, detergents, oils, and grease in wastewater.

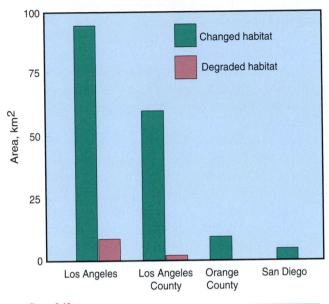

Extent of areas changed or degraded by four major sewage outfalls in the California Bight, 1978–1979. (Data from Mearns 1981.)

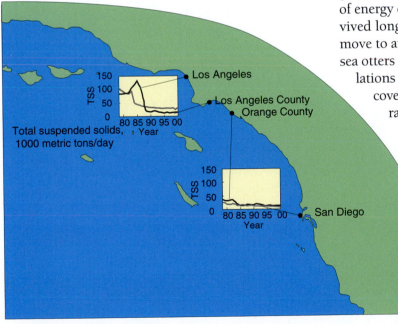

Graphs showing the reduction in discharged total suspended solids (TSS) for the past two decades at the four major sewage outfalls in the California Bight. (Data from Steinberger and Schiff 2002.)

In the United States, most ocean discharges of wastewater are supposed to meet those secondary treatment standards, but many still do not, including some that discharge into southern California coastal waters. Until the mid-1980s, treated sewage containing about a quarter of a million tons of suspended solids was discharged from 4 large and 15 small publicly owned sewage treatment plants each day. These solids are similar to detritus from natural marine sources in its general composition and nutritional value for zooplankton and benthic detritus feeders. Measurable changes in species diversity and biomass of benthic infauna and kelp beds can be found, but these changes depend on the rate of discharge and the degree of treatment before release. Increased abundance of fish and benthic invertebrates have been noted in the vicinity of some outfalls; at others, benthic communities have been noticeably degraded. Of the four major sewer outfalls emptying into the Southern California Bight, two had caused obvious degradation in several square kilometers around the outfall site (FIGURE 8.48). Collectively, the four outfalls significantly changed or degraded nearly 200 km² of seafloor during the 1970s and 1980s.

These energy-rich substances from treated sewage allowed urchin populations to evade the usual consequences that befall animal populations when they overgraze their plant food sources. These alternative sources of energy ensured that large numbers of urchins survived long enough after decimating one kelp bed to move to another. In central California kelp beds that sea otters have recolonized since 1950, urchin populations are now kept low, and kelp forests have recovered throughout most of the otters' geographic range. The kelp beds just off San Diego, however, have made a dramatic recovery since 1960 without sea otters. The recovery there was more likely due to improved urban sewage treatment, especially reducing the amounts of discharged solids (FIGURE 8.49). The activities of other predators, particularly sea stars and the sheephead fish, also played important roles. In an unusual turnabout, these recovered urchin populations, so recently considered pests in need of eradication, are now themselves targets of a rapidly expanding commercial fishery to supply urchin roe to local and international sushi markets.

Summary Points

Temperate Coastal Seas

- More than 90% of marine animals are benthic, living in close association with the seafloor, at the interface with the overlying water, dependent on the characteristics of each and the exchange of substances between the two.

Seafloor Characteristics

- The composition of the sea bottom is determined by the items that it accumulates (such as plankton, wastes, and detritus), the activities of organisms that live on it and within it, and, at least in shallow water, by the amount of energy available in waves and currents.

Animal–Sediment Relationships

- Benthic animals are either epifaunal, living on the sediment, or infaunal, living within the sediment. Infaunal species include large macrofauna that plow through the sediments as they move, interstitial meiofauna that travel between the particles of sediment, and microfaunal creatures that are so small they are able to live on a single sediment particle.
- The water column and the sediment itself are useful sources of food for benthic species. Some, such as selective detritus feeders and generalist deposit feeders, obtain their food from the sediment itself. Others, such as selective suspension feeders and generalist filter feeders, live on or within the sediment but obtain their food from passing waters.

Larval Dispersal

- About 75% of slow-moving, sedentary, or attached animals easily extend their geographic range via the production of tremendous numbers of eggs and sperm that will unite to form larvae that are temporarily planktonic (or meroplanktonic).
- Although bottom type, bottom texture, chemical attractants, current speeds, sounds, light, presence of conspecific adults, and more are suspected of causing meroplanktonic larvae to drop out of the water column, settle on the seafloor, and metamorphose into a juvenile form, our understanding of this important phenomenon is poor.

Intertidal Communities

- Daily fluctuations in tidal heights result in an intertidal, or littoral, zone forming on all shorelines, regardless of slope or texture. This interesting zone typically is inhabited by species of marine origin that undergo significant physiological stress during periods of low tide.
- Distance from mean low water on any shoreline is correlated with tremendous variations in temperature, duration of exposure, wave shock, light intensity, predation pressure, competition for space, wetness, and other factors. Therefore, because characteristic organisms tend to live at preferred distances from mean low water, distinct horizontal bands of zonation are often observed.
- The upper intertidal of rocky shorelines hosts organisms that suffer with frequent desiccation and punctuated food supplies. The middle intertidal is more densely populated with species more troubled by competition for food and space than physical limitations of the environment. The lower intertidal hosts a much more diversified assemblage of plants and animals that are exposed to air for only a short period of time each day.

- Sandy beaches and muddy shores are depositional environments characterized by deposits of unconsolidated sediments and accumulations of detritus. Patterns of vertical zonation do occur, but they are more apparent within the sediment than along its surface.
- Because oil is less dense than seawater, when it is spilled (intentionally or not), most of it ends up in intertidal zones. Although the major source of oil pollution in the sea is seemingly innocuous runoff from urban areas and careless use and storage of petroleum products, tanker accidents and pipeline blowouts gain much more attention from the media.

Shallow Subtidal Communities

- Below the effects of waves and tides, the depositional seafloor is blanketed with soft sediments. Luxurious kelp communities dominate in temperate areas, with North America's west coast hosting a more complex and extensive kelp community than New England.

Topics for Discussion and Review

1. Why do you think most marine animals are benthic? What advantages are there to a benthic existence (as opposed to a pelagic life in the water column)?
2. Distinguish the terms epifaunal and infaunal and give an example of each type of animal. Then do the same for the terms macrofaunal, meiofaunal, and microfaunal.
3. List and discuss the selective advantages of meroplanktonic larval stages for benthic animals living in shallow water.
4. Compare the species diversity of the middle and lower intertidal zones on rocky shores and on sandy beaches. What factors influence or create these differences?
5. Compare and contrast deposit, detritus, filter, and suspension feeding, being sure to discuss the advantages and disadvantages of each, list a representative animal that performs each type of feeding, and relate them to the type of substrate that each animal inhabits.
6. Summarize the major factors that influence the vertical distribution of intertidal plant and animal species and describe how these factors are influenced by tidal fluctuations.
7. Why are benthic epifauna and attached plants seldom found on sandy exposed beaches?
8. Discuss the ecologic relationships between intertidal mussels, barnacles, macroalgae, and sea stars on a temperate rocky coast.
9. Describe the methods currently used to clean oiled beaches after a large spill.
10. Describe the seafloor conditions that lead to more complex kelp communities on the west coast of North America as opposed to the east coast.

Suggestions for Further Reading

Bertness, M. D. 1989. Intraspecific competition and facilitation in a northern acorn barnacle population. *Ecology* 70:257–268.

Bradbury, I. R., and P. V. Snelgrove. 2001. Contrasting larval transport in demersal fish and benthic invertebrates: the roles of behaviour and advective processes in determining spatial pattern. *Canadian Journal of Fisheries and Aquatic Sciences* 58:811–823.

Brown, A. C., and A. McLachlan. 1990. *Ecology of Sandy Shores.* Elsevier, New York.

Carroll, M. L., and R. C. Highsmith. 1996. Role of catastrophic disturbance in mediating *Nucella-Mytilus* interactions in the Alaskan rocky intertidal. *Marine Ecology Progress Series* 138:125–133.

Carson, R. L. 1979. *The Edge of the Sea.* Houghton Mifflin, Boston.

Constable, A. J. 1999. Ecology of benthic macroinvertebrates in soft-sediment environments: a review of progress towards quantitative models and predictions. *Austral Ecology* 24:452–476.

Denny, M. W. 1985. Wave force on intertidal organisms: a case study. *Limnology and Oceanography* 30:1171–1187.

Epifanio, C. E., and R. W. Garvine. 2001. Larval transport on the Atlantic Continental Shelf of North America: a review. *Estuarine, Coastal and Shelf Science* 52:51–77.

Fenchel, T. 1975. Factors determining the distribution patterns of mud snails. *Oecologia* 20:1–17.

Fingas, M. F., and J. Charles. 2000. *The Basics of Oil Spill Cleanup*. CRC Press, Boca Raton, FL.

Gaines, S. D., and M. D. Bertness. 1992. Dispersal of juveniles and variable recruitment in sessile marine species. *Nature* 360:579–580.

Gosner, K. 1978. *A Field Guide to the Atlantic Seashore*. Peterson Field Guide Series. Houghton-Mifflin, Boston.

Hinchee, R. E., B. C. Alleman, R. E. Hoeppel, and R. E. Mann. 1994. *Hydrocarbon Bioremediation*. Lewis Publishers, Inc. Boca Raton, FL.

Hinton, S. 1987. *Seashore Life of Southern California*. University of California Press, Berkeley.

Horn, M., K. Martin, and M. Chotkowski. 1998. *Intertidal Fishes: Life in Two Worlds*. Academic Press.

Jackson N. L., K. F. Nordstrom, I. Eliot, and G. Masselink. 2002 . "Low energy" sandy beaches in marine and estuarine environments: a review. *Geomorphology* 48:147–162.

Kaplan, E. H. 1984. *A Field Guide to Southeastern and Caribbean Seashores: Cape Hatteras to the Gulf coast, Florida, and the Caribbean*. Peterson Field Guide Series. Houghton-Mifflin, Boston.

Koehl, M. A. R. 1982. The interaction of moving water and sessile organisms. *Scientific American* 247:124–134.

Kozloff, E. N. 1987. *Marine Invertebrates of the Pacific Northwest*. University of Washington Press, Seattle.

Lippson, A. J., and R. L. Lippson. 1984. *Life in the Chesapeake Bay*. Johns Hopkins University Press, Baltimore.

Little, C., and J. A. Kitching. 1996. *The Biology of Rocky Shores*. Oxford University Press, Oxford.

Lively, C. M., P. T. Raimondi, and L. F. Delph. 1993. Intertidal community structure: space-time interactions in the northern Gulf of California. *Ecology* 74:162–173.

Menge, B. A. 1975. Brood or broadcast? The adaptive significance of different reproductive strategies in the two intertidal sea stars *Leptasterias hexaxtis* and *Pisaster ochraceus*. *Marine Biology* 31:87–100.

Menge, B. A., B. A. Daley, P. A. Wheeler, E. Dahlhof, E. Sanford, and P. T. Strub. 1997. Benthic-pelagic links and rocky intertidal communities: bottom-up effects on top-down control? *Proceedings of the National Academy of Science* 94:14530–14535.

Morse, A. N. C. 1991. How do planktonic larvae know where to settle? *American Scientist* 79:154–167.

Nienhuis, P. H., and A. C. Mathieson. 1991. *Intertidal and Littoral Ecosystems*. Elsevier Health Sciences.

Paine, R. T. 1990. Benthic macroalgal competition: complications and consequences. *Journal of Phycology* 26:12–17.

Peterson, C. H. 1991. Intertidal zonation of marine invertebrates in sand and mud. *American Scientist* 79:236–248.

Pezeshki, S. R., M. W. Hester, Q. Lin, and J. A. Nyman. 2000. The effects of oil spill and clean-up on dominant US Gulf coast marsh macrophytes: a review. *Environmental Pollution* 108:129–139.

Robles, C., R. Sherwood-Stevens, and M. Alvarado. 1995. Responses of a key intertidal predator to varying recruitment of its prey. *Ecology* 76:565–579.

Smith, D. L., and K. B. Johnson. 1996. *A Guide to Marine Coastal Plankton and Marine Invertebrate Larvae.* Kendall/Hunt Publishing Company.

Underwood, A. J. 2000. Experimental ecology of rocky intertidal habitats: what are we learning? *Journal of Experimental Marine Biology and Ecology* 250:51–76.

Young, C., M. Sewell, and M. Rice. 2001. *Atlas of Marine Invertebrate Larvae.* Academy Press.

CHAPTER 9

A coral reef is a living thing comprising corals, sponges, algae, and other species competing for space, light, and food.

CHAPTER OUTLINE

Tropical and Subtropical Shallow Seas

Ecologists are fond of discussing and determining limiting factors, such as a nutrient that is no longer present in adequate supply. Adding other nutrients, which are already present in sufficient quantity, to this system will not spur renewed productivity. Addition of the limiting nutrient, or factor in more general terms, is likely to stimulate production and growth until some other factor or the same one becomes limiting.

In tropical oceans, hard substrate is widely viewed as an important limiting factor. Most multicellular plants and invertebrates need to attach to the substrate. Yet soft substrates, such as those described in Chapters 7 and 8, provide rather poor anchorage and can be quite abrasive. A hard object on which a plant or animal can live is therefore very desirable, and the presence of such items greatly influences the appearance and diversity of tropical coastlines. In this chapter we emphasize two tropical communities that by virtue of their presence alone provide millions of hectares of firm substrate on which countless plants and animals live, many of which are found nowhere else on Earth. The species that compose coral reefs, mangrove stands, and seagrass beds are inherently interesting in themselves, but the fact that they also create the physical structures for unique biologic communities makes them especially fascinating and important subjects of study.

9.1 Coral Reefs

For many people, thoughts of tropical islands conjure up images of a special type of marine ecosystem, coral reefs. Coral reefs are famous for a diversity of species that rivals rain forests, for the myriad colors exhibited by their inhabitants and for the amazing biologic interactions that have evolved there. Unlike the intertidal communities discussed in Chapter 8, coral reefs are actually produced by the organisms that live on them. The entire reef, which may extend for hundreds of kilometers, is primarily composed of a veneer of tiny sea anemone-like creatures called coral polyps. These small colonial animals slowly produce the massive carbonate infrastructure of the reef itself, on, around, and in which a vast array of other organisms live.

Therein lies a wonderful biologic paradox. Think of any common terrestrial ecosystem, a temperate forest, a tropical jungle, a mid-western plain, or the field adjacent to your house. These areas are dominated by a great variety of plants (the producers) and contain just a handful of animal species, both herbivores and carnivores (the consumers). Conversely, a typical coral reef contains an impressive assemblage of consumers and just a few plants. The coral animals that create the reef feed by removing plankton from the water column, as do the many sponge species that decorate the reef, represent the second most important component of the benthic fauna on coral reefs. Yet tropical seas are virtually devoid of plankton. That is why azure tropical waters are so transparent. So a coral reef can be viewed as one giant animal that is inhabited by hundreds of other animals, such as sponges, snails and clams, squids and octopuses, sea anemones and jellyfishes, shrimps and crabs, worms and fishes. The questions remain: Where are the primary producers on a coral reef? Can an ecosystem violate the second law of thermodynamics by containing more consumers than producers? Why don't planktivorous reef creatures, such as corals and sponges, starve to death in the nearly plankton-free waters that surround them? In this section we attempt to answer these fascinating biologic riddles.

Anatomy and Growth

Coral is a general term used to describe a variety of cnidarian species (see pp. 133–134). Some grow as individual colonies; hence, not all corals produce reefs. And not all reefs are formed by corals; some modern reefs are formed by oysters, annelid worm tubes, red algae, or even cyanobacteria.

Reef-forming corals, the primary species that secrete the $CaCO_3$ matrix of coral reefs, are members of the class Anthozoa (see p. 134). All anthozoans are radially symmetric, a morphology that is adaptive for sessile organisms, such as corals and sea anemones. Anthozoans are subdivided into two subclasses. The subclass Octocorallia, comprising soft corals, sea fans, sea whips, sea pansies, and sea pens, are characterized by the presence of eight pinnate, or feather-like, tentacles. Members of the subclass Hexacorallia possess a mouth that is surrounded by multiples of six smooth tentacles and include four orders of sea anemones (some exist as individuals, some in colonies, and others are tube-dwellers) and three orders of corals (stony corals, false corals, and black corals). One group, the stony corals (order Scleractinia), is responsible for creating coral reefs. Stony corals and most of their cnidarian relatives are carnivores that use tentacles armed with cnidocytes that ring the mouth (FIGURE 9.1) to capture prey and push it into their gastrovascular cavity where it is digested.

Most corals are colonial, built of numerous basic structural units, or polyps (FIGURE 9.2 and see p. 134), each usually just a few millimeters in diameter. Coral polyps sit in calcareous cups, or **corallites**, an exoskeleton secreted by their basal epithelium. Several wall-like **septa** radiate from the sides of each corallite, and a stalagmite-like **columella** extends upward from its floor. Periodically, the coral polyp grows upward by withdrawing itself up and secreting a new basal plate, a partition that provides a new elevated

figure 9.1

Extended polyps of a coral colony. The numerous light-colored spots on the tentacles are batteries of cnidocytes.

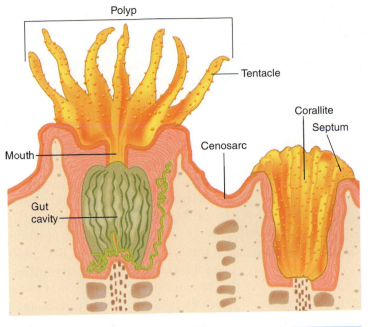

figure 9.2

Cross-section of a coral polyp and a calcareous corallite skeleton. The living coral tissue forms a thin interconnection, the cenosarc, over the surface of the reef.

bottom in the corallite. In addition, the coral colony also increases in diameter by adding new asexually cloned polyps to its periphery. These new polyps secrete their own $CaCO_3$ corallites that share a wall with neighboring polyps. All polyps that make up the colony are interconnected over the lips of their corallites via a thin sheet of tissue called a **cenosarc**. Therefore, touching a living coral colony in any way can easily crush the cenosarc against its own $CaCO_3$ skeleton, thus compromising the colony by leaving it open to infection.

The growth rate of corals is affected by light intensity (which is affected by water motion, depth, and turbidity), day length, water temperature, plankton concentrations, predation, and competition with other corals. Stony corals exhibit a large variety of growth forms that are typically described as encrusting, massive, branching, or foliaceous (FIGURE 9.3). In addition, many species are rather polymorphic, expressing different growth forms in response to differences in wave exposure or depth. Therefore, growth may seem like a simple parameter to measure but it is not. Techniques for monitoring the growth of corals include measuring an increase in weight, or diameter, or surface area, or number of branches, or length of branches. Individual coral colonies may grow continually for centuries or even longer. Some species exceed several meters in size. In general, species with more porous skeletons grow more rapidly than species with denser skeletons, and branching species grow more quickly than massive species. Overall, an entire reef will grow upward as much as 1 mm/yr and spread horizontally 8 mm/yr.

The growth of an individual coral or an entire reef is not simply a function of the local rate of calcification for that species. The persistence of a coral colony or reef depends on a balance between the deposition and removal of $CaCO_3$. In addition to

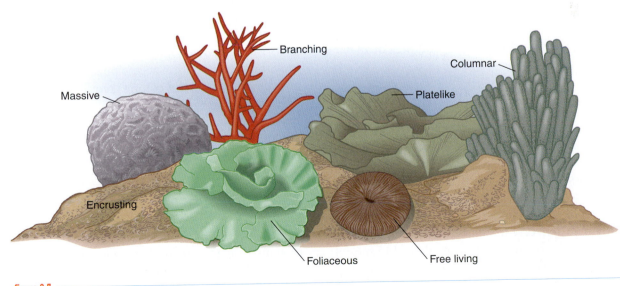

figure 9.3

Corals exhibit a large variety of growth forms.

figure 9.4

Parrotfishes, major grazers of coral skeletal material, use their powerful jaws to produce large amounts of carbonate sand on the reef.

carbonate deposition by corals, several types of other organisms also contribute to the structure of coral reefs, including encrusting and segmented calcareous red and green algae, calcareous colonial hydrozoans, skeletons of crustaceans, bryozoans, and single-celled foraminiferans, mollusk shells, tests and spines of echinoderms, sponge spicules, and serpulid polychaete tubes. The loss or erosion of carbonate is caused by grazers or scrapers, such as sea urchins (echinoids are the major grazers in the Atlantic) and fishes (FIGURE 9.4); etchers, such as bacteria, fungi, and algae (especially *Ostreobium*) that penetrate coral substrates. Infaunal organisms, such as sponges, bivalves, sipunculans and polychaetes also drill or bore into coral skeletons. From this encrusted integrated base of living and dead skeletal remains, coral-reef ecosystems have evolved as the most complex of all benthic associations.

Coral Distribution

Like sea anemones, corals are ubiquitous. Non–reef-forming corals can be found in the deep sea (e.g., black corals) and in temperate zones (such as *Astrangia* on ship wrecks off New England), as well as in the tropics. However, there are several restrictions to the distribution of reef-forming corals, which are more abundant and diverse in the Indo-Pacific (about 700 species) than in the Atlantic Ocean (about 145 species; FIGURE 9.5). First, coral reefs generally are restricted to tropical and subtropical regions (usually below 30° latitude) where the annual sea-surface temperature

averages at least 20°C. Second, coral reefs generally are better developed on the eastern margins of continents. Third, coral reefs generally thrive only in normal-salinity seawater; hence, reefs are rare on the eastern coast of South America because of the enormous outflow of fresh water there from the Amazon River system. Fourth, reef-forming corals are usually found within 50 m of the surface in clear water on exposed surfaces.

These first two biogeographic restrictions suggest that reef-forming corals generally thrive only in warmer water, probably because only in warm waters can the high rates of $CaCO_3$ deposition needed for reef building be achieved. Hence, they are found in low latitudes and on eastern shorelines where coastal upwelling of cold water is less common and where the major ocean gyres direct warm tropical currents (see Figure 1.38). These latitudinal limits of coral-reef development also are often influenced by competition with macroalgae, with macroalgae being favored in higher latitudes because of increased nutrient concentrations, decreased water temperatures, and perhaps decreased grazing pressure.

The third generalized limit to the global distribution of coral reefs suggests that coral animals cannot tolerate low salinity seawater or the sedimentation and high concentration of nutrients associated with rivers and freshwater runoff. The final biogeographic limitation, that coral reefs typically grow within 50 m of the surface in clear water on exposed surfaces, suggests that they need sunlight for their survival and growth. This limitation seems puzzling. Why would an animal (i.e., a coral colony) require sunlight, and why would their growth rates be affected by light intensity and day length, as described above? The answer to this question is also the answer to the apparent paradox posed at the beginning of this chapter.

Coral Ecology

Living intracellularly within the endodermal tissues of all reef-building, or **hermatypic**, corals are masses of symbiotic **zooxanthellae**, unicellular algae that, like all other photosynthetic organisms, require light. Solitary non–reef-building corals, such as *Astrangia* off the coast of New York, do not possess zooxan-

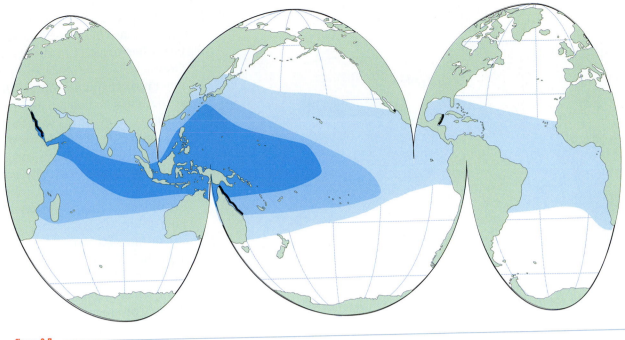

figure 9.5

Distribution of reef-forming corals, by number of genera: light blue, < 20 genera; medium blue, 20–40 genera; dark blue, > 40 genera. Heavy black lines indicate continental barrier reefs.

thellae and are termed ahermatypic. Zooxanthellae is a general term for a variety of photosynthetic cells that are symbiotic with several types of invertebrate species. The most common species is *Symbiodinium microadriaticum*, a member of the protist division Dinophyta (see pp. 73–76). Unlike the dinophytes that appear in Figure 3.16, zooxanthellae lose their flagella and cellulose cell walls. They occur in concentrations of up to one million cells/cm^2 of coral surface and often provide most of the color seen in corals. In fact, corals that grow in bright sunlight are often creamy white, whereas those in deep shade are nearly black. This difference is due to the cellular concentrations of photosynthetic pigments rather than differences in the densities of cells.

Zooxanthellae and corals derive several benefits from each other. Thus, this relationship usually is considered a mutualistic one (see Figure 2.16). Corals provide the zooxanthellae with a constant protected environment and an abundance of nutrients (CO_2 and nitrogenous and phosphate wastes from cellular respiration of the coral). In return, the host corals receive photosynthetic products (O_2 and energy-rich organic substances) from the symbiotic algae by stimulating or promoting their release with specific signal molecules that appear to alter the membrane permeability of the algae. These zooxanthellae photosynthetically produce 10–100 times more carbon than is necessary for their own cellular needs; almost all this excess is transferred to the coral. Nearly all the carbon that is transferred to the coral is respired and not used to build new coral tissue because it is low in nitrogen and phosphorus. This contribution by the zooxanthellae is sufficient to satisfy the daily energy needs of several species of corals (in fact, soft corals are obligate symbionts, having lost the ability to capture and ingest plankton). The total contribution of symbiotic zooxanthellae to the energy budget of the reef is several times higher than phytoplankton production occurring in the waters above many reefs.

Hence, corals are able to construct enormous reefs in plankton-poor waters because they receive a significant supply of food from their algal associates. The coral animals also avoid the necessity of excreting some of their cellular wastes (which the algae absorb and use) and experience greater calcification rates than hermatypic corals that have been experimentally separated from their algal symbionts (**FIGURE 9.6**).

Additional primary production on coral reefs is provided by several types of rather cryptic plants. These include encrusting calcareous red algae, filamentous green algae that invades dead corals, a brown algal turf, photosynthetic symbionts in other reef invertebrates, macroalgae anchored in the sand, seagrass, and phytoplankton cells in the water column over the reef.

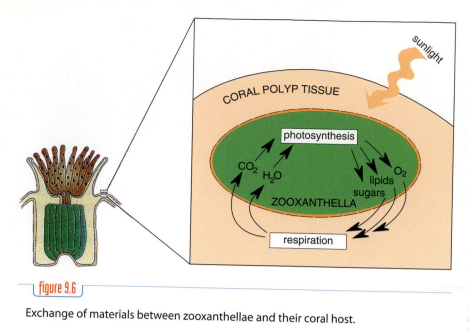

figure 9.6

Exchange of materials between zooxanthellae and their coral host.

Despite the nutritional contribution of zooxanthellae, coral polyps remain superbly equipped to prey upon a variety of external sources of food, and only soft corals depend solely on zooxanthellae. Corals with large polyps and tentacles, such as *Favia* or *Mussa,* feed exclusively on small fishes and larger zooplankton, such as copepods, amphipods, and worms. Species with smaller polyps, such as *Porites* or *Siderastrea,* use ciliary currents to collect small plankton and detritus particles. Most coral polyps are capable of using mesenterial filaments to harvest particulate organic carbon from surrounding sediments. Corals also use their mucus–ciliary system (analogous to the ciliated epithelium in the windpipe of humans) to trap and ingest organic particles as small as suspended bacteria, bits of drifting fish slime, and even organic substances dissolved in passing seawater. Finally, the still controversial concept of endo-upwelling has been suggested as a possible source of additional dissolved nutrients, wherein geothermal heat deep within island reefs drive the upwelling of nutrient-rich water through the reef structure from depths of several hundred meters.

Corals are not the only animals on the reef that possess photosynthetic symbionts. Zooxanthellae also occur in other anthozoans, some medusae (such as the upside-down jellyfish, *Cassiopea*), sponges, and giant clams (see page 264). In addition, it is well documented that sponges possess photosynthetic cyanobacteria. These photosynthetic symbionts are found in about 40% of sponge species from the Atlantic and Pacific Oceans, although their contribu-tion to sponge ecology in the two oceans differs dramatically. On the Great Barrier Reef in the Pacific Ocean, 90% of the sponges on the outer reefs are **phototrophic** (they are flattened and obtain up to half of their energy from cyanobacteria), with 6 of 10 species producing three times as much oxygen as they consume. Virtually none of the sponges studied in the Caribbean Sea is phototrophic. This results in Caribbean sponges consuming 10 times more prey than their Pacific relatives. Perhaps this different reliance on energy from cyanobacteria is because primary productivity in the western Atlantic is higher than in the western Pacific.

Finally, nitrogen fixation, an activity that is light dependent, has recently been found to be associated with bacteria living in the skeletons of various hermatypic corals. These nitrogen-fixing bacteria benefit from organic carbon excreted by the coral tissue. Perhaps this symbiotic association is as important to corals as their mutualism with zooxanthellae.

The living richness of coral reefs stands in obvious contrast to the generally unproductive tropical oceans in which they live. The precise trophic relationships between producers and consumers on the reef are still largely unknown. Coral colonies seem to function as highly efficient trophic systems with their own photosynthetic, herbivorous, and carnivorous components. Critical nutrients are rapidly recycled between the producer and consumer components of the coral colony. Because much of the nutrient cycling is accomplished within the coral tissues, little opportunity exists for the nutrients to escape from the coral production system. Coral colonies, therefore, are able to recycle their limited supply of nutrients rapidly between internal producer and consumer components and keep productivity in coral-reef communities relatively high (up to 5000 gC/m^2/yr) compared with other regions of the ocean (see Table 4.2).

Coral-Reef Formation

Coral reefs occur in two general types, **shelf reefs** that grow on continental margins and **oceanic reefs** that surround islands. Oceanic reefs may be divided into three general subtypes: **fringing reefs, barrier** reefs, and **atolls**. Most shelf reefs are fringing reefs, which form borders along shorelines. Some of the Hawaiian reefs and other relatively young oceanic reefs are also of this type. The longest fringing reef known extends throughout the Red Sea, extending about 400 km. Barrier reefs are further offshore and are separated from the shoreline by a lagoon. The Great Barrier Reef of Australia is by far the largest single biologic feature on Earth, bordering about 2000 km of Australia's northeast coast. Smaller barrier reefs occur in the Caribbean Sea. Atolls are generally ring-shaped reefs from which a few low islands project above the sea surface (FIGURE 9.7). The largest atoll known is Kwajalein Atoll in the Marshall Islands, which has a lagoon 100 km long and 55 m deep.

Charles Darwin is famous for his concept of natural selection, which he proposed as the mechanism by which biologic evolution occurs. Darwin's propensity to view the world with the perspective of geologic time certainly aided him in the development of his revolutionary hypothesis. However, this tendency also influenced his other studies. For example, before he published his *Origin of Species*, he published two works that required this perspective of deep time. One concerned the churning effect that earthworms had on topsoil by steadily eating and defecating while crawling their way through our lawns. The second concerned the development of the three types of coral reefs described above.

Darwin studied the morphology of coral reefs on several islands while serving as a naturalist aboard the *H.M.S. Beagle* during its voyage to circumnavigate the Earth from 1831 to 1836. His observations led him to propose that essentially all oceanic coral reefs were supported by volcanic mountains beneath their surfaces. Fringing reefs, barrier reefs, and atolls, he suggested, were sequential developmental stages in the life cycle of a single reef. Within the tropics, he argued, newly formed volcanic islands and submerged volcanoes that almost reach the sea surface are eventually populated by planktonic coral larvae from other nearby coral islands. The coral larvae settle and grow near the surface close to the shore, forming a fringing reef (FIGURE 9.8, left). The most rapid growth occurs on the outer sides of the reef where food and oxygen-rich waters are more abundant. Waves break loose pieces of the reef and move them down the slopes of the volcano. More corals establish themselves on this debris and grow toward the surface. He reasoned that the weight of the expanding reef and the increasing density of the cooling volcano cause the island to sink slowly. If the upward growth of the reef keeps pace

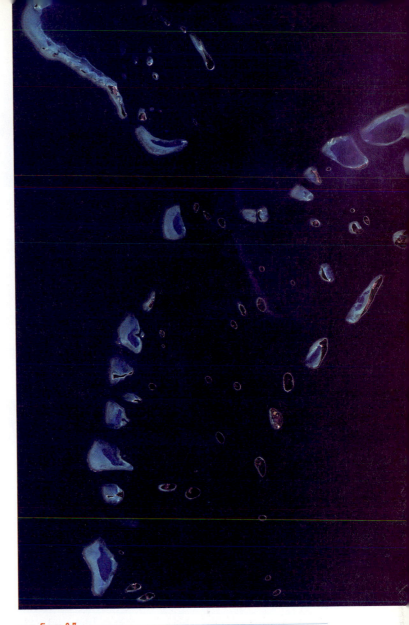

figure 9.7

A satellite view of a portion of the hundreds of atolls that make up the nation of Maldives.

with the sinking island, the coral maintains its position in the sunlit surface waters. If the upward growth of the reef does not keep pace with the sinking island, the reef is pulled into the cold darkness below the photic zone and expires. Such a dead sunken reef is called a **guyot** (pronounced gee-oh).

As the island sinks away from the growing reef, the top of the reef widens. Eventually, this reef crest or flat becomes so wide that many of the corals on the quiet inner edge of the reef die because the water that reaches them is devoid of nutrients and oxygen and contains high concentrations of waste products. The dead corals are soon covered with reef debris and form a shallow lagoon. Delicate coral forms

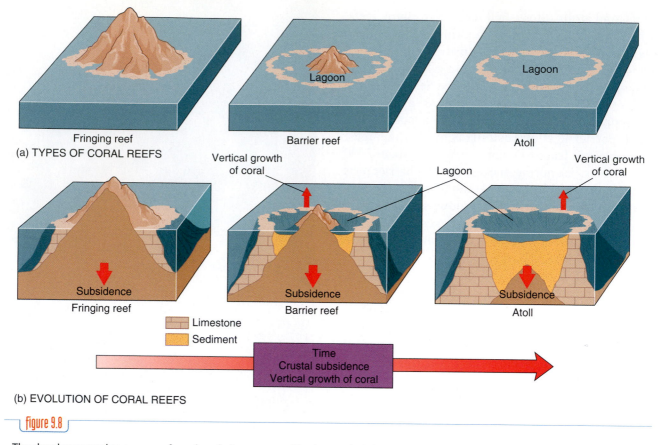

(a) TYPES OF CORAL REEFS

Fringing reef

Barrier reef

Lagoon

Atoll

Lagoon

Vertical growth of coral

Lagoon

Vertical growth of coral

Subsidence
Fringing reef

Subsidence
Barrier reef

Subsidence
Atoll

Limestone
Sediment

Time
Crustal subsidence
Vertical growth of coral

(b) EVOLUTION OF CORAL REEFS

figure 9.8

The developmental sequence of coral reefs, from young fringing reefs (left), to barrier reefs (center), and finally to atolls (right).

survive in the lagoon, protected from the waves by what is now a barrier reef (Figure 9.8, center). With further sinking, the volcanic core of the island may disappear completely beneath the surface of the lagoon and leave behind a ring of low-lying islands supported on a platform of coral debris, an atoll (Figure 9.8, right).

Darwin's concept of coral-reef formation is, with a few modifications, widely accepted today. Test drilling on several atolls has revealed, as Darwin predicted, thick caps of carbonate reef material overlying submerged volcanoes. Two test holes drilled on Enewetak Atoll (the site of US hydrogen bomb tests in the 1950s) penetrated 1268 and 1405 m of shallow-water reef deposits, respectively, before reaching the basalt rock of the volcano on which the reef had formed. For the past 60 million years, Enewetak apparently has been slowly subsiding as its surrounding reef grew around it. Because this transition from a fringing morphology through a barrier morphology to an atoll requires a great deal of time and because the Atlantic Ocean is much younger than the Pacific Ocean, atolls are virtually nonexistent in the Atlantic Ocean.

Some anecdotal information reinforces the scientific data that support Darwin's hypothesis of coral-reef development. For example, British explorer Captain James Cook discovered Hawaii in January 1779, during Makahika, a festival to honor the god Lono. The Hawaiian natives initially thought that Captain Cook was Lono (who was said to come from the sea). After realizing their mistake, they killed him. A monument was soon built in the surf to commemorate his arrival and death. Today, that monument can be found off shore at a depth of 20 m, yet the reef surrounding the island is still growing just under the surface of the sea.

For the past several hundred thousand years, the formation and melting of vast continental glaciers have produced extensive worldwide fluctuations in sea level. Darwin was aware of these fluctuations yet had no means of predicting their effects on coral-reef development. Fifteen thousand years ago, during the last glacial maximum, average sea level was about 150 m below its present level. As the ice melted, sea level gradually rose (about 1 cm/yr) until it reached its present level nearly 6000 years ago. Many coral reefs did not grow upward quickly enough and per-

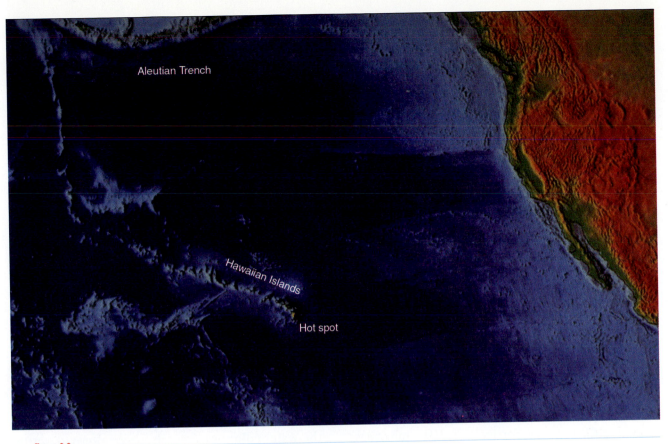

figure 9.9

Chains of volcanoes along the Hawaiian Island–Emperor Seamount are carried, in a conveyer-belt fashion, north into deeper water by the movement of the Pacific Plate. Each volcano was formed over the "hot spot," a continuous source of new molten material presently under Hawaii, and is carried to its eventual destruction in the Aleutian Trench.

ished. Those that did keep up with the rising sea are the living reefs we see today. Coral reefs in the Atlantic Ocean seem most susceptible to glacier-induced changes in their morphology, and barrier reefs are most common in the Atlantic Ocean.

Coral reefs have also been subjected to the effects of global plate tectonics. The Hawaiian Islands and the reefs they support have been transported to the northwest by the movement of the Pacific Plate. Atolls at the northern end of the chain appear to have drowned as they reached the "Darwin Point," a threshold beyond which coral atoll growth cannot keep pace with recent changes in sea level (FIGURE 9.9). At the Darwin Point, only about 20% of the necessary $CaCO_3$ production is contributed by corals.

Reproduction in Corals

Corals reproduce in a variety of ways, both asexually and sexually. Most corals bud off new polyps along their margins asexually as they increase in diameter.

Sometimes, these new polyps sever the cenosarc and initiate a new colony that is a clone of their neighbor. Branching species, such as *Acropora,* are frequently broken by storms or ship anchors into clonal colonies by fragmentation, the production of new colonies from portions broken off established colonies. Fragmentation decreases the risk of mortality of the genotype and avoids the risk of high mortality of larvae and juveniles during sexual reproduction. In addition, fragmentation by species with high growth rates often results in that species dominating certain reef zones (such as the buttress zone discussed below) as well as rapid recolonization after a disturbance. Researchers also have observed "polyp bail-out" in the laboratory, when polyps crawl out of their corallites and drift away. It is not known if these polyps remain viable or if this occurs naturally on coral reefs.

Corals also reproduce sexually, either by brooding fertilized eggs internally or by spawning millions of gametes into the water column for external fertilization. In brooding species, the eggs remain in the gastrovascular cavities of the adults where they are fertilized by motile sperm cells. The developing zygotes and resultant larvae are brooded before they are released to settle nearby. Some evidence suggests that coral species with small polyps have low numbers of eggs combined with internal fertilization and brooding, whereas large-cupped species spawn huge quantities of eggs that are fertilized externally. In addition, the strategy of sexual reproduction used (brooding larvae vs. spawning gametes) is highly correlated with taxonomic affiliation at the family level. Members of the families Agariciidae, Dendrophylliidae, and Pocilloporidae commonly brood, whereas broadcast spawning is predominant in the Acroporidae, Caryophyllidae, Faviidae, and Rhizangidae. Family Poritidae includes both brooders and broadcasters.

Of approximately two hundred species of corals studied on the Great Barrier Reef, 131 were hermaphroditic spawners, 37 were diecious spawners, 11 were hermaphroditic brooders, and 7 were diecious brooders. Hence, spawning by hermaphrodites seems to be the most common method of sexual reproduction among corals. Spawning is usually accomplished during a highly synchronous event known as mass spawning. On the Great Barrier Reef of Australia, mass spawning by corals is a spectacular sight. More than 100 of the 340 species of corals found there synchronously spawn on only one night each year, just a few days after the late spring full moon (FIGURE 9.10). A similar episode of mass spawning has been documented in the Gulf of Mexico, in the evening 8 days after the full August moon. Mass spawning by corals seems to be induced by specific dark periods, and it can be delayed by experimentally extended light periods. Mass

spawning also seems to be broadly influenced by temperature. Such highly seasonal spawning is surprising in the tropics, an area wherein reproduction throughout the year is said to be the norm because of relatively constant climatic conditions.

A few days after spawning, the fertilized eggs develop into a ciliated **planula** larva (FIGURE 9.11). These larvae, each with a supply of zooxanthellae, initially are positively phototactic; they swim toward brighter light. This ensures that they remain near the sea surface where maximal dispersal by surface currents is likely. Then, after a specific interval, they become negatively phototactic and attempt to settle on the sea floor. They thrive only if they encounter their preferred water and bottom conditions. From these planula larvae, new coral colonies develop and mature in 7–10 years. Richmond has reported that the larvae of *Pocillopora damicornis* are capable of reversible metamorphosis. In this species, the planula larva settles and begins to metamorphose into a juvenile. It forms a $CaCO_3$ exoskeleton, a mouth, and tentacles. However, if it is stressed within the first 3 days of settling, it will sever its attachments to its carbonate exoskeleton, revert back into a planula larva, and reenter the water column to search for an alternate settlement site. During their planktonic phase, coral larvae are capable of settling new volcanic islands some distance from their island of origin. When they do, the form of the reef they eventually create depends on existing environmental conditions and the prior developmental history of reefs in the area.

It is unclear why corals spawn synchronously, and why this event occurs just several nights after the full moon. One advantage to mass spawning is

(a)

(b)

 figure 9.10

Spawning corals. (a) Female staghorn coral releasing eggs; (b) male mushroom coral releasing sperm.

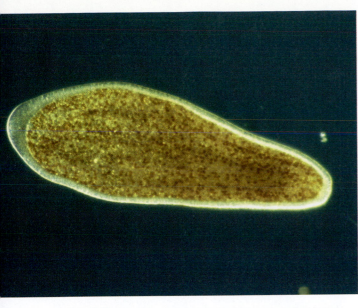

Micrograph of a planula larva of the coral *Acropora*.

that the chance of fertilization will increase greatly for one species. It is unclear, however, why mass spawnings are multiple species events, in that simultaneous spawning by many species may increase the risk of gamete loss via hybridization. Perhaps such an epidemic spawning event overwhelms (and satiates) active predators and filter feeders in the area, increasing the likelihood of gamete survival. However, these species also risk big losses by spawning on just a few nights each year. A sudden rain storm and subsequent drop in salinity of surface waters during a mass spawning event around Magnetic Island in November 1981 destroyed the entire reproductive effort of those corals for that year. Mass spawning is not a universal behavior of reef corals; in the northern Red Sea, none of the major species of corals reproduces at the same time as any of the other major species.

Interestingly, calcareous green algae in the Caribbean also exhibit mass spawning. Nine Caribbean species in five genera participate in a predawn episode, with a total of 17 species of green algae exhibiting highly synchronous reproductive patterns. Unlike the coral phenomenon described above, closely related algal species broadcast their gametes at different times and the environmental or biologic triggers of these events remain unknown. In all cases, gametes from both sexes remain motile for 40–60 minutes after release but sink quickly after combining to form a zygote.

Zonation on Coral Reefs

Environmental conditions that favor some coral-reef inhabitants over others in a particular habitat depend a great deal on wave force, water depth, temperature, salinity, and a host of biologic factors. These conditions vary greatly across a reef and provide for both horizontal and vertical zonation of the coral and algal species that form the reef. FIGURE 9.12, a cross-section of an idealized Indo-Pacific atoll, includes the major features and zones of the reef.

The living base of a coral reef begins as deep as 150 m below sea level. Between 150 and 50 m on outer reef slopes, a few small fragile species, such as *Leptoseris*, exist despite the low level of sunlight that penetrates to these depths. Above 50 m and extending up to the base of vigorous wave action (approximately 20 m) is a transition zone between deep and shallow water associations. In this zone, the corals and algae receive adequate sunlight yet are sufficiently deep to avoid the adverse effects of surface waves. Several of the delicately branched species commonly found in the protected lagoon waters also occur in this transition zone.

From a depth of about 20 m to just below the low-tide line is a rugged zone of spurs, or buttresses, radiating out from the reef. Interspersed between the buttresses are grooves that slope down the reef face. This windward profile of alternating buttresses and grooves is useful in dissipating some of the energy of waves that crash into the face of the reef, but damage to the reef and its inhabitants is inevitable. The grooves

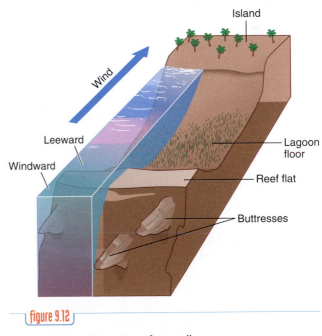

Cross-sectional zonation of an atoll.

drain debris and sediment produced by wave impacts off the reef and into deeper water. Continual heavy surf has limited detailed studies of the buttress zone, but it is known to be dominated by several species of encrusting coralline algae and by rapidly growing branching coral species (such as some *Acropora*) that repair damage quickly and thrive when fragmented. Small fishes seem to be in every hole and crevice on the reef, and many of the larger fishes of the reef—sharks, jacks, and barracudas—patrol the buttresses in search of food.

Most coral reefs are swept by the broad reaches of the trade winds. The waves generated by these winds crash as thundering breakers on the windward sides of reefs. Windward reefs are usually characterized by a low jagged algal ridge. The algal ridge suffers the full fury of incoming waves. In this high-energy habitat, a few species of calcareous red algae, especially *Porolithon*, *Hydrolithon*, *Goniolithon*, and *Lithothamnion*, flourish and produce the ridge, creating new reef material as rapidly as the waves erode it. A few snails, limpets, and urchins (FIGURE 9.13) also can be found wedged into surface irregularities. Slicing across the algal ridge are surge channels that flush bits and fragments of reef material off the reef and down the seaward slope.

Extending behind the algal ridge to the island (or, if the island is absent, to the lagoon) is a reef flat, a nearly level surface barely covered by water at low tide in the Atlantic (Indo-Pacific reef flats are intertidal). The reef flat may be narrow or very wide, may consist of several subzones, and may have an immense variety of coral species and growth forms. In places where the water deepens to a meter or so, small raised microatolls occur. Microatolls are produced by a half dozen different genera of corals and, with other coral growth forms, provide the framework for the richest and most varied habitat on the reef. Burrowing sea urchins are common, and calcareous green algae and several species of large foraminiferans thrive and add their skeletons to the sand-sized deposits on the reef flat. The sand, in turn, provides shelter for other urchins, sea cucumbers, and burrowing worms and mollusks.

One of the most spectacular animal of the reef flat is the giant clam, *Tridacna*. The largest species of this genus occasionally exceeds a meter in length and weighs greater than 100 kg. Some tridacnids sit exposed atop the reef platform; others rock slowly to work themselves into the growing coral structure beneath (FIGURE 9.14). Like corals and many other invertebrates, tridacnid clams house dense concentrations

figure 9.13

Echinometra, a common tropical sea urchin.

figure 9.14

A giant clam, *Tridacna*, amid mixed corals from Fiji. Note the siphon opening and extended mantle tissue between open shell edges.

of zooxanthellae in specialized tissues, particularly the enlarged mantle that lines the edges of its shell. When the shell is open, the pigmented mantle tissues with their zooxanthellae are fully exposed to the energy of the tropical sun.

Tridacnid clams were long thought to "farm" their zooxanthellae in blood sinuses within the mantle and then transport them to the digestive glands, there to be digested by single-celled amoebocytes. Using elaborate staining and electron microscope techniques, Fankboner demonstrated that the digestive amoebocytes of *Tridacna* selectively cull and destroy old or

Variation in coral growth forms from the Solomon Islands: (a) plate coral, *Acropora*; (b) brain coral, *Diploria*; and (c) staghorn coral, *Acropora*.

degenerate zooxanthellae. Healthy zooxanthellae are maintained to provide photosynthetic products to their hosts in dissolved rather than cellular form.

The tranquil waters of the lagoon protect two general life zones: the lagoon reef and the lagoon floor. The lagoon reef is a leeward reef. It forms the shallow margin of the lagoon proper and is usually free of severe wave action. It lacks the algal ridge characteristic of the windward reef and in its place has a more luxuriant stand of corals (FIGURE 9.15). Other algae, some specialized to burrow into coral, and uncountable species of crustaceans, echinoderms, mollusks, anemones, gorgonians, and representatives of many other animal phyla flourish in the lagoon reef. In this gentle protected environment, single coral colonies of *Porites* and *Acropora* may achieve gigantic proportions. Branching bush and tree-like forms extend several meters from their bases. The plating, branching, and overtopping structures common in the protected lagoon are most likely structural adaptations evolved in response to competition for particles of food and sunlight, two resources vital to the survival of reef-forming corals.

Coral Diversity and Catastrophic Mortality

The great diversity of species on coral reefs is legendary and rivals that of tropical rain forests. The classic explanation for this high diversity was that the uniform and predictable conditions on coral reefs promoted high diversity by enabling species to become increasingly specialized. (This steady-state hypothesis has also been used to explain the high diversity of rain forests and the deep sea.) Recently, this view has been challenged by an argument that suggests that the high diversity of coral reefs is a non-equilibrium state whose diversity can persist only if

it is disturbed. Like some rocky intertidal communities discussed in Chapter 8, coral reefs are subject to severe disturbances often enough that equilibrium, or a climax stage, may never be reached, and high diversity is maintained by frequent catastrophic mortality. The catastrophic mortality of corals and coral-reef species that has been observed in the past 10–15 years can be viewed as natural perturbations of these communities rather than abnormal events. One common natural cause of catastrophic coral mortality is storm waves from hurricanes and typhoons. At Heron Island on the Great Barrier Reef, for example, the highest number of species of corals occurs on the crests and outer slopes that are constantly exposed to damaging waves. In fact, it has been reported that the most significant factor determining the spatial and temporal organization of Hawaiian coral-reef communities is physical disturbance from waves. Nevertheless, there is still some cause for concern. The recent rate of loss of coral reefs worldwide may be higher than ever before. When coupled with natural causes of reef mortality, these human-induced mortality events may exceed a reef's ability to recover.

Coral reefs fringe about one sixth of the world's coastlines and are estimated to house about 25% of known marine species. Sadly, over half of those reefs are now threatened by human activities (with the Caribbean region being hardest hit). Moreover, the forecast for future mortality is even more dire. In 2000, the Global Coral Reef Monitoring Network estimated that 27% of Earth's coral reefs are now dead, an additional 14% will be lost in the next 10 years, and an expected 18% will die within 30 years or less.

The human activities that are implicated as causes for such unprecedented reef mortality include agricultural activities, deforestation, and coastal development, all of which introduce sediments, excessive nutrients, and assorted pollutants into coastal areas (similar to the problems experienced by many estuaries discussed in Chapter 7). A coating of sediment on a coral colony can smother it, clog its feeding structures, and increase the colony's energy expenditure by causing its mucus–ciliary system to work overtime to rid its surface of sediment particles. Sediments also decrease the photosynthetic output of zooxanthellae by shading them and reducing their light absorption.

Elevated nutrient levels occur when runoff from agricultural areas injects excess quantities of fertilizers in the waters that bathe coral reefs. Sewage runoff also supplies unnaturally high concentrations of nutrients to coastal areas. These increased concentrations of nitrogen and phosphorus enhance algal growth and enable algae to dominate corals in their competition for space on the reef. It also results in phytoplankton blooms that cloud the water and further handicap zooxanthellae. These two common human-made causes of coral mortality, increased sedimentation and nutrification, are thought to be responsible for the devastation of corals in the Florida Keys, in parts of Hawaii, and elsewhere that has occurred in recent years.

Other reef herbivores seem to be equally important in helping corals maintain their dominance over rapidly growing seaweeds. A water-borne pathogen killed large numbers of a ubiquitous, long-spined, black sea urchin (*Diadema antillarum*) in the Caribbean Sea in 1983. It is estimated that 93% of the urchins in an area of 5 million square kilometers perished during what Knowlton called "the most extensive and severe mass mortality ever reported for a marine organism." The rapid extermination of this urchin, an important grazer of algae, enabled algal populations to overgrow corals in their competition for space on reefs. Some Caribbean reefs still have not recovered and remain green fuzzy remnants of their former beauty. Although the cause of this epidemic remains undetermined, some speculate that the Panama Canal enabled a virulent Pacific pathogen to make its way into the Caribbean Sea to cause these urchin deaths. If this is the case, humans are once again to blame for the resultant coral mortality.

Moreover, many reefs are badly overfished, and this removal of the majority of herbivorous teleosts from a reef by overfishing also enables macroalgae to overgrow corals quickly. Yet this is not the only impact that overfishing has on reefs. The methods used

for removing desirable teleosts, either for food or for aquarium display, also result in reef destruction. Because the structural complexity of coral reefs provides countless homes and hiding places for reef fishes, traditional low-impact fishing methods (such as hooks, lines, and nets) are inefficient ways to capture reef teleosts. So fishers and shell collectors on reefs often turn to much more destructive methods, including dynamite, crowbars, and poisons, to obtain their quarry. Dynamite and crowbars destroy the physical structure of the reef, and poisons (cyanide is used most commonly) can stun fishes that then become available for the live fish trade (popular among Asian restaurants and aquarium hobbyists). It is estimated that the use of cyanide results in the unintentional deaths of about 50% of the fishes exposed on the reef and the subsequent deaths of about 40% of captured fishes during transport. Thus, this 1.2 billion dollar industry centered in southeast Asia, although lucrative, is very damaging and incredibly wasteful. It is estimated that only about 4% of Philippine reefs and less than 7% of Indonesian reefs remain unaffected by cyanide use.

Overfishing can also lead to outbreaks of coral predators, with massive mortality of corals being the obvious consequence. Over the past four decades, there have been several outbreaks of the coral-eating crown-of-thorns sea star, *Acanthaster planci* (FIGURE 9.16a), in the western Pacific Ocean. In places, coral mortality adjacent to aggregations of the sea stars approached 100%. During the first outbreak in the 1960s, ecologists speculated that these sudden occurrences of large populations of this damaging sea star were the result of human activities, specifically the disappearance of its major predator, the Pacific triton, *Charonia tritonus* (Figure 9.16b). A large and beautiful snail, the Pacific triton had been nearly exterminated by shell collectors. Others suspect that the population increases are due to natural causes, such as unusually frequent storms.

Birkeland showed that more recent *Acanthaster* outbreaks occurred about 3 years after periods of unusually abundant rainfall. Perhaps abnormally high rainfall causes nutrient runoff, which leads to plankton blooms that feed *Acanthaster* larvae, resulting in subsequent increases in their successful settlement. Others suggest that outbreaks of *Acanthaster* are a recent phenomenon that is caused by the overfishing of prawns (a major predator of juvenile *Acanthaster*). Although still poorly understood, outbreaks of *Acanthaster* and resultant reef mortality may have been

(a)

(b)

figure 9.16

(a) The predatory seastar, *Acanthaster*, and (b) its major predator, the Pacific triton, *Charonia*.

augmented by the overfishing of tritons and prawns by humans.

Previously unknown coral diseases also have begun plaguing coral reefs on global scales. Porter documented a fourfold increase in disease at 160 reef sites that he has monitored in Florida since 1996. His data show that 37% of all coral species in Florida have died since his study began, and 85% of the elkhorn coral has expired. Elkhorn coral, sometimes described as "the sequoias of the reef," is a beautiful, orange, 3-m-tall branching species that so dominates Caribbean reefs that the seaward-facing Palmata Zone is named in recognition of its ubiquitous occurrence (its scientific name is *Acropora palmata*). *Serratia marcescens,* a bacterium common in human feces and sewage, is the cause of whitepox disease, the affliction causing the death of elkhorn coral. The disappearance of this majestic species is perhaps the greatest esthetic loss suffered to date.

Of the many new diseases currently plaguing corals, most are named by the appearance of the affected coral tissue (as in whitepox above); black-band disease, white-band disease, white plague, and bleaching are most common. **Black-band disease**, first reported in the 1970s in Belize and Bermuda, is now causing high mortalities in susceptible corals worldwide. This disease is characterized by a band of blackened necrotic tissue that advances several millimeters a day around coral colonies (FIGURE 9.17). Black-band disease is caused by *Phormidium corallyticum,* a sulfate-

reducing cyanobacterium that invades corals, attacks their zooxanthellae, feeds on dying coral tissues, and grows as a densely interwoven mat that separates the cenosarc from the coral's skeleton. This tissue damage eventually results in death because of the invasion of a consortium of opportunistic pathogens, such as *Beggiatoa* and *Desulfovibrio*. This consortium of cells, several of which are known only from humans

figure 9.17

External symptoms of black-band disease on coral.

(and their sewage), creates a sulfide-rich environment that prevents photosynthesis by zooxanthellae.

White-band disease was first reported in the late 1970s in Caribbean species of *Acropora* (elkhorn and staghorn corals are the most well-known members of this genus). By 1989, 95% of the elkhorn coral in St. Croix had succumbed to this disease, which also involves a suite of pathogenic agents. **White plague** (or just plague) is a disease that resembles white-band disease, only it moves and kills much more quickly. It first appeared in the Florida Keys in the 1980s. Although this disease, like most tissue-sloughing coral ailments, is poorly understood, the 1995 plague in Florida seems to have been caused by *Sphingomonas*, a common bacterium that causes infections, septicemia, and peritonitis in humans. Once again, sewage transport of this pathogen is suspected.

A team of scientists from the US Geological Survey believe that none of the many hypotheses offered to explain coral deaths around the world is adequate to explain the vast distribution of coral diseases (nor coral's inability to recover after a die-off). They suggest that the hundreds of millions of tons of dust (see Figure 3.33) carried by winds to the Americas from Africa and Asia each year may transport viable pathogens, nutrients, trace metals, and other organic contaminants that could contribute to reef deaths worldwide.

Perhaps the most well-studied cause of episodic coral mortalities is **bleaching**, a recently characterized phenomenon first observed in the mid-1980s. Bleaching occurs when physiologically stressed pathogen-free corals expel their mutualistic zooxanthellae (FIGURE 9.18). This results in a whitening of the colony (due to the appearance of the $CaCO_3$ skeleton of the coral visible through its now pigment-free cenosarc) and perhaps its death. Bleaching events have been correlated with increased sea surface temperatures, such as those that occur in the tropical eastern Pacific Ocean during El Niño–Southern Oscillation (ENSO; see pp. 117–119). For example, nearly all the living coral in the Galapagos Islands bleached and died after the severe 1982–1983 ENSO episode. Some species completely disappeared during this event in the eastern Atlantic. Dennis reported that the even more intense ENSO event of 1997–1998 resulted in the deaths of fully one sixth of our planet's coral via bleaching. This event enabled the bacterium, *Vibrio shiloi,* to invade Mediterranean corals, and the Maldives were so badly impacted that virtually none of their corals survived. There is now some concern

that the 1200 atolls that constitute the Maldives nation (Figure 9.7) are no longer protected from erosion, which may lead to the complete disappearance of this archipelago over time.

The first Caribbean bleaching event occurred in 1987–1988 and affected all species living down to 30 m depth (only *Madracis* and *Acropora* seem minimally affected). A second event occurred in 1990. Unlike Pacific episodes that are usually attributed to increased water temperatures, both mass bleaching events in the Caribbean Sea were not readily explained by temperature alone. Recent studies suggested that decreased temperatures, increased levels of ultraviolet radiation, increases sediment loads, changes in salinity, or toxic chemicals may play additional roles in Caribbean bleaching episodes during periods of calm clear water that occur during ENSO events.

However, the actual cause and mechanism of coral bleaching remain unknown. Do corals evict malfunctioning algae, thus hurting themselves in the process? Or do the zooxanthellae voluntarily leave stressed coral polyps? Some have even suggested that latent viral infections are induced by the above coral stressors.

Recently, Baker speculated that bleaching may be the natural response of corals to rapidly changing environmental conditions (changes that are the result of human activities). He viewed bleaching as an ecologic gamble that corals take to rid themselves of suboptimal algal symbionts that no longer thrive in

figure 9.18

Effect of coral bleaching on a Caribbean coral head.

our human-modified environment. Baker suggested that the coral strategy may be to sacrifice their health and growth in the short term for long-term benefits. If bleaching enables corals to recruit ENSO-tolerant algal partners, it may help them survive the severe warming trends that are predicted by contemporary climate models.

9.2 Coral-Reef Fishes

Associated with the reef and lagoon but with the mobility to escape the limitations of a benthic ex- istence are thousands of species of reef fishes (FIGURE 9.19). These fishes find protection on the reef, prey on the plants and animals living there, and sometimes nib- ble at the reef itself. These assemblages of shallow- water coral-reef fishes are easily observed by divers and have been intensively studied for decades. Less well known are the fishes of the deeper portions of coral-reef communities (below 100 m). Submersible- based studies recently demonstrated that as one works down the reef face into deep water, the same general assemblages are present, but individual numbers and species diversity both diminish.

Reef face Algal Ridge Reef flat

figure 9.19

General habitats of some common reef fishes on a tropical Caribbean reef: 1. nurse shark (*Ginglymostoma*), 2. reef shark (*Carcharhinus*), 3. barracuda (*Sphyraena*), 4. surgeonfish (*Acanthurus*), 5. butterflyfish (*Chaetodon*), 6. angelfish (*Pomacanthus*), 7. hawkfish (*Amblycirrhitus*), 8. grouper (*Mycteroperca*), 9. moray eel (*Gymnothorax*), 10. stingray (*Dasyatis*), 11. grunt (*Haemulon*), 12. soldierfish (*Myripristis*), 13. porcupinefish (*Diodon*).

Coral-Reef Sharks and Rays

Sharks are often described as large voracious predators. Yet about 80% of known species are less than 2 m in length, half of all shark species are less than 60 cm long, and some species barely attain a length of 30 cm. Many of these smaller sharks are only found on coral reefs.

Without question, nurse sharks, carpet sharks, wobbegongs, and bamboosharks in the order Orectolobiformes dominate coral reefs. Some carcharhinid sharks, such as reef sharks, blacktips and whitetips, lemons, bulls, and tigers, also frequent coral reefs. All these reef-dwelling sharks contradict an oft-repeated myth by coasting to a stop and resting on the sea floor for many hours at a time. Thanks to their inshore existence, they have developed the ability to flex their muscular gill slits and create the necessary flow of water over their gills even while stationary.

Most sharks that inhabit coral reefs also fail to fit the standard view of sharks as apex predators. Some consume large invertebrates, such as conchs, sea urchins, and clams from the sea floor. Caribbean nurse sharks suck sleeping wrasses from the sand under which they sleep. Many cryptic species are ambush predators, launching themselves from the reef when a prey species swims nearby. The numerous dermal flaps on the jaw margin of wobbegongs may function as lures to "bait" prey near their mouths (FIGURE 9.20). Nurse sharks and some rays perch on extended pectoral fins in an attempt to attract prey to the cave-like space that they create just under their chins. Some reef sharks are masters when it comes to extracting prey from reef crevices or using their snouts to flip coral rubble to reveal hidden crustaceans or annelids. And filter-feeding whale sharks and manta rays routinely visit reefs to consume the reproductive products of spawning corals and fishes.

figure 9.20

Dermal flaps around the mouth of a wobbegong, a benthic shark.

Coral-Reef Teleosts

Several groups are common to all the major regions characterized by coral reefs (FIGURE 9.21). These include grunts, snappers, cardinalfishes, moray eels, porcupinefishes, butterflyfishes, squirrelfishes, groupers, triggerfishes, gobies, blennies, wrasses, parrotfishes, surgeonfishes, and seahorses. Many of these fishes are thought to be major importers of important limiting nutrients to local reef systems by foraging on pelagic prey during the day and then defecating at night while resting on the reef. Others feed in surrounding seagrass meadows at night and defecate on the reef while resting during the day. The results of this off-reef predation are converted through detritus food chains to dissolved nutrients usable by plants, phytoplankton, and the coral-based zooxanthellae.

Symbiotic Relationships

Excellent examples of all the types of symbiosis (summarized in Figure 2.16) can be found in many of the abundant animal groups of the coral reef. Our discussions are limited to some of the better-known symbiotic relationships involving coral-reef fishes. These relationships span the entire spectrum of symbiosis, from very casual commensal associations to highly evolved parasitic relationships.

Remoras (FIGURE 9.22) associate with sharks, billfishes, parrotfishes, sea turtles, and even the occasional dolphin in a mutualistic symbiosis. The remora's first dorsal fin is modified as a sucking disc and is used to attach itself to its host. From its attached position, it feeds on scraps from the host and often cleans the host of external parasites. Thus, the remora gains food, a free ride, and protection via proximity (a special benefit of symbiosis called **inquilinism**), whereas the host rids itself of many ectoparasites. A similar association in the open ocean exists between sharks and pilotfishes (*Naucrates*). The pilotfishes swim below and in front of their hosts and scavenge bits of food from the shark's meal. It has been speculated that pilotfishes may attract prey species to the shark.

It is common for smaller defenseless fishes to live on or near better-defended species of reef invertebrates. For example, shrimpfishes often hover vertically in a head-down position among the long sharp spines of sea urchins in a commensal symbiosis (FIGURE 9.23).

Close-up of a small coral head with some of its many associated fish species.

Two remoras, *Echeneis*, with modified dorsal fins accompanying a nurse shark, *Ginglymostoma*.

The shrimpfishes acquire protection from that sea urchin without affecting it. Brightly colored clownfishes and anemonefishes find equally effective shelter by nestling among the stinging tentacles of several species of sea anemones (FIGURE 9.24). This relationship, somewhat more complex that those just described, is also probably a mutualistic one. In return for the protection they obtain, clownfishes assume the role of "bait" and lure other fishes within reach of the anemone. They occasionally collect morsels of food and, in at least one observed instance, catch other fishes and feed them to the host anemone. Clownfishes, however, are not immune to the venomous cnidocytes of all sea anemones. Although some clownfishes are innately protected from some anemone species (i.e., their protection results from their normal development rather than from contact with chemical, visual or mechanical stimuli from an anemone), Elliot and Mariscal demonstrated that some clownfishes must acclimate to some anemone species. Moreover, they also reported

figure 9.23

Shrimpfish, *Aeoliscus*, seeking shelter amid the spines of a sea urchin.

figure 9.24

Clownfishes, *Amphiprion*, hovering near their host anemone.

that other clownfishes are unable to acclimate to a particular species of anemone.

The increased popularity of skin diving and scuba diving has revealed some remarkable cleaning associations involving a surprising number of animals. Cleaning symbiosis is a form of mutualism; one partner picks external parasites and damaged tissue from the other. The first partner gets the parasites to eat; the other partner has an irritation removed.

The behavioral and structural adaptations of cleaners are well developed in a half dozen species of shrimps (FIGURE 9.25) and several groups of small fishes. Tropical cleaning fishes include juvenile butterflyfishes, angelfishes, and damselfishes, but only some neon gobies in the Atlantic (*Elacatinus*) and cleaner wrasses in the Pacific (*Labroides*) are cleaning spe-

cialists throughout their lives (FIGURE 9.26). All tropical cleaning fishes are brightly marked, are equipped with pointed pincer-like snouts and beaks, and occupy a cleaning station around an obvious rock outcrop or coral head. Most are solitary; a few species, however, live in pairs or larger breeding groups.

Host fishes approach cleaning stations, frequently queuing up and jockeying for position near the cleaner. Often, they assume unnatural and awkward poses similar to courtship displays. As the cleaner fish moves toward the host, it inspects the host's fins, skin, mouth, and gill chambers and then picks away parasites, slime, and infected tissue.

In the Bahamas, Limbaugh tested the cleaner's role in subduing parasites and the infections of other reef fishes. Two weeks after he had removed all known cleaners from two small reefs, the areas were vacated by nearly all but territorial fish species. Those species that remained had an overall ratty appearance and showed signs of increased parasitism, frayed fins, and ulcerated skin. Limbaugh concluded that symbiotic cleaners were essential in maintaining healthy fish populations in his study area.

Losey conducted similar studies on a Hawaiian reef. In this situation, the small cleaner wrasse, *Labroides phthirophagus* (the major cleaner on the reef), was excluded from the study site for more than 6 months. During that time, no increase in the level of parasite infestation was observed. This result suggests that for some cleaner–host associations the role of the cleaner is not crucial. The cleaner may be dependent on the host for food, but the host's need for the cleaner seems to be variable.

The fine line separating mutualistic cleaning of external parasites and actual parasitism of the host fish is occasionally crossed by cleaner fishes. In addition to unwanted parasites and diseased tissue, some cleaners take a little extra healthy tissue or scales or graze on the skin mucus secreted by the host. Thus, the total range of associations displayed by cleaning fishes encompasses mutualism, commensalism, and parasitism.

Because parasitism is such a widespread way of life in the sea, few fishes avoid contact with parasites at some time in their lives. The groups notorious for creating parasitic problems in humans—viruses, bacteria, flatworms, roundworms, and leeches—also plague marine fishes. Despite the bewildering array of parasites that infest fish, very few fishes become full-time parasites themselves. A remarkable exception is pearlfishes. They find refuge in the intestinal tracts of sea cucumbers, the stomachs of certain sea stars, the body cavities of sea squirts, and the shells of clams. Once this association is established, some pearlfishes assume a parasitic existence, feeding on and seriously damaging the host's respiratory structures and gonads. When seeking a host sea cucumber, pearlfishes detect a chemical substance from the cucumber and then orient themselves toward the respiratory current coming from the cucumber's cloaca. (Sea cucumbers draw in and expel water through their cloacae for gas exchange.) The fish enters the digestive tract tail first via the cloaca. The hosts are not willing participants in this relationship. They sometimes eject their digestive and respiratory organs in an attempt to rid themselves of the symbiont. In fact, sea cucumbers of the genus *Actinopyga* have evolved five teeth on their cloacal margin, perhaps as an anti-pearlfish mechanism.

Coloration

Against the colorful background of their coral environment, reef fishes have evolved equally brilliant hues and color patterns. The colors are derived from skin or internal pigments and from iridescent surface features (like those of a bird's feathers) with optical properties that produce color effects. Most fishes form accurate visual color images of what they see but, like humans, they are susceptible to misleading visual images and camouflage.

Our interpretation of the adaptive significance of color in fishes falls into three general categories: concealment, disguise, and advertisement. Some

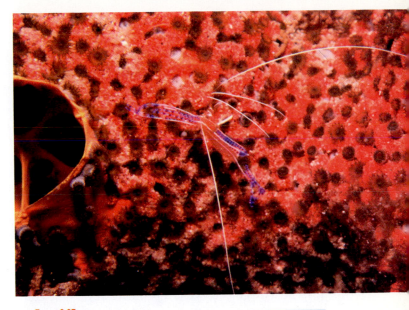

figure 9.25

A nearly transparent cleaner shrimp, *Periclimenes*, on a Caribbean sponge.

figure 9.26

A small wrasse, *Labroides*, cleaning external parasites from a lionfish, *Pterois*.

seemingly conspicuous fishes resemble their coral environment so well that they are nearly invisible when in their natural setting. Extensive color changes often supplement their basic camouflage when they are moving to different surroundings. These rapid color changes are accomplished by expanding and contracting the colored granules of pigmented cells (chromatophores) in the skin and are governed by the direct action of light on the skin, by hormones, and by nerves connected to each chromatophore. As the chromatophore pigments disperse, the color changes become more obvious (FIGURE 9.27). When the granules are contracted, the pigment retreats to the center of the cell, and little of it is visible. Other cells, called iridocytes, contain reflecting crystals of guanine. Iridocytes can produce an entire spectrum of colors within a few seconds.

Several distinctive fishes conceal themselves with color displays reminiscent of disruptive coloration, or dazzle camouflage. Bold contrasting lines, blotches, and bands tend to disrupt the fish's image and draw attention away from recognizable features such as eyes. Eyes are common targets for attack by predators, so a disguised eye is a protected eye. One common strategy masks the eye with a dark band across the black staring pupil so that it appears to be con-

tinuous with some other part of the body (FIGURE 9.28). To carry the deception even further, masks around the real eyes are sometimes accompanied by fake eyespots on other parts of the body or fins. Eyespots, intended as visual attention-getters, are usually set off by concentric rings to form a bull's eye. Presumably, predators are drawn away from the eyes and head and drawn to less vital parts of the body.

The flashy color patterns of cleaning fishes serve different functions. If the fishes are to attract any business, they must be conspicuous. So they advertise themselves and their location with bright startling color combinations. These bold advertisement displays are also useful for sexual recognition. One or both sexes of certain species assume bright color patterns during the breeding period. The colors play a prominent role in the courtship displays, which lead to spawning. During this period, the positive value gained from sexual displays must offset the adverse

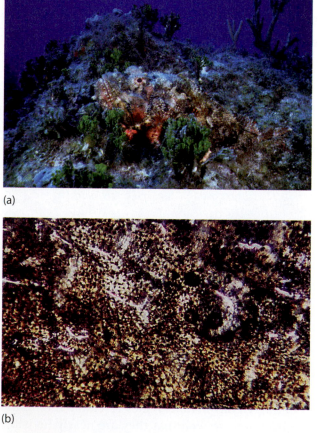

(a)

(b)

figure 9.27

A well-camouflaged scorpionfish, *Scorpaena* (a), with magnified chromatophores from a section of skin (b). The black and brown pigments of some are expanded and diffused; others are densely concentrated in small spots.

figure 9.28

Disruptive coloration patterns of two species of butterflyfish, *Chaetodon*.

impact of attracting hungry predators. Between breeding periods, these fishes usually assume a drab, less conspicuous appearance.

Advertisement displays are also used to warn potential predators that their prey carry sharp or venomous spines, poisonous flesh, or other features that would be painful or dangerous if eaten. Predatory fishes recognize the color patterns of unpalatable fishes and learn to avoid them.

Occasionally, a species capitalizes on the advertisement displays of another fish by closely mimicking its appearance. The cleaner wrasse (*Labroides*, the small fish in Figure 9.26) is nearly immune to predation because of the cleaning role it performs for its potential predators. Over much of its range, *Labroides* lives close to a small blenny (*Aspidontus*). The blenny so closely resembles *Labroides* in size, shape, and coloration (FIGURE 9.29) that it fools many of the predatory fishes that approach the wrasse's cleaning station. Not content to share *Labroides*' immunity to predation, the blenny also uses its disguise to prey on fishes that mistakenly approach it for cleaning. This ability to disguise and thereby be protected is known as mimicry.

Only in the clear waters of the tropics and subtropics does color play such a significant role in the lives of shallow-water animals. In the more productive and turbid waters of temperate and colder latitudes, light does not penetrate as deeply nor is the range of colors available. In coastal waters and kelp beds, monotony and drabness of appearance, not brilliance, are the keys to camouflage. In the deep ocean, color is even less important. Without light to illuminate their pigments, it matters little whether deepwater organisms appear red, blue, black, or chartreuse when viewed at the surface. In the abyss, they would all assume the uniform blackness of their surroundings were it not for bioluminescence (discussed in Chapter 10).

Spawning and Recruitment

Coral-reef teleosts are a very diverse lot, yet most share a common life-history strategy (most adult reef teleosts are benthic fishes that spawn in the water column). Only about 20–30% of reef species (damselfishes, gobies, and triggerfishes) deposit 1-mm-long benthic eggs that stick to the substrate until they hatch after 1–4 days (longer in some species). Of these, damselfishes are the most conspicuous, and the courtship and mating rituals of some species are well known (FIGURE 9.30). For example, bicolor damselfish, *Stegastes partitus*, mate between full and new moons. The male builds a nest and then attempts to persuade females to deposit their eggs in his nest by performing a series of dips in the water column. Females typically choose the male that performs the most dips in a given time. Presumably, the female uses his dance to assess his health and fitness. Because the male must guard her eggs for 3–4 days (FIGURE 9.31), the female uses the rate of dipping during courtship to determine which male in her vicinity has the most energy stored as fat (energy that will be very useful while he guards her eggs relentlessly before their hatching).

figure 9.29

A cleaner wrasse, *Labroides* (above), and its mimic, *Aspidontus* (below).

figure 9.30

Two bicolor damselfish mate inside a discarded PVC pipe on a Caribbean reef.

Yet most reef teleosts are pelagic spawners. Many species, 30 or more at any given time, will assemble around the same coral promontory to broadcast as many as 50,000 eggs apiece into the water column (FIGURE 9.32) during the course of an hour or so. The coral pinnacle that is selected is not obviously different from neighboring outgrowths, but the fishes seem to

understand the difference. In fact, if one removes all female blueheads from a reef, the new set of replacement females that arrives will pick many (maybe completely) new spawning sites, and these sites will become the new "traditional" sites on that reef. This is one of the few examples of "culture" in fishes. Wrasses are known to travel 1.5 km to a spawning site (a distance equal to 15,000 body lengths, or the equivalent of a 55-km roundtrip for a human), and Nassau groupers, *Epinephelus striatus*, may travel up to 240 km to spawn.

After fertilization, these pelagic eggs drift away from the reef and disperse for a period of time that ranges from 1 day to a year or more. This time period, known as the **pelagic larval duration**, influences dispersal greatly and is of intense interest because of the influence of pelagic larval durations on the potential efficacy of National Marine Sanctuaries.

In 1972, a century after the United States established the first national park at Yellowstone, legislation was passed to create the National Marine Sanctuaries Program. The intent of this legislation was to provide similar protection to selected coastal habitats as we have for land areas designated as national parks. The designation of an area as a marine sanctuary says to all that, like our national parks, this is a safe refuge where people can observe organisms in their natural environment, but nothing may be harmed or removed. Three decades later, only 15 marine sanctuaries have been designated (FIGURE 9.33).

The National Marine Sanctuary Program should be viewed as a crucial part of new management practices in which whole communities, and not just individual species, are offered some degree of protection from habitat degradation. Nevertheless, because of the interactions between and vagaries of the pelagic larval durations of different teleosts and surface currents, one can never be certain that protection of a spawning population on one reef will ensure augmented recruitment of juveniles into that sanctuary. It seems just as likely that one's efforts at one location will result in huge benefits in terms of larval recruitment somewhere down current.

Therefore, a great deal of research effort is also being directed to the study of settlement or the passage from the pelagic existence of a larva to the benthic life of a juvenile reef fish. Hypothetical recruitment factors that are being investigated include active mechanisms, such as larvae being attracted to reef sounds (such as waves breaking on the reef crest or the snapping of shrimp) or reef smells (in a mechanism analogous to the homing behavior of salmon) as well as passive mechanisms that may

figure 9.31

A male green damselfish (*Abudefduf*) guards its purple egg mass in Hawaii.

figure 9.32

Most reef teleosts spawn in the water column above the reef.

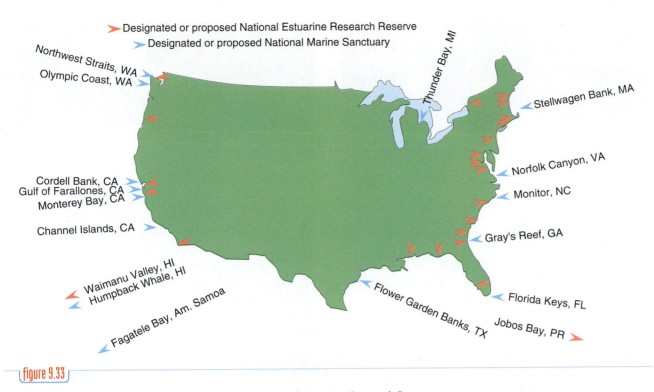

figure 9.33

Locations of US National Marine Sanctuaries and National Estuarine Research Reserves.

enable the settlement of postlarval juveniles (such as retention in gyres around some Hawaiian islands).

It is estimated that only about 1% of the eggs spawned survive to produce an individual that will settle on a reef. But successful settlement does not mean that the intense mortality is over. About 90% of those teleost larvae that settle on a reef are eaten during their first night. If a larva is lucky enough to survive its first night on the reef, it then must metamorphose into a benthic juvenile within a day or two. Only 50% of those that begin metamorphosis survive to complete the transformation. Once they acquire the morphology, appearance, and behavior of a juvenile, mortality decreases greatly.

Sexual Systems in Reef Fishes

Reef teleosts display a great variety of sexual systems, from relatively straightforward gonochorism with separate males and females (as seen in grunts, snappers, and most damselfishes) to complex systems involving hermaphroditic individuals that play the roles of both genders during their adult lives. Simultaneous hermaphrodites function as males and females at the same time, whereas sequential hermaphrodites are born as one gender and then change sex during their life.

The best-known simultaneous, or synchronous, hermaphrodite on the reef is the hamlet (a relative of groupers in family Serranidae). All hamlets are monogamous, forming faithful pairs that often last throughout the breeding season. During courtship, one member of the pair will "act male" and will release sperm during a stereotypical "clasping" in the water column above the reef (FIGURE 9.34). Its mate will play the female role and release eggs. Immediately after spawning, the couple will reverse roles and recourt, with the first individual now acting female and releasing eggs and the second fish acting male and releasing sperm. They then repeat this ritual 2–15 times, trading roles repeatedly, until they are both spent and unable to court with another fish.

Sequential hermaphrodites come in two varieties, those that begin their adult life as a male (**protandry**) and those that begin their adult life as a female (**protogyny**). Both types change sex as they age. The best-known protandrous hermaphrodites on the reef are clownfishes (see Figure 9.24), which all begin their adult lives as a male. In each clownfish community (often a single sea anemone), a pecking order exists wherein the largest individual is an adult female, the second largest is an adult male, and the remaining individuals are all undifferentiated juveniles. If the alpha female is removed via death or predation, oppression on the large male is released and he becomes a female. Soon after, the individual that formerly was third in line differentiates and matures into an adult male.

figure 9.34

Clasping hamlets above a reef.

figure 9.35

Male and female "bluehead" wrasses in their initial yellow phase.

Beginning adult life as a female is much more common among reef fishes, and wrasses and parrotfishes in the family Labridae are perhaps the best-known protogynous hermaphrodites. The bluehead, *Thalassoma bifasciatum,* of Caribbean coral reefs has a particularly interesting sexual system that is greatly influenced by the number of wrasses in a given area. All blueheads settle on the reef as tiny yellow individuals. Under conditions of high wrasse density, the young females grow to become larger yellow females and the young males grow to become larger yellow males (FIGURE 9.35). As adults, they participate in group spawning events wherein a sphere of yellow fish appear over the reef, one large female races toward the surface and spawns while being closely followed by 10–15 adult males that streak toward the surface and compete for the opportunity to fertilize her eggs. This ritual is very costly in that leaving the safety of the reef makes these fishes extremely vulnerable to predation. (One estimate suggests that a single adult jack will consume about 1 ton of fishes, many of which are spawning wrasses, during its life.) Nevertheless, this seemingly careless behavior may be important in helping the female choose the fittest male in the group.

When bluehead wrasses are not abundant in a given area, such that females can "choose" to mate with the bigger single males, blueheads will engage in pair spawning instead, and the "real" bluehead will appear. The largest female will double her size and change sex, thus transforming herself into a terminal-phase male (sometimes called a "supermale") with a blue head, a greenish body, and an "oreo" around his neck (so named because the two black bands flanking a single white band closely resembles the popular cookie). This blueheaded terminal-phase male (FIGURE 9.36) is responsible for this species' common name, and only he is involved with pair-spawning events. Each day, this Darwinian champion will establish a territory and spawn with as many as 100 females. And with the exception of small yellow "streaker" males that dart in to fertilize a tiny portion of the eggs liberated by some females, only the territory-holding male experiences very high reproductive success (because he will spawn with many females each day, while being streaked only a few times). This helps explain why the mating system of wrasses is density dependent. If there are a lot of wrasses on the reef, the terminal-phase bluehead will be streaked so often that he will lose the benefit of being large and maintaining a territory (i.e., high reproductive success).

Unfortunately, once a large female becomes a terminal-phase male, his days are numbered. His exceptional reproductive effort is very expensive, and his propensity to run the extremely risky reproductive gauntlet up to 100 times each day almost always results in a lifespan that ends after just a few weeks. As soon as he dies, the process starts again,

Terminal-phase bluehead male surveying his territory.

tributes half the genetic information to be passed on to the next generation. In the event the dominant male of a *Labriodes* population dies or is removed, the most dominant of the remaining females immediately assumes the behavioral role of the male. Within 2 weeks, the dominant individual's color patterns change and the sex transformation to a male is complete. In this manner, males are produced only as they are needed, and then only from the most dominant of the remaining members of the population.

Before ending our discussion of hermaphroditic reef fishes, it is helpful to consider the benefits of such a sexual system. Why should an individual change its sex during its life, and under what circumstances should a reef fish begin life as a male (as opposed to a female)? Biologists hypothesize that changing sex is advantageous when reproductive success is closely tied to body size, and when being one sex (as opposed to the other) results in increased reproductive success at a given body size (this is known as the size-advantage model). FIGURE 9.37a shows the relative reproductive success of male and female clownfishes from the time they are small adults to sometime later when they are older and larger. Note that body size does not influence the reproductive success of male clownfishes. It remains the same as they grow because they produce countless sperm cells at all body sizes and they do not compete with other males for access to females (they simply mate with the adult female inhabiting their sea anemone). Yet female reproductive success among clownfishes is greatly influenced by body size because eggs are very large and

wherein the largest female in the area begins her 2-week transition into a terminal-phase male.

The tropical cleaner fish *Labroides* (also in the family Labridae; see Figure 9.26) is also protogynous. This inhabitant of the Great Barrier Reef of Australia occurs in small social groups of about 10 individuals. Each group consists of one dominant male and several females existing in a hierarchical social group. This type of social and breeding organization is termed **polygyny**. Only the dominant most aggressive individual functions as the male and, by himself, con-

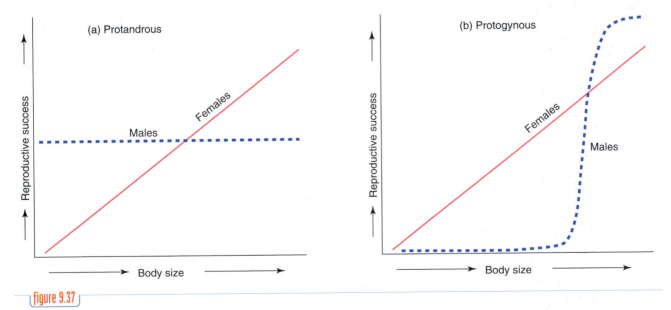

Relative reproductive success experienced by males and females of protandrous clownfishes (a) and protogynous wrasses (b).

expensive to produce. Therefore, young (small) females do not produce nearly as many eggs as larger (older) females. Hence, it is advantageous for clownfishes to be male while young (and small) and then change into females once older (and larger). Via this sex-changing strategy, they achieve maximal reproductive success at all sizes.

Figure 9.37b presents the relative reproductive success achieved by protogynous fishes, such as the blueheaded wrasses described above. Female wrasses are very similar to female clownfishes in that they only produce a large number of eggs when they possess a large body. Unlike male clownfishes, small (young) male wrasses experience very limited reproductive success because they are too small to compete for access to females and their eggs. Hence, their reproductive success is essentially zero while young and small and then suddenly sky-rockets once they become large enough to gain access to eggs in the water column. Hence, unlike the situation experienced by clownfishes, it is better for a wrasse to begin life as a small female (because small males do not reproduce at all) and later change into a male after becoming large and competitive.

But the size-advantage model does not explain all changes of sex. The coral goby, *Paragobiodon echinocephalus*, changes sex both ways. Coral gobies are small black fish with orange heads that live on the branching coral, *Stylophora pistillata*. On each coral colony, only the two largest fish (one male and one female of similar sizes) breed monogamously. The reproductive success of the pair is positively correlated with size in both sexes; larger males are more successful at guarding benthic eggs, and larger females can produce more eggs. Hence, the size-advantage model does not explain the factors that induce sex change in this species. Instead, change in social rank determines the direction of sex change. When a coral goby loses its mate, it prefers to change sex in either direction to form a mating bond with the nearest adult goby (as opposed to traveling a great distance to locate a heterosexual mate).

9.3 Some Tropical Marine Tetrapods

Teleost fishes are clearly the dominant group of vertebrates associated with coral reefs. Even so, most marine reptiles and a few marine mammals are also common inhabitants of the warm waters of corals reefs and lagoon and reef-flat seagrass beds. Like reef-forming corals and tropical mangroves, countless seagrass blades growing in all tropical lagoons provide a prodigious surface area on which other organisms (**epibionts**) can attach and grow. In St. Ann's Bay, Jamaica, Maciá determined that, on each square meter of seafloor, seagrass blades provide an average of nearly 300 m^2 of surface on which epibionts can attach. This vast expanse of surface area does not go unnoticed by local organisms. About 175 species of plants and animals have already been observed living attached to blades of turtle grass in the Caribbean region, including various algae, sponges, hydroids, sea anemones, amphipods, bryozoans, tunicates, annelids, and snails. These same seagrass beds support the foraging activities of a few species of unusual marine tetrapods that are unusual simply because they are herbivores.

If you were to visit a tropical seagrass meadow with a mask and snorkel, your first impression would likely be that there is very little life in seagrass beds (FIGURE 9.38). This is because, other than the epibionts, which simply appear as a whitish fuzz on older blades, you would encounter very few animals. The primary reason for the paucity of animals among seagrass is that this habitat experiences a great deal of sedimentation. Waves arrive from the open sea and break upon the reef flat, creating sediment-laden currents that stream into the lagoon. In the lagoon, the currents slow as they are forced to meander through millions of seagrass blades. This decreased current velocity is insufficient to transport larger sediment particles, and they begin to settle onto the sea floor among the seagrass. Many organisms can not endure this high rate of sedimentation because it interferes

A typical view of a seagrass-covered lagoon floor.

with feeding, it hinders respiration by clogging gills, it easily abrades soft tissues, and it buries smaller organisms in an avalanche of particles. However, the high sedimentation rates that repel many potential seagrass residents is actually attractive to deposit feeding sea cucumbers and mojarras, silvery fishes that make a living by straining mouthfuls of sediment through their gill rakers in search of organic morsels.

Macroalgae such as *Halimeda, Penicillus, Acetabularia,* and *Caulerpa* are often very common in tropical lagoons, and because some of these also precipitate $CaCO_3$ that they have extracted from seawater (like corals), they contribute additional carbonate sediment to the sea floor after the die. About 85% of the biomass of the Caribbean's *Halimeda* is $CaCO_3$, and this genus alone can contribute three kilograms of carbonate sand per square meter in just 1 year.

The expansive meadow of seagrass and macroalgae that grows in most tropical lagoons is an irresistible source of food to several common herbivores. In addition to high concentrations of herbivorous parrotfishes and surgeonfishes that leave the safety of the adjacent reef at night to forage in seagrass, a few species of large marine reptiles and mammals rely extensively on tropical seagrass beds and are important components of seagrass communities.

Tropical Marine Reptiles

Marine reptiles such as sea turtles, sea snakes, the marine crocodile, and the marine iguana are, like other reptiles, ectothermic (see pp. 24–25). Therefore, they are largely restricted to tropical latitudes. Sea snakes move up the east coast of Asia by staying within the warm northward flowing Kuroshio Current, and some sea turtles occur off Long Island because of the influence of the warm Gulf Stream in the region. But cold shock is a common cause of strandings in sea turtles that stray into higher latitudes, and most marine reptiles prefer to remain near the equator.

A single species of lizard, the marine iguana of the Galapagos Islands, is marine (FIGURE 9.39, and see Figure 6.22a). In contrast to the herbivorous habits of marine iguanas, the 21 species of crocodilians existing today are all carnivorous. Most, including alligators, caimans, and gharials, inhabit freshwater river and lake systems. Crocodiles (family Crocodylidae, FIGURE 9.40a) are more euryhaline than their relatives, and two species routinely enter the sea. The American crocodile is quite at home in the sea, occurring from the southern tip of Florida through the Caribbean to northern South America. And the

figure 9.39

Several marine iguanas basking on a rock in the Galapagos.

(a)

(b)

figure 9.40

(a) The saltwater crocodile, *Crocodylus,* in a north Australia coastal river. (b) Beach warning sign on Australia's northeast coast.

saltwater crocodile, *Crocodylus porosus*, inhabits tropical estuaries and mangrove swamps around islands of the Indo-Australian archipelago from Asia to Australia. Although it is not uncommon to see this largest living crocodilian at sea hundreds of kilometers from land, it is more frequently encountered near shore where it poses some danger to swimmers and divers (Figure 9.40b).

Sea snakes are derived from the highly venomous elapids, a terrestrial family that includes cobras, coral snakes, and kraits. Although some biologists lump sea snakes into family Elapidae with the cobras and coral snakes, we prefer to recognize two families of sea snakes, the laticaudids, which lay 4-inch-long sticky eggs supratidally, and the more highly derived hydrophiids, which are helpless on land. Hydrophiid sea snakes never leave the sea and retain their eggs to give birth to live young there, much like the ovovivparous sharks described in Chapter 6.

Like their terrestrial relatives, sea snakes are highly venomous, although they are difficult to excite, and bites on human divers are quite rare. The 100 or so extant species all inhabit Indo-Pacific waters, with the exception of *Pelamis platurus* (see Figure 6.22b) in the eastern tropical Pacific. There are no sea snakes in the Atlantic Ocean, and all reports of sightings there stem from sightings of moray and snake eels.

Sea snakes are extremely specialized for life in the ocean. The distal end of their tail is laterally flattened into an oar-like structure, the large ventral scales common in terrestrial snakes are reduced or absent in most species, their valved nostrils are dorsally located on their snouts, and uptake of oxygen through their skin while under water has been demonstrated. Like all advanced snakes, sea snakes possess only a right lung, but unlike terrestrial species, the single lung of sea snakes extends back to the cloaca and is not highly vascularized, suggesting that it serves a hydrostatic function like the swim bladders of teleost fishes.

The seven species of sea turtles that survive today are classified into two families, six shelled species in family Chelonidae and the shell-less leatherback sea turtle, *Dermochelys coriacea* (FIGURE 9.41). Leatherbacks are adapted to colder water than other sea turtles (and marine reptiles in general) and are the most widely distributed of all sea turtles as a result. They routinely travel far from the

tropics, easily reaching the North Sea, Barents Sea, and New Foundland in the Atlantic and ranging from New Zealand to Alaska in the Pacific. Their ability to thrive in very cold water is poorly understood, yet several hypothetical attributes that may enable leatherbacks to enter cold waters have been proposed. These include protection gained from their thick and oily dermis, behavioral thermoregulation by basking in the sun at the surface and absorbing solar energy through their black backs, and heat retention via circulatory countercurrent heat exchangers in their flippers. In addition, many biologists credit their cold-water tolerance to gigantothermy, a phenomenon wherein large organisms with relatively little surface area are able to stay warm for very long periods of time (once they successfully heat up their large volume via surface basking). Leatherbacks are the largest living sea turtles and are one of the largest living reptiles. The largest specimen ever measured, an individual that stranded on the coast of Wales in 1988, had carapace and flipper lengths of nearly 260 cm and a body weight that exceeded 900 kg. Nevertheless, most species of marine turtles are tropical and subtropical and are often common in reefy areas.

All sea turtles except the green sea turtle are carnivores, and most consume a variety of reef-dwelling benthic invertebrates. Loggerheads seem to prefer large mollusks, the two ridley turtles target crabs, and Indo-Pacific flatback sea turtles emphasize sea cucumbers in the diets. Hawksbill sea turtles have been

figure 9.41

The leatherback sea turtle, *Dermochelys coriacea*.

described as having a "diet of glass" in that they frequently consume sponges with siliceous spicules. Because leatherback sea turtles are highly oceanic, only approaching land during their breeding season, they feed mainly on gelatinous epipelagic invertebrates such as medusae, siphonophores, and tunicates. Unfortunately, common items in their diet closely resemble plastic debris in the sea, and leatherbacks are highly prone to ingesting discarded plastic trash that ultimately clogs their digestive tract and kills them.

Sea turtles are well known for their need to return to land to lay their eggs in nests dug in sandy beaches above the high tide lines of tropical and subtropical shores. Temperature-dependent sex determination, a phenomenon wherein the sex of some reptiles is determined not by genetics but rather by the nest temperature that the eggs experience during development, occurs in all crocodilians and many turtles, including sea turtles. In sea turtles, warmer eggs yield more female hatchlings; the opposite is true among crocodilians.

The ability of sea turtles to navigate over thousands of kilometers of open ocean to return to nesting beaches year after year is legendary. The most striking example of this ability is displayed by green sea turtles in the Atlantic Ocean. Only the green sea turtle, *Chelonia* (see Figure 6.22c), is herbivorous, preferring to graze on seagrasses (although it does consume macroalgae and kelps). Some Atlantic sea turtles feed along the coast of Brazil and nest on Ascension Island, a tiny volcanic peak only 20 km in diameter located about 2200 km offshore.

These turtles lay their eggs in the warm sandy beaches along the north and west coasts of Ascension Island. Immediately after hatching, the young turtles instinctively dig themselves out of the sand, scurry into the water, and head directly out to sea. During this very short period, they are heavily preyed upon by seabirds and large fishes. Once they are beyond the hazards of shoreline and surf, they presumably are picked up by the South Atlantic Equatorial Current and are carried toward Brazil at speeds of 1–2 km/h. Less than 2 months are needed to passively drift to Brazil, yet nothing is known of the young turtles' whereabouts or activities during their first year.

As they mature, the turtles congregate along the mainland coast of Brazil, where they graze on turtle grass, *Thalassia*, and other seagrasses in shallow flats. Features of the nesting migration back to Ascension Island are not well known, but the adult turtles do show up there in great numbers during the nesting season. Mating apparently occurs only near the nesting ground. Either the males accompany the females on their migration from Brazil to Ascension or they make a precisely timed, but independent, trip on their own. Either way, the males get to the nesting area and can be seen just outside the surf zone competing for the attentions of females.

Females go ashore several times during the nesting season and deposit about a hundred eggs each time (FIGURE 9.42). These egg-laying episodes provide the only opportunity researchers have to capture and tag large numbers of green turtles at their nesting sites. (Because males do not leave the water, almost nothing is known about their migratory behavior.) Tagging results indicate that the females leave Ascension

figure 9.42

The green sea turtle, *Chelonia mydas*, laying eggs in a sandy beach nest.

Threats to the Diversity of Living Sea Turtles

Although sea turtles first appeared during the early Cretaceous period (about 140 million years ago), this unusual evolutionary success did not prepare them for the arrival of *Homo sapiens,* the most efficient predator ever to hunt on Earth. Our relentless population growth and ubiquitous coastal developments have destroyed many nesting beaches, and our incessant need for food has caused us to steal their eggs and kill unknown numbers of sea turtles for their meat, fat, shells, and skin. Careless fishing methods have resulted in the unintentional deaths of thousands more. In fact, sea turtles may hold the unenviable distinction of being the only marine animal to suffer tremendous human-induced mortality both on land and in the sea. The obvious result is that all seven living species of sea turtles described in Chapters 6 and 9 are declining in number, and only two of seven species are not listed officially as endangered.

The olive ridley and flatback sea turtles are the lucky ones; although there is evidence their numbers are declining, they are not yet listed as endangered. The olive ridley is found throughout the world (with the exception of Hawaii and the Caribbean Sea) and may be the most abundant sea turtle alive today. The only species to nest on the east coast of India, the olive ridley used to arrive on the beaches of Gahirmatha in numbers that approached 250,000 females. But like other species, the olive ridley is not immune to the trawls of shrimpers, gill nets, lighted beaches, coastal development, and other disturbances. The expected massive nesting of female olive ridleys at Gahirmatha did not occur at all in 1997 or 1998.

The flatback sea turtle of Australia, Indonesia, and Papua New Guinea, so named because the scutes on its back are poorly defined, is unique among sea turtles. First, nesting females deposit only 50–60 eggs per clutch (half that produced by other species). Second, this is the only species without a pelagic stage in its life cycle in that most flatbacks do not migrate. Rather, they spend their entire life swimming around the northern coast of Australia. All other sea turtles disappear upon hatching and do not return until they are about 30 cm long. Except for loggerheads in the North Atlantic, which may ride the oceanic gyre for several years, the whereabouts of most young sea turtles during these "lost years" are unknown. Although the range of flatbacks is somewhat limited (relative to most species of sea turtles), the fact that their hunt is limited mostly to Australian Aborigines may reduce the human impact on this species.

The same cannot be said for the other species of "ridley" turtle, Kemp's ridley of the Caribbean Sea and Gulf of Mexico, which is the rarest of all sea turtles (it is estimated that only 1500 individuals exist today). Kemp's ridley (the origin of the term "ridley" is unknown) is one of the smallest species of sea turtles and is extremely vulnerable for two reasons. First, they nest mostly dur-ing the day (just like Australian flatbacks), making them easy prey to human hunters (Figure B9.1). Second, they nest only on one beach, Rancho Nuevo, about 160 km south of Texas on the Gulf coast of Mexico, which makes the entire population extremely susceptible to disturbances there. A film shot in 1947 shows about 40,000 female Kemp's ridley turtles coming ashore at Rancho Nuevo simultaneously in an incredible phenomenon known as an *arribada* (Spanish for "arrival"). In 1995, only 580 nests were counted at Rancho Nuevo. Historically, the eggs of Kemp's ridley turtles were dug up and sold in local bars (the patrons would drop the raw egg into a beer and drink the concoction thinking that it was an effective aphrodisiac). Even though Kemp's ridley turtles have received full legal protection from the Mexican government since 1960, high mortality occurs in the nets of Gulf of Mexico shrimp trawlers when turtles accidentally drown before the trawl gear is retrieved. In 1989, all U.S. shrimp boats became required by law to install and use turtle excluder devices (or TEDs) in their trawl nets. In theory, sea turtles are large enough and strong enough to push the TEDs open (the little shrimp are not) such that they function as an escape hatch. In practice, some TEDs malfunction and little enforcement is present. Trawls kill between 1000 and 10,000 sea turtles (mostly Kemp's ridleys and loggerheads) each year.

Green sea turtles may be the most commercially valuable living reptile. They are killed for their skin, meat, shells, and cartilage (called calipee), their most valuable commodity, which is cut from between the bones of their bottom shell and used in soup. Green sea turtles were pivotal in educating us that artificial lights affect the reproductive success of sea turtles. Under normal circumstances, the sea at night is brighter than adjacent land (due to the reflection of celestial bodies off its surface). Postnesting females and emerging hatchlings are attracted to this brightness and normally use it to find the sea. Unfortunately, artificial lights, such as street lights and other sources of shoreline illumination (such as high-rise hotels), overwhelm the dim reflections of the ocean's surface and cause nesting females and hatchlings to head for land

rather than the sea. There they soon die due to predation, exhaustion, dehydration, or impacts by motor vehicles. This source of mortality was so great in Volusia County, Florida that citizens sued the county in 1997 to decrease artificial illumination along nesting beaches during the breeding season and to prevent people from driving or parking cars on nesting beaches. Their suit was contested all the way to the Supreme Court, and the private citizens won. Today, Volusia County is a model for turtle protection and conservation.

More recently, green sea turtles have suffered from an epidemic of fibropapillomatosis (50% of wild turtles in Florida and Hawaii are infected). Afflicted turtles are anemic and emaciated, fibrous tumors develop in multiple organs, and buoyancy problems, bowel stoppage, renal failure, and pressure necrosis results. Death is common. To date, the cause is unknown but is believed to be a tumor-causing retrovirus, herpesvirus, or papillomavirus.

Hawksbill sea turtles, which are found in all tropical seas, have their own unique problem. Unlike other sea turtles, they do not shed their skin by exfoliation. Therefore, the epidermal scutes that overlay the bones of their shell are unusually thick and in demand commercially. When harvested, the scutes

Threats to the Diversity of Living Sea Turtles

are treated and sanded smooth. This is the source of "tortoiseshell," used for eyeglass frames, shoehorns, combs, the backs of hand-held mirrors, and as decorations on innumerable other products. The introduction of plastics in the 1930s temporarily reduced the pressure on hawksbills, but today people are once again craving real tortoiseshell, and fishers who participate in black-market trade are happy to meet the demand.

Finally, leatherback sea turtles, the largest living sea turtle and the heaviest reptile on Earth, is projected by James Spotila of Drexel University to be extinct in the Pacific Ocean by 2010, where they fall victims to gill nets set by fishers from several Asian nations. Clearly, the ancient lineage of sea turtles is vulnerable worldwide to a variety of serious threats, and only public education and altered human behavior can save them from extinction.

Additional Reading
Ellis, R. 2003. *The Empty Ocean*. Island Press/Shearwater Books, Washington.

www.bioscience.jbpub.com/marinebiology
For more information on this topic, go to this book's website at http://www.bioscience.jbpub.com/marinebiology.

Figure B9.1   This Kemp's ridley sea turtle's newly laid eggs are collected for a captive rearing program.

Island after laying the eggs and return to the Brazilian coast.

It has been hypothesized that adult green turtles use a straightforward navigation system using the sun on their spawning migration back to their Ascension Island nesting sites. Ascension Island lies due east of Brazil at 8° S latitude. Adult turtles may use the height of the noonday sun to judge latitude, swim to the east at 8° S latitude, and eventually make landfall on Ascension island. Island finding by the turtles might be improved if, once they were close to Ascension Island on the down-current side, they detected a characteristic chemical given off by the island. No one is sure how well green turtles can use celestial cues (if at all) or how well they can detect chemicals dissolved in water. Until these aspects of turtle biology are studied further, the guidance system of green turtles must remain hypothetical.

Green sea turtles used to be incredibly abundant. It has been estimated that about 33 million green sea turtles lived in the Caribbean region in the early 1800s. Harvesting of turtles for meat and raiding nests for their eggs have so devastated the Caribbean population that today less than 5000 green turtles are thought to exist in the Caribbean basin. Consequently, herbivorous fishes and sea urchins have become the dominant consumers of seagrasses there. Formerly, before they were hunted to near extinction, sirenians (manatees in the Atlantic and dugongs in the Indo-Pacific) used to be very abundant in lagoonal systems, grazing like "sea cows" in an aquatic meadow.

Mammalian Grazers of Seagrasses

Manatees and dugongs (order Sirenia) are the only marine mammals that are herbivores. Manatees and dugongs consume a wide variety of tropical and subtropical seagrasses, including *Enhalus, Halophila, Halodule, Cymodocea, Thalassia, Thalasodendron, Syringodium,* and *Zostera.* Algae also are eaten but only in limited amounts if seagrasses are abundant. At the very southern extent of their range, dugongs in subtropical Moreton Bay, Australia also consume sessile benthic invertebrates including sea squirts and polychaete worms, presumably to augment the protein content of their otherwise low-protein diet of seagrasses.

The strongly down-turned snout of the dugong causes its mouth to open almost straight downward, and it is virtually an obligate bottom feeder subsisting on seagrasses less than 20 cm high. Manatees, in contrast, have only a relatively slight deflection and are generalists, feeding at any level in the water column from bottom to surface, and are able to easily take floating vegetation. Manatees graze on a large variety of coastal and freshwater vegetation, including several species of submerged seagrasses, floating freshwater plants (*Hydrilla* and water hyacinths), and even the leaves and shoots of emergent mangroves.

Sirenians are the only marine mammals that have a prehensile snout. The short muscular snout of manatees is covered with modified vibrissae that have a prehensile function to bring vegetation to the mouth (FIGURE 9.43). During feeding, dugongs gouge visible tracks into seagrass stands and bottom sediments while grubbing seagrasses from the bottom. Each track follows a serpentine course and appears to represent the continuous feeding effort of a single dive. Like terrestrial mammalian herbivores, the cheek teeth of both manatees and dugongs are adapted for grinding cellulose-rich vegetation.

Even though seagrass meadows are very productive, green sea turtles, manatees, and dugongs are the only large herbivores to graze on them commonly. The low energy value and protein content of a seagrass diet for manatees and dugongs has been suggested as contributing factors to their slow and sluggish behavior, low metabolic rates relative to other marine mammals, and the consequent need for these endotherms to remain in tropical and warm subtropical waters. Several other species of marine mammals (described in Chapter 10) also occupy tropical waters, but none is as completely restricted to warm and shallow waters as these large, slow, herbivorous sirenians.

figure 9.43

The Caribbean manatee manipulating plant food with its snout vibrissae and flippers.

Summary Points

Tropical and Subtropical Shallow Seas

- Hard substrate is an important limiting factor in the tropical ocean.
- Coral reefs, mangrove stands, and seagrass meadows provide homes and attachment sites for countless marine organisms.

Coral Reefs

- Coral reefs are created by many species of colonial cnidarians. These anemone-like polyps produce a $CaCO_3$ skeleton in a great variety of sizes and shapes.
- Living coral reefs usually are located within 30° latitude of the equator, in water that averages at least 20°C, on the eastern side of most continents, and at depths no greater than 50 m.
- Reef-building corals contain a mutualistic single-celled dinophyte that provides photosynthetic products to the coral to aid in its survival and growth. These tiny zooxanthellae are also associated with medusae, sponges, and giant clams.
- Coral reefs occur in three morphologies: fringing reefs, barrier reefs, and atolls. Charles Darwin was the first to suggest that these reef designs are sequential developmental stages in the life cycle of a single reef.
- Corals reproduce in a great variety of ways, both asexually and sexually. Most sexually reproducing corals are hermaphroditic spawners.
- Wave force, water depth, temperature, salinity, and a host of biologic factors favor some corals and reef inhabitants over others. These conditions vary greatly across a reef and result in both horizontal and vertical zonation of the species that form the reef.
- Although biologists acknowledge that occasional catastrophic mortality may be beneficial to reefs in that it helps maintain their extraordinary biodiversity, it is clear that reefs worldwide are threatened by human activities. One group predicts that nearly 60% of the Earth's coral reefs will die within 30 years, succumbing to pollution, destructive fishing practices, bleaching, and a host of diseases.

Coral-Reef Fishes

- Reefs worldwide are dominated by benthic orectolobid sharks (nurses, wobbegongs, and bamboosharks) and more typical pelagic carcharhinid sharks (blacktips, whitetips, tigers, and reefs sharks).
- About 50% of all living vertebrates are teleost fishes, and many of these fishes inhabit coral reefs. This great diversity of fishes, living in close association with each other and numerous reef invertebrates, results in numerous symbiotic relationships such as inquilinism and cleaning behaviors.
- The brightly colored patterns of coral-reef fishes illustrate the advertisement, disguise, and concealment roles of brilliant coloration in a coral-reef environment.
- About one fourth of all reef-fish species place sticky benthic eggs in a guarded nest on the reef, whereas most reef teleosts are pelagic spawners that release many thousands of gametes into the water column.
- The recruitment of postlarval reef fishes to a coral reef is essential to maintaining the health and diversity of the reef itself. Postlarval fishes that do settle on reefs are offspring either of residents of that reef or of fishes living upcurrent on adjacent reefs. It is crucial for managers and conservationists to determine the relative contributions made by each source of recruited larvae.
- The great diversity of reef fishes results in an equally great diversity of sexual systems, from more typical species with separate sexes to complex systems involving simultaneous hermaphrodites that produce both eggs and sperm to sequential hermaphrodites that change sex.

Some Tropical Marine Tetrapods

- The 100 or so species of sea snakes are relatives of the highly venomous cobras, coral snakes, and kraits. Most sea snakes are highly derived and are able to complete their entire life cycle at sea, remaining underwater for 8 hours or more.
- Most sea turtles (seven species are known) frequent coral reefs and seagrass meadows, often navigating over great distances to return to preferred nesting beaches year after year.

STUDY GUIDE

- Manatees and dugongs are the only herbivorous marine mammals. They use their prehensile snouts to graze on a wide variety of sea grasses and the occasional macroalga.

Topics for Discussion and Review

1. Summarize the limitations to coral-reef distribution and then explain why coral reefs do not form at all latitudes and depths.
2. Tally the pros and cons experienced by each member of the symbiotic relationship between corals and zooxanthellae and demonstrate that this relationship is mutualistic.
3. Why is hermaphroditic spawning the most common method of sexual reproduction in reef-building corals?
4. What can be done to slow or end the global destruction of coral reefs that we are experiencing currently?
5. Describe the relationship between clownfishes and sea anemones, listing the benefits and disadvantages experienced by each.
6. Generate a list of all potential cues that postlarval reef fishes could use to locate the coral reef on which they eventually settle.
7. Summarize the factors that lead to protogynous hermaphroditism (as opposed to protandry) in coral-reef fishes.
8. Summarize all differences between terrestrial and marine snakes. Then suggest additional adaptations that would benefit sea snakes in their ability to live in the sea.
9. Generate a list of all potential cues that sea turtles could use while navigating over great distances to return to their preferred nesting beaches.
10. Why are manatees and dugongs restricted to tropical and subtropical waters even though their preferred food (seagrasses) occurs commonly at most latitudes?

Suggestions for Further Reading

Birkeland, C. 1997. *Life and Death of Coral Reefs.* Chapman and Hall, London.

Bortone, S. A. 1999. *Seagrasses: Monitoring, Ecology, Physiology, and Management.* CRC Press, Boca Raton, FL.

Brown, B. E. 1997. Coral bleaching: causes and consequences. *Coral Reefs* 16:S129–S138.

Dubinsky, Z., and P. L. Jokiel. 1994. Ratio of energy and nutrient fluxes regulates symbiosis between zooxanthellae and corals. *Pacific Science* 48:313–324.

Dunson, W. A. 1975. *The Biology Of Sea Snakes.* University Park Press, Baltimore.

Fautin, D. G. 1991. The anemonefish symbiosis: what is known and what is not? *Symbiosis* 10:23–46.

Heatwole, H. 1987. *Sea Snakes.* New South Wales University Press, Kensington.

Hemminga, M. A., and C. A. Duarte. 2001. *Seagrass Ecology.* Cambridge University Press, England.

Jackson, J. B. C., M. X. Kirby, W. H. Berger, K. A. Bjorndal, L. W. Botsford, B. J. Bourque, R. H. Bradbury, R. Cooke, J. Erlandson, J. A. Estes, T. P. Hughes, S. Kidwell, C. B. Lange, H. S. Lenihan, J. M. Pandolfi, C. H. Peterson, R. S. Steneck, M. J. Tegner, and R. R. Warner. 2001. Historical overfishing and the recent collapse of coastal ecosystems. *Science* 293:629–638.

Jones, G. P., M. J. Milicich, M. J. Emslie, and C. Lunow. Self-recruitment in a coral reef fish population. *Nature* 402:802–804.

Kuta, K. G., and L. L. Richardson. 1996. Abundance and distribution of black band disease on coral reefs in the northern Florida Keys. *Coral Reefs* 15:219–223.

Lohmann, K. J., and C. M. F. Lohmann. 1998. Migratory guidance mechanisms in marine turtles. *Journal of Avian Biology* 29:585–596.

Lutz, P. L., and J. A. Musick. 1996. *The Biology of Sea Turtles.* CRC Press, Boca Raton, FL.

Márquez, M. R. 1990. *Sea Turtles of the World.* An annotated and illustrated catalogue of sea turtle species known to date. FAO Fisheries Synopsis, no. 125, vol. 11, Rome, FAO.

Nakashima, Y., T. Kuwamura, and Y. Yogo. 1995. Why be a both-ways sex changer? *Ethology* 101:301–307.

Richardson, L. L. 1998. Coral diseases: what is really known? *Trends in Ecology and Evolution* 13:438–443.

Sale, P. F. *The Ecology of Fishes on Coral Reefs.* Academic Press, San Diego.

Sapp, J. 1999. *What Is Natural? Coral Reef Crisis.* Oxford University Press, New York.

Spalding, M. D., C. Ravilious, and E. P. Green. 2001. *World Atlas of Coral Reefs.* Prepared at the UNEP World Conservation Monitoring Centre. University of California Press, Berkeley.

Swearer, S. E., J. E. Caselle, D. W. Lea, and R. R. Warner. 1999. Larval retention and recruitment in an island population of a coral-reef fish. *Nature* 402:799–802.

Wolanski, E., R. Richmond, L. McCook, and H. Sweatman. 2003. Mud, marine snow, and coral reefs. *American Scientist* 91:44–51.

Wooton, R. J. 1990. *Ecology of Teleost Fishes.* Chapman and Hall, London.

CHAPTER 10

The open sea, or epipelagic zone, is the largest habitat on Earth.

The Open Sea

Life in the pelagic division of the marine environment exists in a three-dimensional nutritionally dilute medium. Microscopic protists and the major groups of small herbivores and many of their larger predators live in near-surface waters. The distribution of pelagic animals reflects their dependency on the primary producers of the sea for food. Away from continental shelves, reefs, and coastal upwelling, primary production rates tend to be low. Pelagic animals congregate in or near the photic zone, yet they typically are less abundant than animal populations in temperate or tropical coastal seas. At greater depths, population densities diminish rapidly, but animal life never completely disappears even in the deepest parts of the ocean.

The pelagic division is a realm that presents few obvious ecologic niches for its inhabitants. The small local variations that do exist in temperature and chemical characteristics tend to be smoothed out by turbulent mixing and diffusion processes, so variations in light intensity, water temperature, and food availability change only on horizontal scales of several kilometers.

10.1 Inhabitants of the Pelagic Division

The pelagic division of the environment is home to two major ecologic groups of marine animals, the zooplankton and nekton. Zooplankton are represented by **holoplankton**, the permanent planktonic forms, and temporary **meroplankton**, including larval stages of shallow-water broadcast-spawning invertebrates and fishes. Most nektonic animals also begin their lives as meroplankton. As they grow and improve their swimming capabilities, they eventually graduate to the status of nekton. Meroplankton tend to be concentrated in near-shore neritic provinces over continental shelves and near shallow banks, reefs, and estuaries where the adults that produced them live. Their patterns of abundance are strongly related to the seasonal distribution and productivity cycles of local phytoplankton communities (see Chapter 4).

More than 5000 species of holoplankton have been described. Prominent among these are members of all three protozoan phyla as well as cnidari-ans, ctenophores, chaetognaths, crustacean arthropods, and invertebrate chordates (see Chapter 5). Holoplankton use flotation and buoyancy techniques similar to those found in phytoplankton. Because most holoplankton are characteristically small, they increase their frictional resistance to the water by having high surface area-to-body volume ratios. A profusion of spines, hairs, wings, and other surface extensions also increases frictional resistance to sinking (FIGURE 10.1). Because of their very small cell sizes, microscopic protists have great difficulty in overcoming the viscous forces between water molecules by swimming. They experience almost no glide in their microscopic world; when they stop swimming, they instantly stop moving, so these microscopic cells must swim continuously if they are to move at all.

A large variety of gelatinous zooplankton exists, including numerous jellyfish medusae, siphonophores, pelagic mollusks (FIGURE 10.2a), ctenophores (Figure 10.2b), and tunicates. The tunicates include barrel-shaped salps, with life cycles alternating between solitary individuals and colonial clones (Figure 10.2c), as well

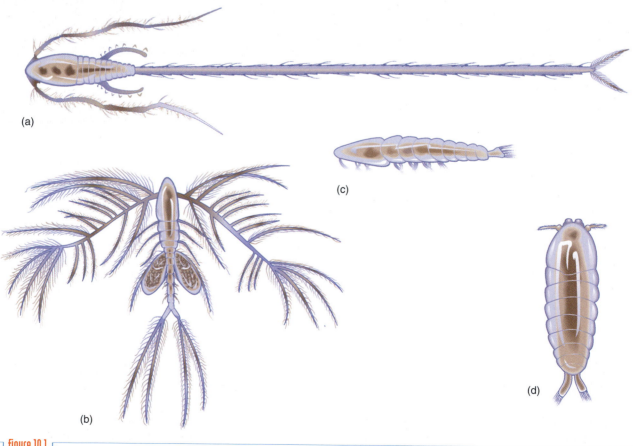

(a)

(b)

(c)

(d)

figure 10.1

Some planktonic copepods exhibiting structural adaptions for floatation: (a) *Aegisthus,* (b) *Oithona,* (c) side and (d) top views of *Sapphirina.* (Redrawn from Sverdrup, Johnson, and Fleming, 1970.)

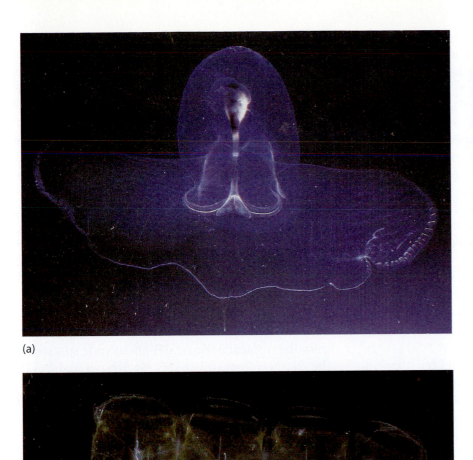

(a)

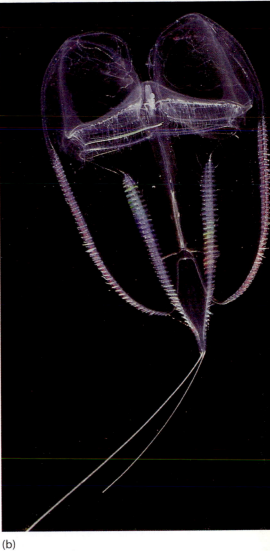

(b)

(c)

figure 10.2

Some large gelatinous zooplankton. (a) A pelagic mollusk, *Corolla*. (b) A ctenophore, *Eurhamphea,* swimming with eight rows of ciliated combs. (c) A short chain of seven salps (*Pegea*) cloned from a single parent. All photos approximately life-size.

as the smaller appendicularians. Because the bodies of all these gelatinous species contain at least 95% water and their body densities are very close to that of seawater, buoyancy is not a problem. Their small proportion of organic material accounts for their low metabolic rates and for their body sizes, which range from about 1 mm to several meters. Their relatively large sizes (in comparison with most other zooplankton) and nearly transparent appearance in water confer some protection from larger pelagic predators, such as sea turtles.

Crustaceans are the most numerous and widespread species of holoplankton. Copepods, euphausiids, amphipods, and decapods (Figure 10.1, and see Figure 5.32) all contribute substantially to near-surface plankton communities. Calanoid copepods, such as

Calanus, account for the bulk of herbivorous zooplankton in the 1- to 5-mm size range. Euphausiids are the giants of the planktonic crustaceans, yet they seldom exceed 5 cm.

The number of holoplankton species living in the pelagic province of the world ocean is matched by an equal number of nektonic species roaming the same waters. These animals represent most of the taxonomic groups that have achieved the large body sizes and well-developed swimming powers needed to exploit the pelagic realm. Absolute body size is crucial; once a well-muscled animal exceeds a few centimeters in body length, the viscous forces of water that limit continuous swimming by zooplankton begin to diminish, and efficient swimming becomes possible. Kilometer-scale distances can be covered in minutes rather than hours or days, and horizontal migrations to improve conditions for survival become possible and even feasible.

Most nekton are vertebrates, and most marine vertebrates are teleost fishes. Of the numerous groups of marine invertebrates that live in the sea, only squids and a few species of shrimps are truly nektonic. In some regions, vast numbers of small squids less than 1 m in length form important intermediate links in pelagic food webs. The giant squid, *Architeuthis,* lives at greater depths. These large animals occasionally reach 18 m in length and weights of about 2 tons, with tentacles approximately 12 m long and eyes as large as soccer balls.

10.2 Geographic Patterns of Distribution

Away from the influences of continental borders, life in the upper 200 m of the world ocean drifts along in the large semienclosed current gyres described in Chapter 1. This upper layer of the oceanic province, the **epipelagic zone**, is approximately coincident with the photic zone. In marked contrast to the numerous life zones available to animals on the sea bottom, the epipelagic zone is partitioned into only a few major habitats reflecting the major marine climatic zones shown in Figure 1.27. Each major epipelagic habitat is broadly defined by its own unique combination of water temperature and salinity characteristics and is occupied by a suite of species that, over long periods of time, have adapted to that set of environmental conditions.

These geographic patterns of distribution are nicely illustrated by six closely related species of planktonic euphausiids (FIGURE 10.3) yet are indicative of the general large-scale distribution of many other epipelagic animal species. First, there is a tendency for each species to be distributed in a broad latitudinal band across one or more oceans. Well-defined patterns of tropical (*Euphausia diomediae*), subtropical (*E. brevis*), and south polar (*E. superba*) distribution are evident. Some species, such as *E. diomediae* and *E. brevis,* are broadly tolerant to their environmental regimes and occupy wide latitudinal bands. Other species (*E. longirostris, Thysanoessa gregaria,* and *E. superba*) occupy narrower latitudinal ranges. Similar regimes in both the northern and southern hemispheres are frequently inhabited by the same species. The subtropical *E. brevis* and the temperate-water *T. gregaria* exhibit this **antitropical distribution.** Often these lower-latitude species extend into all three major ocean basins (*E. brevis* and *T. gregaria*), but occasionally they do not (*E. diomediae* is absent from the tropical Atlantic). High-latitude species in the southern hemisphere (*E. longirostris* and *E. superba*) also extend around the globe, aided by extensive oceanic connections between Antarctica and the other southern continents. Similar circumglobal distributions are more difficult to accomplish in the higher latitudes of the northern hemisphere (*T. longipes* is present only in the North Pacific).

The boundaries of these zones overlap slightly, but analyses of the distribution patterns of numerous other species of zooplankton confirm that these boundaries define, in a very real way, the major epipelagic habitats of the oceanic province. However, smaller-scale variations of environmental features do exist and do influence the structure of pelagic communities by contributing additional texture to the physical–chemical terrain, creating potential niches for occupation. Zooplankton, like phytoplankton, have patchy distributions at virtually all levels of sampling, from several kilometers down to microscopic distances. Some of this patchiness develops from the local effects of grazing or predation, some from responses to chemical and physical gradients, and some from the social responses (such as aggregation or avoidance) of planktonic species to each other.

10.3 Vertical Distribution of Pelagic Animals

Although the epipelagic zone accounts for less than 10% of the ocean's volume, most pelagic animals are found there. Most epipelagic nekton are carnivorous predators of the higher trophic levels of pelagic food webs. They are typically large in size

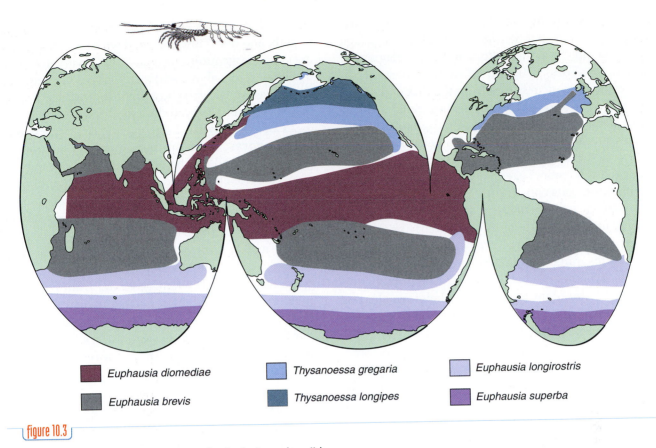

figure 10.3

The global distribution of six species of epipelagic euphausiids.

■ *Euphausia diomediae*	■ *Thysanoessa gregaria*	■ *Euphausia longirostris*
■ *Euphausia brevis*	■ *Thysanoessa longipes*	■ *Euphausia superba*

when compared with zooplankton, are effective swimmers, and some accomplish impressive feats of migration to locate food or to improve their chances for successful reproduction. Their ability to move rapidly over appreciable distances tends to erase the sharper distributional boundaries exhibited by zooplankton.

Epipelagic animals of the open ocean seldom exhibit the bright coloration so common in coral-reef fishes and invertebrates. Instead, **countershading** is a common pattern of coloration (FIGURE 10.4). Many abundant fishes, whales, and squids have dark, often green or blue, pigmentation on their dorsal surfaces, with silvery or white pigmentation on their ventral surfaces. When viewed from above, the pigmented upper surfaces of countershaded fishes blend with the darker background below. From beneath, the lighter undersides may be difficult to distinguish from the ambient light coming from the sea surface. From either view, these fishes tend to blend visually into, rather than stand out against, their watery background. The functions of countershading may be many, from protecting prey species from visual detection by their predators to hiding predators from their prey. Flashing of silvery bellies or white side

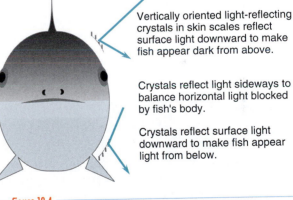

Vertically oriented light-reflecting crystals in skin scales reflect surface light downward to make fish appear dark from above.

Crystals reflect light sideways to balance horizontal light blocked by fish's body.

Crystals reflect surface light downward to make fish appear light from below.

figure 10.4

Expected patterns of light reflection from the surface of a countershaded fish.

stripes during abrupt turns may alert individuals in a school to the maneuvers of their immediate neighbors. Finally, skin pigments may have additional structural properties or physiologic functions completely independent of the visual appearance of the animal.

Below the sunlit waters of the epipelagic zone lies the **mesopelagic zone**, a world where animals live in very dim light and depend on primary production from the photic zone above. The mesopelagic zone extends from the bottom of the epipelagic zone down to about 1000 m. Most members of the mesopelagic zone rely totally on the flux of particles from above for food. This downward transport is accelerated by the conversion of dispersed microscopic cells to fecal pellets (FIGURE 10.5) and organic aggregates. The fecal pellets of calanoid copepods, for instance, sink about 10 times faster than do the individual phytoplankton cells constituting the pellet.

Macroscopic particles produced by incorporating living and dead material into irregularly shaped organic aggregates are known as "marine snow." These aggregates, ranging from one to several millimeters in size, are composed of living and dead phytoplankton cells, abundant bacteria, exoskeletons shed by crustaceans, and other detrital material. Sometimes these aggregates also include fecal pellets, so the distinction between the two types of particles blurs somewhat. Both types of particles serve as sites of additional aggregation by other members of the plankton community. Bacteria inoculate the particles and initiate processes of decomposition. Dinophytes exploit the nutrients released by the activity of the bacteria, and grazing herbivores are attracted by the concentration of energy-rich organic material, both living and dead.

Considerably less is known about the biology of mesopelagic animals than about epipelagic animals, and less still is known about animals living below 1000 m. Fishes living in the mesopelagic zone are typically much smaller than fishes of the epipelagic

zone. Mesopelagic fishes seldom exceed 10 cm in length, and many are equipped with well-developed teeth and large mouths (FIGURE 10.6a). Only very dim light penetrates from above, and many species have evolved large eyes sensitive to low light intensities (Figure 10.6b) to detect prey and predators alike.

Correlated with large eyes is the presence of **photophores**, light-producing organs most commonly arranged on the ventral surface of the body (FIGURES 10.6 and 10.7). The position and arrangement of photophores suggest two likely functions. The light produced by the ventral photophores usually approximates the intensity of the background light found at the normal daytime depths of these fishes. The light from the photophores may disrupt the visual silhouette of the fish when observed from below by making the fish's silhouette visually blend with the faint background light from above in a manner similar to countershading in near-surface fishes. Elaborate arrangements of photophores are species

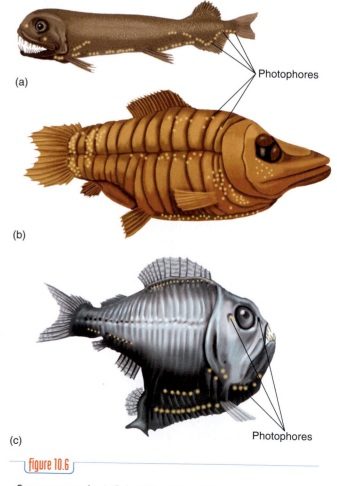

(a)

(b)

(c)

Photophores

Photophores

figure 10.6

Some mesopelagic fish: (a) loosejaw, *Aristostomias;* (b) spookfish, *Opisthoproctus;* and (c) hatchetfish, *Argyropelecus.* All are 5–20 cm in length.

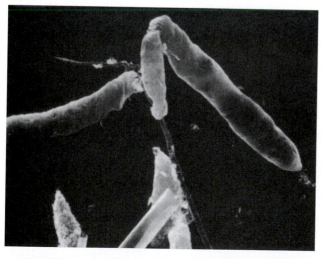

figure 10.5

An SEM of a copepod fecal pellet containing fragments of phytoplankton cells.

figure 10.7

A midwater lantern fish, *Bolinichthys*. Small white dots are light-producing photophores.

unique, suggesting that photophores are also used for species identification. With little to be seen at these depths except the pattern of photophores, appropriate mate selection may depend on the existence of species-specific patterns of photophores.

Below the mesopelagic zone, light from the surface is so dim that it cannot be detected with human eyes nor does it stimulate the visual systems of most deep-sea fishes. The light seen at depths below 1000 m comes largely from photophores. At these depths, photophores are used as lures for prey, as species-recognition signals, and possibly even as lanterns to illuminate small patches of the surrounding blackness. Most fishes found at these depths are not vertical migrators. Instead, some depend on the unpredictable sinking of food particles from the more heavily populated and productive waters above. These fishes are typically small and have flabby, soft, nearly transparent flesh supported by very thin bones (FIGURE 10.8). Others feed on mesopelagic fishes, often engulfing fishes that are nearly their own size.

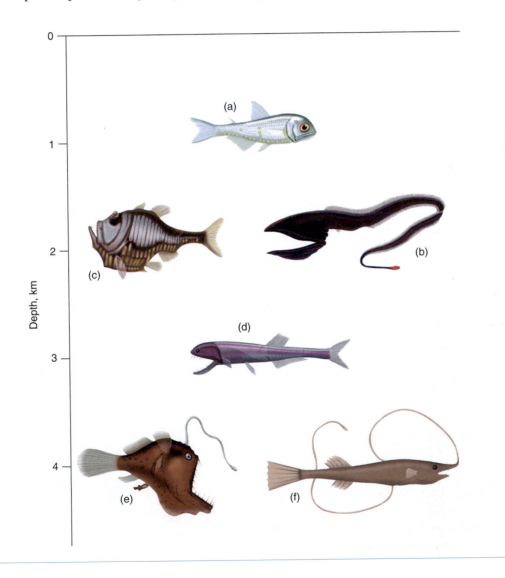

figure 10.8

A few fish of the deep sea, shown at their typical depths. Most have reduced bodies, large mouths, and lures to attract prey. (a) A lantern fish, *Bolinichthys;* (b) a gulper, *Eurypharynx;* (c) a hatchetfish, *Argyropelecus;* (d) a bristlemouth, *Cyclothone;* (e) a female anglerfish, *Melanocetus,* with an attached male; and (f) another anglerfish, *Gigantactis.*

10.4 Vertical Migration: Tying the Upper Zones Together

Zooplankton and small nekton, such as lanternfishes (Figure 10.7), are poor long-distance travelers. They can, however, experience very different environmental conditions by vertically moving modest distances of a few tens of meters. Water temperature, light intensity, pressure, and food availability all change markedly as the distance from the sea surface increases.

The mesopelagic zone offers some distinct advantages when compared with life nearer to the sea surface. Prey species are more difficult to detect by their predators in the dim light. Decreased water temperatures at middepths lower the metabolic rates and the food and oxygen needed to maintain those rates. The cold water, with its increased density and viscosity, also slows the sinking rates of food particles. The most abundant supply of food particles, however, still resides in the epipelagic zone just above.

To exploit the benefits of both zones better, large numbers of mesopelagic animals periodically migrate upward to feed in near-surface waters. The most common pattern of vertical migration occurs on a daily cycle. At dusk, these midwater animals ascend to the photic zone and feed throughout the night. Before daybreak, they begin migrating to deeper darker waters to spend the day. The following evening, the pattern is repeated. In antarctic waters, the daily pattern of vertical migration often breaks down; the animals generally remain in the photic zone during the summer and in deeper waters during the winter.

This pattern of daily, or **diurnal, vertical migration** has been deduced from numerous sources of information. Net collections of animals from several depths at different times throughout the day have shown that more animals are near the surface at night than during the day (FIGURE 10.9). Direct observations from submersible vehicles support these conclusions.

Ship-mounted sonar devices also are used to study the behavior of vertically migrating animals because the sonar sound signal is partially reflected by concentrations or layers of midwater ani-

mals. These **deep sound-scattering layers (DSSL)** ascend nearly to the surface and disperse at dusk (FIGURE 10.10). At daybreak, the layers reform and descend to their usual daytime depths (200–600 m). Often, three or more distinct layers are discernible, extending over broad oceanic areas.

The species composition of the DSSL is still an unsettled question. Most inhabitants of the mesopelagic zone are too small or too sparsely distributed to reflect sound signals strongly. Net tows and observations from manned submersibles suggest three groups of animals that cause the deep scattering layers: euphausiids, small fishes (primarily lanternfishes, Figure 10.7), and siphonophores (FIGURE 10.11). Euphausiids and small fishes are often abundant members at depths where the DSSLs occur. The strong echoes of sound pulses may be due in part to the resonating qualities of the gas-filled swim bladders of lanternfishes and air floats of siphonophores. Relatively few swim bladders or air floats are necessary to produce strong echoes at certain sound frequencies. Whatever the composition of the DSSLs, they are merely sound-reflecting indicators of much more extensive vertically migrating assemblages of animals not detected by sonar. Undoubtedly, members of the DSSLs graze on smaller vertical migrators and are in turn preyed upon by larger fishes and squids.

These vertical migrations, only a few hundred meters in extent, occur over short time periods. The copepod *Calanus,* for example, is only a few mil-

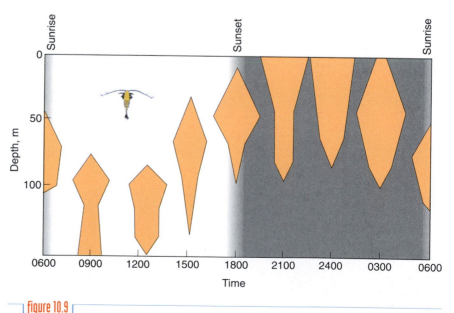

figure 10.9

A generalized kite diagram of net collections of adult female copepods, *Calanus finmarchicus,* during a complete one-day vertical migration cycle. The width of each part of each "kite" represents relative numbers of animals. Night hours are shaded.

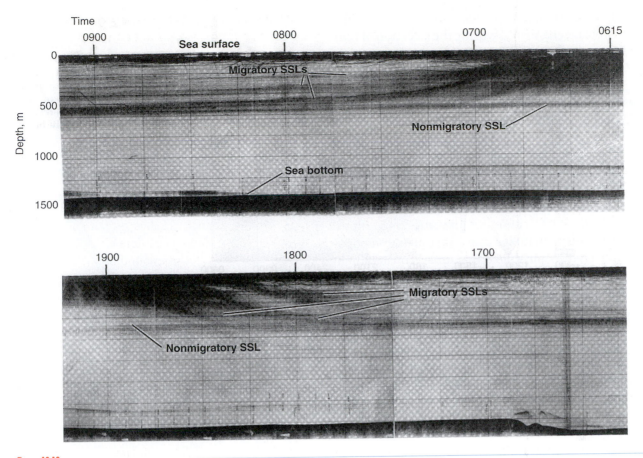

0900   0800   0700   0615

Sea surface

0

Migratory SSLs

Depth, m

500

Nonmigratory SSL

1000

Sea bottom

1500

1900   1800   1700

Migratory SSLs

Nonmigratory SSL

figure 10.10

A SONAR record of several distinct sound scattering layers at a station near the Canary Islands. Except for slight drift corrections, the recording ship remained stationary. The time scale reads from right to left. Multiple layers begin migrating downward at daylight (0700 hr), remain near 500 m throughout the day, then migrate upward in the evening (1800–1900 hr).

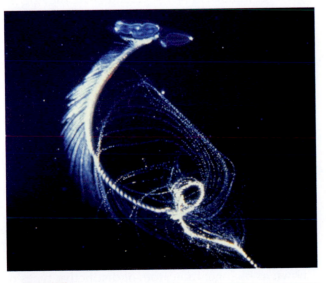

figure 10.11

A mid-water siphonophore with a small gas-filled pneumatophore at the upper end.

limeters long, yet it can swim upward at 15 m/h and descend at 50 m/h. Larger 2-cm euphausiids swim in excess of 100 m/h. If diurnal vertical migrations are foraging trips from below into the productive photic zone, why do these animals descend after feeding? Why don't they remain in the photic zone? Many explanations for the adaptive value of vertical migration have been offered.

One explanation is that diurnal vertical migration enables animals to capitalize on the more abundant food resources of the photic zone in the dark of night and to escape visual detection by predators in the refuge of the dimly lit mesopelagic zone during the day. But vertical migration is useful in other ways. Lower water temperatures at the deeper depths reduce an animal's metabolic rate and its energy requirements. The energy conserved may be sufficient to offset the lack of food during the day and the energy expenditures incurred during the actual migration. Each of these explanations offers plausible mechanisms favoring the selection of individuals possessing the genetic information required to accomplish vertical migration.

Vertical migrations of zooplankton and small nekton also occur on seasonal time scales. While in the late copepodite stage, *Calanus finmarchicus* spends the winter months of low primary productivity in the North Atlantic at depths near 1000 m. When the spring diatom bloom develops, it molts to the adult form and begins to rise into the photic zone.

Daily or seasonal changes in light intensity seem to be the most likely stimulus for vertical migrations. Experiments with mixed coastal zooplankton populations have demonstrated that under constant light and temperature conditions, some species of copepods maintained their diurnal migratory behavior as a circadian rhythm for several days without relying on external cues such as light intensity. Other species did not migrate and apparently require light or another external stimulus to initiate vertical migration behavior. Electric lights lowered into the water at night and bright moonlight can drive near-surface DSSLs downward. Even midday solar eclipses influence vertical migrators by causing them to move toward the surface for the duration of an eclipse. Collectively, these responses demonstrate a sensitivity to light intensity by natural populations of vertical migrators. Each DSSL follows an isolume (a constant light intensity) characteristic of the top of the layer at its normal daytime depth (FIGURE 10.12). As the sunlight intensity decreases in late afternoon, the isolume moves toward the sea surface, and the DSSL follows with a precision seldom seen in natural populations.

10.5 Feeding on Dispersed Prey

In most marine communities, individual prey items typically become larger in size but fewer in number and biomass at successively higher trophic levels. Two generalized food webs are shown in FIGURE 10.13 to illustrate some fundamental differences in trophic linkages between oceanic regions of high and low productivity. Food webs of subtropical waters leading to tuna-sized predators that are low in biomass begin with very small phytoplankton and bacteria and include numerous trophic levels. The antarctic upwelling system, in contrast, is characterized by relatively large (though still microscopic) primary producers, larger filter-feeding herbivores, extremely large carnivores, and fewer trophic levels. High-latitude food webs are described in Chapter 12.

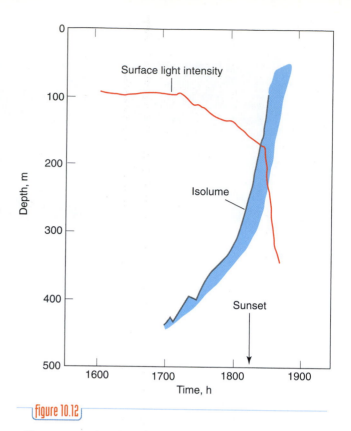

figure 10.12

The upward migration of a scattering layer (colored portions of the graph) at sunset. Note the very close correspondence between the isolume and the top of the scattering layer, with both rising as the surface light intensity diminishes. (Redrawn from Boden and Kampa 1967.)

In tropical and subtropical pelagic environments, small and dispersed prey items insufficient to support large numbers of large predators are the rule. To secure adequate food supplies, larger pelagic animals here either must capture large but widely scattered prey items or be able to harvest very large numbers of smaller more abundant prey efficiently. The foraging activities of large filter-feeding baleen whales, basking sharks, and whale sharks are mostly limited to more productive coastal or high-latitude waters. Single-prey predators range in size from large tunas, billfishes, and dolphins over a meter or two in length to centimeter-long voracious chaetognaths (FIGURE 10.14) and barely visible filter-feeding copepods. Chaetognaths are found throughout the world ocean in numbers sufficient to decimate whole broods of young fishes. In addition to their significant role in pelagic food webs, some species of chaetognaths are well-known as biologic indicators of distinctive types of surface ocean water.

Copepods are common prey of chaetognaths. All adult calanoid copepod species are similar in body

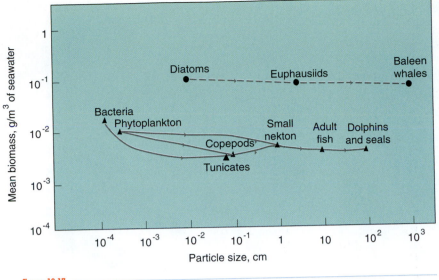

figure 10.13

The relationship between food particle size and biomass in two pelagic food chains. Note that the biomass in the Antarctic (dashed lines) is about ten times higher at all trophic levels than those in subtropical gyres (solid lines), and the biomass of each food chain decreases at higher trophic levels. (Redrawn from Steele 1980.)

figure 10.14

A small chaetognath, *Sagitta,* capturing and consuming a fish larva its own size.

bacteria through the common types of phytoplankton to organic aggregates and large centric diatoms. This size spectrum presents several opportunities for small versatile particle grazers to adopt feeding strategies that select for optimal-sized food items.

Although copepods prey on large phytoplankton cells when they are available, calanoid species can capture the smaller more abundant microplankton with a basket-like filtering mechanism derived from their complex feathery feeding appendages (FIGURE 10.15). The hair-like setae on the appendages of *Calanus* are fine enough to retain nanoplankton-sized food particles larger than 10 μm. Laboratory observations have revealed that food particles are carried into the filter basket by currents generated by the feeding appendages and five pairs of more posterior thoracic swimming legs. Special long setae on the feeding appendages remove the trapped particles and direct them to the mouth. With this filtering mechanism, *Calanus* species are capable of exploiting a wide size range of food particles.

Even though *Calanus* and similar copepods can shift rapidly from one food particle size to the other, they prefer larger food particles to smaller ones. Studies using high-speed photomicrography techniques have shown that calanoid copepods efficiently capture sparsely scattered, single, large food items such as protozoans, small fish eggs, and large diatoms. As the copepods' feeding actions drive them forward in the water, their antennae extend laterally to function as an array of sensory receptors to detect minute disturbances surrounding larger food items (FIGURE 10.16). If the food particle is detected near the end of an antenna, the animal quickly adjusts its swimming direction to bring the particle within reach of an extended feeding appendage. The mouthparts then seize and manipulate the particles before eating them. Because filtering and large-particle seizure cannot operate simultaneously, this mode of feeding is interrupted when the copepods are filtering small particles.

Such particle-size selectivity by copepods is likely an important factor in stabilizing phytoplankton populations. When phytoplankton populations of a particular cell size become more abundant through growth

form and general feeding behavior, suggesting the evolution of a very successful functional form. Copepods and other small pelagic particle grazers are typically exposed to a wide spectrum of food particle sizes. Food particles range from abundant minute

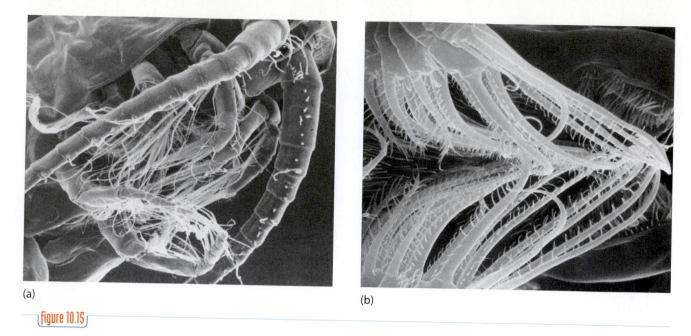

(a)   (b)

figure 10.15

(a) A scanning electron micrograph of the thorax and filter-feeding mechanism of *Calanus*, shown in side view. (b) Higher magnification ventral view of *Calanus*, showing the filtering basket formed by the second maxillae.

figure 10.16

Copepod detection and capture behavior of individual diatoms (green). Sensory cells arranged in arrays on large antennae provide information to detect prey and guide the response until the capture is made (right).

and reproduction, they attract increased grazing pressures as more copepods shift feeding strategies to concentrate on them. It is unlikely, however, that the phytoplankton population would be grazed to extinction. Several species exhibit ingestion rates that are dependent on the concentration of phytoplankton cells (FIGURE 10.17). For a certain food particle size, ingestion rates increase with increasing particle density to some crucial maximum. Beyond the maximum particle density, some aspect of the copepod's food-processing system appears to become saturated, and no further increase in ingestion rates occurs. Conversely, ingestion rates decrease with decreasing phytoplankton densities; copepods most likely will shift to another more optimal concentration of food particles before the first population is exhausted completely.

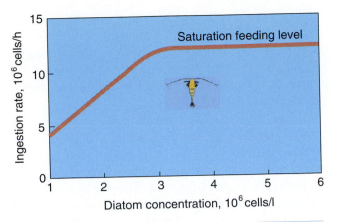

figure 10.17

The ingestion rate of a copepod, *Calanus*, as a function of the concentration of its food (in this case, the diatom *Thalassiosira*, shown in figure 3.8). The ingestion rate peaks near 3000 diatom cells per ml, and no further increase is seen even at much higher concentrations. (Redrawn from Frost 1972.)

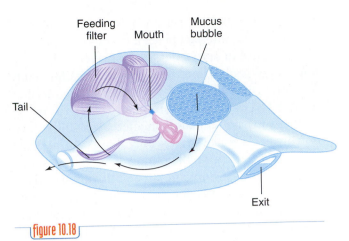

figure 10.18

The appendicularian *Oikopleura,* within its mucous bubble. Arrows indicate path of water flow.

In contrast to the rigid filter devices of crustaceans, some gelatinous herbivores rely on nets or webs of mucus to ensnare food particles. One unusual example is *Gleba*, one of the few planktonic gastropod mollusks. When feeding, *Gleba* secretes a mucous web that often exceeds 2 m in diameter. The free-floating web spreads horizontally as it is produced, maintaining a single point of attachment at the animal's mouth. As the animal and its web slowly sink, bacteria and small phytoplankton become trapped in the mucus. The web, with its load of food, is formed into a mucous string and ingested. *Gleba* then swims upward to repeat the behavior.

Another elaborate mucous feeding system is found in appendicularians, small tadpole-shaped invertebrate chordates. Like *Oikopleura* (FIGURE 10.18), most appendicularians live enclosed within delicate transparent mucous bubbles. Food-laden water, pumped by the tail beat of the occupant, enters the bubble through openings at one end. These openings are screened with fine-meshed grills to exclude large phytoplankton cells. Smaller cells enter the bubble and are trapped on a complex internal mucous feeding screen. Every few seconds, the animal sucks the particles off the screen and into its mouth. When the grill in the bubble wall becomes clogged or the interior is fouled with feces, the entire bubble is abandoned and a new one is constructed, sometimes in as little as 10 minutes. With this feeding mechanism, even bacteria-sized particles can be harvested efficiently, because these ani-

mals achieve filtering rates several times higher than those of calanoid copepods. The larger gelatinous salps are even better; a small chain of colonial salps (such as the seven shown in Figure 10.2c) is capable of filtering as much water as 3000 copepods can in the same amount of time.

The mechanisms used by zooplankton to glean small diffuse food particles from the water reflect the crucial roles these animals play in pelagic food webs. Planktonic tunicates are the largest herbivorous grazers to exploit the very small phytoplankton found in tropical and subtropical areas of low productivity. Because their filters clog quickly when they encounter the high phytoplankton densities of upwelling areas or temperate spring blooms, they are less successful competitors in these areas. It is in these regions of high productivity that planktonic crustaceans, particularly the calanoid copepods and euphausiids, thrive.

10.6 Buoyancy

Living and moving in three dimensions well above the seafloor creates some buoyancy problems for pelagic animals because the bone and muscle tissues needed for locomotion in these animals are more dense than seawater. Stored fats and oils, which are less dense than water, are common buoyancy devices used by some pelagic marine animals. Whales, seals, and penguins maintain thick blubber layers just under the skin. Many sharks and a variety of teleosts store large quantities of oils in their livers and muscle tissues. In fact, in some species of sharks, the liver accounts for about one third of the body weight.

Fats and oils, however, are only slightly less dense than seawater. This fact poses a serious challenge for many small but active nektonic species. They cannot energetically afford to carry around a huge oily liver or a thick blubber layer nor can they sacrifice muscle and bone to lighten their load. The solution for many marine animals is an internal gas-filled flotation organ. At sea level, air is only about 0.1% as dense as seawater, and a small volume of air can provide a large amount of buoyancy. The buoyancy derived from a volume of gas depends on the volume of seawater the gas displaces. Unlike fats or oils, gases are compressible; they occupy different volumes at different pressures and depths. At sea level, the pressure created by the Earth's envelope of air is about 1 kg/cm^2 or 1 atm. Below the sea surface, the water pressure increases about 1 atm for each 10-m increase in depth. Thus, the total pressure experienced by a fish at 5000 m is 501 atm (more than 3.5 $tons/in^2$).

A few genera of colonial cnidarians maintain positive buoyancy by secreting gases into a float, or **pneumatophore**. *Velella* (sometimes called by-the-wind sailor, Jack-by-the-wind, or, simply, purple sail; FIGURE 10.19a) and the larger Portuguese man-of-war, *Physalia* (Figure 10.19b) have large pneumatophores and float at the sea surface. The pneumatophore acts as a sail to catch surface breezes and transport the colony long distances. Both *Velella* and *Physalia*, with only one species in each genus, have worldwide distributions.

Other siphonophores with gas floats (Figure 10.11) are neutrally buoyant and can easily change their vertical position in the water column by swimming. A gas gland within the pneumatophore secretes gas into the float. Excess gases are vented through a small pore that is opened and closed by a muscular valve, or sphincter, and the siphonophore's buoyancy is adjusted accordingly.

Gas is also used for buoyancy by a small planktonic nudibranch, *Glaucus*. It produces and stores intestinal gases to offset the weight of its shell. Another planktonic snail, *Janthina*, forms a cluster of bubbles at the surface and clings to it. These adaptations are apparently related to the preference of the nudibranch and snail for feeding on the soft parts of *Velella* (which also floats at the surface). These zooplankton adapted to live permanently at the sea surface are known as **neuston**. (Another example of a neustonic animal is the water strider shown in Figure 1.15.)

Air in the lungs of mammals, reptiles, and birds also can provide some buoyancy. However, it is the teleost fishes that have the most sophisticated system for using air to solve their buoyancy problems. Many teleosts, especially active species with exten-

(a)

(b)

figure 10.19

(a) Purple sail, *Velella*, and (b) the Portuguese man-of-war, *Physalia*, both floating at the sea surface. The trailing tentacles of *Physalia* may reach 50 m in length.

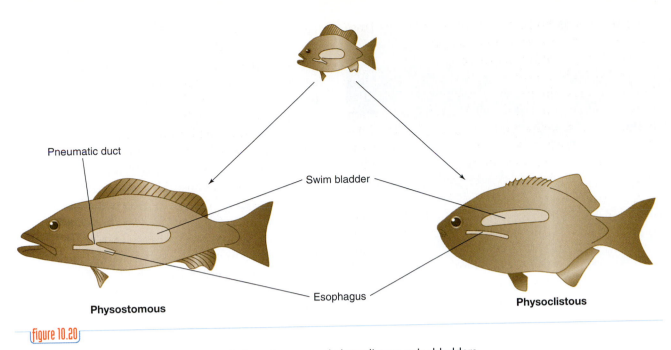

Pneumatic duct

Swim bladder

Esophagus

Physostomous

Physoclistous

figure 10.20

The development and relative positions of physostomous and physoclistous swim bladders.

sive muscle and skeletal tissue, have body densities about 5% greater than that of seawater. To achieve neutral buoyancy, many of these fishes have an internal swim bladder filled with gases (mostly N_2 and O_2, the most abundant gases in the atmosphere and also dissolved in seawater). The swim bladders of bony fishes develop embryonically from an out-pouching of the esophagus (FIGURE 10.20). The densely woven fibers that make up the bladder wall are embedded with a layer of overlapping crystals of guanine to make the bladder wall nearly impermeable to gases.

The connection between the esophagus and swim bladder, called the **pneumatic duct**, is present during the larval or juvenile stages of all teleosts. In some species, the pneumatic duct remains intact in the adult. This is the **physostomous** swim bladder condition. In other species, the duct disappears as the fish matures to create a **physoclistous** swim bladder (Figure 10.20). Nearly half of the more than 25,000 species of teleosts, however, lose not only the pneumatic duct but also the entire swim bladder when they mature. Swim bladders are notably lacking in some bottom fishes and in active continuously swimming fishes such as tuna.

Swim bladders are not rigid structures; the volume of water they displace is subject to changing water pressures at different depths. To maintain neutral buoyancy at different depths, the volume of a fish's swim bladder must remain constant. A fish that swims downward experiences greater external water pres-

sure, which squeezes its swim bladder and reduces the bladder's volume and the fish's buoyancy. The fish must then increase the quantity of gas in the bladder to compensate for the volume change.

An ascending fish has the opposite problem: It must get rid of swim bladder gases as rapidly as they expand. Some shallow-water physostomous fishes fill their swim bladders simply by gulping air at the sea surface. They also release excess gases through their pneumatic duct and eventually out the mouth or gills. Fishes with physoclistous swim bladders, however, lack a pneumatic duct, and their ability to rapidly add or remove bladder gases to compensate for a rapid depth change is limited. If a deep-water fish with a physoclistous swim bladder ascends rapidly, the decreased water pressure enables the gases within the somewhat elastic swim bladder to expand and reduce the overall density of the fish. The density decrease may be so great that the fish is unable to descend for some time. This is well illustrated by the appearance of fishes brought to the surface (unwillingly, of course) from deep water on fishing lines or in trawls. It is not unusual for the swim bladders of such fishes to expand so much that severe internal organ damage occurs (FIGURE 10.21). During a slower natural ascent, excess gases are reabsorbed back into the bloodstream, but this takes some time. This is accomplished at a specialized region of the swim bladder, the oval body (FIGURE 10.22), that is richly supplied with blood vessels for resorption of gases. The oval body is isolated from

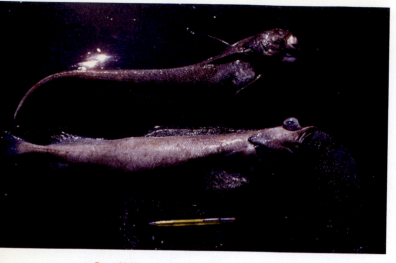

figure 10.21

Two deep-sea fishes on the deck of a ship after being hauled up from a depth of 800 m. Both fish were seriously damaged and distorted by the rapid expansion of gases in their swim bladders as they were brought to the surface.

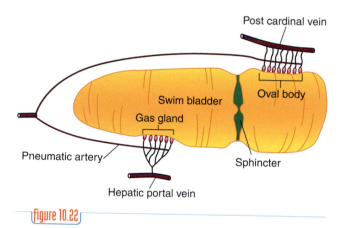

Post cardinal vein

Swim bladder

Oval body

Gas gland

Pneumatic artery

Sphincter

Hepatic portal vein

figure 10.22

A physoclistous swim bladder and associated blood vessels. The area of the gas gland is diagrammed in greater detail in Figure 10.23.

the remainder of the swim bladder by a sphincter that controls the flow of bladder gases to the oval body.

Both types of fish swim bladders have gas glands that regulate the secretion of gas from the blood into the bladder when these fishes are below the sea surface and have no access to air. Because the process of filling the swim bladder is the same in both types, only the physoclistous swim bladder is described. Fishes with gas-filled swim bladders have been collected from depths as great as 7000 m. The gas pressure needed within the swim bladder to balance the water pressure at that depth is about 700 atm (5 tons/in²). Such extreme gas pressures

are achieved by a dramatic increase in the O_2 concentration of the bladder gases. Oxygen commonly accounts for more than 50%, and occasionally exceeds 90%, of the gas mixture of the swim bladders of deep-ocean fishes.

How are O_2 and N_2, which are dissolved in seawater at pressures no greater than 1 atm, concentrated in swim bladders at pressures as great as 700 atm? Below the sea surface, fishes must fill their swim bladders with gases absorbed from seawater by their gills. These gases are transported in the blood to the gas gland of the swim bladder and then are secreted into the bladder at pressures equal to external water pressures. When highly oxygenated hemoglobin reaches the gas gland of a swim bladder (Figure 10.22), the O_2 must be induced to leave the hemoglobin and diffuse into the swim bladder, often in the face of high O_2 pressures within the bladder. When a fish descends and its swim bladder volume must be increased, stretch receptors in the swim bladder wall stimulate the gas gland to produce lactic acid. The lactic acid diffuses into the blood vessels and lowers the pH of the blood. Lower pH conditions reduce the oxygen-carrying capacity of hemoglobin and induce it to unload a large portion of its O_2. The free O_2, which has not yet left the blood, is now no longer associated with the hemoglobin. The total effect of lactic acid on hemoglobin is sufficient to produce about 2 atm of O_2 pressure at the gas gland of the swim bladder.

Eventually, the O_2 will diffuse into the swim bladder if the O_2 pressure there is not greater than 2 atm. This mechanism alone, however, is capable of producing swim bladder gas pressures useful only to relatively shallow fishes. Fishes with gas-filled swim bladders living below about 20 m rely on another structure, an extensive rete mirabilia, associated with the gas gland (FIGURES 10.22 and 10.23) to achieve that extra swim bladder gas pressure. A typical rete system may contain a few hundred or as many as 200,000 tiny capillary-sized blood vessels. Each tiny rete vessel approaches the gas gland carrying oxygen-rich hemoglobin and then doubles back on itself without penetrating the swim bladder. These complex rete systems form another countercurrent exchange system that operates on the same principle as that described for fish gills in Chapter 6 to concentrate O_2. Here, too, this simple system works by creating and manipulating O_2 diffusion gradients.

Until the O_2 unloaded from hemoglobin by lactic acid at the gas gland is concentrated sufficiently to match the pressure in the swim bladder, it will sim-

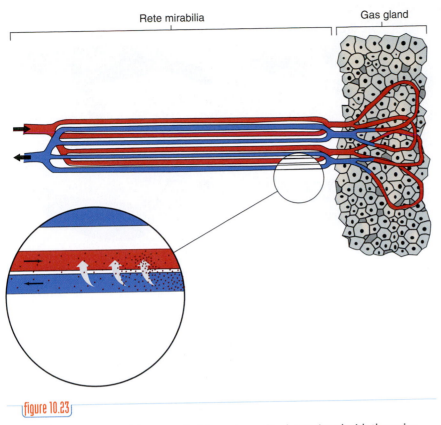

Rete mirabilia   Gas gland

figure 10.23

A simplified diagram of the rete mirabile and gas gland associated with the swim bladders of many bony fishes. Inset illustrates the countercurrent arrangement of blood flow (small arrows) and the diffusion of O_2 from outgoing to incoming blood vessels (broad arrows). (Adapted from Hoar 1983.)

the lack of a specialized transport system for N_2 (as hemoglobin is for O_2) relegates N_2 to the role of a minor gas in swim bladders, especially at great depths.

As the pressure of gases inside swim bladders increases, so do their densities. At 7000 m, the greatest depth at which gas-filled swim bladders have been found, the gas within a swim bladder is so compressed that its density is almost as great as that of fat. For some fishes at great depths, the constant energy expenditures needed to maintain a full swim bladder become unrealistic, and swim bladder gases are replaced with fat. Fat-filled swim bladders provide almost as much buoyancy as gases do at great depths but are much simpler to maintain because fat will not compress at high pressures. Fat-filled swim bladders are also found in many vertically migrating fish species, such as lanternfishes (Figure 10.7), that move through pressure changes of 10 to 40 atm twice each day.

ply be carried through the turn of the rete capillary and start to leave the gas gland. However, its path is adjacent to and parallel with a capillary carrying blood toward the gas gland. As the concentration of the dissolved O_2 of the blood leaving the gas gland is higher than that of the incoming capillary, the O_2 will diffuse across the capillary walls and back into the incoming blood of adjacent capillaries. The O_2 forced off the hemoglobin at the gas gland is thus trapped in this countercurrent diffusion loop as it tries to leave the gas gland. Given sufficient time and enough cycles around this loop, the pressure of O_2 in the capillaries will surpass even very high pressures of the swim bladder, and O_2 will diffuse from the gas gland into the bladder.

As one might expect, a long rete is capable of concentrating more O_2 at the gas gland than is a short one. Still, the rete does not need to be unmanageably long. A rete only 1 cm long can secrete O_2 at pressures up to 2000 atm, well in excess of the swim bladder pressures needed in the deepest parts of the sea. The rete mirabilia concentrates N_2 as well as O_2; however,

10.7 Locomotion

In the pelagic environment, animals often must move long distances to find food or mates or to otherwise improve conditions for their survival. Structural or behavioral adaptations that allow animals to swim with reduced energy expenditures enable them to divert more energy to growth and reproduction and contribute to the potential success of an individual.

Nekton are large and fast animals. Consequently, their energetic costs of locomotion are expensive and represent a major expenditure of their available resources. The cost of swimming is affected by swimming speed, the flow patterns of water around the swimmer, and by two physical properties of water, density and viscosity. Water is greater than 800 times more dense than air and at least 30 times more viscous. Thus, the frictional resistance to moving through water is considerably greater than in air. As a consequence, movement through water imposes severe limitations on speed and energetic performance for nekton. There are, however, advantages to moving in water. Propulsive forces are easier to generate in

figure 10.24

Power and glide strokes of three pectoral-swimming tetrapods.

Power stroke

Glide stroke

water than in air. And because the body densities of most nekton are similar to that of water, most marine nekton are approximately neutrally buoyant. So swimming animals can maintain their vertical position in the water with little energy expenditure because they do not need to support their weight during locomotion as do terrestrial animals.

Most pelagic fishes use side-to-side motions of their caudal fins as their chief source of propulsion (see Chapter 6), as do tetrapods as diverse as seals and sea snakes. Whales move their caudal flukes in vertical motions to achieve their swimming power. Other tetrapods use paddling (turtles) or underwater flying motions (penguins, sea lions, and many pelagic rays), using their pectoral flippers to do most of the work (FIGURE 10.24). A few invertebrate nekton are also excellent swimmers. Squids and other cephalopods take water into their mantle cavities and then expel it at high speeds through a nozzle-like **siphon**. The siphon can be aimed in any direction for rapid course corrections and for maneuvering purposes. Squids and cuttlefishes also use their undulating lateral fins in much the same manner as benthic skates and rays.

One useful approach to gaining a general understanding of the energy costs associated with locomotion is to evaluate an animal's **cost of transport (COT)**. COT comparisons can be made between different modes of locomotion and for animals of different sizes (see Box 10, Dolphin Swimming, pp. 310–311). All species in all modes of locomotion have some preferred speed at which their COT is minimum. When this COT_{min} is calculated and plotted as a function of

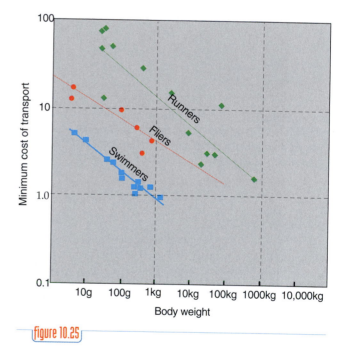

figure 10.25

Relationship between COT_{min} and body size for different modes of locomotion. (Adapted from Tucker 1959.)

body mass, it is clear that both flying and swimming impose lower COT_{min} than does walking or running (FIGURE 10.25). Flying is energetically demanding but covers long distances in a short time, so the COT_{min} is low. Swimmers, regardless of size, do not need to support the weight of their bodies, so their COT_{min} is lower still. In contrast to fliers, though, which cannot be larger than about 40 kg, swimmers can be extremely large and still move efficiently.

Body Shape

To maintain high swimming velocities or to swim in an energetically efficient manner, a swimmer must overcome (or at least reduce) several different components of the total hydrodynamic drag on the body of the swimmer. Each of these components is influenced by the swimmer's body size and shape. The body shape of a fast swimmer, such as a tuna or dolphin, is a compromise between different hypothetical body forms, each of which reduces some component of the total drag and enables the animal to slip through the water with as little resistance as possible. **Frictional drag** is a function of the extent of wetted surface an animal has in contact with the water and the density and viscosity of the water. Frictional drag is lower if the flow of water over the body surface is smooth, or laminar, and high if it is turbulent. **Pressure drag**, sometimes called form drag, is the consequence of displacing an amount of water equal to the swimmer's largest cross-sectional area (from a head-on view). Another drag component is **induced drag** created by the fins, flukes, or flippers that swimmers use to produce their thrust. Finally, an additional **wave drag** is created by the production of surface waves when swimming at or near the sea surface. Because all marine tetrapods must surface frequently to breathe, wave drag can contribute substantially to the total drag that they must overcome to swim.

Frictional and form drag of a fast swimmer are reduced with a streamlined body form that is roundly blunt at the front end, tapered to a point in the rear, and round in cross-section. The **fineness ratio** is the ratio of an animal's body length to its maximum body diameter. For efficient swimmers, fineness ratios range from 3 to 7, with the ideal near 4.5. Most cetaceans exhibit fineness ratios near 6–7, with only killer whales and right whales having nearly ideal ratios of 4.0–4.5. Most tunas also have nearly ideal fineness ratios. This streamlined shape of most fast nekton, excluding their fins, is the best possible body shape to reduce the several components of drag and to slip through the water with as little resistance as possible.

Unlike fishes, marine birds and mammals must necessarily surface to breathe more frequently when swimming at higher speeds. Because small dolphins or penguins must surface to breathe more frequently than large whales, it is more difficult for them to escape beneath the high drag associated with the sea surface. Only by remaining at least 2.5 body diameters below the sea surface can they minimize wave drag. At high speeds, the high wave drag associated with surfacing can be partly avoided by leaping above the water–air interface and gliding airborne for a few body lengths. The aerial phase of this type of **porpoising** or leaping locomotion sends the animal above the high-drag environment of the water surface while simultaneously providing an opportunity to breathe. The velocity at which it becomes more efficient to leap rather than to remain submerged is known as the **crossover speed**. The crossover speed is estimated to be about 5 m/s for small spotted dolphins and increases with increasing body size until leaping becomes a prohibitively expensive mode of locomotion in cetaceans longer than about 10 m.

Speed

Several species of nekton are noted for their amazingly fast swimming speeds. The oceanic dolphin *Stenella* has been clocked in controlled tank situations at better than 40 km/h (approximately 25 mi/h). Top speeds of killer whales are estimated to be 40 to 55 km/h. For comparison, human Olympic-class swimmers achieve sprint speeds of only 4 to 5 km/h. Using specially designed fishing poles to measure speeds at which line is stripped from a reel, a 1-meter-long barracuda has been clocked at 40 km/h. Brief sprints of similarly sized yellowfin tunas and wahoos have been clocked at more than 70 km/h, and large bluefin tunas may be capable of speeds in excess of 110 km/h (70 mi/h). This estimate may not be as farfetched as it seems because some bluefins reach lengths of 4 m and presumably would be much faster than a fish only 1 m long, but it has yet to be confirmed.

What enables dolphins, tunas, and similar fishes to swim so fast? In addition to having nearly optimal streamlined body shapes (FIGURE 10.26a), many cetaceans are faster than most fishes because they are endothermic and so can maintain high and continuous rates of power output. But what about rapidly swimming fishes (Figure 10.26b)? The exceptional swimming abilities of tuna and tuna-like fishes go beyond simply having a streamlined body form and an efficient caudal fin. Their streamlined body form is complemented by other friction-reducing features, including small and smooth scales, nonbulging eyes, fins that can be retracted into slots and out of the path of water flow when not needed for maneuvering, and numerous small median **finlets** on the dorsal and ventral sides of the rear part of the body function to reduce turbulence in that region.

Most of the caudal flexing of a tuna is localized in the region of the **caudal peduncle**, the region where

Dolphin Swimming

Swimming and diving represent major energetic expenditures in marine mammals. In wild bottlenose dolphins, locomotor activities can represent more than 82% of the animal's daytime activity budget. In a 1975 article, Tucker summarized and calculated the minimum costs of locomotion for a large variety of animals spanning several orders of magnitude in body size (see Figure 10.25). The minimum cost of transport (COT_{min}) is defined as the minimum power required to transport an animal's weight over some distance. This is analogous to determining the minimum amount of fuel necessary to drive an automobile between two cities. For both animals and automobiles, the power-to-speed curve is U shaped, with the power requirements reaching a minimum at some intermediate speed. The COT_{min} occurs at the speed at which the power requirements are minimum.

It is apparent from Tucker's figure that within any one mode of locomotion (running, flying, or swimming), the COT_{min} decreases with increasing body size and is essentially independent of taxonomic affiliation. Conspicuously absent from Tucker's summary were estimates of the COT for cetaceans or other marine mam-

mals. The relationship between body size and COT_{min} suggested to several researchers that large swimming animals should have exceedingly low COT_{min}, but experimental evidence to test that prediction was not available until recently. Measuring power output rates requires that the metabolic rate of a subject animal be measured, and that is difficult to do with a large animal swimming in the open ocean (see Box 2, p. 54). To overcome some of these problems, Williams and coworkers trained two Atlantic bottlenose dolphins to swim in open water beside a pace boat and to match its speed. Heart and breathing rates, previously calibrated to oxygen consumption rates, were monitored and recorded continuously during each 20- to 25-minute test session to estimate metabolic rate. Blood samples

for lactate (a product of anaerobic respiration) analysis were collected immediately after each session.

The results of this study indicate that the COT is minimum for these swimmers at speeds of 2.1 m/s and that their COT increased 100% at 2.9 m/s. However, when speeds were increased above approximately 3 m/s, the dolphins invariably switched to wave riding, a behavior that is best described as surfing the stern wake of the pace boat (Figure B10.1). When the dolphins were wave riding at 3.8 m/s, their COT was only 13% higher than the minimum at 2.1 m/s. The large energy savings that accompanies wave riding at higher speeds explains dolphins' common practice of riding the bow or stern waves of ships and even of large whales, apparently with little effort.

Figure B10.1   Dolphin surfing the wake of a boat.

How does the COT of dolphins compare with the COT of other swimmers? Williams and colleagues demonstrated that dolphins are efficient swimmers, with COT_{min} about an order of magnitude lower than that of pinnipeds and even lower than humans or of other surface swimmers (Figure B10.2). Yet the dolphin COT_{min} is still several times higher than a hypothetical fish of comparable size. The additional costs incurred by dolphins are presumably associated with the mammalian re-

quirement of maintaining high body temperatures. In essence, this is the overhead cost of keeping the motor warmed up and running, a cost not experienced by poikilothermic fish. The COT_{min} of dolphins is substantially lower than that of pinnipeds and is comparable (when differences in body sizes are accommodated) with the estimated COT_{min} of migrating adult gray whales.

The other energy-robbing condition experienced by air-breathing marine tetrapods occurs when they

Dolphin Swimming

approach the sea surface to breathe. There they encounter additional wave drag due to gravitational forces associated with creating waves at the air–sea interface. At depths below 2.5 times an animal's greatest body diameter, these gravitational forces are negligible, but as a swimming mammal ascends, its total drag rapidly increases to a maximum of fivefold at the sea surface due to energy lost in the formation of surface waves. Any mammal can reduce its COT by swimming at depths below 2.5 body diameters. However, it must return to the surface periodically to breathe. At very high swimming speeds, porpoising (described on p. 309) becomes a practical behavior to avoid some of the high drag encountered at the sea surface (Figure B10.1).

Additional Reading

Weihs, D. 2002. Dynamics of dolphin porpoising revisited. *Integrative and Comparative Biology* 42:1071–1078.

Williams, T. A., W. A. Friedl, M. L. Fong, R. M. Yamada, P. Sedivy, and J. E. Haun. 1992. Travel at low energetic cost by swimming and wave-riding bottlenose dolphins. *Nature* 355:821–23.

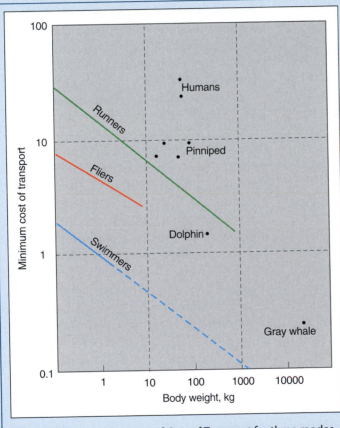
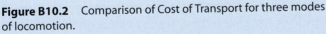

Figure B10.2 Comparison of Cost of Transport for three modes of locomotion.

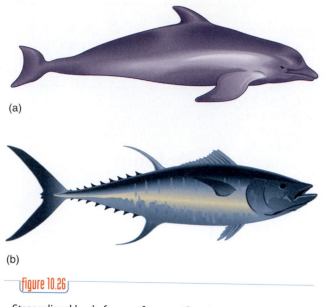

figure 10.26

Streamlined body forms of two swift pelagic animals:
(a) bottlenose dolphin, *Tursiops*; (b) tuna, *Thunnus*.

the caudal fin joins the rest of the body. The caudal peduncle, flattened in cross-section, produces little resistance to lateral movements. The rigid caudal fin is lunate and has a high aspect ratio (usually greater than 7; see Figure 6.17). The tail beats rapidly with relatively short strokes. This type of caudal fin creates a large thrust with little drag but also provides very little maneuverability.

Nearly 75% of the total body weight of a tuna is composed of swimming muscles. In tuna, each myomere (see Figure 6.16) overlaps several body segments and is anchored securely to the vertebral column. Tendons extend from the myomeres across the caudal peduncle and attach directly to the caudal fin. Tuna swimming muscles consist of segregated masses of red and white muscle fibers. Structurally, red muscle fibers are much smaller in diameter (25–45 µm) than white muscle fibers (135 µm) and are rich in myoglobin, the red pigment with a strong chemical affinity for O_2 (similar to that of hemoglobin). The small size of the red muscle cells provides extensive surface area which, in conjunction with myoglobin, greatly facilitates O_2 transfer to the red muscle cells. Physiologically, red muscle cells respire aerobically and white muscle cells respire anaerobically, converting glycogen to lactic acid.

The metabolic rate (and power output) of tuna red muscle, and probably of red muscle in other fishes, is about six times as great as that of white muscle. The relative amount of red and white muscle a fish has is related to the general level of activity the fish experiences. An ambush predator such as a grouper has almost no red muscle, but its large mass of white muscle fibers can power short fast lunges to capture prey or elude predators. At the other extreme are tunas, with over 50% of their swimming muscles composed of red muscle fibers. Electrodes monitoring activity of muscle tissues indicate that at slow normal cruising speeds, only red muscles of tunas contract. White muscles come into play only during sprints. Top speeds of about 10 body lengths per second can be maintained for about 1 second, but cruising speeds of 2 to 4 body lengths per second can be maintained indefinitely (**FIGURE 10.27**). The power for continuous swimming comes from the red muscle masses, with white muscle being held in reserve for peak power demands. White muscle does not require an immediate O_2 supply; it can operate anaerobically and accumulate lactic acid during stress situations. The lactic acid can be converted back into glycogen or some other substance when the demand for O_2 has diminished. Tuna, with a greater proportion of red muscle, are able to maintain a faster cruising speed than most other fishes indefinitely.

Fishes are generally considered to be ectothermic animals. The heat generated by metabolic processes within the body may elevate the body temperature slightly above the ambient water temperature, but the heat gain is quickly lost to the surrounding seawater (**TABLE 10.1**, left column). A few exceptionally fast fishes, however, have red muscle masses that are much warmer than the surrounding water (Table 10.1, right column). The magnitude of muscle temperature elevation above the water temperature is usually consistent for each species. The one well-studied exception is the bluefin tuna (*Thunnus thynnus*), which has a consistently high red muscle temperature regardless of water temperature. In water of 25°C, for example, the core muscle temperature of the bluefin tuna is near 32°C and declines only slightly to 30°C when the animal moves to seawater with a temperature of 7°C.

Within certain limits, metabolic processes, including muscle contractions, occur more rapidly at higher temperatures. Consequently, the power output of a warm muscle is greater than that of a cold muscle. Tuna and some sharks, including porbeagles, makos, threshers, and white sharks, all possess fascinating heat-conserving anatomic features. The swimming muscles of most bony fishes receive blood from the dorsal aorta just under the vertebral column. The major blood source for the red mus-

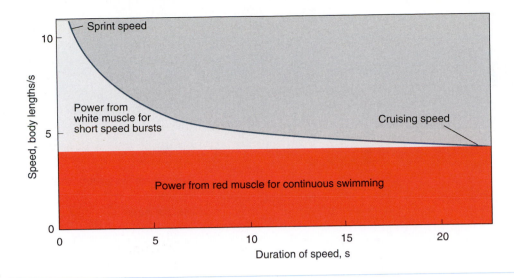

figure 10.27

Duration of swimming speeds for white and red muscles. White muscle is used for short bursts at sprint speeds and fatigues rapidly; red muscle maintains continuous cruising speeds. (Adapted from Bainbridge 1960.)

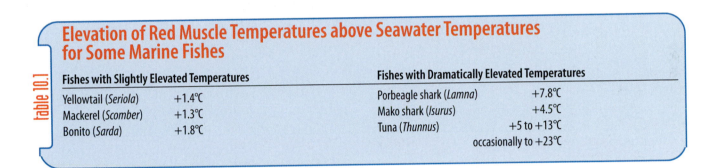

table 10.1

Elevation of Red Muscle Temperatures above Seawater Temperatures for Some Marine Fishes

Fishes with Slightly Elevated Temperatures		Fishes with Dramatically Elevated Temperatures	
Yellowtail (*Seriola*)	+1.4°C	Porbeagle shark (*Lamna*)	+7.8°C
Mackerel (*Scomber*)	+1.3°C	Mako shark (*Isurus*)	+4.5°C
Bonito (*Sarda*)	+1.8°C	Tuna (*Thunnus*)	+5 to +13°C
			occasionally to +23°C

cle masses of these sharks and most tuna is a cutaneous artery under the skin on either side of the body (FIGURE 10.28). The blood flows from the cutaneous artery to the red muscle and then returns to the cutaneous vein. Between the cutaneous vessels and the red muscles are extensive countercurrent heat exchangers that facilitate heat retention within the red muscle. Cold blood enters the countercurrent system and is warmed by the blood leaving the warm red muscle. As a result, little of the heat generated in the red muscles is lost.

All the previously described features collectively function to provide some sharks, tunas, billfishes, and other similar fishes with the ability to cruise continually at moderate speeds and with the opportunity to be the efficient pelagic predators they are. TABLE 10.2 summarizes these features and compares tuna with a normally noncruising fish (a rockfish) that typically lies in wait for its prey.

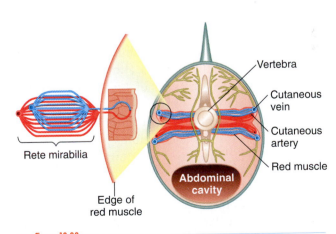

figure 10.28

Cross section of a tuna showing the position of the red muscles (shaded) and the countercurrent system of small arteries and veins serving the red muscles. (After Carey 1973.)

Table 10.2

Functional Comparison of Some Features That Influence the Swimming Speeds of a Noncruising Fish (Rockfish) and a Specialized Cruiser (Tuna)

Characteristic	Rockfish	Tuna
Body feature		
Shape		
Front view		
Rigidity	Flexible body	Rigid body
Scales	Abundant large scales	Small scales
Eyes	Bulging eyes	Nonprotruding eyes covered with adipose lid
% of thrust by body	50%	Almost none
Dorsal fin	Broad-based and high	Small, fits into slot
Caudal peduncle		
Form	Compressed	Depressed
Cross-sectional shape		
Keels	Absent	Present
Finlets	Absent	Present
Caudal fin		
Aspect ratio	Low, 3	High, 7–10
Rigidity	Flexible	Rigid
Maneuverability	Good	Poor
Tail-beat frequency	Low	High
Tail-beat amplitude	Large	Small
Swimming muscles		
% of body weight	50–65%	75%
% red muscle	20%	50% or more
Body temperature	Ambient	Elevated

Modified from Fierstine and Walters 1968.

Schooling

Successful feeding techniques used by large whales, numerous fishes, and even a few birds and seals are dependent on the presence of abundant and dense aggregations of smaller animals. Hundreds of species of smaller fishes, a few types of squids, and several species of dolphins exist in well-defined social organizations called **schools**. Fish schools vary in size from a few individuals to enormous populations extending over several square kilometers (FIGURE 10.29). Schools usually consist of a single species, with all members similar in size or age. Larger fishes swim faster than smaller ones, and mixed populations quickly sort themselves out according to their size. The spatial organization of individuals within a school remains remarkably constant as the school moves or changes direction. Individuals typically line up parallel to each other, swim in the same direction, and maintain fixed spacings between individuals. When the school turns, it turns abruptly, and the animals on one flank assume the lead. The spatial arrange-

ment within schools seems to be maintained with the use of visual or vibrational cues.

Why do some species band together in schools to be so conveniently eaten by larger predators (as indicated in FIGURE 10.30)? A part of the answer seems to be that for small animals with no other means of individual defense, schooling behavior provides a degree of protection. Most of our present understanding of the survival value of schooling behavior is based on conjecture because experiments with natural populations are exceedingly difficult to conduct and evaluate. Predatory fishes have less chance of encountering prey if the prey are members of a school because the individuals of the prey species are concentrated in compact units rather than dispersed over a much larger area. Moreover, once a predator encounters a school, satiation of the predator enables most members of the school to escape unharmed. (Because commercial fishermen are immune to satiation, our predatory activities often result in rapid overexploitation of schooling species; see Chapter 13.) Large numbers of fishes in a school may achieve additional

figure 10.29

A small portion of a large school of spotted dolphins (*Stenella*) sometimes found with schools of tuna.

survival advantages by confusing predators with continually shifting and changing positions; they might even discourage hungry predators with the illusion of an impressively large and formidable opponent.

Schooling also can serve as a drag-reducing behavior as individuals draft behind leading individuals, much as race car drivers do. Laboratory studies with fishes that instinctively school also indicate that if these fishes are isolated at an early age and prevented from schooling, they learn more slowly, begin feeding later, grow more slowly, and are more prone to predation than their siblings who are allowed to school. Schooling behavior also serves as a mechanism to keep reproductively active members of a population together. Many schooling species reproduce by broadcast spawning, and dense concentrations of ma-

figure 10.30

A skipjack (*Katsuwonus*) in a school of baitfish.

ture individuals spawning simultaneously ensure a high proportion of egg fertilization and probably greater larval survival.

Several species of pelagic tunas, especially bigeye and yellowfin, tend to congregate under near-surface schools of spinner or common dolphins. The behavioral reasons for the large mixed schools of pelagic tunas and small dolphins shown in Figure 10.29 are not understood; the consequences to the dolphins are described in Chapter 13.

Migration

In the sea, only the larger and faster swimming nekton are capable of accomplishing regular long-distance migrations, and a few examples are described below. These migrations commonly serve to integrate the reproductive cycles of adults into local and seasonal variations in the patterns of primary productivity. Many species of nekton participate in regular and directed migratory movements that are several orders of magnitude larger in both time and space scales than are the patterns of vertical migration described earlier in this chapter. Some migrators require several months of each year to accomplish their oceanic treks. In general, these migrations are adaptations to better exploit a greater range of resources for feeding or for reproduction. For example, the food available in spawning areas may be appropriate for larval and juvenile stages, but it might not support the mature members of the population. So the adults congregate for part of the year in productive feeding areas elsewhere that may be unsuitable for the survival of the younger stages.

Migratory patterns of marine animals often exhibit a strong similarity to patterns of ocean surface currents. Juvenile stages of some species may be carried long distances from spawning and hatching areas by ocean currents. Although adults may use currents for a free ride, many types of larvae and juvenile fishes are absolutely dependent on current drift for their migratory movements. The down-current drift of these young may require the adults to make an active compensatory return migration against the current flow to return to the spawning grounds.

Migrating teleosts typically move below the sea surface and well away from the coast, making it difficult or even impossible for us to observe directly their migratory behavior. Most of our understanding of oceanic migrations has been inferred from studies using visual or electronic tags and from distributional patterns of eggs, larvae, young individuals, and adults of a species. When a general progression of developmental stages from egg to adult can be found extending from one oceanic area to another, a migratory route between those areas may be inferred.

Animals marked with visual tags can yield valuable information about their migratory routes and speeds, but only if the tags are repeatedly observed during migration or are recovered after the migration is completed. The application of tagging programs is thus limited to animals that can be recaptured in large numbers (usually commercially harvested species) or to animals whose tags can be observed frequently at the sea surface. Newer techniques, such as continuous tracking of individual animals fitted with radio or sound transmitters, have added considerably to our knowledge of oceanic migration patterns. Radio tags attached to tetrapods that must surface for air periodically can be monitored by orbiting satellites. These efforts are revealing more of the fine details of oceanic movements by obtaining nearly continuous global coverage of swimming, diving, and migratory behaviors of reptiles, birds, and mammals, even when they migrate into extremely remote parts of the world ocean.

Examples of Extensive Oceanic Migrations

Four salmon species in the genus *Oncorhynchus* live only in the North Pacific. All are **anadromous**; they spend much of their lives at sea and then return to freshwater streams and lakes to spawn. They deposit their eggs in beds of gravel, and the eggs remain there through the winter. After spawning, the adult salmon die. Because the migratory patterns of the various types of salmon are similar, only the patterns of the sockeye salmon are described here. After hatching in the spring, the young sockeye remain in freshwater streams and lakes for about 2 years as they develop to a stage known as **smolts.** The smolts then migrate downstream and into the sea and enter a period of heavy feeding and rapid growth.

Sockeye salmon, as well as other salmon species, follow well-defined migratory routes, usually 10 to 20 m deep, during the oceanic phase of their migrations. These migrations closely follow the surface current patterns in the North Pacific subarctic current gyre (FIGURE 10.31, top), but the sockeye move faster than the currents. After several years at sea and several complete circuits of the gyre, the sockeye approach sexual maturity, move toward the coast, and seek out freshwater streams. Strong evidence supports a home-stream hypothesis that each salmon returns to precisely the same stream and tributary in which it was spawned. There it spawns for its only time and dies.

Tuna also have extensive migrations. Skipjack tuna are widely distributed in the warm latitudes of

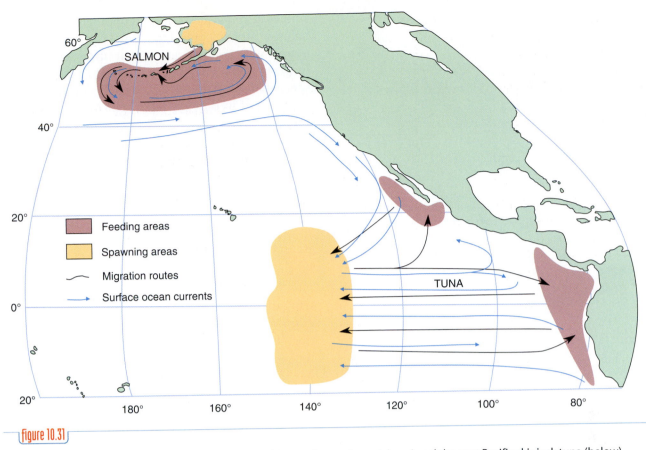

figure 10.31

The general oceanic migratory patterns of the Bristol Bay sockeye salmon (above) and the east Pacific skipjack tuna (below). Note the apparent relationship between these migratory patterns and surface ocean currents. These currents are identified in Figure 1.38. (Adapted from Royce, Smith, and Hartt 1968.)

the world ocean. Several genetically distinct populations probably exist, but we examine only the eastern Pacific population. These tuna spawn during the summer in surface equatorial waters west of 130° W longitude (Figure 10.31, bottom). For several months, the young tunas remain in the central Pacific spawning grounds. After reaching lengths of approximately 30 cm, they either actively migrate or are passively carried to the east in the Pacific Equatorial Countercurrent. These adolescent fishes remain in the eastern Pacific for about 1 year as they mature. Two feeding grounds, one off Mexico and another off Central America and Ecuador, are the major centers of skipjack concentrations in the eastern Pacific.

As the skipjack approach sexual maturity, they leave the Mexican and Central–South American feeding grounds and follow the west-flowing equatorial currents back to the spawning area. After spawning, the adults retrace the Equatorial Countercurrent they followed as adolescents. However, the feeding adults are seldom found as far to the east. Subsequent re-

turns to the spawning area follow the general pattern established by the first spawning migration.

The Atlantic eel, *Anguilla*, exhibits a migratory pattern just the reverse of the Pacific salmon. This eel also migrates between fresh and salt water. But, in complete contrast to salmon, Atlantic eels are **catadromous.** They hatch at sea and then migrate into lakes and streams, where they grow to maturity. Two species of the Atlantic eels exist: the European eel and the American eel. The distinction between the species is based on their geographic distribution and anatomic and genetic differences of the adults (FIGURE 10.32). Both species spawn deep beneath the Sargasso Sea of the central North Atlantic. Their eggs hatch in the spring to produce a leaf-shaped transparent **leptocephalus larva** about 5 cm long (FIGURE 10.33). The leptocephalus larvae, drifting near the surface, float out of the Sargasso Sea and move to the north and east in the Gulf Stream. After 1 year of drifting, American eel larvae metamorphose into young **elvers** that move into rivers along the eastern coast of North

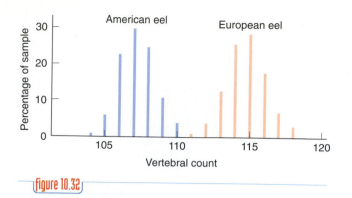

figure 10.32

Variation in vertebral counts of anguillid eels collected in America and in Europe. (Redrawn from Cushing 1968.)

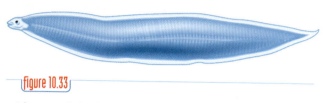

figure 10.33

A leptocephalus larva of the eel *Anguilla*.

America. The European eel larvae continue to drift for another year across the North Atlantic to the European coast (FIGURE 10.34). There, most enter rivers and move upstream. The remainder of the European population requires still another year to cross the Mediterranean Sea before entering fresh water.

After several years (sometimes as many as 10) in fresh water, the mature eels (now called yellow eels) undergo physical and physiologic changes in preparation for their return to the sea as silver eels. Their eyes enlarge, and they assume a silvery and dark countershaded pattern characteristic of midwater marine fishes. Then they migrate downstream and presumably return to the Sargasso Sea, where they spawn and die. Very few adult silver eels have been captured in the open sea, and none has been taken from the spawning area itself. Thus, the spawning migration back to the Sargasso Sea is still a matter of some speculation. To avoid swimming against the substantial current of the Gulf Stream, European eels likely follow the south-flowing Canary Current after leaving European rivers, then west in the North Atlantic Equatorial Current, and eventually to the region of the Sargasso Sea (Figure 10.34). The American eels apparently swim across the Gulf Stream to their spawning area.

Studies of variation in the structure of mitochondrial DNA confirm that American and European eels are genetically isolated separate species. These studies also revealed a hybrid population of eels inhabiting streams in Iceland. But despite repeated attempts with sophisticated sonar, underwater video cameras, and high-speed nets, no adult eels of either species (or of their hybrids) have yet been observed or captured in the presumed spawning areas of the Sargasso Sea.

Example of Migration Integrating Feeding and Reproduction

One of the best-studied foraging patterns among pinnipeds is that of the northern elephant seal (FIGURE 10.35). Adults of this species range over much of the North Pacific Ocean, making them nearly permanent members of the pelagic realm. They stay at sea for 8–9 months each year to forage, interrupted by two round-trip migrations to near-shore island breeding rookeries. With the development of microprocessor-based dive time and depth recorders, details of the migratory and foraging behavior of these deep-diving seals have been fairly well-documented. By attaching time and depth recorders to adults as they leave their island rookeries, dive behaviors of individual animals have been recorded continuously for several months until their return to their rookeries with their time and depth recorded data.

Foraging

When ashore for breeding or for molting, adult elephant seals fast, and each fasting period is followed by prolonged periods of foraging in offshore waters of the North Pacific Ocean. The postbreeding migration averages 2.5 months for females and 4 months for males. After returning to their breeding rookery for a short molting period of 3–4 weeks, during which they shed their old guard hairs, they again migrate north and west into deeper water. During this postmolt migration, females forage for about 7 months and males for about 4 months (FIGURE 10.36). Both males and females then return again to their rookeries for breeding. Individuals return to the same foraging area as during postbreeding and postmolt migrations.

While at sea, both males and females dive almost continuously, remaining submerged for about 90% of the total time. Dives for both sexes average 23 min, with males having slightly longer maximum dive durations. These dive durations probably are near the aerobic dive limit for this species (see Chapter 6). Elephant seals of both sexes spend about 35% of their dive time near maximum depth, between 200 and 800 m. In general, daytime dives are 100–200 m

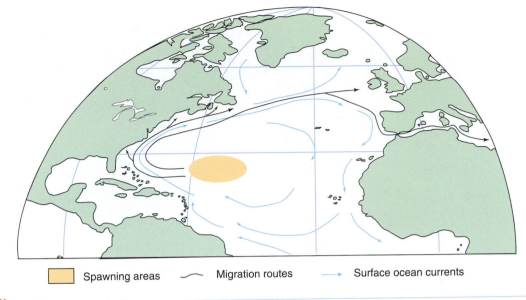

| Spawning areas | ⌒ Migration routes | → Surface ocean currents |

figure 10.34

Migratory routes of the larvae and young of anguillid eels. The return migrations of their adults have been omitted for clarity.

figure 10.35

Northern elephant seals hauled out on a California beach.

deeper than nighttime dives. The principal prey of elephant seals are mesopelagic squid and fishes that spend daylight hours below 400 m and ascend nearer the surface at night. This upward nighttime movement reflects the pattern of diurnal vertical migration of these prey species and is likely the basis for the difference in maximum depths between day and night dives of elephant seals.

Foraging elephant seals are widely dispersed over the northeast Pacific Ocean and exhibit strong gen- der differences for preferred foraging locations. Females tend to remain south of 50° N latitude, particularly in the subarctic Frontal Zone (Figure 10.36). Adult males transit farther north, through the feeding areas of females to subarctic waters of the Gulf of Alaska and the offshore boundary of the Alaska Current flowing along the south side of the Aleutian Islands. These areas of aggregation are regions of high primary productivity and contain abundant fish and squid communities.

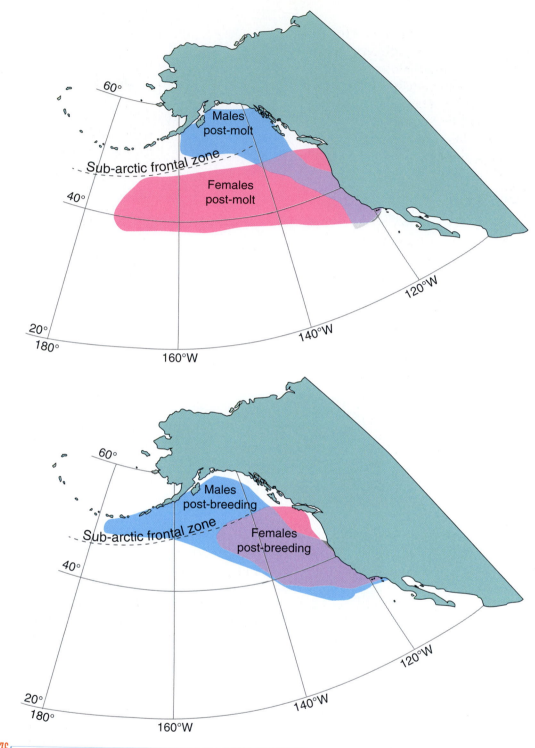

figure 10.36

Geographical distribution of male and female elephant seals during post-molt (top) and post-breeding migrations. (Adapted from Stewart and DeLong, 1993.)

Breeding

Twice each year, adult elephant seals return from foraging at sea to the same breeding beach on which they were born, once to breed and again several months later to molt. Although many vertebrates migrate long distances between breeding seasons, elephant seals are the best known to make a double migration each year. Their individual annual movements of 18,000–21,000 km rival those of gray and humpback whales in terms of the greatest distance traveled.

Most pinniped species breed and pup on land, whereas the remainder mate and give birth on fast ice or pack ice (ice breeders are discussed in Chapter 12). Species like elephant seals that pup on land select their pupping and mating rookeries on islands free of large terrestrial predators or on isolated mainland beaches and sandbars not easily accessible to such predators. Consequently, available rookery space may become limiting, and females become densely aggregated in large breeding colonies. These dense aggregations establish conditions favorable to a polygynous mating system. Almost all pinniped species that breed on land, including elephant seals, are extremely polygynous and strongly sexually dimorphic, with males showing obvious secondary sex characteristics. Breeding male northern elephant seals are five to six times larger than adult females, have an elongated proboscis, enlarged canine teeth, and thick cornified skin on the neck as secondary sexual characteristics (FIGURE 10.37).

Mammals are predisposed for polygyny because of the disparity in the number of gametes produced by males and females and of the relatively minor role males usually play in the successful rearing of offspring. Males can mate with many females during a breeding season, but females, although able to mate numerous times, can only achieve one pregnancy during each breeding season. Provisioning offspring until they are nutritionally independent is the most energetically expensive component of reproduction. With very few exceptions, care and feeding of marine mammal offspring is the sole responsibility of the offspring's mother. In many species, this disparity between the reproductive best interests of males and females of the same species is associated with **sexual dimorphism**, an obvious and often dramatic difference in the size, appearance, and behavior of adult males and females.

Adult male elephant seals arrive at rookery sites in early December, and successful breeders remain while fasting until late February or early March. Ar-

figure 10.37

Male and female elephant seals display obvious sexual dimorphism. The females lack the elongated nose, enlarged canines, thickened neck, and large size characteristic of sexually mature males.

rival of these males displaces the juveniles who used the rookeries as haul-out sites during the nonbreeding season. Late-pregnant females arrive in late December and aggregate in preferred rookery sites already occupied by dominant males. Single pups are born and begin nursing about 1 week after females arrive.

Adult females fast through their 4-week lactation, remaining on the beach near their pups. Elephant seal milk is low (about 10%) in fat in the early stages of lactation but increases to about 40% through the last half of lactation. The gradual substitution of fat for water is probably an adaptation of the mother to conserve water late in her fast; it also contributes to the developing blubber store of the pup before weaning. Pups weigh about 40 kg at birth and gain an average of 4 kg each day before weaning. Overall, nearly 60% of the total energy expenditure of a lactating elephant seal mother is in the form of milk for her pup.

At weaning, most elephant seal pups have quadrupled their birth weights; however, a few may have increased their birth weights as much as sevenfold by sneaking milk from other lactating females after their own mothers have departed to feed at sea. Weaned elephant seal pups do not immediately commence foraging for their own food. Rather, they remain on the pupping beaches for a postweaning fast lasting an average of 8–10 weeks. Only when pups have lost about 30% of their weaned body mass and are at least 3 months old do they enter shallow waters to feed for the first time.

Lactating elephant seal females enter estrus and mate about 2 weeks after pupping. Adult male elephant seals compete for reproductive control of these females by establishing dominance hierarchies to exclude other males from breeding activities. This type of mating system often is referred to as female (or harem) defense polygyny. Elephant seal males fast for the duration of the 3-month breeding season, and the increased lipid stores associated with large male body mass confer clear advantages in successfully withstanding these extended periods without food. The most dominant (or alpha) male in a dominance hierarchy defends nearby females against incursions by subordinate males. In areas crowded with females, an alpha male can dominate the breeding of about 50 females.

Subdominant, usually younger, males use other strategies to gain access to at least a few estrus females. Some male elephant seals occupy less preferred breeding beaches that attract fewer males with whom they must compete. However, these beaches also attract few females. In more crowded rookeries, subdominant males may accomplish occasional matings by sneaking into large harems in which alpha males are more likely to be distracted or by mobbing a female as she departs from the rookery at the end of the breeding season. Almost all these departing females are already pregnant by a dominant male, so these late copulations by subdominant males are unlikely to lead to paternity.

Mating strategies of female pinnipeds have been less studied than those of males. This is in part due to the crowded rookery conditions of highly polygynous species. It is sometimes difficult to distinguish a female's behavior directed at defending her pup from strictly mating-related behavior. Females may move away from or vocally protest mounting attempts by subdominant males, thus attracting the notice of higher ranking males. A female's position within a group, especially a large group, also may influence with whom she will mate. Older or more dominant females near the center of a harem will be more likely to mate with the dominant male, leaving less-dominant females to be pushed to the edge of the harem where they are more exposed to mating forays by subdominant males.

After the last of the females have departed the rookery, emaciated adult males begin to leave in February or March. Adult females and juveniles return to the rookery for a 1-month molting fast between mid-March and mid-May. Adult males return to the rookery in June and July for their molt. Then the juvenile animals return to use the beach as a haul-out site until the beginning of the next breeding season.

The duration of gestation in placental mammals ranges from 18 days in jumping mice to 22–24 months in elephants. In terrestrial species, gestation duration is roughly related to the size of the fetus: Larger fetuses require longer gestation periods to develop. For elephant seals, both parturition and fertilization occur in the same geographic location and at the same time of year, but in successive years, and gestation periods must be about 1 year in duration regardless of body size. Consequently, these and all other species of pinnipeds must stretch it out to prolong gestation for a year.

Adjustments to fit gestation periods of less than 1 year into an annual time frame are accomplished by a reproductive phenomenon known as **seasonal delayed implantation** (FIGURE 10.38). In seasonal delayed implantation, the zygote undergoes several cell divisions to form a hollow ball of a few hundred cells, the blastocyst, which then remains inactive in the female's uterus for several months. After this delay, the blastocyst becomes implanted into the inner wall of the uterus, a placental connection develops between the embryo and the uterine wall, and normal embryonic growth and development resume through the remainder of the gestation period. In effect, delayed implantation substantially extends the normal developmental gestation period to enable mating when adults are aggregated in rookeries rather than when dispersed at sea. By providing flexibility in the length of gestation, this reproductive strategy confines birth and mating to a relatively brief period of time ashore and enables the young to be born when conditions are optimal for their survival. Because female elephant seals are at sea and inaccessible when blastocyst implantation occurs, the physiologic and hormonal mechanisms controlling delayed implantation are not well understood.

The geographic segregation of foraging areas by male and female northern elephant seals may reflect the preferences of males for larger and more oil-rich species of squid found in higher latitude subarctic waters, even though they must travel farther through the foraging areas of the females to get to these high-latitude foraging grounds. The smaller females of this species also may have different energetic requirements that encourage them to undertake shorter migratory distances while giving up access to energy-rich

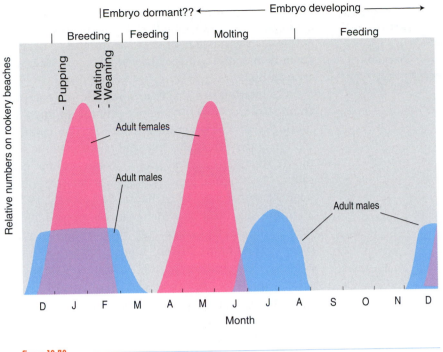

Embryo dormant?? ←————— Embryo developing —————→

| Breeding | Feeding | Molting | Feeding |

- Pupping
- Mating
- Weaning

Adult females

Adult males

Adult males

Relative numbers on rookery beaches

D J F M A M J J A S O N D

Month

figure 10.38

Seasonal patterns of rookery use by adult elephant seals. Periods of embryo dormancy and of active fetal growth in pregnant females are indicated.

prey at higher latitudes. Strikingly similar patterns of latitudinal segregation by gender in North Pacific foraging areas are exhibited by sperm whales. Their prey are also primarily mesopelagic squid, and their seasonal migrations and diving patterns may also reflect the geographic and vertical distribution of their preferred squid prey.

The differential migrations of male and female elephant seals appear to develop during puberty, when growth rates of males are substantially greater than those of females. These patterns are well established by the time males are 4–5 years old. Although body mass will vary substantially on a seasonal basis in those species that experience prolonged postweaning or seasonal fasts, body length tends to increase regularly until physical maturity is reached. For marine mammals, growth to physical maturity continues for several years after reaching sexual maturity, so old sexually mature individuals are often much larger than younger but still sexually mature individuals of the same gender. Second, in polygynous and sexually dimorphic species of pinnipeds and odontocetes (such as killer and sperm whales), patterns of male growth exhibit a delay in the age of sexual maturity to accommodate a period of accelerated growth into body sizes much larger than those of females (FIGURE 10.39). Before polygynous males can compete successfully for breeding territories or establish a high dominance rank, they must achieve a body size substantially larger than that of females. Males of all polygynous species delay both sexual and physical maturity as compared with females.

The longer wait to sexual maturity comes at a substantial cost reflected in overall mortality. For example, in northern elephant seals using central California rookeries, life histories of individual males collected between 1967 and 1986 were similar; only 7–14% of males survived to 8–9 years of age, when they could first successfully compete for females. Females, however, must only survive to about 4 years of age before they can successfully breed. The period of accelerated growth in males often corresponds to the age at which females of the same species achieve sexual maturity. Although the life history of male northern elephant seals is geared toward high mating success late in life, the chance of living to that age of high mating success is small. Fewer than 10% of a study group of males managed to mate at all during their lifetimes, whereas the most successful male of that group mated with 121 females.

The mortality experienced by elephant seals during their first year is caused by several agents, ranging from starvation to parasitic lungworm infections to predation by white sharks. White sharks are not the capricious or mindless killers portrayed by the "Jaws" movies. Rather, they are skilled predators of pinnipeds. About 1 m long at birth, young white sharks initially feed on small teleost fishes. By the time they are 3 m long, they begin to shift to the adult preference for pinnipeds, especially young seals. This shift in preferred prey coincides with movements to higher latitudes (in both hemispheres) where aggregations of seals are abundant.

The predatory behavior of white sharks on seals involves a stealthy approach along the sea floor in shallow areas where seals enter the water. White sharks exhibit strong countershading patterns, and their approach must be difficult for seals to detect from above. When the victim is located, likely by silhouetting it against the lighter background of the

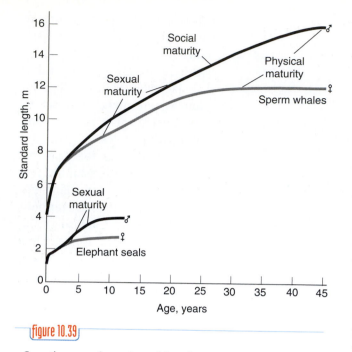

figure 10.39

Growth curves for male and female northern elephant seals (bottom) and sperm whales (top). (Adapted from Clinton, 1994 and Lockyer, 1981.)

sea surface, the shark lunges upward from below and bites the seal to cause profuse bleeding (FIGURE 10.40). The bleeding seal is then either carried underwater to drown in the shark's jaws or is left at the surface until it dies.

From analyses of stomach contents, large white sharks appear to prefer seals and whales over other kinds of prey, such as birds, sea otters, or fishes. This selective preference for marine mammals that use fat-rich blubber for insulation may be related to the high metabolic demands of maintaining elevated muscle temperatures and high growth rates in the temperate waters occupied by their seal prey.

10.8 Orienting in the Sea

How do pelagic migratory species know where they are and where they are going when there are few obvious "landmarks" for them to follow? Before an animal can successfully accomplish a directed movement from one place to another, it must orient itself both in time and in space. Biologic clocks operating on circadian and longer period rhythms (see Chapter 8) are important factors in the orientation process. A variety of environmental factors serves as cues to adjust or reset the timing of these rhythms. Well known among these timing factors is the day length, which changes with predictable regularity through the seasons. Day length, water temperature, and food availability might serve as useful cues for following the passage of the seasons and as triggers for seasonal migrations of gray whales and other marine animals.

Orientation in space is somewhat more complex than orientation in time, especially for animals migrating below the sea surface where directional information derived from the apparent position of the sun, moon, or stars is unavailable. It is known that eels, salmon, sharks, and many other fishes have extremely keen olfactory senses (see Chapter 6). Since the 1970s, an impressive body of evidence has been gathered to support the idea that salmon use olfactory cues to guide them to their home stream. The sequential odor hypothesis of migrating salmon postulates that young salmon smolt are imprinted with a sequence of stream odors during their downstream trip to the ocean. When returning back to their home stream as adults, the remembered cues are played back in reverse to act as a sequence of sign stimuli that release a positive response to swim upstream. The chemical nature of the characteristic odors in stream water remains largely unidentified. However, these or similar odors are not likely to be concentrated sufficiently in the open ocean to guide the oceanic phase of the salmon's migration. What then are the guideposts available to nekton migrating across huge expanses of open ocean well below the sea surface?

Currents are among the most stable structural features of the open ocean. The migratory patterns of many fishes and other marine animals seem closely associated with surface current patterns. But how can a fish detect the direction or speed of an ocean current if it can see neither the surface nor the bottom? The sharp temperature and salinity gradients sometimes found at the edges of ocean currents might be detected by some fishes, but only if they venture across the edge of the current. However, available evidence suggests that salmon, tuna, and possibly adult eels migrate within currents, not along their edges.

Current speeds and directions are difficult to detect from the surface if the observer is being carried along by the current. But animals below the surface may be able to detect ocean surface currents by visually observing the speeds and direction of horizontally moving debris and plankton (FIGURE 10.41). In the ocean, the water velocity generally decreases with depth (see Figure 1.36). Swimmers near the bottom of the current should see particles above them moving in the direction of the current. Slower moving particles below a swimmer would appear to move

figure 10.40

Typical attack behavior of a white shark on a seal. The shark approaches the seal from below to attack (left), then may carry the seal underwater or let the seal float to bleed.

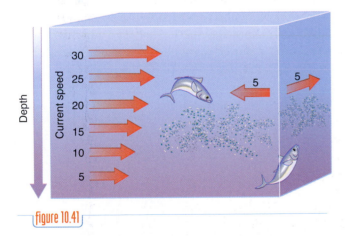

figure 10.41

Possible speed and direction cues for fish in an ocean current. To a drifting fish above an accumulation of debris and plankton (shaded region), the debris appears to move backward. From below, the debris appears to be carried forward in the direction of the current. Numbers indicate relative current speeds.

backward as the swimmer is carried forward by the current. In this manner, animals could determine the current direction and orient their swimming motions either in the same direction or directly against it.

When charged ions of seawater are moved by the ocean's currents through the magnetic field of the Earth, a weak electrical potential is generated in a process that is similar to the operation of an electrical generator. These ocean current potentials have been measured with ship-towed electrodes and are used to compute current speeds. Some preliminary laboratory evidence suggests that at least the Atlantic eel, *Anguilla*, and the Atlantic salmon, *Salmo*, are sensitive to electrical potentials of the same magnitude as those generated by ocean currents. In addition, they are most sensitive to the electrical potential when the long axes of their bodies are aligned with the direction of the current. Sharks, rays, and green turtles also have a demonstrated ability to sense and respond to the Earth's magnetic field.

10.9 Echolocation

To echo a theme introduced in Chapter 1, seawater is not very transparent to light, but it is an excellent transmitter of sound energy. Marine mammals have very good hearing (described in Chapter 6), and they have taken advantage of the sound-conducting properties of water to compensate partially for the generally poor visibility conditions found below the sea surface. Many marine mammals can obtain some information about their surroundings simply by listening to the environmental sounds that surround them. Others, however, have evolved systems for actively producing sounds to illuminate targets acoustically for detailed examination.

Soon after the first hydrophone was lowered into the sea, it became apparent that whales and pinnipeds could generate a tremendous repertoire of underwater vocalizations. Many of the moans, squeals, and wails that we can hear are evidently for communication. Bottlenose dolphins, *Tursiops,* produce a large variety of whistle-like sounds, and captive individuals have been shown to understand complex linguistic subtleties. Other sounds, especially those of the humpback whale, *Megaptera,* have a fascinating musical quality. The songs of each humpback whale population are identifiably different from the songs of other populations, are probably produced exclusively by adult males advertising to females, and are culturally transmitted from one individual to another within each population. Each song is composed of numerous phrases, some of which are repeated several times. During each breeding season the songs evolve; some phrases are modified and others are added or deleted.

About 20% of all mammal (and even a few bird) species have overcome the problems of orienting themselves and locating objects in the dark or underwater with **echolocation**. Echolocation, also re-

ferred to as biosonar, consists of animals producing sharp sounds and listening for reflected echoes as the sounds bounce off target objects. Bats are well-known echolocators, but so too are many marine mammals, particularly toothed whales. To echolocate successfully, an animal must be able to produce an appropriate sound signal, detect its echo, and then mentally process that signal to extract meaningful information about its immediate environment.

The sounds most useful for echolocation are neither squeals nor songs but trains of broad-frequency clicks of very short duration. Much more is known about the echolocating capabilities of the smaller whales, such as *Tursiops,* because they are frequently maintained in captivity for convenient study. *Tursiops* use clicks consisting of sound frequencies audible to humans as well as higher frequency clicks, often exceeding 150 kHz (FIGURE 10.42a). Because our human ears are sensitive to air-borne sound frequencies between about 18 vibrations per seconds (or hertz) and 18,000 Hz (=18 kHz), most of the energy in dolphin echolocation clicks is well beyond the upper range of human hearing. These frequencies are referred to as **ultrasonic**. Each click lasts only a fraction of a millisecond and is repeated as often as 600 times each second (Figure 10.42b). As each click strikes a target, a portion of its sound energy is reflected back to the source. Click repetition rates are adjusted to allow the click echo to return to the animal during an extremely short silent period between outgoing clicks. The time required for a click to travel from an ani-

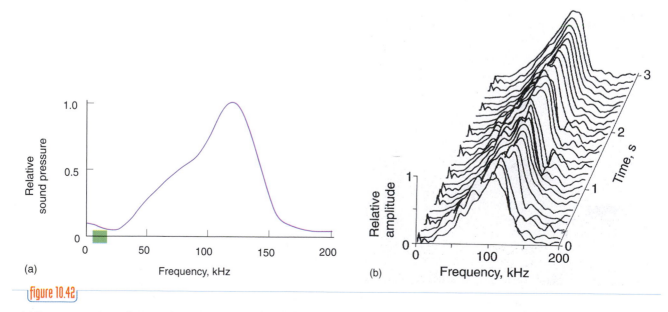

figure 10.42

(a) Power spectrum of a typical echolocation click, and (b) display of the power spectra of a single series of echolocation clicks of a bottlenose dolphin. Green band in (a) spans the frequency range of human hearing.

mal to the reflecting target and back again is a measure of the distance to the target. As that distance varies, so will the time necessary for the echo to return. Continued reevaluation of returning echoes from a moving target can indicate the target's speed and direction of travel.

Relying solely on their echolocating abilities, captive blind-folded bottlenose dolphins have repeatedly demonstrated aptitudes for discriminating between objects of a similar nature: two fishes of the same general size and shape, equal-sized plates of different metals, and pieces of metal differing only slightly in thickness. In the wild, these animals must acoustically survey their surroundings, while simultaneously distinguishing their own echolocation clicks from the cacophony of other sounds so frequently present in large herds of wild dolphins (Figure 10.29).

How do whales produce the sounds involved in echolocation? The larynx of toothed whales is well muscled and complicated in structure, yet it lacks vocal cords and is not used in sound production. The elongated tip of the larynx extends across the esophagus into a common tube leading to the blowhole to separate the pathways for food and air completely. Just inside the blowhole are a pair of heavily muscled valves, the nasal plugs. Associated with the nasal plugs and a complex of air sacs branching from the nasal passage are the dorsal bursae, which drive a pair of **phonic lips** that vibrate to produce the clicks (see Figure 6.47). Clicks produced here are directed forward by the concave front of the skull and then focused by the fatty lens-shaped **melon,** the rounded forehead structure so characteristic of toothed whales (FIGURE 10.43), to concentrate the clicks into narrow directional beams. Recent research indicates that some species of toothed whales may also stun fish prey with intense blasts of sound energy, presumably using the same sound production system used for echolocation.

To make this acoustic picture even more complicated, at the same time a dolphin is producing a train of rapidly repeated echolocation clicks, it can simultaneously produce frequency-modulated tonal whistle signals that vary in pitch from 2 to 30 kHz and direct those emitted sound signals forward from the melon in different directional beam patterns. It can do this while continually varying the frequency content of the clicks to adjust to the changing background noise or to the acoustic characteristics of the target.

Sperm whales are the largest tooth whales. They are notable for their massive and very distinctive foreheads and extremely powerful and structurally complex echolocation signals. Inside the forehead is a highly specialized organ, the **spermaceti organ,** which may occupy 40% of the whale's total length and 20% of its body weight. This organ is filled with a fine-quality liquid, or spermaceti oil, once prized by whalers for candle making and for burning in lanterns. The spermaceti organ is encased within a wall of extremely tough connective tissue just above another fat-filled organ of similar size, the **junk.** The junk is thought to be homologous with the melon of smaller toothed whales. The entire structure sits in the hollow of the rostrum and the amphitheater-like front of the skull (FIGURE 10.44).

Between the anterior ends of the junk and spermaceti organ are a pair of large opposed phonic lips (FIGURE 10.45a) that, as in dolphins, are the origin of the click sounds of these whales. Although impossible to test in live sperm whales, anatomic evidence suggests that air from the larynx is forced through the right nasal passage to open the phonic lips, which then snap shut with a loud clap. Most of that sound energy is reflected backward by the frontal air sac (acting as an acoustical mirror) and channeled through the spermaceti organ to be reflected again

figure 10.43

A bottlenose dolphin (*Tursiops*) with a prominent melon.

figure 10.44

A sperm whale skeleton. Note the concave shape of the skull, which in life is filled by the spermaceti organ.

(a)

Figure 10.45

(a) Phonic lips of a sperm whale, and (b) a cutaway view of the complex structure of a sperm whale head. (Adapted from Norris and Harvey 1972.)

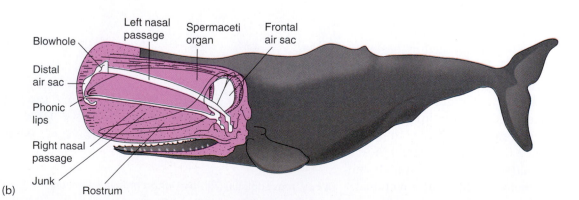

Blowhole

Left nasal passage

Spermaceti organ

Frontal air sac

Distal air sac

Phonic lips

Right nasal passage

Junk

Rostrum

(b)

by another acoustical mirror, the distal air sac (Figure 10.45b). From there, the click is focused through the junk and projected forward through the front of the head. The resulting echolocation clicks are repeated more slowly and at lower frequencies than those of *Tursiops*.

This description conflicts somewhat with the traditional view of sperm whale echolocation clicks, which have been described as reverberant pulses lasting about 24 ms and consisting of a series of several rapidly reverberating individual clicks. The reverberant click sounds long reported in the literature appear to represent "leakage" of some of the click energy from the sides of the whale's head as the click signal bounces repeatedly between the two acoustical mirrors at either end of the spermaceti organ. Mohl and his associates recently demonstrated that when a recording hydrophone is directly in the frontal sound beam of an echolocating sperm whale, there is no click reverberation, and the sound intensity level of the single focused click is the loudest sound yet recorded from any natural biologic source.

These intense clicks of sperm whales travel for several kilometers in the sea. The powerful long-range echolocation systems of sperm whales may partially explain their success as efficient predators of larger mesopelagic squid. Try to visualize these whales cruising along at the sea surface with a constant supply of air, periodically scanning the unseen depths below with a short burst of echolocation click pulses. Only when a target worthy of pursuit is detected and its location pinpointed does the whale depart from its air supply and go after its meal (FIGURE 10.46).

How common is echolocation in marine mammals? Presently, it is uncertain because it is difficult to establish whether or not wild populations are indeed using the echolocation-like clicks for the purposes of orientation and location. If judgments can be made from the types of sounds produced, then echolocation is assumed to occur in all toothed whales, some pinnipeds, and possibly a few baleen whales. Click series with echolocation-like qualities have been recorded in the presence of gray whales in the North Pacific and blue and minke whales in the North Atlantic. It is not unreasonable to assume that these animals use these sounds, as well as any other sensory means they possess, to find food, locate the bottom, and evaluate the nonvisible portion of their surroundings. Some of these possible uses of intentionally produced sounds by baleen whales are considered in Chapter 12.

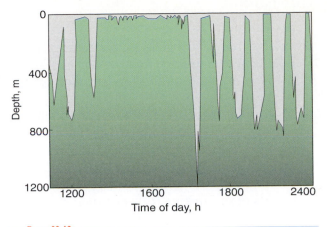

figure 10.46

Continuous 13-hour trace of 9 deep dives by a single adult sperm whale. Depths during the dives were recorded by a digital time-depth recorder.

STUDY GUIDE

Marine Biology On Line

Connect to this book's web site: **http://www.bioscience.jbpub.com/marinebiology** *The site features eLearning, an online review area that provides quizzes, chapter outlines, and other tools to help you study for your class. You can also follow useful links for more in-depth information.*

Summary Points

The pelagic realm is a three-dimensional nutritionally dilute habitat wherein rates of primary production tend to be low and few obvious ecologic niches are present.

Inhabitants of the Pelagic Division

- The pelagic division is home to two major ecologic groups, the zooplankton and nekton. Zooplankton are represented by more than 5000 species of permanent holoplankton (including all three protozoan phyla, cnidarians, ctenophores, chaetognaths, crustaceans, and invertebrate chordates) and larval stages of many invertebrates and fishes (the temporary meroplankton).

- An equal number of nektonic species roam the same waters; most nekton are vertebrates, and most marine vertebrates are teleost fishes. Of the numerous groups of marine invertebrates that live in the sea, only squids and a few species of shrimps are truly nektonic.

Geographic Patterns of Distribution

- Within the center of the large, semienclosed, oceanic current gyres is the epipelagic, or photic, zone. Each major epipelagic habitat, broadly defined by its own unique combination of water temperature and salinity characteristics, is nicely delineated by six closely related species of krill, whose distributions are indicative of the general large-scale distribution of many other epipelagic animal species.

- However, smaller scale variations in environmental features, as well as the local intensity of grazing or predation, chemical and physical gradients, and behaviors (such as aggregation or avoidance), do result in patchy distributions at virtually all levels of sampling, from several kilometers down to microscopic distances.

Vertical Distribution of Pelagic Animals

- Although the epipelagic zone accounts for less than 10% of the ocean's volume, most pelagic animals are found there. Most are counter-shaded carnivores acting at higher trophic levels and effective swimmers, which enables them to erase the sharper distributional boundaries exhibited by zooplankton.

- From the bottom of the sunlit epipelagic zone to about 1000 m lies the mesopelagic zone, a world where animals live in very dim light and depend on primary production from the photic zone above. Mesopelagic fishes seldom exceed 10 cm in length, and many are equipped with well-developed teeth, large mouths, highly sensitive eyes, and photophores, light-producing organs most commonly arranged on the ventral surface of the body. The light from photophores may disrupt the visual silhouette of the fish when observed from below and may also used for species recognition.

- Below the mesopelagic zone, light comes largely from photophores, which are used as lures for prey, as species-recognition signals, and possibly even as lanterns at these great depths.

Vertical Migration: Tying the Upper Zones Together

- Zooplankton and small nektonic species can experience very different environmental conditions by moving upward modest distances, because temperature, light intensity, and food availability all increase markedly as the distance from the sea surface decreases. Because the mesopelagic zone also offers some distinct advantages, large numbers of mesopelagic animals, such as euphausiids, small fishes (primarily lanternfishes), and siphonophores, periodically migrate vertically to enjoy the benefits of both zones.

- Daily or seasonal changes in light intensity seem to be the most likely stimulus for vertical migrations. The animals follow an isolume (a constant light intensity); as sunlight decreases in late afternoon, the isolume moves toward the sea surface, and the migrating community follows with a precision seldom seen in natural populations.

Feeding on Dispersed Prey

- Copepods and other small pelagic particle grazers are typically exposed to a wide spectrum of food particle sizes, which presents several opportunities for these small versatile particle grazers to adopt feeding strategies that selects for optimal-sized food items. The mechanisms used by zooplankton to glean small diffuse food particles from the water reflect the crucial roles these animals play in pelagic food webs.

Buoyancy

- Living and moving in three dimensions well above the seafloor creates some buoyancy problems for pelagic animals because the bone and muscle tissues needed for locomotion in these animals are more dense than seawater. Stored fats and oils, which are only slightly less dense than seawater, or an internal gas-filled flotation organ are common buoyancy devices used by pelagic marine animals.

- The swim bladders of bony fishes develop embryonically from an outpouching of the esophagus. This connection with the esophagus may remain intact in the adult (the physostomous condition) or disappear as the fish matures (a physoclistous swim bladder).

- Both types of swim bladders have gas glands that regulate the secretion of gas from the blood into the bladder when these fishes are below the sea surface and have no access to air. The gas gland and associated countercurrent rete mirabilia of bony fishes are capable of concentrating gases from the blood into their swim bladders at high pressures.

Locomotion

- Nekton are large and fast animals that often must move long distances to find food or mates or to otherwise improve conditions for their survival. Because water is greater than 800 times more dense than air and at least 30 times more viscous, their energetic costs of locomotion are expensive and represent a major expenditure of their available resources.
- Most pelagic fishes, seals, and sea snakes use side-to-side motions of their bodies as their chief source of propulsion; whales move their flukes in vertical motions; turtles paddle; and penguins, sea lions, and many pelagic rays use underwater flying motions. Squids and other cephalopods take water into their mantle cavities and then expel it at high speeds through a nozzle-like siphon.
- The body shape of a fast swimmer, such as a tuna or dolphin, is a compromise between different hypothetical body forms, each of which reduces some component of the total drag and enables the animal to slip through the water with as little resistance as possible.
- Dolphins, tunas, and some sharks are able to swim very quickly. The exceptional speed of tuna-like fishes is due to their streamlined body, small and smooth scales, nonbulging eyes, retractable fins, and numerous small median finlets that reduce turbulence. Moreover, nearly 75% of the total body weight of a tuna is composed of swimming muscles. The major blood source for the red muscle masses of most tuna is a cutaneous artery that is associated with extensive countercurrent heat exchangers that facilitate heat retention within the red muscle. As a result, little of the heat generated in the red

muscles is lost and the red muscle masses are much warmer than the surrounding water.
- Hundreds of species exist in well-defined social organizations called schools for protection, as a means of reducing drag while swimming, or to keep reproductively active members of a population together.
- Larger and faster nekton participate in regular and directed migrations that serve to integrate the reproductive cycles of adults into local and seasonal variations in patterns of primary productivity. Migration routes, which can be inferred by using visual or electronic tags and from distributional patterns of eggs, larvae, and variously aged individuals of a species, often correlate well with patterns of ocean surface currents.
- Some oceanic migrations are extensive. Six species of salmon in the North Pacific are anadromous, spending much of their lives far at sea and then returning to freshwater streams and lakes to spawn. Atlantic eels are catadromous, hatching at sea and then migrating into lakes and streams where they grow to maturity.
- One of the best-studied migration patterns among pinnipeds, and its correlation with foraging and breeding, is that of the northern elephant seal. Adults of this species stay at sea for 8–9 months each year to forage and make a double migration each year to near-shore islands for breeding or for molting.

Orienting in the Sea

- An animal must orient itself both in time and in space to migrate successfully. Biologic clocks are important factors in the timing aspect of navigation, and a variety of environmental factors, such as day length, water temperature, and food availability, serves as cues to adjust or reset the timing of these clocks.
- Orientation in space is somewhat more complex, especially for animals migrating below the sea surface where directional information derived from celestial objects is unavailable. Salmon use olfactory cues, many fishes and other marine animals seem to associate with surface current patterns, Atlantic eels and salmon are sensitive to electrical potentials of the same magnitude as those generated by ocean currents, and sharks, rays, and green turtles can sense and respond to the Earth's magnetic field.

Echolocation

- To compensate for reduced visibility and their inability to smell under water, odontocetes and some other groups have a sophisticated system of echolocation for target detection and orientation.

Topics for Discussion and Review

1. Mesopelagic fishes differ from more familiar epipelagic and coastal fishes in many ways. Summarize them.
2. Summarize the hypothetical uses of photophores in marine animals.
3. The exceptional speed of tuna-like fishes is by many unique anatomic features. Describe them.
4. Summarize the common buoyancy devices used by pelagic marine animals.
5. Discuss the hypothetical advantages of vertical migration for mesopelagic species.
6. Review the mechanisms used by zooplankton to collect diffuse food.
7. What are some of the hypothetical advantages of schooling behavior?
8. Why do some marine animals migrate such enormous distances?
9. Summarize the hypothetical cues used by marine animals to orientation in space during their long migrations.
10. How do toothed whales echolocate?

Suggestions for Further Reading

Alldredge, A. L., and L. P. Madin. 1982. Pelagic tunicates: unique herbivores in the marine plankton. *Bioscience* 32:655–663.

Alldredge, A. L., and M. Silver. 1988. Characteristics, dynamics and significance of marine snow. *Progress in Oceanography* 20:41–82.

Azam, F., T. Fenchel, J. G. Gray, L. A. Meyer-Reil, and T. Thingstad. 1983. The ecological role of water column microbes in the sea. *Marine Ecology Progress Series* 10:257–265.

Block, B. A., J. R. Finnerty, A. F. R. Stewart, and J. Kidd. 1993. Evolution of endothermy in fish: mapping physiological traits on a molecular phylogeny. *Science* 260:210–214.

Block, B. A., and E. D. Stevens. 2001. *Tuna: Physiological Ecology and Evolution.* Academic Press, San Diego.

Briggs, J. C. 1974. *Marine Zoogeography.* McGraw-Hill, New York.

Bundy, M. H., and H. A. Vanderploeg. 2002. Detection and capture of inert particles by calanoid copepods: the role of the feeding current. *Journal of Plankton Research* 24:215–223.

Carey, G. F., J. M. Teal, J. W. Kanwisher, K. D. Lawson, and J. S. Beckett. 1971. Warm-bodied fish. *American Zoologist* 11:137–145.

Duró, A., and E. Saiz. 2000. Distribution and trophic ecology of chaetognaths in the western Mediterranean in relation to an inshore–offshore gradient. *Journal of Plankton Research* 22:339–361.

Emlet, R. B., and R. R. Strathman. 1985. Gravity, drag, and feeding currents of small zooplankton. *Science* 228: 1016–1017.

Enright, J. T. 1977. Diurnal vertical migration: adaptive significance and timing. I. Selective advantage: A metabolic model. *Limnology and Oceanography* 22:873–886.

Folt C. L., and C. W. Burns. 1999. Biological drivers of zooplankton patchiness. *Trends in Ecology and Evolution* 14:300–305.

Francis, L. 1991. Sailing downwind: aerodynamic performance of the *Velella* sail. *Journal of Experimental Biology* 158:117–132.

Frost, B. W. 1972. Effects of size and concentration of food particles on the feeding behavior of the marine planktonic copepod *Calanus pacificus. Limnology and Oceanography* 17:805–815.

Gibbons, M. J. 1993. Vertical migration and feeding of *Euphausia lucens* at two 72h stations in the southern Benguela upwelling region. *Marine Biology* 116:257–268.

Graham, W. M., S. MacIntyre, and A. L. Alldredge. 1999. Diel variations of marine snow concentration in surface waters and implications for particle flux in the sea—Monterey Bay, central California. *Deep Sea Research Part I: Oceanographic Research Papers* 47:367–395.

Hardy, A. 1971. *The Open Sea: Its Natural History. Part I: The World Of Plankton.* Houghton Mifflin, Boston.

Hastings, J. W. 1971. Light to hide by: ventral luminescence to camouflage the silhouette. *Science* 173:1016–1017.

Inada, Y., and K. Kawachi. 2002. Order and flexibility in the motion of fish schools. *Journal of Theoretical Biology* 214:371–387.

Lampitt. R. S., B. J. Bett, K. Kiriakoulakis, E. E. Popova, O. Ragueneau, A. Vangriesheim, and G. A. Wolff. 2001. Material supply to the abyssal seafloor in the Northeast Atlantic. *Progress in Oceanography* 50:27–63.

Le Boeuf, B. J., and R. M. Laws. 1994. *Elephant Seals: Population Ecology, Behavior, and Physiology*. University of California Press, Berkeley.

Lohmann, K. J., and C. M. F. Lohmann. 1998. Migratory guidance mechanisms in marine turtles. *Journal of Avian Biology* 29:585–596.

Martinussen, M. B., and U. Bamstedt. 1999. Nutritional ecology of gelatinous planktonic predators. Digestion rate in relation to type and amount of prey. *Journal of Experimental Marine Biology and Ecology* 232:61–84.

Mauchline, J., A. J. Southward, J. H. Blaxter, and P. Tyler. 1998. *Advances in Marine Biology: The Biology of Calanoid Copepods*. Academic Press, San Diego.

O'Dor, R. K., and D. M. Webber. 1986. The constraints on cephalopods: why squid aren't fish. *Canadian Journal of Zoology* 64:1591–1605.

Ohman, M. D. 1990. The demographic benefits of diel vertical migration by zooplankton. *Ecological Monographs* 60:257–281.

Paffenhöfer, G.-A., and M. G. Mazzocchi. 2003. Vertical distribution of subtropical epiplanktonic copepods. *Journal of Plankton Research* 25:1139–1156.

Piatt, J. F., and D. A. Methuen. 1993. Threshold foraging behavior of baleen whales. *Marine Ecology Progress Series* 84:205–210.

Purcell, J. E., and M. N. Arai. 2001. Interactions of pelagic cnidarians and ctenophores with fish: a review. *Hydrobiologia* 451:27–44.

Rome, L. C., D. Swank, and D. Corda. 1993. How fish power swimming. *Science* 261:340–343.

Ryer, C. H., and B. L. Olla. 1998. Shifting the balance between foraging and predator avoidance: the importance of food distribution for a schooling pelagic forager. *Environmental Biology of Fishes* 52:467–475.

Saito, H., and T. Kiørboe. 2001. Feeding rates in the chaetognath *Sagitta elegans*: effects of prey size, prey swimming behaviour and small-scale turbulence. *Journal of Plankton Research* 23:1385–1398.

Sanderson, S. L., and R. Wassersug. 1990. Suspension-feeding vertebrates. *Scientific American* 262:96–101.

Shanks, A. L. 2002. The abundance, vertical flux, and still-water and apparent sinking rates of marine snow in a shallow coastal water column. *Continental Shelf Research* 22:2045–2064.

Shanks, A. L., and M. L. Reeder. 1993. Reducing microzones and sulfide production in marine snow. *Marine Ecology Progress Series* 96:43–47.

Suzuki, K., T. Takagi, and T. Hiraishi. 2003. Video analysis of fish schooling behavior in finite space using a mathematical model. *Fisheries Research* 60:3–10.

Tamura T., and Y. Fujise. 2002. Geographical and seasonal changes of the prey species of minke whale in the Northwestern Pacific. *ICES Journal of Marine Science* 59:516–528.

Thomas, J. A., R. A. Kastelein, and A. Y. Supin. 1992. *Marine Mammal Sensory Systems*. Plenum Press, New York.

Todd, C. D., M. S. Laverak, and G. Boxshall. 1996. *Coastal Marine Zooplankton: A Practical Manual for Students*. Cambridge University Press, Cambridge.

Turner, J. T., A. Ianora, F. Esposito, Y. Carotenuto, and A. Miralto. 2002. Zooplankton feeding ecology: does a diet of *Phaeocystis* support good copepod grazing, survival, egg production and egg hatching success? *Journal of Plankton Research* 24:1185–1195.

Vourinen, I., M. Rajasilta, and J. Salo. 1983. Selective predation and habitat shift in a copepod species—support for the predation hypothesis. *Oecologia* 59:62–64.

Waggett, R., and J. Costello. 1999. Capture mechanisms used by the lobate ctenophore, *Mnemiopsis leidyi*, preying on the copepod *Acartia tonsa*. *Journal of Plankton Research* 21:2037–2052.

Webb, P. W. 1984. Form and function in fish swimming. *Scientific American* 251:72–84.

Yen, J., B. Sanderson, J. R. Strickler, and A. Okubo. 1991. Feeding currents and energy dissipation by *Euchaeta rimana*, a subtropical pelagic copepod. *Limnology and Oceanography* 36:362–369.

CHAPTER
11

Biologists deploy a bottom dredge to sample the inhabitants of the deep-sea floor.

CHAPTER OUTLINE

The Deep-Sea Floor

Beyond the continental shelves, the seafloor descends sharply down continental slopes to the perpetual dark and cold abyssal zone of the deep sea. Underlying the oceanic province of the world ocean are vast expanses of nearly flat and featureless abyssal plains and hills, plus a scattering of tectonically active ridges, rises, and trenches. Three fourths of the ocean bottom (the abyssal and hadal zones of Figure 1.46) lies at depths below 3000 m. A large portion of the deep-ocean basin consists of broad, flat, sediment-covered abyssal plains. Abyssal plains typically extend seaward from the bases of continental slopes and oceanic ridge and rise systems at depths between 3 and 5 km below the sea surface. The long narrow floors of marine trenches distributed around the margins of the deep-sea basins are generally deeper than 6 km. Trenches cover only 2% of the seafloor and are characterized by the most extreme pressure regimes experienced by living organisms anywhere on this planet.

Interrupting the continuity of abyssal plains are mountainous linear ridge and rise systems encircling the globe. The axes of these ridge and rise systems are the source of new oceanic crust, created as the global forces of plate tectonics rift plates of the Earth's crust apart. Growth of new crust is slow, about as fast as your fingernails grow, yet it causes about 60% of the Earth's surface (or 85% of the seafloor) to be recycled every 200 million years. Several remarkable discoveries associated with ridge and rise spreading centers have been made in the past two decades. These discoveries are radically changing our understanding of life in the deep sea and are clarifying how metal-rich mineral deposits are formed from chemical interaction between seawater and the Earth's crust.

11.1 Living Conditions on the Deep-Sea Floor

It is not possible to establish a precise global depth boundary between the animals of the deep sea and the shallow-water fauna of the continental shelves. Generally, the boundary exists as a vague region of transition on the continental slopes bordering the deep-sea basin. However, animals of the "deep sea" commonly extend into shallower water in polar seas and, on occasion, even extend to the inner portions of high-latitude continental shelves.

Most of the sea bottom is covered with thick accumulations of fine sediment particles, skeletons of planktonic organisms, and other debris (FIGURE 11.1) that have settled from the surface. Marine sediments are derived from several sources. A few minerals precipitate from their dissolved state in seawater to produce irregular deposits on the seafloor. Manganese nodules (FIGURE 11.2) are a well-known example of this type of deposit. Such deposits have potential commercial importance as a source of minerals, but they also may have an important influence on the structure of some benthic communities, and their removal in mining operations may cause catastrophic mortality in harvest areas.

Sedimentary materials found in the deep-ocean basins away from continental margins are composed of the mineralized skeletal remains of planktonic or-

figure 11.2

Manganese nodules scattered on the surface of the seafloor in the Pacific Ocean.

ganisms. These deposits, known as **oozes**, are characterized by their chemical composition. Siliceous oozes contain cell walls of diatoms and the internal silicate skeletons of planktonic radiolarians (see Figure 5.3). The skeletons of other planktonic protozoans, the foraminiferans, constitute most of the extensive calcareous oozes found on the ocean floor. These oceanic oozes accumulate very slowly, approximately 1 cm of new sediment every 1000 years. In the deep sea, oceanic oozes exhibit distributional patterns that reflect the surface abundance of their biologic sources. These patterns are shown in FIGURE 11.3.

The deep-sea bottom is one of the most rigorous and inaccessible environments on Earth. Below 3000 m, the water is cold, averaging 2°C and dipping slightly below 0°C in polar regions (TABLE 11.1). Water temperatures at these depths vary little on time scales of years to decades, creating an extremely constant thermal environment for the inhabitants of the deep sea. Pressures created by the overlying water are tremendous, ranging from 300 to 600 atmospheres (atm) on the abyssal seafloor and exceeding 1000 atm in the deepest trenches. Laboratory studies have confirmed that metabolic rates of deep-sea bacteria are lower at pressures normally experienced on the seafloor than they are at sea surface pressures. Less clear is the response of multicellular organisms to high pressures. Several studies have suggested that pressure-induced reductions in metabolic rates may lead to lowered growth rates, lowered reproductive rates, and increased

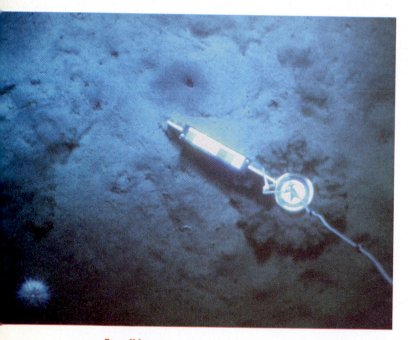

figure 11.1

Fine-grained bottom sediments off the Oregon coast disturbed by the impact of a current-direction indicator.

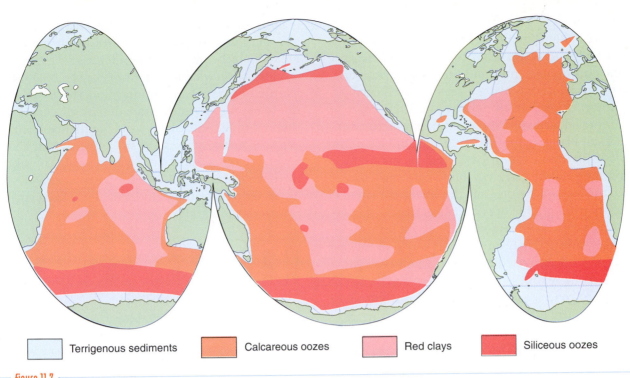

| Terrigenous sediments | Calcareous oozes | Red clays | Siliceous oozes |

figure 11.3

Distribution of ocean-bottom sediments. (Adapted from Tait [1968] and Sverdrup, Johnson, and Fleming [1942].)

table 11.1

Characteristics of a Typical Abyssal-Plain Habitat at 3000 m

Water pressure	300 atm
Water temperature	1–2°C
Salinity	34.5–35‰
Dissolved oxygen	5 ppm
Light	Bioluminescence only
Current speed	Slow, <1 cm/s or 0.7 km/day
Sediment	
Type	Soft fine oozes or clay
Deposition rate	<0.01 mm/yr
Organic content	0–0.5%

figure 11.4

A group of large crustacean amphipods feeding on bait in the Peru-Chile Trench at a depth of 7000 m. For scale, the cable at the right is just over 6 cm in diameter.

life spans in the deep sea, culminating in occasional examples of deep-sea gigantism (FIGURE 11.4). Other recent studies have found that the depth-related decline in metabolic rates of crustaceans can be explained as metabolic adjustments to temperature declines with increasing depth and not to a separate depth or pressure effect. Because the effects of high pressures on growth rates and maximum sizes of deep-sea animals are not yet very well understood, this question will not be resolved without more study.

The effects of intense pressure are greatest in trenches, where water depths typically exceed 6000 m.

Individual trenches exist as "islands" of extreme-pressure habitats for animal life, isolated from each other by broad expanses of abyssal plains. Because of the enormous depths and pressures characterizing trenches, studies of deep-sea life have been fragmentary at best, inhibited by the high costs of sampling so far below the surface and by the difficulties of bringing healthy deep-sea animals to the surface for study. Captured animals encounter extreme temperature and pressure changes when hauled from the bottom. By the time they reach the surface, they are usually dead or seriously damaged. Consequently, we presently know more about features on the back side of Earth's moon than we do about vast areas of its deep-sea floor. In fact, we have visited Venus about 30 times, but we have been to the deepest part of the ocean only twice (and the bottom of the deepest trench is just 11 km from the surface).

Most sampling of living animals from the deep-sea floor is accomplished with dredges, trawls, or sampling scoops (FIGURE 11.5) operated from surface ships at the ends of extremely long cables. It is a bit like trying to sample animals of a rain forest by blindly trailing a scoop on a rope from a balloon 4000 m above the canopy. Some luck is involved, and the sample returned to the ship is typically biased against animals fast enough to evade the sampler and those too fragile to survive the sampling experience and the trip to the surface. In the last quarter century, this kind of blind remote sampling has been supplemented with direct and video observations from manned submersibles and unmanned remotely operated vehicles.

Despite the difficulties inherent in sampling the deep-sea floor, two general and contrasting pictures of animal life there are emerging. Abyssal plains are vast areas of soft mud habitats generally deeper that 3000 m and extending over distances measured in hundreds or even thousands of kilometers. They are sparsely populated by animal communities with low biomass but surprisingly high species richness. These animals are completely dependent on the rain of energy-rich material from the photic zone far above. In almost complete contrast to abyssal plains are deep-sea hydrothermal vent communities. They occupy small and localized rocky habitats on the axes of ridge and rise systems and are characterized by a relatively small number of species existing in an almost overwhelming abundance of biomass. One feature of these vent communities that absolutely floored biologists at the time of their discovery was their complete nutritional independence from the primary productivity of the surface waters far above. Several other types of small vent or seep communities have since been discovered, also existing in similar states of nutritional independence from surface primary productivity.

11.2 Transfer of Oxygen and Energy from the Epipelagic Zone to the Deep Sea

The nonvent animal communities of the deep sea rely on phytoplankton in the photic zone far above for their oxygen and their food. Dissolved O_2 needed to support the metabolic activities of deep-sea animals must come from the surface. Diffusion and sinking of cold dense water masses are the chief mechanisms of O_2 transport into the deep-ocean basin, with diffusion providing O_2 from the surface downward and sinking water masses filling the deeper ocean basins with O_2-rich water. As it diffuses downward from surface waters, dissolved O_2 is slowly diminished by animals and bacteria, leaving an O_2 minimum zone at intermediate depths (usually between

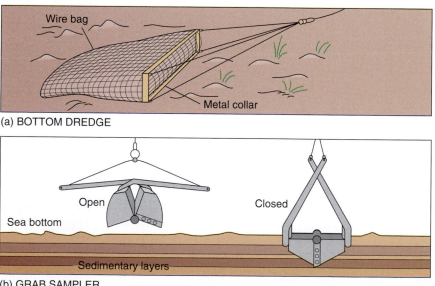

(a) BOTTOM DREDGE

(b) GRAB SAMPLER

figure 11.5

Two types of seafloor samplers: (a) bottom dredge, which skims the surface of the sediment, and (b) grab sampler, which removes a quantitative "bite" of sediment and its inhabitants.

MAY 1

JUNE 29

figure 11.6

Seafloor images showing the deposition of phytodetritus before (a) and 2 months after (b) a phytoplankton bloom in the photic zone above.

1000 and 2000 m; see Figure 1.32). Below this zone, dissolved O_2 gradually increases to just above the sea bottom. The presence of the O_2 minimum zone reduces the abundance and activity of mid-depth consumers. One study in the Pacific of a steep-sided volcanic seamount that penetrated the O_2 minimum zone showed that the number and abundance of benthic species present were much greater just below the O_2 minimum zone than within it.

The other crucial and variable resource in the deep sea is food. Because there is no sunlight, there are no photosynthesizers. Food for deep-sea benthic communities, by necessity, comes from above. Only recently have studies determined the rate and condition of the food's arrival at the sea bottom. Several studies have demonstrated that, contrary to earlier assumptions, a tight coupling exists between near-surface primary productivity and the eventual consumption of that production by inhabitants of the deep-sea floor (FIGURE 11.6). In most areas of the deep ocean, seasonal variations in the organic content of surface sediments reflect the seasonality of primary production occurring in the photic zone directly above. Such seasonal pulses of organic matter are often detectable several centimeters below the sediment surface as well, as infaunal animals rework this material through their burrowing and feeding activities.

Sinking rates of food-rich particles are largely a function of their size, with smaller particles sinking more slowly. As they sink slowly through the photic zone, much of the near-surface phytoplankton are aggregated into larger gelatinous blobs or are compacted into fecal pellets of zooplankton, which settle rapidly to the bottom. This fallout of fecal particles and gelatinous aggregates accelerates the transport of organic material to the abyss, falling from the surface waters in a few days rather than the weeks or months of settling time necessary for smaller phytoplankton particles. Thus, the time lag between the initial production of organic material in the photic zone and its arrival on the deep-sea floor may be a few days or as much as several weeks. On the way down, sinking particles of organic material are repeatedly consumed by pelagic scavengers and colonized by bacteria. Consequently, sinking particles, especially the very small ones, lose much of their nutritive value by the time they settle to the bottom. The amount of photic zone productivity that actually reaches the sea bottom is typically only a few percent, and it is even less at greater depths.

11.3 Life on Abyssal Plains

Abyssal plains are flat and stable places occasionally interrupted by low abyssal hills. They are inhabited by a distinctive group of animals, although few exhibit structural adaptations that make them appear notably different from their shallow-water relatives. Instead, a shift in dominant taxonomic groups occurs in deeper water. Echinoderms (especially sea cucumbers and crinoids), polychaete worms, pycnogonids, and isopod and amphipod crustaceans become abundant, whereas bivalves and other mollusks and sea stars decline in number. Some taxonomic groups are virtually absent until relatively great depths are reached. Most species of pogonophorans, for instance, are found below 3000 m, and 30% of known species are restricted to trenches below 5000 m.

Low temperatures, high pressures, and a limited food supply led to an early and widespread belief that the rigorous and specialized climate of the deep sea would not support a highly diversified assemblage of animals. It was assumed that only a few highly adapted animals could succeed in the abyss. Repeated sampling did reveal a marked decline in both density and biomass of organisms at greater depths, both presumed responses to the limiting effects of decreasing food and O_2 with depth. However, improvements in sampling equipment and sample analysis techniques in the past quarter century have shown that deep-sea assemblages contain a diversity of species comparable with or even exceeding the diversity of soft-bottom communities in shallow inshore waters.

Using an "epibenthic sledge," a sampling device combining the best attributes of a bottom trawl and a dredge (FIGURE 11.7), Hessler and Sanders demonstrated an increase in species diversity with increasing depth along a transect from shallow New England coastal waters to deep-ocean basins near Bermuda nearly 5000 m deep. Other researchers studying different deep-sea regions have not always found this pattern of increasing diversity with increasing depth; it seems to be strongly dependent on the choice of study locations. Collectively, these studies dispel the basic tenet first established by the *Challenger* Expedition almost a century earlier that animal diversity decreases with increasing depth in the deep sea (TABLE 11.2). At the same time, these studies confirm another *Challenger* observation: that the biomass of animal life in the sea becomes more sparse at greater depths. So although species diversity is surprisingly high at great depths, the average number of individuals and their average body sizes do tend to decrease consistently with depth. Yet so much of the deep-sea floor remains unsampled that we might yet find thousands more species if it is ever explored with the intensity that we currently direct to more accessible ecosystems, such as tropical rain forests.

To explain the relatively high species diversity of abyssal plains (FIGURE 11.8), researchers have proposed several complementary explanations. First, the enormous extent of abyssal plains (about 200 million km^2) provides large areas with few barriers to block dispersal, so species exhibit broad geographic ranges. When large areas of the deep sea are sampled at the same depth, approximately one additional species is added

figure 11.7

An epibenthic sledge for sampling epifauna and near-surface infauna.

table 11.2

Diversity of Major Animal Phyla (Indicated by Number of Species and Families) Collected from a Small Sampling Area at a Depth of 2100 m Off the New England Coast

Phylum	No. of Families	No. of Species
Annelida	49	385
Arthropoda	40	185
Mollusca	43	106
Echinodermata	13	39
Cnidaria	10	19
Pogonophora	5	13
Sipuncula	3	15
Echiura	2	4
Nemertea	1	22
Hemichordata	1	4
Priapulida	1	2
Brachiopoda	1	2
Bryozoa	1	1
Chordata	1	1

Adapted from Gage, J. D., and P. A. Tyler. 1991.

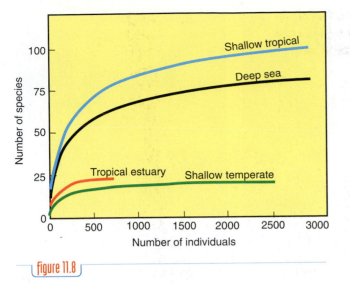

Comparison of deep-sea species diversity (for polychaete annelids and bivalve mollusks) with three other marine environments. (Adapted from Sanders, 1968.)

for each additional square kilometer sampled. When different depths are sampled, the rate at which additional species are acquired is even greater. These trends suggest very high species diversity in the deep sea and also point out one of the basic difficulties in clarifying the pattern of that diversity: Enormous areas must be sampled (at enormous costs) before we have an accurate picture of the number and distribution of species in the deep sea.

Second, within large areas of nearly constant deep-sea environments, small-scale variations in food availability and physical or biologic disturbance create new opportunities for colonizing by additional species. Events such as food-falls of large animal carcasses or the sorting and modification of sediment deposits when animals form tubes and compact sediments to produce fecal pellets and castings may serve to maintain some spatial diversity in what otherwise seems a very uniform environment.

A portion of the smaller particles of organic material settling to the bottom is immediately claimed by suspension feeders. These are often epifauna clinging to rocky outcrops, shells, or manganese nodules. Other epifauna use stilt-like appendages to keep their bodies from sinking into the sediments (see Figure 5.31). Most benthic animals in the deep sea, however, are infaunal deposit feeders. Analyses of their stomach contents indicate that many infauna indiscriminately engulf sediments containing smaller in-

fauna and bacteria as well as dead organic material. It has been estimated that 30–40% of the organic material available at the bottom is first absorbed by benthic bacteria which, in turn, are consumed by larger animals. An important component of the organic material is the chitinous exoskeletons shed by planktonic crustaceans; these cannot be digested by most animals but are decomposed and used by bacteria.

Large rapidly sinking particles, including dead squids, fishes, and an occasional whale or seal, may provide a significant, although unpredictable, supply of food to inhabitants of the deep-sea floor. When large food parcels do contribute significantly to the energy budget of the deep sea, they are rapidly consumed by various wide-ranging scavenger–predators. Most scavenger–predators seem swift enough to evade bottom trawls and are best known from photographs taken by underwater cameras baited with food (FIGURE 11.9). They seem to be food generalists, capable of rapidly locating and dispatching large food items as they arrive on the bottom. Some of the food is eventually dispersed as detritus and fecal wastes to other benthic consumers.

In abyssal plains, as in shallow-water soft-bottom communities, **deposit feeding** is common. A variety of worms, mollusks, echinoderms, and crustaceans obtains their nourishment by ingesting accumulated detritus and digesting its organic material. These deposit feeders engulf sediments and process them through their digestive tracts. They extract nourishment from the organic material in the sediment in much the same manner as earthworms do. Some deposit feeders indiscriminately ingest any available sediments; others select organically rich substrates for consumption.

Several animal species are capable of extracting sufficient nourishment from sediments by conducting digestive processes outside their bodies. These animals absorb the products of digestion either through specialized organs or across the general body wall. Deep-sea pogonophorans, small worms related to the large tube worms recently discovered in association with deep-sea hot springs (see Figure 11.13), depend on this type of **absorptive feeding**, as do numerous echinoderms. Sea stars are usually carnivorous, but a few species are quite opportunistic. When feeding on bottom sediments, some sea stars extrude their stomachs outside their bodies to digest and absorb organic matter from the sediments. The omnipresent bacteria also depend on extracellular digestion. As they absorb nutrients and their population grows, they in turn become a significant source of particulate food for deposit feeders.

The term **cropper** has been applied to deep-sea animals that have merged the roles of predator and deposit feeder (FIGURE 11.10). These croppers, by preying heavily on populations of smaller deposit feeders and bacteria, may be responsible for reducing competition for food and for encouraging coexistence between several species sharing the same food resource. Thus, although a deep-sea community as a whole is food limited, populations within the community need not be as long as they are heavily preyed upon. With competition for food reduced by cropping, fewer species are pushed to extinction (on a local scale) and species diversity remains high. The disruptive effects of large croppers on smaller ones can be compared with predators, ice, or logs as disturbance mechanisms in rocky intertidal communities. By nonselectively reducing competition, they lessen the possibility of a species being excluded by resource competition.

In the physically stable environment of the deep sea, patterns of reproduction are thought to differ appreciably from the reproduction patterns of shallow-water benthic animals. Few deep-water benthic species produce planktonic larvae because the chances of larvae reaching the food-rich photic zone several kilometers above and successfully returning to the ocean floor for permanent settlement are extremely remote.

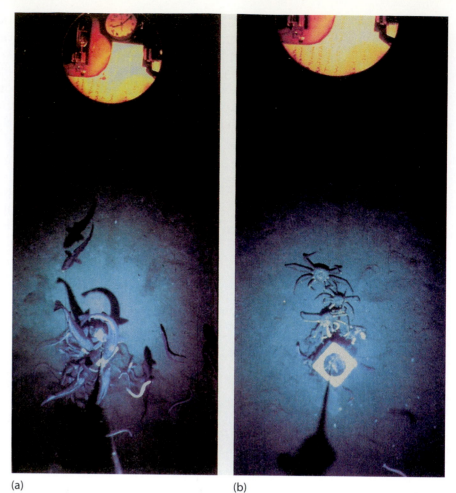

(a)   (b)

figure 11.9

A bait can lowered to the seafloor at a depth of 1390 m off the northern Baja California coast quickly attracts mobile scavenging fishes (a). Several hours later (b), the same scene is dominated by slower invertebrates, including the tanner crabs seen here.

figure 11.10

An aggregation of deep-sea cucumbers, *Scotoplanes,* and scattered brittle stars feeding on bottom deposits at a depth of approximately 1000 m.

To compensate for the absence of a dependable external food supply, fewer and larger (lecithotrophic) eggs are produced. The larvae are supplied with adequate yolks so they can develop to fairly advanced stages before hatching. Brood pouches and other similar adaptations further enhance the chances of survival by protecting the eggs until they hatch. Consequently, the larvae have a reasonable chance of completing their development in one of the most severe environments on Earth.

11.4 Vent and Seep Communities

The dispersed nature of mostly small animals on and in soft sediments of abyssal plains contrasts sharply with the compact, even exuberant, nature of deep-sea vent and seep communities. Vent communities are organized around hot-water outflows vented from cracks in the seafloor. Most are located on the axes of oceanic ridge or rise systems; a few others are associated with oceanic island-arc systems. Seep communities tend to be more dispersed in areas where hydrocarbons, particularly methane or other natural gases, are percolating up through sediments to the surface of the seafloor (FIGURE 11.11).

Hydrothermal Vent Communities

The discovery of deep-sea animals intimately associated with hydrothermal vents has radically changed our understanding of life in the deep sea. The hydrothermal vents at the centers of these communities are the geologic plumbing that channel hot metal-enriched water from deeper within the crust to the surface of the seafloor. Because most of these vents are located on the axes of seafloor ridge or rise structures, they exist in very young rocks that form as crustal plates spreading apart (see Figure 1.6). Rising molten magma cools, solidifies, and creates a new oceanic crust made of black volcanic basalt. If you could approach a hydrothermal vent from an adjacent abyssal plain, you would observe the fine sediment blanket that covers the basaltic ocean crust thinning until it disappears at the ridge or rise axis, exposing the rough rugged surfaces of freshly cooled and solidified basalt. It is this zero-age oceanic crust, too young to have accumulated any sediment, that creates conditions for the development of hydrothermal vents and the animal and bacterial communities associated with them (FIGURE 11.12)

Deep-sea hydrothermal vent communities were first directly observed in 1977 when a team

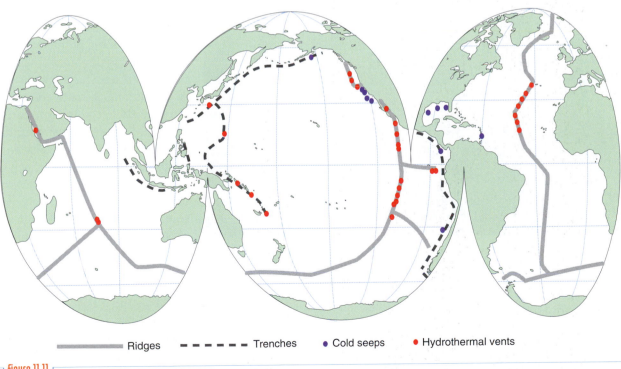

figure 11.11

Approximate locations of confirmed hydrothermal vent communities (red dots) and cold seeps (blue dots). Most are associated with trenches or with actively spreading ridge systems.

Ridges — Trenches - - - - • Cold seeps • Hydrothermal vents

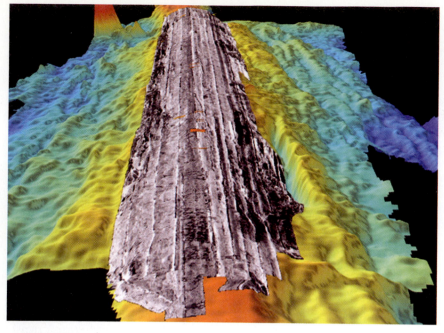

figure 11.12

Sidescan sonar image (gray scale) overlaid onto multibeam bathymetry (color) for the East Pacific Rise (9° 25'N to 9° 57'N) at a depth of 2500 m. The sidescan image covers an area about 31 × 7 km and resolves seafloor features that are as small as 2 m.

figure 11.13

Red-plumed tube worms, *Riftia*, with a few other members of this unusual deep-sea community.

in small areas of shimmering hot water pouring from the seafloor. Water temperature often exceeds 100°C at vent sites, in sharp contrast to the near-uniform 2°C abyssal water just a few meters away. Even though some of these vent animals are adapted to high water temperatures, it is the altered chemistry of the water, not the heat, that provides the basis of life in vent communities.

These hot-water plumes pouring from seafloor vents are the end products of seawater circulating through the many cracks and fissures of the new crust as it forms along the axes of the rise system (FIGURE 11.14). It has been estimated that a volume of water equivalent to the volume of all Earth's oceans circulates through these ridge crack systems approximately every 8 million years. As seawater percolates through these cracks, crustal rock temperatures above 350°C drive some complex chemical reactions to alter several properties of the circulating seawater. In addition to being heated well above the boiling point at sea level, vent water is devoid of dissolved O_2, is very acidic, and is enriched in metals such as manganese, iron, copper, zinc, and silicon. Sulfate (SO_4^{2-}), the third most abundant ion in seawater, reacts with water when heated to form hydrogen sulfide (H_2S). Despite being toxic to most animals, H_2S plays a central role in subsequent biologic processes as the heated water emerges from the seafloor vent.

Some of the H_2S reacts with iron and other metal ions to form metal sulfides in concentrations several million times greater than those found in average seawater. At the vent opening, this super-enriched metal sulfide solution quickly mixes with cold surrounding seawater, and a dense black "smoke" of iron sulfide particles is produced (FIGURE 11.15). The massive quantities of black particles cre-

of geologists studying the Galapagos Rift Zone west of Ecuador discovered several remarkable assemblages of animals living in water nearly 3 km deep (FIGURE 11.13). Dense aggregations of large mussels, clams, giant tube worms, and crabs were found clustered

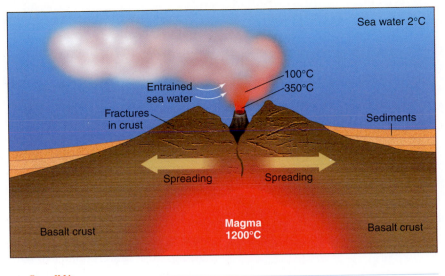

figure 11.14

Cross-section of a ridge axis and the plumbing connected to a vent chimney. Seawater flows through numerous fissures in the hot basalt crust of the seafloor.

figure 11.15

A black smoker on the Galapagos Rift Zone. Particles in the "smoke" are major sources of metal deposits around vent chimneys.

ated at chimney openings atop these vents are known as "black smokers." These particles either settle immediately around the vent or are oxidized by oxygen from surrounding nonvent water to produce thick deposits of metal oxides in the vicinity of the vent.

Mixing of vent and nonvent waters in proximity of vent chimneys creates strong gradients in dissolved O_2, pH, water temperature, and H_2S extending outward from the eye of the vent. Clouds of free-living bacteria, so abundant near vent openings that they make the water appear milky, are important primary producers in hydrothermal vent communities. This form of primary production, however, is a form of **chemosynthesis** rather than photosynthesis characteristic of the sunlit portions of the ocean. Like near-surface primary producers, these bacterial communities also experience large cyclic fluctuations in productivity rates. Rather than being seasonally driven as are phytoplankton communities, vent bacteria periodically thrive and bloom in response to changes in flow rates of vent water in response to local shifts in crustal plates.

Vent bacteria use dissolved O_2 from the surrounding water to oxidize the abundant H_2S back to SO_4^{2-} to obtain energy to synthesize complex and energy-rich organic compounds (FIGURE 11.16). In this manner, bacterial primary production occurs locally, and the vent community is independent of surface productivity for their food. Vent-based chemosynthesis is not totally independent of photosynthesis at the sea surface, because the O_2 used in chemosynthesis must originate from phytoplankton in the photic zone.

Bacterial chemosynthesis is not a newly discovered form of primary productivity; it was first described in the late 19th century. Vent-based chemosynthesis, however, is unusual in that it is the principal energy source for the entire base of hydrothermal vent food webs. The unusual chemosynthetic bacteria of hydrothermal vent communities are the focus of intensive biochemical, physiologic, and genetic studies. It is generally agreed that several species of bacteria tolerate extremely high water temperatures (100–115°C) and that these "thermophiles" belong to the division Archeobacteria (see Chapter 2 and Figure 2.7). The most abundant species of chemosynthetic bacteria have adapted to one of two markedly different vent community habitats: as suspended bacteria in and around black-smoker fluids or as symbiotic bacteria living on or within tissues of vent-dwelling animals.

Suspended bacteria may be consumed by dense aggregations of large clams and mussels found nowhere else in the sea (FIGURE 11.17). By trapping bacteria on their mucus-lined gills (see Figure 5.24) as do

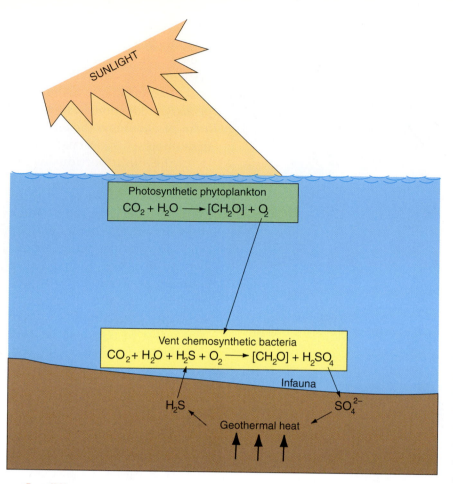

figure 11.16

Comparison of primary production in photosynthetic and chemosynthetic systems. Transfer of O_2 from the surface to vent bacteria materially link these two marine primary-production systems.

their shallow-water suspension-feeding relatives, these large vent-community bivalves may capitalize on an abundant, yet extremely small, supply of food particles. Some evidence also exists that these vent bivalves also maintain endosymbiotic chemosynthetic bacteria within their gills and harvest them much as giant clams do in coral-reef communities (see p. 264).

The large red-plumed vestimentiferid worm, *Riftia*, shown in Figure 11.13, completely lacks a digestive tract and is unable to harvest suspended bacteria. Instead, it maintains internal bacterial symbionts (endosymbionts) that oxidize H_2S within a specialized organ, the **trophosome**. Major blood vessels link the trophosome to the bright red plume, where gas exchange occurs (FIGURE 11.18). The blood of *Riftia* contains two proteins to deal with the high concentrations of dissolved H_2S and small amounts of dissolved O_2 in the deep-sea hot-spring environment. The first protein is a rare sulfide-binding protein used to transport large quantities of H_2S from the plume, where it is absorbed, to the trophosome, where it is used by bacterial symbionts. Hemoglobin, the other blood protein, is used to bind and transport O_2 within these animals. By binding O_2 and H_2S to different blood proteins, the high toxicity usually associated with H_2S is avoided and the O_2 is prevented from prematurely oxidizing the H_2S before both are delivered to the endosymbionts in the trophosome.

figure 11.17

Aggregations of large vent clams, *Calyptogena*, with a clump of *Riftia* tube worms in the background.

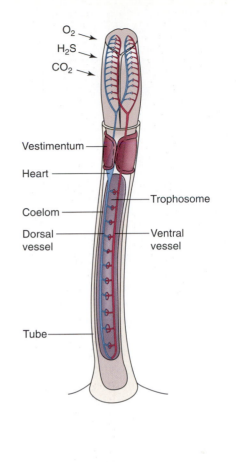

figure 11.18

External appearance (left) and internal anatomy (right) of the tube worm, *Riftia*.

In these deep-sea vent communities, bacterial use of O_2 from the water in the immediate vicinity of the vents is extensive. To compete effectively with bacteria for the limited available O_2 and to sequester O_2 away from H_2S, larger animals need some type of blood pigment with a strong affinity for O_2. Hemoglobin is abundant in *Riftia* and in large bivalve mollusks, including the large clam, *Calyptogena* (Figure 11.17). In the crabs most commonly found in vent communities (*Bythograea*, FIGURE 11.19), the blue pigment hemocyanin is used instead. The blood pigments of these vent animals are relatively insensitive to changing temperatures. Such thermal insensitivity is unusual and likely has evolved to accommodate the extremely wide range of water temperatures experienced over very short distances around vents. These pigments thus assist the larger deep-sea vent animals to extract the little O_2 left over by the abundant bacteria. With these adaptations, these populations of benthic animals have tapped into an unusual energy

figure 11.19

Aggregations of large vent crabs, *Bythograea*, with blue-pigmented blood visible through their carapaces.

Deep Submersibles for Seafloor Studies

For humans to explore the deeper parts of the seafloor, they must either use remotely operated unmanned vehicles (ROVs) or work from inside submersibles designed to protect human passengers from the enormous pressures and to provide oxygen, light, and other life support. Both ROVs and manned submersibles have their roles in deep-sea research. In this sense, there are close parallels between the vessels we use to explore the deep ocean and the ones used to explore the far reaches of space.

The best-known manned deep submersible vehicle used for ocean research is a high-tech but awkward-looking 7-m-long minisubmarine dubbed *Alvin,* named for Allyn Vine of Woods Hole Oceanographic Insti-

tution (WHOI), one of the early proponents for a deep submergence vehicle for seafloor studies. *Alvin* was delivered to Woods Hole in 1964. The submarine's first tender was crafted from a pair of surplus Navy pontoons and was named *Lulu* after Al Vine's mother. *Alvin* has room inside its 3-m-diameter titanium hull for a pilot and two researcher/passengers and for a myriad of sampling and recording instruments outside its hull (Figure B11.1). Typical dives to *Alvin*'s maximum depth limit of 4500 m takes about 8 hours. For the past 40 years, it has been the workhorse of manned deep-sea research efforts in the United States. A few highlights of its long and successful career are described here.

The first real test of the underwater capabilities of *Alvin* came in 1966, when a U.S. Air Force B-52 collided with an air tanker over Spain and dropped one of its H-bombs in the Mediterranean off the Spanish coast. The bomb was located and then lost during an attempt to attach lift lines. The bomb was relocated 2 weeks later and eventually recovered by a military submersible.

During the launch for Dive 308 in 1968, *Alvin*'s support cables failed

and *Alvin* sank to the seafloor in 1500 m of water. *Alvin* remained on the bottom until recovered in late 1969 by another submersible, the deep submergence vehicle *Aluminaut*. The sinking caused *Alvin* very little structural damage. Lunches left behind by the rapidly departing crew almost a year earlier were soggy but otherwise unchanged because of the near-freezing temperatures and absence of oxygen in the deep sea.

In 1977, *Lulu* carried *Alvin* through the Panama Canal and into the Pacific Ocean for the first time. Near the Galapagos Islands, *Alvin* discovered abundant, and previously unknown, animal life closely associated with hydrothermal vents 2500 m below the sea surface (see p. 343). This discovery ushered in entirely new areas of research in marine sciences.

By 1980, *Alvin* had completed its 1000th dive during another visit to the Galapagos Rift Zone. Three years later, *Alvin* was refitted to operate from a new support ship, the research vessel *Atlantis II*. With its new mother ship, *Alvin* could operate in higher latitudes and in rougher weather. In 1986, *Alvin* made 12 dives to the wreck of RMS *Titanic* in the

source—the heat and chemicals of the Earth's crust—and have achieved a nutritional emancipation from the fall-out products of the photic zone above.

Larval Dispersal of Hydrothermal Vent Species

How are newly formed hydrothermal vents initially colonized when the nearest vent community with reproductively mature animals may be hundreds of miles distant? Little is known about the spawning behaviors or early developmental stages of most vent animal species. *Riftia* is perhaps the best studied of these, and it is still not known if fertilization is internal or

external in this tube worm. Free-swimming larvae of most vent community members remain elusive, and larval durations (if any) are unknown. Reproductive strategies of members of these small and isolated vent communities must somehow balance the demands of retaining some of their progeny within their existing vent communities with the conflicting need to disperse larval forms to colonize newly formed vents.

Several "bridging" mechanisms to disperse larvae and explain the presence of the same species at different hot springs have been proposed. Larvae may drift passively with bottom currents that flow parallel to ridge or rise axes, eventually reaching a new settlement site. Alternatively, nonswimming larvae

North Atlantic to test a new ROV called *Jason Jr.* and to document the wreck photographically.

New calculations and tests made in 1994 (*Alvin's* 30th birthday) demonstrated that *Alvin's* titanium pressure hull was capable of withstanding pressures to the current limit of 4500 m. That increase of its depth rating of 500 m allows *Alvin* to explore an additional 20% of the seafloor. By this time, every individual part of the original *Alvin* has been replaced a piece at a time.

Figure B11.1 *Alvin.*

Alvin's support ship, *Atlantis II,* was retired in 1996 and was replaced with a new *Atlantis,* outfitted with capabilities for supporting a new generation of ROVs to operate in conjunction with *Alvin.*

Jason II is one of a new generation of ROV capable of routine operation to depths of 6500 m and communicating its data back to shore via the Internet. The others are the towed imaging system *Argo II* and the DSL-120 side-scan sonar system. All three systems are part of the U.S. National Deep Submergence Facility operated by WHOI for the U.S. ocean sciences community. These are the ROVs the new *Atlantis* was designed for, allowing investigators to use the complete suite of manned and unmanned underwater tools from the same platform anywhere in the world.

For future work, a concept development study for building a new *Alvin* has been funded. The planned

Deep Submersibles for Seafloor Studies

capabilities and features of a new *Alvin* will include increased depth capability and bottom time, improved fields of view for researchers and pilot, increased science payload, and reduced physical and chemical disturbances to science study areas. These improvements should produce a new submersible with increased scientific and operational capabilities, including greatly improved access to most of the abyssal regions of the world's oceans.

www.bioscience.jbpub.com/ marinebiology
For more information on this topic, go to this book's website at http://www.bioscience.jbpub.com/ marinebiology.

---

may be transported as hitchhikers on other community members capable of walking (crabs) or swimming (fishes) between vent sites.

Another intriguing possibility came to light as an outgrowth of a modest observation made off the California coast in 1987 from the deep-sea research submersible *Alvin.* A decomposing, but still identifiable, whale carcass on the seafloor was surrounded by numerous clams of two species previously seen only at vent sites. Dense mats of sulfide bacteria were also found. This suggests that hot-spring community members may "hop" from carcass to carcass, eventually moving long distances along the sea bottom. Such larval hops, if they occur, may be assisted by the vertical rise and eventual horizontal drift of warm plume water from an active vent. It is not yet known if such "whalefalls" occur frequently and closely enough to provide the larval-hopping network of food-rich material necessary to bridge the distances that exist between these remarkable deep-sea communities.

Cold-Seep Communities

As the search for more deep-sea hydrothermal vent communities continues in all the major ocean basins, a scattering of other densely populated animal communities dependent on chemosynthetic bacteria are being located (Figure 11.11). In addition to whale-falls, these sites include cold-water brine seeps in the Gulf of Mexico and in several crustal subduction zones, methane seeps in several ocean basins, and earthquake-disturbed sediments of the Laurentian fans in the eastern North Atlantic. All these sites support localized dense communities of benthic animals and chemosynthetic bacteria similar to those found at deep-sea hydrothermal vent sites. All these sites echo the fundamental biologic theme of hydrothermal vent sites: It is the generation of sulfide by whatever means available that supports chemosynthetic bacteria as the community's primary producers. These unexpected discoveries suggest that we have only begun to understand the biology of this rigorous and fascinating environment.

STUDY GUIDE

Summary Points

Living Conditions on the Deep-Sea Floor

- The deep-sea floor descends sharply down continental slopes to the dark, cold, and featureless abyssal plains, which are punctuating only by a scattering of tectonically active ridges, rises, and trenches.
- Most of the seafloor is covered with thick accumulations of fine sediment particles, mineralized skeletal remains of planktonic organisms, known as oozes, that accumulate very slowly (about 1 cm/1000 yr).

Transfer of Oxygen and Energy from the Epipelagic Zone to the Deep Sea

- The diffusion and sinking of cold dense water masses are the chief mechanisms of O_2 transport into the deep sea, although dissolved O_2 is slowly diminished by animals and bacteria, leaving an O_2 minimum zone at intermediate depths. Below this zone, dissolved O_2 gradually increases to just above the sea bottom.
- Food for deep-sea benthic communities sinks from above at rates that are tightly coupled with primary productivity at the sunlit surface.

Hence, the seasonality of primary production occurring in the photic zone directly above results in seasonal pulses of organics supplied to the deep sea.
- Sinking rates of food-rich particles correlate well with size and are accelerated after the phytoplankton are aggregated into larger gelatinous blobs or are compacted into fecal pellets of zooplankton. Thus, the time lag between the initial production of organic material in the photic zone and its arrival on the deep-sea floor is greatly shortened, although smaller sinking particles lose much of their nutritive value by the time they settle to the bottom.

Life on Abyssal Plains

- A shift in dominant taxonomic groups occurs in deeper water, with echinoderms, polychaete worms, pycnogonids, and isopod and amphipod crustaceans becoming abundant, whereas mollusks and sea stars decline in number.
- Although both density and biomass of organisms decline markedly at greater depths, species diversity is comparable with or even exceeds that of soft-bottom communities in shallow inshore waters.
- Although rapidly sinking dead squids, fishes, and an occasional whale or seal may provide a significant if unpredictable supply of food and a portion of the smaller particles settling to the bottom is immediately claimed by suspension feeders, most benthic animals in the deep sea are infaunal deposit feeders. Deposit feeders engulf sediments and process them through their digestive tracts, extracting nourishment from the sediment in much the same manner as earthworms.

- Additional deep-sea feeding strategies include animals that absorb the products of digestion either through specialized organs or across the general body wall and croppers that have merged the roles of predator and deposit feeder by preying heavily on populations of smaller deposit feeders and bacteria.

Vent and Seep Communities

- Deep-sea hot springs, recently discovered along the axes of ridge and rise systems, support unique communities of deep-sea animals. Dissolved H_2S emerging from seafloor cracks is used as an energy source by chemosynthetic bacteria, which are the source of nutrition for dense populations of the unique animals clustered around these springs.
- Additional densely populated animal communities dependent on chemosynthetic bacteria, including cold-water brine seeps, methane seeps, and earthquake-disturbed sediments of the Laurentian fans, also have been discovered.

Topics for Discussion and Review

1. Summarize ambient conditions present on the seafloor at a depth of 3000 m.
2. Characterize the sediments of the deep sea in terms of composition, source, and particle size.
3. What is the oxygen minimum zone? Where is it? How does it form?
4. Describe the different sources of food available to animals of the deep sea.
5. Why does the nutritional quality of surface particles decrease as they sink through the water column on their way to the abyssal plains?
6. Compare and contrast the benthic invertebrate fauna of the deep sea and the shallow subtidal.
7. If the deep sea is so inhospitable (see Discussion Topic no. 1 above), why is the diversity of deep-sea animals similar to or even greater than that of shallow subtidal communities?
8. Why are infaunal deposit feeders so abundant in the deep sea? How do they feed?
9. Why is broadcast spawning less common in deep-sea animals than in shallow-water animals?

10. Where in the Atlantic Ocean would you predict that additional deep-sea hot springs and their associated chemosynthesis-powered communities might be found? Why?

Suggestions for Further Reading

Albert, O. T., and O. A. Bergstad. 1993. Temporal and spatial variation in the species composition of trawl samples from a demersal fish community. *Journal Fish Biology* 43(A):209–222.

Angel, M. V., and T. L. Rice. 1996. The ecology of the deep sea and its relevance to global waste management. *Journal of Applied Ecology* 33:915–926.

Arp, A. J., and J. J. Childress. 1983. Sulfide binding by the blood of the hydrothermal vent tube worm *Riftia pachyptila*. *Science* 219:295–297.

Barry, J. P., H. G. Greene, D. L. Orange, C. H. Baxter, B. H. Robison, R. E. Kochevar, J. W. Nybakken, D. L. Reed, and C. M. McHugh. 1996. Biologic and geologic characteristics of cold seeps in Monterey Bay, California. *Deep Sea Research Part I: Oceanographic Research Papers* 43:1739–1762.

Billett, D. S. M., R. S. Lampitt, A. L. Rice, and R. F. C. Mantoura. 1983. Seasonal sedimentation of phytoplankton to the deep-sea benthos. *Nature* 302:520–522.

Butman, C. A., J. T. Carlton, and S. R. Palumbi. 1995. Whaling effects on deep-sea biodiversity. *Conservation Biology* 9:462–464.

Childress, J. J., ed. 1988. Hydrothermal vents, a case study of the biology and chemistry of a deep-sea hydrothermal vent of the Galapagos Rift. *Deep-Sea Research* 35:1677–1849.

Childress, J. J. 1995. Are there physiological and biochemical adaptations of metabolism in deep-sea animals? *Trends in Ecology and Evolution* 10:30–36.

Childress, J., H. Felback, and G. Somero. 1987. Symbiosis in the deep sea. *Scientific American* 256:106–112.

Corliss, J. B., and R. D. Ballard. 1977. Oases of life in the cold abyss. *National Geographic* 152:441–453.

Eder, W., M. Schmidt, M. Koch, D. Garbe-Schönberg, and R. Huber. 2002. Prokaryotic phylogenetic diversity and corresponding geochemical data of the brine–seawater interface of the Shaban Deep, Red Sea. *Environmental Microbiology* 4:758–763.

Edmond, J. M., and K. Von Damm. 1983. Hot springs on the sea floor. *Scientific American* 249: 78–93.

Ernst, W. G., and J. G. Morin. 1982. *The Environment of the Deep Sea.* Prentice-Hall, Englewood Cliffs, NJ.

Gage, J. D. 1996. Why are there so many species in deep-sea sediments? *Journal of Experimental Marine Biology and Ecology* 200:257–286.

Gage, J. D., and P. A. Tyler. 1991. *Deep-Sea Biology: A Natural History of Organisms of the Deep-Sea Floor.* Cambridge University Press, Cambridge.

Gatehouse, L. N. L., A. Shannon, E. P. J. Burgess, J. T. Christeller, M. Sibuet, and K. Olu. 1998. Biogeography, biodiversity and fluid dependence of deep-sea cold-seep communities at active and passive margins. *Deep Sea Research Part II: Topical Studies in Oceanography* 45:517–567.

Gooday, A. J. 2002. Biological responses to seasonally varying fluxes of organic matter to the ocean floor: a review. *Journal of Oceanography* 58:305–332.

Grassle, J. F. 1991. Deep-sea benthic biodiversity. *Bioscience* 41:464–469.

Haedrich, R. L., and N. R. Merrett. 1990. Little evidence for faunal zonation or communities in deep sea demersal fish faunas. *Progress in Oceanography* 24:239–250.

Hecker, B. 1985. Fauna from a cold sulfur-seep in the Gulf of Mexico: comparison with hydrothermal vent communities and evolutionary implications. *Bulletin of the Biological Society of Washington* 6:465–473.

Higgins, R. P., and H. Theil. 1988. *Introduction to the Study of Meiofauna.* Smithsonian Institution Press, Washington, DC.

Hill, T. M., J. P. Kennett, and H. J. Spero. 2003. Foraminifera as indicators of methane-rich environments: a study of modern methane seeps in Santa Barbara Channel, California. *Marine Micropaleontology* 49:123–138.

Hollister, C. D., A. R. M. Nowell, and P. A. Jumars. 1984. The dynamic abyss. *Scientific American* 251: 42–53.

Jannasch, H. W. 1989. Serendipity in deep-sea microbiology: lessons from the *Alvin* lunch. *Oceanus* 31:28–33.

Jones, A. T., and C. L. Morgan. 2003. Code of practice for ocean mining: an international effort to develop a code for environmental management of marine mining. *Marine Georesources and Geotechnology* 21:105–114.

Kiorboe, T., K. Tang, and H.-P. Grossart. 2003. Dynamics of microbial communities on marine snow aggregates: colonization, growth, detachment, and grazing mortality of attached bacteria. *Applied and Environmental Microbiology* 69:3036–3047.

Knittel, K., A. Boetius, A. Lemke, H. Eilers, K. Lochte, O. Pfannkuche, P. Linke, and R. Amann. 2003. Activity, distribution, and diversity of sulfate reducers and other bacteria in sediments above gas hydrate (Cascadia Margin, Oregon). *Geomicrobiology Journal* 20:269–294.

Lorensona, T. D., K. A. Kvenvoldena, F. D. Hostettlera, R. J. Rosenbauera, D. L. Orange, and J. B. Martinc. 2002. Hydrocarbon geochemistry of cold seeps in the Monterey Bay National Marine Sanctuary. *Marine Geology* 181:285–304.

Lutz, R. A. 1991. The biology of deep-sea vents and seeps. *Oceanus* 34:75–83.

Lutz, R. A., D. Jablonski, and R. D. Turner. 1984. Larval development and dispersal at deep-sea hydrothermal vents. *Science* 226:1451–1454.

Lutz, R. A., T. M. Shank, D. J. Fornari, R. M. Haymon, M. D. Lilley, K. L. Von Damm, and D. Desbruyeres. 1994. Rapid growth at deep-sea vents. *Nature* 371:663–664.

Macpherson, E., and C. M. Duarte. 1991. Bathymetric trends in demersal fish size: is there a relationship? *Marine Ecology Progress Series* 71:103–112.

Mullineaux, L. S., P. H. Wiebe, and E. T. Baker. 1991. Hydrothermal vent plumes: larval highways in the deep sea? *Oceanus* 34:64–68.

Page, H. M., C. R. Fisher, and J. J. Childress. 1990. Role of filter-feeding in the nutritional biology of a deep-sea mussel with methanotrophic symbionts. *Marine Biology* 104:251–257.

Pimenov, N. V., A. S. Savvichev, I. L. Rusanov, A. Y. Lein, and M. V. Ivanov. 2000. Microbiological processes of the carbon and sulfur cycles at cold methane seeps of the North Atlantic. *Microbiology* 69:709–720.

Rex, M. A., C. T. Stuart, R. R. Hessler, J. A. Allen, H. L. Sanders, and G. D. F. Wilson. 1993. Global-scale latitudinal patterns of species diversity in the deep-sea benthos. *Nature* 365:636–639.

Sayles, F. L., W. R. Martin, and W. G. Deuser. 1994. Response of benthic oxygen demand to particulate organic carbon supply in the deep sea near Bermuda. *Nature* 371:686–689.

Smith, C. R. 1985. Food for the deep-sea: utilization, dispersal and flux of nekton falls at the Santa Catalina Basin floor. *Deep-Sea Research* 32:417–422.

Smith, C. R., H. Kukert, R. A. Wheatcroft, P. A. Jumars, and J. W. Deming. 1989. Vent fauna on whale remains. *Nature* 341:27–28.

Tourova, T. P., T. V. Kolganova, B. B. Kuznetsov, and N. V. Pimenov. 2002. Phylogenetic diversity of the archaeal component in microbial mats on coral-like structures associated with methane seeps in the Black Sea. *Microbiology* 71:196–201.

Tunnicliffe, V. 1991. The biology of hydrothermal vents: ecology and evolution. *Oceanography and Marine Biology Annual Reviews* 29:319–407.

Tunnicliffe, V. 1992a. Hydrothermal-vent communities of the deep sea. *American Scientist* 80:336–349.

Tunnicliffe, V. 1992b. The nature and origin of the modern hydrothermal vent fauna. *Palaios* 7:338–350.

Tunnicliffe, V., and C. M. R. Fowler. 1996. Influence of sea-floor spreading on the global hydrothermal vent fauna. *Nature* 379:531–533.

Van Dover, C. L. 2000. *The Ecology of Deep-Sea Hydrothermal Vents*. Princeton University Press, New Jersey.

Van Dover, C. L., P. Aharon, J. M. Bernhard, E. Caylor, M. Doerries, W. Flickinger, W. Gilhooly, S. K. Goffredi, K. E. Knick, S. A. Macko, S. Rapoport, E. C. Raulfs, C. Ruppel, J. L. Salerno, R. D. Seitz, B. K. Sen Gupta, T. Shank, M. Turnipseed, and R. Vrijenhoek. 2003. Blake Ridge methane seeps: characterization of a soft-sediment, chemosynthetically based ecosystem. *Deep Sea Research Part I: Oceanographic Research Papers* 50:281–300.

Zal, F., E. Leize, F. H. Lallier, A. Toulmond, A. Vandorsselaer, and J. J. Childress. 1998. S-Sulfhemoglobin and disulfide exchange—the mechanisms of sulfide binding by *Riftia pachyptila* hemoglobins. *Proceedings of the National Academy of Sciences* 95:8997–9002.

CHAPTER 12

Pacific walrus hauled out on an Alaskan beach.

Marine Birds and Mammals in Polar Seas

The two polar ends of the Earth share several environmental characteristics that distinguish them from other marine environments so far considered in this text. Both experience long winter nights without sunlight. Low levels of sunlight keep sea surface temperatures hovering around 0°C even in summer. Large parts of both polar marine environments remain perpetually covered by permanent sea ice known as **fast ice**. Even larger areas freeze over in winter to form **pack ice** that thaws and disappears each summer. For the purposes of this chapter, polar seas are defined as those areas of the ocean characterized by a cover of either permanent fast ice or seasonal pack ice. The approximate geographic extent of fast and pack ice is shown in FIGURE 12.1. Sea ice is a solid physical structure encountered in no other marine ecosystem. It acts as a barrier to insulate seawater from continued chilling effects of the atmosphere in winter, so sea ice never more than a few meters thick forms in even the most extreme winter temperature regimes. Sea ice also provides a stable and nearly predator-free platform on which some birds and mammals can raise their young, yet it also effectively bars many of those same animals from moving easily from the ice surface to forage the water below.

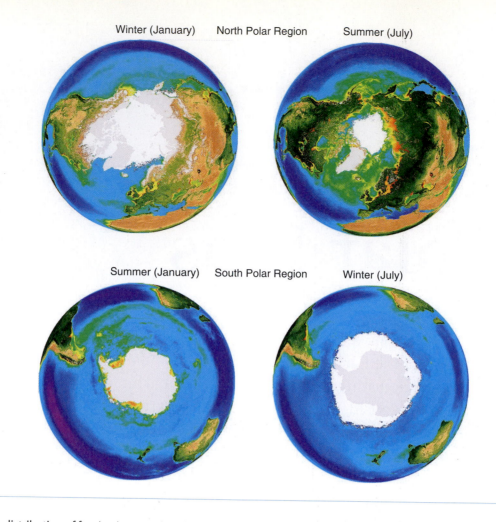

figure 12.1

Approximate distribution of fast ice (summer) and pack ice (winter) in the north and south polar regions.

In addition to these similarities, some strong contrasts exist between north and south polar marine environments. In Chapter 4, we emphasized that the Arctic is an ice-covered ocean surrounded by continents; the Antarctic is a frozen continent surrounded by ice-covered seas. Much of the Arctic Ocean is permanently covered by fast ice, and the annual average productivity rate there is low (about 25 g C/m²/yr). Around the Antarctic continent, however, upwelling of deep nutrient-rich water supports very high summertime primary production rates and annual productivity rates of over 100 g C/m²/yr. This uninterrupted zone of upwelling extends around the entire continent to support a band of high phytoplankton productivity from the ice edge north to the Antarctic Convergence (Figure 3.40). Comparable latitudes in the Arctic are interrupted by land masses to form the small regional Bering, Baffin, Greenland, and Norwegian Seas. Some important features of the two polar regions are compared in TABLE 12.1.

In this chapter, much of the emphasis is on marine homeotherms, the birds and mammals, that have evolved means of remaining active and even reproducing while exposed to the near-freezing temperatures of seawater and the much colder air temperatures found at high latitudes. Maintaining high core body temperatures in such cold water is energetically costly, yet the benefits of occupying these extreme habitats are offset by the availability of seasonally abundant food in the water and by access to near predator-free ice platforms for reproduction. Several additional features characterize many of the homeotherms that successfully exploit polar or subpolar environments. They have large bodies with space available to store lipids for extended food-poor seasons, and their low costs of transport (see Chapter 10) facilitate long-distance seasonal migrations to and from these areas of intensive summer feeding. Whales, pinnipeds, and penguins sport an insulating blanket of **blubber**, a sheath of fat and connective tissue of varying thickness that

Comparison of Arctic and Antarctic Oceanographic Features

Table 12.1

Feature	Arctic	Antarctic
Shelf	Broad, two narrow openings	Narrow, open to all oceans
River input	Several	None
Nutrients in photic zone	Seasonally depleted	High throughout year
Icebergs	Small, irregular, not in arctic basin	Abundant, large, tabular
Pack ice		
Maximum area ($\times 10^6$ km^2)	13	22
Age	Mostly multiyear	Mostly 1 year
Thickness	3.5 m	1.5 m

Adapted from Hempel 1991.

shields the deeper body musculature from the cold of the surrounding water. The insulative value of blubber is a function of both its thickness and its lipid content, so it is not effective as an insulator for flying birds, but works well for large swimming animals where a thick blubber layer does not seriously distort an animal's body shape or proportions.

12.1 The Arctic

Phytoplankton productivity patterns at higher latitudes are marked by large swings between long winters of essentially no productivity and short bursts of high summertime production. The phytoplankton cells produced in these summer blooms tend to be individually much larger than those in lower latitudes. These in turn feed relatively large planktonic copepods and benthic consumers, particularly amphipod crustaceans and bivalve mollusks. The stage is then set for short food chains supporting very abundant, although seasonally more variable, aggregations of larger marine mammals and seabirds. Common responses of these homeotherms to cope with the seasonal variability in their food supplies include extended fasting periods and long migratory excursions to lower latitude waters in winter.

Arctic Mammals

At birth, the young of cetaceans are capable swimmers and instinctively surface to take their first breath. Polar bear and pinniped pups are unable to swim at birth, so their birth must occur on land or on ice floes. The newborn of marine mammals are large; blue whale calves weigh about 3 tons at birth. Still, these newborn mammals are smaller than their parents, so they have higher surface-area-to-volume ratios, and their insulating layers of blubber or fur are not usu-

ally well developed. Several features compensate for the potentially serious problem of heat loss and body temperature maintenance in newborn marine mammals. Terrestrial pupping in pinnipeds provides some time for growth before the pups must face their first winter at sea. The larger cetaceans spend their summers feeding in cold polar and subpolar waters and then undertake long migrations to their calving grounds in tropical and subtropical seas (FIGURE 12.2). In these warm waters, their calves have an opportunity to gain considerable weight before migrating back to their frigid summer feeding grounds.

The growth rates of the young of some marine mammal species are truly astounding. Nursing blue whales grow from 3 tons at birth to 23 tons when weaned a scant 7 months later (an average weight gain of almost 100 kg a day). Hooded seal pups gain over 7 kg each day of their extremely short 4-day nursing period before weaning (FIGURE 12.3). These prodigious growth rates are supported by an abundant supply of high-fat milk. Cetacean milk is 25–50% fat (cow's milk ranges from 3 to 5% fat), and several species of seals produce milk that is 50–60% fat. The daily milk yield of a blue whale has been estimated at nearly 600 liters. In pinnipeds, 2–5 liters are more typical. In both cetaceans and pinnipeds, species occupying colder waters consistently produce milk with a higher fat content.

The energy demands made on females to produce a large offspring and then to supply it with large quantities of fatty milk until it is weaned (usually a few weeks to several months) are exceedingly high. Even the water that goes into the milk imposes additional osmotic costs that must be paid with further energy expenditures. Marine mammals have long gestation periods (several months to more than a year), tend to reproduce not more than once a year, and, with the exception of polar bears, produce only

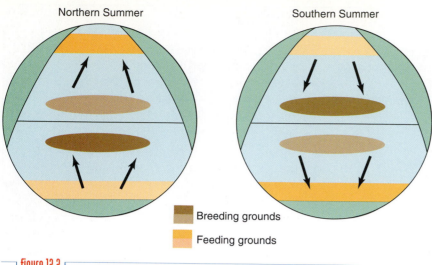

Northern Summer   Southern Summer

Breeding grounds

Feeding grounds

figure 12.2

Generalized migratory patterns of large whales between summer feeding and winter breeding grounds. Northern and southern populations follow the same migratory pattern but do so 6 months out of phase with each other. Consequently, northern and southern populations of the same species remain isolated from each other, even though both populations approach equatorial latitudes. (Adapted from Mackintosh 1966.)

figure 12.3

Hooded seal mother and pup hauled out on pack ice in background; an adult male is in foreground.

a single offspring each time they reproduce. In this section, several different arctic species are compared to provide a sense of how marine mammals solve specific challenges in this extreme and sometimes extremely harsh environment.

Pinnipeds

Several species of seals, including ringed, harp, and hooded seals, breed on arctic fast or pack ice. These species all show a reduced level or complete absence of polygyny that is so strongly developed in elephant seals (described in Chapter 10) and other species of

pinnipeds that breed in lower latitudes. This difference seems to be due to the difficulty males have in gaining and controlling access to multiple females in the physically unstable environment of pack ice. The extensive distribution of pack ice encourages breeding females to disperse widely while still providing the same protection from most predators that island rookeries do for elephant seals. Yet the temporary and unpredictable nature of the location, extent, and breakup time of typical pack ice restricts pupping to a short time period when the ice is most stable. Females of ice-pupping species cannot expect to return to the same "place" for pupping year after year, as elephant seal females do, so males are forced to search widely for dispersed females.

In monogynous seals, males can succeed reproductively by locating an estrus female and mating with her. With population gender ratios approximating 1:1, there will be approximately as many females available as there are males competing for them. These males apparently are limited to monogyny simply because the nonaggregated distribution of females does not allow them to access more than one female at a time or to exclude other males from access to females other than the one he is attending. The larger size and other sexually dimorphic characteristics seen in polygynous males confer no selective advantage, and monomorphism or even slight reverse sexual dimorphism becomes the norm for monogynous species.

Mating in the water is common in these species, and this behavior provides males with even fewer opportunities to control the mating choices of females. To compensate somewhat, some ice-breeding seals have evolved complex vocalizations underwater to enhance their attractiveness and advertise their interest in mating. Mating in water is especially difficult for human observers to study, and it is much more so when it occurs under polar ice.

Like pack ice, the vast extent of arctic fast ice (Figure 12.1) also allows females breeding on fast ice to disperse widely, although access to cracks and holes in the ice usually promotes some clumping of individuals. In the Arctic, ringed seal females give birth to pups in isolated and hidden snow caves above

figure 12.4

Female ringed seal nursing her pup on pack ice.

breathing holes in the fast ice to avoid polar bears. Because females are isolated and dispersed, individual male ringed seals typically have access to only one female and are considered monogynous. They have a broad arctic distribution and are found wherever openings in fast ice occur, even as far north as the Pole. Under the ice, they feed on small fish, krill, and planktonic amphipods. This small seal (FIGURE 12.4) is the most abundant pinniped in arctic waters, with a current population exceeding 2.5 million.

Harp seals are almost as numerous as ringed seals, although their distribution is more limited, extending eastward from Hudson Bay around Greenland to northern Siberia. A dorsal harp-like "V" and black hood is worn by adults. They exploit a broad range of prey species, concentrating on capelin and arctic and polar cod. This species is probably best known as the target species for human hunters of the white pelts of very young pups (FIGURE 12.5). These pups continue to be hunted each spring in the Canadian Atlantic before the break-up of the pack ice. The documented annual kill between 1996 and 1999 was 460,000 pups.

Hooded seals are unusual ice seals for many reasons. When females are ready to pup, they haul out on pack ice, loosely positioning themselves about 50 m apart. After their pup is born, their 4-day nursing period is the shortest of any pinniped (and probably the shortest of any large mammal). Soon after weaning, the female enters the water where she mates with an attendant male. The male soon returns to the ice to continue his search for other estrous females. Individual males may mate with as many as six to eight females in one breeding season. They are the only polygynous arctic ice seal, and the appearance of adult males reflects the strong sexual dimorphism that is associated with polygyny. Male hooded seals sport an unusual two-part nasal "hood" ornament that is inflated during competitive displays. When inflated by closing the nostrils, one part, the hood, enlarges to cover the face and top of the head (FIGURE 12.6a). Males can also inflate a very elastic nasal septum to form a large membranous pink balloon that extends forward from one nostril (Figure 12.6b).

Walruses are the only species in the pinniped family Odobenidae. They are found all around the arctic basin where ice is sufficiently thin to break through to feed and where the sea bottom is not more than 80 m deep (FIGURE 12.7). Walruses are easily identified by their long paired tusks (FIGURE 12.8) that are present in both sexes but are shorter and more slender in females. Tusks of males are larger, occasionally reaching 1 m in length. Male walruses use their tusks mainly in dominance displays directed toward other males; females use them to defend themselves and their young from aggressive males and from marauding polar bears. On occasion, both sexes use their tusks to pull themselves onto ice floes. Tusks are not usually used in feeding, although rarely they are used to kill seal prey by stabbing.

Although some walruses occasionally eat ringed and bearded seals and birds using their tongue vacuum pump to suck away the skin, blubber, and intestines, their typical prey are several species of benthic clams. Walruses forage for clams by swimming along the sea bottom in a head-down position, stirring up sediments with their snout and its very sensitive vibrissae (FIGURE 12.9). When a clam is located, it is either sucked into the mouth directly or is first excavated with jets of water squirted from the mouth. The soft tissue of large clams is sucked out of the shell by the

figure 12.5

Harp seal pup, long a target of fur hunters, on Canadian pack ice.

(a)

(b)

figure 12.6

Hooded seal. (a) Hood; (b) balloon.

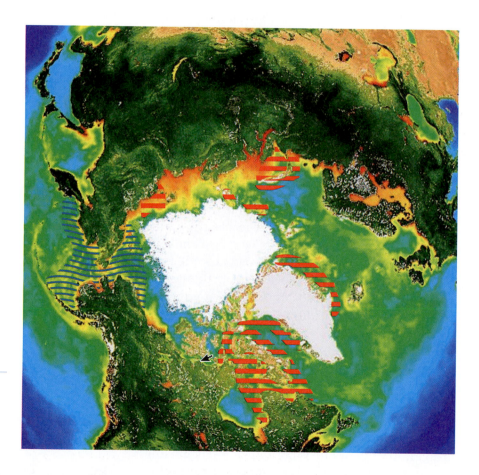

figure 12.7

Geographic distribution of the Atlantic (red) and Pacific (blue) subspecies of walrus, *Odobenus rosmarus*.

tongue. Individual walruses consume large quantities of shallow-water benthic clams. They can harvest clams at an impressive rate of 40–60 clams per dive for a total of several thousand clams each day. During their foraging activities, walruses create long furrows and pits on the sea bottom and can be important agents of disturbance in shallow soft-bottom arctic benthic communities.

Walruses are polygynous breeders, with adult males typically growing to about twice the size as adult females. Walruses mate in water, giving females a selective advantage that does not occur on land: directly selecting the male with whom they will copulate. Pacific walruses have been described as having a **lek** or lek-like mating system. During the breeding season, female walruses haul out on ice or rest in the

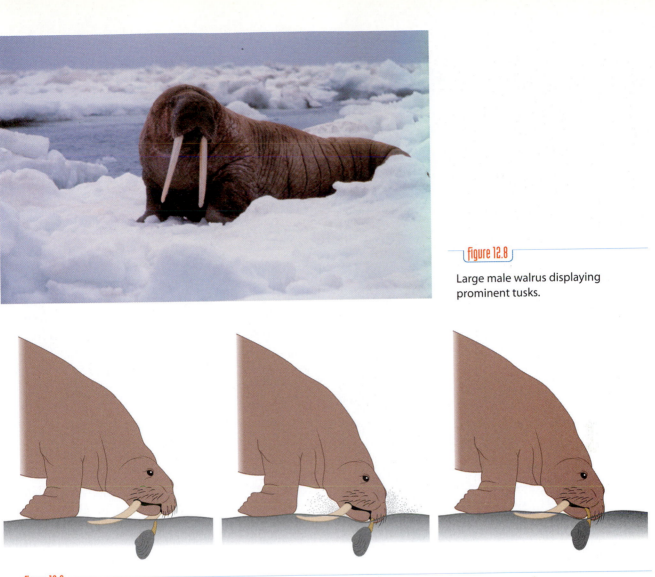

figure 12.8

Large male walrus displaying prominent tusks.

figure 12.9

Benthic suction-feeding behavior of walruses used to capture buried bivalves.

water while one or more adult males station themselves in the water nearby to perform underwater visual displays while producing a series of amazingly bell-like or gong-like sounds to attract females. These ritual courtship displays are similar to the well-known lekking behaviors of some African antelopes. In all lekking species, including walruses, males only aggregate and display; females make the choice of their mating partners. The elaborate display behaviors of males allow females an opportunity to make the best genetic deal possible by getting males to advertise their own fitness before selecting a mate.

The gestation period for walruses is 15 months, so the annual reproductive schedules characteristic of other pinnipeds cannot be maintained. Walrus calves are nursed for about 1 year and are weaned gradually during their second year as they improve their diving abilities and foraging skills. Adult females, consequently, mate and produce a calf only every 3 years.

Polar Bears

The polar bear, *Ursus maritimus,* is the animal icon of the arctic (FIGURE 12.10a). Polar bears have completely white fur, slightly longer necks, and larger body sizes than other bears. They are broadly distributed on ice-covered waters throughout the arctic. In Hudson Bay and other regions that become completely ice free in summer, they are forced to spend several months on land fasting until the fall freeze-up.

Like most other terrestrial mammals, polar bears have few specialized adaptations for efficient swimming. They have large feet that form flat plates oriented perpendicular to the direction of motion,

(a)

(b)

figure 12.10

(a) A polar bear mother and her twin cubs at rest.
(b) Hair-covered paws of a swimming polar bear.

producing inefficient drag-based thrust to move the animal forward. Polar bears swim with a stroke like a crawl, pulling itself through the water with its forelimbs while its hind legs trail behind. When traveling on land, their huge paws help distribute their body weight while their hair-covered foot pads increase friction between the feet and the ice (Figure 12.10b).

Polar bears prey on ice seals (ringed, bearded, and ribbon, and sometimes harbor and harp seals), with walruses, beluga whales, and even birds occasionally taken. Bear predation of ice seals occurs primarily in unstable ice conditions where the bears can stalk seals resting out of water. Seal pups are especially vulnerable to bear predation. In the Canadian Arctic, ringed seals give birth to pups in snow caves above breathing holes in the ice. At weaning, these pups are 50% fat, and it is from these pups that polar bears secure most of their year's energy supply during the short period just after the ringed seal pups are weaned.

The extent of bear predation on walruses varies in different areas of the Arctic, with higher rates of predation in the Canadian Arctic than in the Bering Sea. Bears frequently fail to make a kill by stalking hauled-out walrus herds, yet they often manage to frighten herds into the water. In the ensuing stampede, walrus pups are sometimes injured or isolated, making them easy prey for the bears. It is not known whether bears commonly prey on walruses or seals underwater.

Like some other large marine mammals, polar bears endure prolonged seasonal fasts as the availability of food declines or when reproductive activities prevent foraging. When bears are forced to stay onshore in summer as the ice melts, they are forced to fast for up to 4 months until late autumn when the formation of fall ice permits them to regain access to the newly formed pack ice.

Polar bears breed in late spring when large males spend days tracking potential mates across large expanses of ice. Males compete aggressively for access to these females, because females typically breed only once every 3–4 years. Female polar bears are induced ovulators, so a female must mate repeatedly before she ovulates and can be fertilized. Once pregnant, delayed implantation blocks fetal development until fall, when females dig maternity dens to spend the winter and give birth. (Most polar bears except pregnant females do not occupy winter dens.) Usually, two cubs are born about 2 months into the winter denning period. By March or April, the 3-month-old cubs have increased their birth weights 20-fold, and they are ready to leave their den with their mother. They remain with her until weaned 2 years later when she will mate again and prepare for her next pregnancy.

Mysticete Whales

Although polar bears, emperor penguins, and a few species of pinnipeds remain in these summer-intensive polar and subpolar production systems year-round, mysticete whales provide examples of the more common approach to exploiting these high-latitude production systems, with intensive summer feeding followed by long-distance migrations to low latitudes in winter months.

Mysticete whales are among the largest animals on Earth, and all are filter feeders. All except the gray whale feed on planktonic crustaceans or small shoaling fishes. The character of the baleen, as well as the size and shape of the head, mouth, and body, differ markedly between species of baleen whales (see Figure 6.34). All mysticetes show reverse sexual size dimorphism, with females growing to larger maximum sizes than males. This may provide females with larger lipid stores to help offset the energetic costs associated with rapid fetal growth and lactation.

Several distinctive types of feeding behaviors have been described for baleen whales, depending on the type of whale as well as on its prey. Bowhead whales are large bulky animals with very long and fine baleen plates well adapted to collect copepods, euphausiids, and other small planktonic crustaceans using a surface feeding behavior known as skimming (FIGURE 12.11). Swimming slowly with their mouths slightly agape, water and prey items flow into the mouth and then through the baleen where the small prey are trapped. They also use this feeding behavior well below the sea surface and sometimes even near the sea floor. Bowhead whales live all year along the edge and in leads of pack ice around much of the arctic basin, migrating with the growth and retreat of the pack ice.

Gray whales exhibit the most distinctive feeding behavior of all mysticetes and also have the coarsest and shortest baleen (FIGURE 12.12). In their shallow summer feeding grounds of the Bering and Arctic Seas, these medium-sized whales feed on bottom invertebrates, especially amphipod crustaceans. Direct observations of gray whale feeding behavior demonstrate that these animals roll to one side and suck their prey into the side of the mouth and expel water out the

other side, flushing the mud out through the filter basket formed by the coarse baleen while trapping their invertebrate prey inside the baleen.

Long-distance fasting migrations to low latitudes in winter months is typical of large mysticetes in both hemispheres, with both mating and calving occurring in warm and often protected waters. A general picture of the relationship between these migrations and reproductive timing and behavior is well-illustrated by gray whales. The annual gray whale migration has been extensively studied and is the best known of the large whale migrations. These whales migrate an impressive 18,000 km round-trip each year. Most gray whales spend the summer in the Bering Sea and adjacent areas of the

figure 12.11

Skimming feeding behavior of bowhead whales. (Adapted from Pivorunas 1979.)

figure 12.12

An adult gray whale in its winter breeding lagoon. The baleen plates attached to the upper jaw of its open mouth are evident.

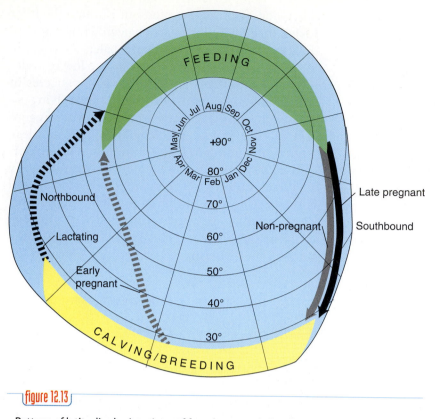

FEEDING

+90°

Northbound

Lactating

Early pregnant

CALVING/BREEDING

80°
70°
60°
50°
40°
30°

May Jun Jul Aug Sep Oct Nov Dec Jan Feb Mar Apr

Late pregnant

Non-pregnant

Southbound

figure 12.13

Pattern of latitudinal migrations of female gray whales through a complete 2-year reproductive cycle. (Adapted from Sumich 1986.)

Arctic Ocean as far north as the edge of the pack ice. Their habit of feeding on bottom invertebrates and their limited capacity to hold their breath (4–5 minutes) restrict their feeding activities to the shallow portions of these seas (usually less than 70 m).

The annual migration of gray whales exists as two superimposed patterns related to the reproductive states of adult females (FIGURE 12.13). Each year approximately one half of the adult females are pregnant. These near-term pregnant females depart the Bering Sea at the start of the southbound migration 2 weeks before other gray whales (Figure 12.13, black solid line). Nonpregnant adult females start south a little later accompanied by adult males, although both groups overlap somewhat the earlier departure of pregnant females.

The southward migration is initiated in autumn, possibly in response to shortening days or to the formation of sea ice in arctic waters. The migration is a procession of gray whales segregated according to age and sex. After they pass through the Aleutian Islands, gray whales follow the long curving shoreline of Alaska (FIGURE 12.14). South of British Columbia, gray whales travel reasonably close to the shoreline, usu-

ally in water less than 200 m deep, within sight of land and likely within hearing of breaking waves on shore. The average speed of southbound gray whales is about 7 km/h. At that speed, most of the whales reach the warm protected coastal lagoons of Baja California by late January.

It is in these lagoons that the pregnant females give birth to a 1-ton calf that is 4–5 m long. The new mothers remain with their calves in the lagoons for about 2 months. During that time, the nursing calves more than double their birth weight in preparation for the rigors of a long migration back to the chilly waters of the Bering Sea. While in the lagoons and adjacent coastal areas, most lactating cows with new calves maintain a spatial separation from other age and sex groups by occupying the inner more protected reaches of the lagoons until the other animals depart. These females, accompanied by their calves, are the last to leave calving lagoons the following spring.

Accompanied by adult males, nonpregnant adult females arrive later than pregnant females and are also the first to leave the lagoon after approximately 30 days. They mate during the southern portion of their coastal migration or after they arrive in their winter lagoons. Both are environments of limited underwater visibility, so direct observations of mating behavior have not been made. Based on above-surface observations of courting activities, mating appears to be promiscuous as it is in most other baleen whale species, although actual paternity is impossible to establish from direct observations of their very active and vigorous courting encounters (FIGURE 12.15). Courting/mating groups including as many as 17 individuals have been observed; however, the intensity of the physical interactions during these courting bouts makes it impossible to determine the gender of all group participants, their interwhale contact patterns, or even accurate assessments of group sizes.

In the absence of direct underwater observations of gray whale mating behavior, arguments for promiscuous mating must be based on other evidence. In comparison with most other baleen whales, gray, bowhead, and right whale males have substantially

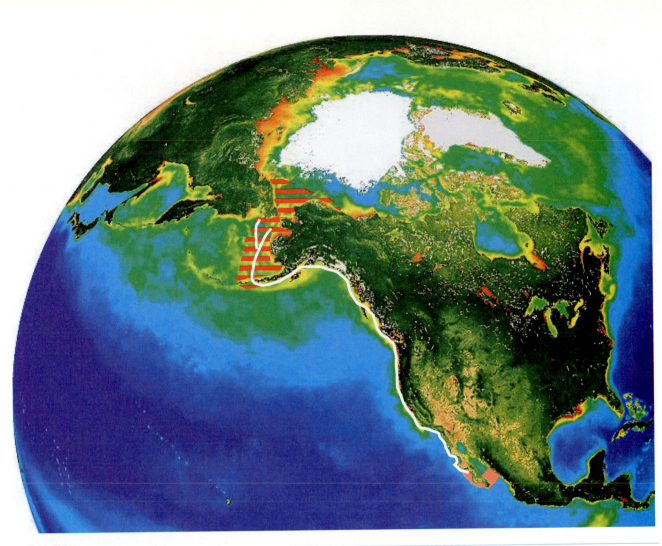

figure 12.14

Migratory route (white line) of the Northeast Pacific gray whale, with summer feeding (red hatching) and winter breeding areas (pink) indicated.

larger testes weight-to-body weight ratios (FIGURE 12.16) and relatively larger penises. Larger testes are presumed to produce greater quantities of sperm. This has been interpreted as evidence of **sperm competition** (a behavior well studied in several species of birds), with copulating males competing with previous copulators by attempting to displace or dilute their sperm within a female's reproductive tract, thus increasing the probability of being the male to fertilize the female's single ovum.

The reproductive cycle of gray whales and other large baleen whales consists of three parts (FIGURE 12.17): a 12- to 13-month gestation period followed by 6 months of lactation, then another 6 months of rest to prepare for the next pregnancy. In early spring, the northward migration begins and is much the reverse of the previous southbound trip, with nursing females and their

calves the last to leave the lagoons. While still nursing, calves and their mothers migrate back to their polar feeding grounds. Traveling at a more leisurely pace than when going south, the whales reach their arctic feeding grounds in the late spring or early summer. With food abundant there, calves are weaned and females begin to replenish their fat and blubber reserves before migrating back to the winter breeding grounds to mate and begin the cycle again.

Toothed Whales

Arctic waters are inhabited or at least visited by three species of toothed whales: belugas, narwhals, and killer whales. Belugas and the closely related narwhals are medium-sized whales intimately associated with pack-ice habitats. Belugas, also known as white whales, are actually born gray and gradually become creamy

figure 12.15

Several courting gray whales in their winter lagoon. One is clearly a male.

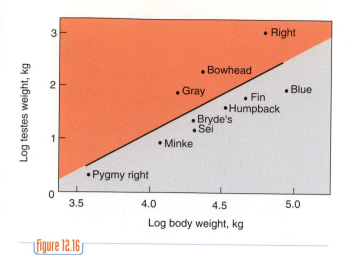

figure 12.16

General relationship between testis size and body size for 10 species of mysticete whales. Gray, bowhead, and right whales (above the diagonal line) have relatively larger testes and are suspected of participating in sperm competition.

white as adults (FIGURE 12.18a). Adaptations to ice-covered waters include a very flexible neck, small flippers, and a thick layer of insulating blubber. In place of a dorsal fin, belugas have a stout dorsal ridge used to break through ice from below. They feed on seasonally and locally abundant fish and invertebrates, both in the water column and on the sea floor.

The most obvious and dramatic feature of narwhals is their long spiral tusk (Figure 12.18b). It is thought by some to be the basis for unicorn legends. A narwhal tusk is actually an elongated upper tooth projecting through the lip. Tusk shapes and sizes are variable and are found on both males and females, although male tusks are longer. The prominent tusk of this species is celebrated in its taxonomic name, *Monodon monoceros* (one tooth, one horn).

Details of the biology of narwhals remain spotty because of their preference for remote polar areas over deep water covered with heavy pack ice in winter. Their general range extends over most of the Atlantic portion of the Arctic Ocean, with only occasional stragglers seen in the Pacific side. They feed mostly on polar and arctic cod, two species of fish often associated with the undersides of pack ice. Narwhals are excellent divers, sometimes to depths greater than 1000 m, where they occasionally feed on bottom fish and midwater squid.

The striking white and black color patterns of killer whales make them the most easily recognized of all the whales (FIGURE 12.19). Killer whales are cosmopolitan in distribution, from tropical waters to ice edges of both polar regions. Captive animal displays, hit movies, and worldwide boat- and shore-based

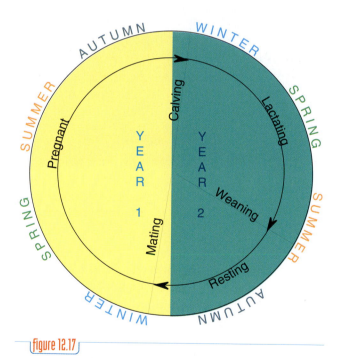

figure 12.17

The reproductive cycle of female gray whales (*Eschrichtius robustus*) including the accompanying growth of fetus and calf during its period of dependency on the female. (Adapted from Berta and Sumich 1999.)

whale-watching activities have made this species a common sight to large numbers of people. And long-term research programs using photographic identification of individual whales in some populations have been crucial to improving our understanding of the biology of this species.

Within their broad geographic range, smaller social groups of killer whales, known as **pods**, occupy smaller individual ranges. Individual killer whale pods consist of a mature female, her offspring, and her daughters' offspring. Thus, pods are affiliations of very closely related individuals, and are **matrilineal**, with each pod member a direct descendent of the oldest female in the pod. In contrast to many other social mammals, killer whale pods include both male and female siblings that remain together all their lives.

The broad ecologic and latitudinal range of killer whales reflects their wide variety of prey species, including fish, cetaceans, pinnipeds, birds, cephalopods, sea turtles, and sea otters. Killer whales are the only cetacean known to attack and consume other marine mammals consistently. Although killer whales have been described as feeding opportunists or generalists, they also can behave as feeding specialists, with the necessary flexibility to respond to variations in the type and abundance of their preferred prey.

In Icelandic coastal waters, for example, some pods of killer whales feed on marine mammals and then switch to fish prey at other times of the year. In both Antarctic and Pacific Northwest waters, prey specialization between killer whale populations occurs. In Washington and British Columbia coastal waters, a resident population feeds almost exclusively on fishes, whereas a nonresident or transient population preys principally on marine mammals, especially harbor seals and sea lions. Sightings of foraging resident pods are strongly associated with seasonal variations in their most common prey, salmon. During months when salmon are unavailable, residents switch prey to herring or bottom fish and may disperse northward several hundred kilometers. In contrast, the prey of transient groups is available year-round, and despite their label, these whales tend to remain in a small area for extended periods of time (see Table 2.1 for a summary of some of the important differences between these two populations).

Different populations use different foraging strategies, even when in close proximity to each other. These strategies reflect different prey types and possibly different pod traditions. Resident whales typically swim in a flank formation when hunting, possibly to maximize the likelihood of detecting prey, although individual hunting is sometimes seen at the periph-

(a)

(b)

figure 12.18

Two arctic toothed whales: (a) beluga, *Delphinapterus leucas*, and (b) narwhal, *Monodon monoceros*.

figure 12.19

A killer whale, *Orcinus orca*, surfacing amid arctic ice.

figure 12.20

Transient killer whale attacking a Dall's porpoise in Alaskan coastal waters.

ery of the main pod group. Resident pods appear to prefer foraging at slack tidal periods when salmon tend to aggregate.

Echolocation and other vocalizations are common when resident pods feed on fish. These vocalizations are less frequent in transient groups preying on other marine mammals, possibly because mammals are more likely to detect vocalizations and exhibit avoidance responses. Mammal-eating killer whales are cooperative foragers and take advantage of an ability, probably unique among marine mammals, to capture prey larger than themselves. The degree of foraging cooperation is variable, depending on the type of prey; it can involve cooperative prey encirclement and capture, division of labor during an attack, and sharing of prey after capture.

Relative to residents, transients use short and irregular echolocation trains composed of clicks that appear structurally variable and low in intensity, more closely resembling random noise. The contrasting black and white coloration patterns of killer whales may facilitate visual coordination and communication within feeding groups of transients. Transient groups commonly encircle their prey, taking turns hitting it with their flippers and flukes. This extended handling of prey may continue for 10–20 minutes before the prey is killed and consumed. When hunting alone, transient whales often use repeated percussive tail-slaps or ramming actions to

kill their prey before consuming it (FIGURE 12.20). Foraging by transient groups occurs at high tides when more of their pinniped prey are in the water and vulnerable to predation.

Differences in the foraging behaviors of the two overlapping populations in the Pacific Northwest persist through time, with each group expressing different vocalizations and different social group sizes as well as different prey preferences and foraging behaviors. Some researchers have suggested that the two populations in the Pacific Northwest are in the process of speciation.

An interesting consequence of this prey specialization is the pattern of indirect trophic interactions that have evolved between these two populations of Orcas. Resident whales compete directly for fish that are also the prey of pinniped species consumed by transient killer whales. Any increase in the carrying capacity of residents will have the effect of decreasing the carrying capacity of transients, because residents compete with the pinniped prey of transients for available fish resources. Any increase in the carrying capacity of transients will have the opposite effect on residents by reducing competitive pressures from pinnipeds on fish resources. These indirect trophic interactions and resulting trophic level efficiencies suggest that residents should be more numerous than transients, and present estimates of the sizes of the two populations support this.

Arctic Birds

As a group, seabirds share two important characteristics: They obtain all their food from the sea, and they use enlarged nasal salt glands (see Figure 6.23) to excrete excess salt ingested during feeding. Only a small number (~3%) of bird species are considered seabirds (this category does not include shore or wading birds), and fewer still occupy high latitudes for at least part of each year. All species of seabirds are from three of the orders of birds listed in Table 6.1, Charadriiformes (gulls, skuas, jaegers, terns, auks, and puffins), Procellariiformes (petrels, fulmars, and albatrosses), and Sphenisciformes (penguins).

Most of the seabirds (the emperor penguin is the obvious exception) described here nest at high latitudes during summer, often in enormous and crowded colonies, and then leave to spend winter months at lower latitudes elsewhere. In contrast to land birds, fewer species of seabirds feed and nest in northern polar regions than in southern regions. Only northern fulmars, ivory gulls, and Ross's gulls consistently depend on arctic ice-associated prey in summer, although some ivory gulls remain near ice-free openings throughout the long polar winter. In slightly lower latitudes, several species of auks feed on capelin, herring, and sand eels.

Almost all high-latitude sea birds must forage while raising their chicks, mandating frequent and often long feeding trips from nesting areas to the sea. Fulmars and auks exemplify two common approaches to foraging exhibited by seabirds in general. Fulmars are closely related to petrels and albatrosses (FIGURE 12.21). All three are pelagic predators, flying long distances, often hundreds of kilometers from nest sites, to forage on near-surface prey. Their ability to exploit long-distance food resources is enhanced by several adaptations: They are adept fliers with long wings to help conserve energy by gliding and slope soaring the backs of ocean waves. Core body temperatures 2–3°C lower than those of most other birds also conserve energy. Subdermal fat and stomach oil reserves allow them to endure long periods without food. Foraging at night when vertically migrating prey are nearer the sea surface is facilitated by their well-developed olfactory sense. And finally, their eggs are less sensitive to chilling when the adults are away.

Auks are the northern hemisphere's answer to penguins of the southern hemisphere. Auks are poor fliers and never make long flights comparable with those of fulmars or petrels. Auks, however, are excellent swimmers and divers. When below the sea

figure 12.21

A northern fulmar, *Fulmarus glacialis,* in search of prey.

figure 12.22

Auk swimming underwater.

surface, their short wings provide thrust in a flying motion (FIGURE 12.22) similar to that of sea lions or penguins. Prey is captured with this sort of subsurface pursuit swimming rather than by plunge diving from the surface.

The arctic tern resembles a small gull and, in summer, ranges north to within 8° of the North Pole, but nests in coastal areas all around the arctic basin. As the northern autumn approaches, arctic terns begin their transequatorial migration to similar latitudes in the Antarctic. This, the longest known annual migration of any bird, maintains these terns in endless summer conditions. It also causes them to overlap the distribution of a southern hemisphere sister species, the antarctic tern. This is another example of the bipolar geographic distribution of a species or of two closely related species described for euphausiids in Chapter 10. The northern fulmar also has a sister species, the southern fulmar, which occupies a similar habitat in the southern hemisphere.

12.2 The Antarctic

The high primary production that occurs around the Antarctic continent during the short polar summer supports a trophic system quite unlike any found in northern latitudes. Essentially all antarctic communities depend directly or indirectly on one species of dominant pelagic herbivore, *Euphausia superba* (FIGURE 12.23). This 6-cm-long, 1-g, superb euphausiid is the largest of all euphausiid species (collectively, these crustaceans are commonly referred to as krill), and it occurs in enormous swarms around much of the antarctic continent. This herbivore is adaptable in its taste for phytoplankton, shifting between filtering abundant pelagic diatoms and scraping ice algae from the undersides of pack ice formations. These swarms of *E. superba* represent standing stocks of $200–250 \times 10^6$ metric tonnes, equivalent to about 2×10^{12} individual animals.

In its central trophic role in antarctic food webs, *E. superba* is the staple prey for benthic invertebrates, pelagic fish and squid, seabirds, and several species of marine mammals, including the largest carnivores on Earth, the blue and fin whales. One of the keys to the ecologic success and dominance of *E. superba* is its ability to survive several long polar winter nights without food during its 5-year lifetime. Diurnal vertical migration is completely suppressed, and *E. superba* descend to their winter depths of 250–500 m and cease feeding. By reducing their metabolic rates during winter, they can survive for more than 7 months without food. During such prolonged periods of no food, these crustaceans metabolize some of their structural body proteins and stored lipids, causing a reduction in body size and eliminating the need to molt and produce new exoskeletal material.

Antarctic Birds

Seventeen species of penguins are widely distributed in the southern hemisphere, from antarctic fast ice in the south to rocky shores of the Galapagos islands on the equator. Two species, the Adelie and emperor penguins, are abundant in antarctic pack-ice habitats. Penguins are flightless birds with streamlined bodies and wings adapted as narrow flippers for efficient lift-based propulsion underwater. On land, their short legs are clumsy for walking on ice or snow, and they commonly flop down on their bellies and "toboggan" while pushing with their flippers and feet. When seeking nesting sites, Adelie and emperor penguins both penetrate pack ice to at least 77° S latitude.

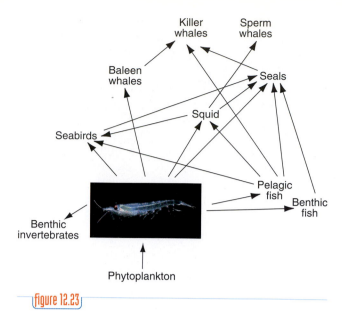

figure 12.23

Antarctic food web with krill, *Euphausia superba*, occupying a central position on the second trophic level.

Adelie penguins are moderate-sized birds, standing about 75 cm tall when adult (FIGURE 12.24). During winter months, they are dispersed around the antarctic continent north of the winter pack-ice edge where they feed on *E. superba*. The latitudinal position of the pack ice, however, is shifting slowly southward because climate warming causes the northern edge of the pack ice to melt earlier in spring and loosen what previously was hard pack ice further south.

In early spring, adults leave the sea to return to their traditional nesting areas in rocky coastal areas blown free of drifting snow but where pack ice persists well into late spring. To reach their nesting areas, Adelie penguins must walk or toboggan over as much as 100 km of sea ice. There, males and females establish pair bonds that may persist for several years, mate, and construct a crude nest of stones.

Typically, two eggs are laid a few days apart. After the second egg is laid, the female departs to resume feeding near the edge of the pack ice, which has migrated southward during the spring melt. Males remain on the nests to incubate the eggs until the females return several weeks later. By the time they resume feeding, males have been ashore without food for as many as 40 days. Through the summer, both parents are kept busy foraging on *E. superba*, filling their crops, and returning to the nest site to feed their chicks by regurgitating partially digested krill. Seldom does the second chick to hatch in a nest survive, because it is unable to overcome the advantage in size and weight of its older sibling. By the end of summer, Adelie chicks have grown to the size of their parent, developed adult plumage, and are ready to begin foraging for themselves near the ice edge. As winter pack ice forms, they move farther north and remain at sea until mature 3 years later.

Emperor penguins are the largest living penguins, exceeding 1.1 m in height (see Figure 6.26a). Emperor penguins have a reproductive pattern similar to that of Adelie penguins, with one crucial exception: They nest on pack ice. For their chicks to grow to adult size before the pack ice melts out from under them, adults cannot nest near the ice edge. Instead, they walk far south of the edge of the pack ice in the dead of winter to start their breeding and nesting season in areas where the pack ice remains until they can complete their breeding season. A single egg is laid, and then the male takes over its incubation, and the female walks across as much as 100 km of pack ice to break her fast in open water.

To attend an incubating egg in air temperatures as low as −70°C and isolate it from the cold ice on which they are standing, male emperor penguins carry their eggs on their feet and cover them with a flap of bare belly skin that is richly supplied with blood vessels to provide warmth for the egg. Here the egg develops for 64 days as the male continues to fast. During incubation, males huddle tightly together in large groups for additional protection against the biting winter winds. Over time, individuals near the edge of the huddle move toward the center, so that everyone spends some time on the exposed edges of the huddle.

With the return of the spring sun, well-fed adult females also return to the nest site to relieve their starving mates, who have been fasting for more than 100 days and face their own long march to open water before they can begin to feed. For the remainder of the summer, both parents forage for themselves and their chicks. In the water, male and female emperor penguins demonstrate exceptional diving abilities as they forego krill to pursue midwater squids and fishes. During feeding bouts, these animals make deep dives, sometimes exceeding 500 m while staying submerged for up to 15 minutes.

Another antarctic nesting bird that is restricted to the southern hemisphere is the antarctic prion, or whale-bird (FIGURE 12.25a). It is the largest of the prions, with a wing span of 55–60 cm. Of the several very similar species of prions, this is the only one that nests on the antarctic continent as well as antarctic islands to the north. In their breeding colonies, antarctic prions construct nests on exposed rocky cliff faces, in cavities under boulders, or in short burrows on grassy slopes and then lay a single egg. After 45 days of incubation, the hatchling is fed by both parents at night. Nighttime feeding of young is also a characteristic of closely related shearwaters.

Prion chicks and adults depart the nests in March, about 50 days after hatching, and move north into subantarctic waters for the winter months. At sea, they are very gregarious in flocks containing thousands of individuals. Antarctic prions feed on *E. superba* and other crustaceans by running/flying along the surface of the water with outstretched wings and bill submerged in the water to scoop their food. Although they occasionally make shallow dives to capture prey, their bills are adapted to skim and strain water with comb-like lamellae on either side of the bill (Figure 12.25b). The lamellae function in much the same manner as whale baleen does.

Antarctic Mammals

The vast apron of sea ice around the antarctic continent is the permanent home of several species of seals (Weddell, crabeater, leopard). In summer months,

figure 12.24

An Adelie penguin on antarctic pack ice.

(a)

(b)

figure 12.25

(a) Antarctic prion. (b) Antarctic prion beak showing the bill lamellae used for filtering krill.

the disintegrating pack ice is visited by several species of large whales (rorquals, killer, sperm, and bottle-nosed whales) that forage for the abundant krill or for larger prey dependent on krill (Figure 12.23). Each year, baleen whales in the Antarctic consume an estimated 30–50 million tons of krill, and seals probably eat a similar amount. These vast shoals of krill probably support half the world's biomass of seals and a large proportion of its baleen whales and seabirds as well.

Pinnipeds

The most abundant seal on Earth is the crabeater seal; its total population probably exceeds 10 million an-

imals. Despite their common and taxonomic names (*Lobodon carcinophaga* = lobe-toothed crab-eater), crabeater seals forage almost exclusively on *E. superba*. This species has highly modified teeth (FIGURE 12.26a) that are effective as strainers to collect krill. Crabeater seals are circumpolar, spending the entire year on pack ice. They migrate on a seasonal schedule as the ice edge advances and retreats with the pattern of freezing and thawing.

Like arctic pack-ice seals described earlier in this chapter, crabeater seals are monogynous. Male crabeater seals have been observed to guard a female on the ice for 1 or 2 weeks after her pup is weaned, presumably after mating had occurred. Mate guarding by a male may enhance his prospects for paternity by preventing other males from access to that female after (and possibly even before) he has mated with her.

Eighty percent of crabeater seal pups die before their first birthday. This surprisingly high mortality rate seems to be due to predation by another ice seal, the leopard seal. For those crabeater seals that survive past their first birthday, most exhibit extensive scarring from previous leopard seal attacks. Leopard seals have teeth very much like those of crabeater seals, adapted for straining krill from water. About half their total diet comes from krill (Figure 12.26b). However, these large and supple inhabitants of antarctic pack-ice regions are also major predators of penguins and crabeater seal pups.

Weddell seals are found associated with cracks and leads in fast ice around the entire antarctic continent (FIGURE 12.27). These openings in the ice must be maintained ice free as breathing holes (see Figure 6.42) by constant abrasion by the hole's occupant. Each animal has a preferred breathing hole and returns to it after each dive. This behavior has been used by a succession of researchers to equip free-swimming Weddell seals with electronic equipment to monitor physiologic responses during dives without restraining them and still have a reasonable expectation that the seal will return to the same breathing hole with its expensive equipment package.

The major contributors to the diets of Weddell seals are antarctic cod and other large bottom fish and squid. To harvest these, Weddell seals are extremely good divers. Some dives exceed 80 minutes in duration and 600 m in depth. These exceptional breath-holding abilities permit male Weddell seals to maintain under-ice breeding territories near breathing holes by excluding other males. Each male is limited to mating only with the females that congregate

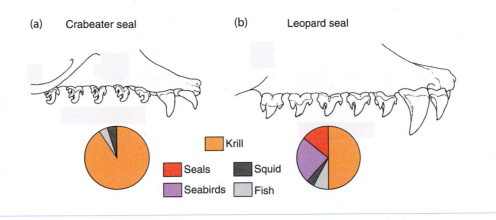

(a) Crabeater seal   (b) Leopard seal

Krill
Seals   Squid
Seabirds   Fish

figure 12.26

Crabeater seal (a) and leopard seal (b) teeth specialized for filtering krill, with the proportion of krill in their respective diets indicated.

figure 12.27

Several Weddell seals on fast ice around a permanent lead or opening.

around his ice hole. Because males must display and mate underwater, their smaller size in comparison with females may make males more agile swimmers and possibly more attractive as potential mates. Males may enhance their attractiveness as mates with long and complex trilling vocalizations (**FIGURE 12.28**). As Weddell seals mate underwater, it is unclear which gender is responsible for mate choice, but female choice is certainly suspected.

Mysticete Whales

The rorquals (family Balaenopteridae) occupying antarctic waters to feed in summer include blue, fin, humpback, and minke whales (see Figure 6.34). All are fast streamlined swimmers equipped with baleen that is intermediate in length between bowhead and gray whale baleen. The mouths of all rorqual species are enormous, extending posteriorly nearly half the total length of the body (**FIGURE 12.29**). All members of this family have 70–80 external grooves in the floor of the mouth and throat. During feeding, this grooved mouth floor expands like pleats to four times its resting size. Alternating longitudinal strips of muscle and blubber interspersed with an elastic protein facilitate this extension. As the mouth floor is extended, small blood vessels in this tissue dilate to give the throat a reddish color (hence the name "rorqual," or red whale, for balaenopterids). As the mouth inflates, it fills with an amount of water equivalent to two thirds the animal's body weight; a mature blue whale might engulf as much as 70 tons of water in a single mouthful.

Water and the small prey it contains enters the open mouth by negative pressure produced by the backward and downward movement of the tongue and by the forward swimming motion of the feeding animal. This method of prey capture, in which large

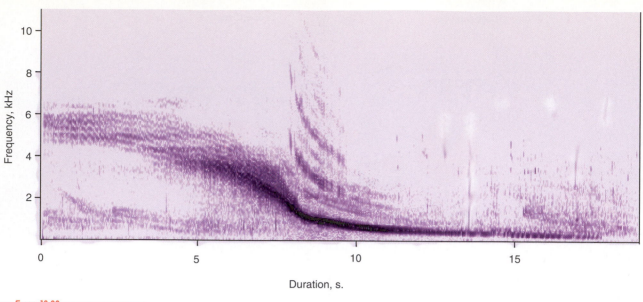

figure 12.28

Sonogram of a 19-second-long descending trill of an underwater Weddell seal, with complex harmonics between 8 and 10 seconds.

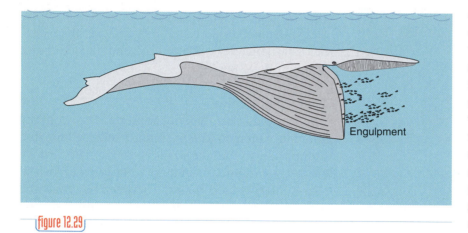

figure 12.29

Engulfment feeding behavior of fin whales and other rorquals. (Adapted from Pivorunas 1979.)

volumes of water and prey are taken in, is referred to as engulfment feeding (Figure 12.29). After engulfing entire shoals of small prey in this manner, the lower jaw is slowly raised to close around the mass of water and prey. Then the muscular tongue acts in concert with contraction of the muscles of the mouth floor (and sometimes with vertical surfacing behavior) to force the water out through the baleen and to assist in swallowing the trapped prey.

In addition to the engulfment behavior of other rorquals, humpback whales often produce a large 10-m diameter curtain or net of ascending grapefruit-sized bubbles of exhaled air, referred to as **bubble-net feeding** (FIGURE 12.30). This net of bubbles is produced by

one member of a group of several cooperating whales that form long-term foraging associations. They seem to confuse prey or cause them to clump into tight balls for easier capture by the whales. While one whale produces the bubble curtain, other members of the group dive below the target prey school and force it into the curtain, and then lunge through the confused school from below. Low-frequency calls that are sometimes produced as the whales lunge through the school of prey may aid in orienting whales during this complex maneuver.

In 1992, the US Navy began to make available to marine mammal scientists the listening capabilities of its Integrated Undersea Surveillance System (IUSS). IUSS was part of the US submarine defense system developed over three decades ago to detect and track Soviet submarines acoustically. The IUSS study has provided a wealth of acoustical information on vocalizations of large baleen whales, especially blue, fin, and minke whales. The vocalizations of blue whales are very loud low-frequency pulses between 15 and 20 Hz, mostly below the range of human hearing, whereas those of fin whales are only slightly higher at 20–30 Hz. Their function is not known, but two plausible explanations have been put forward.

figure 12.30

Final phase of humpback whale bubble-net feeding. Circular trace of the bubble net is still visible.

It is reasonable to conclude that if we can detect these sounds at long distances, other whales should be able to as well. They may therefore function in long distance communication across hundreds or thousands of kilometers of open ocean. These loud, low-frequency, patterned sequences of tones propagate through water with less energy loss than do the higher-frequency whistles or echolocation clicks of small toothed whales described in Chapter 10. In addition to identity calls, these low-frequency pulses may serve an echolocation function, although a very different one than that described for small toothed whales. The low frequency of blue and fin whale tones have very long wavelengths, ranging from 50 m for 30-Hz fin whale calls to 100 m for 15-Hz blue whale calls. If they are used for echolocation, these sound frequencies cannot resolve target features smaller than their respective wavelengths of 50–100 m. Some researchers have speculated that these tone pulses might be used by large mysticetes to locate very large-scale oceanic features, such as continental shelves or islands, sharp differences in water density associated with upwelling of cold water, and possibly even large swarms of *E. superba*.

Despite the differences in body sizes and specializations in feeding behaviors and mechanisms displayed by rorquals, all the rorqual species foraging in antarctic waters exploit a single prey species, the enormously abundant *E. superba*. Their summertime distribution in antarctic waters closely mimic the distribution of krill concentrations. Before the advent of pelagic commercial whaling in antarctic waters at the end of the 19th century, this plentiful food supply supported large populations of all rorqual species. Today, blue, fin, and humpback whale populations still remain depleted as a result of the intense commercial whaling during the first half of the 20th century. For example, 29,000 blue whales were harvested in a single year, 1929, and a total of 1.4 million whales (71 million tons of whale biomass) were removed from antarctic waters by 1960. Today, a total of only 2000 blue whales feed in antarctic waters.

With the larger whale species so efficiently removed from antarctic food webs, the numbers of smaller and commercially less attractive minke whales, Adelie penguins, crabeater and leopard seals, and fish that rely on *E. superba* have responded to the reduced competition with changes in their reproductive rates.

Marine Sanctuaries

Agencies at state, national, and international levels are recognizing the importance of conserving marine biodiversity, and marine protected areas, whether as sanctuaries, parks, or estuarine reserves, play important roles in conserving diversity. In 1979, responding to a proposal by the Republic of Seychelles, the International Whaling Commission (IWC, see p. 403 in Chapter 13) established the Indian Ocean Sanctuary (Figure B12-1). The Sanctuary was initially created for a period of 10 years and was extended indefinitely in 1992, subject to review in 2002. Within this Sanctuary, all pelagic whaling by IWC member nations was prohibited. Several species of baleen whales migrate to the Indian Ocean in summer to breed and calve, and blue and humpback whales are year-round residents. These whales are particularly valuable to the ecotourism industry of the Republic of Seychelles, which is based partly on whale watching.

To extend the effective conservation reach of the Indian Ocean Whale Sanctuary, the IWC created the Southern Ocean Sanctuary in 1994 (Figure B12.1). This Sanctuary includes all waters surrounding Antarctica and bans whaling within

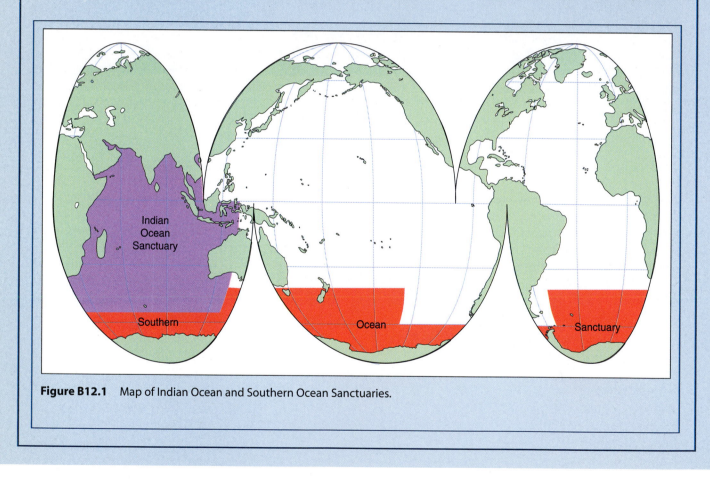

Figure B12.1   Map of Indian Ocean and Southern Ocean Sanctuaries.

The age at sexual maturity of minke whales, for instance, has declined from over 15 years in the 1930s to about 7 years at present, and females have shifted from reproducing every 2 years to every year. As a consequence, the antarctic population of this small rorqual species has exploded to 750,000 animals.

Some of the other complicating effects of commercial whaling are discussed in Chapter 13.

Toothed Whales
Antarctic waters have no permanent toothed whale inhabitants comparable with belugas or narwhals of

the Sanctuary. In doing so, it protects most of the world's baleen whales in their summer feeding grounds. The IWC created this sanctuary to prohibit legal whaling in waters surrounding the Antarctic continent.

The Antarctic minke whale is the world's only remaining large population of baleen whales that has not been seriously depleted by whaling. Protecting the summer feeding grounds of these whales is critical to maintaining their populations.

The protection provided to large whales by this sanctuary status is not complete, however. Japan, a member state of the IWC, consistently refuses to comply with the intent of IWC-sponsored conservation decisions regarding the Southern Ocean Sanctuary and each year unilaterally issues permits to kill several hundred minke whales each year in Antarctic waters for the purpose of "scientific research" (see Figure 13.14).

The National Marine Sanctuaries Program in the United States is intended to create a national system of marine protected areas to conserve, protect, and enhance their biodiversity and ecological integrity (see Figure 9.33). It is administered by the National Oceanic and Atmospheric Administration, a branch of the Department of Commerce. These sanctuaries range in size from the tiny (less than 1 km²) Fagatele Bay National Marine Sanctuary in American Samoa to the Monterey Bay National Marine Sanctuary, extending over 15,744 km².

Only in this way can a reasonable degree of marine species diversity be maintained in a setting that also maintains the natural interrelationships that exist among those species. Several other types of marine protected areas exist in the United States and other countries. A federally managed National Estuarine Research Reserve System (see Figure 9.33) includes 23 designated and protected coastal estuaries. In addition, most coastal states, partially funded by the Federal Coastal Zone Management Act, have developed their own coastal management programs. Abroad, marine protected area programs exist as marine parks, reserves, and preserves. Over 100 designated areas exist around the periphery of the Caribbean Sea. Others range from the well-known Australian Great Barrier Reef Marine Park to little-known parks in developing countries such as Thailand and Indonesia, where tourism is placing growing pressures on fragile coral reef systems.

A United Nations-sponsored international agreement, signed by more than 175 nations so far, was adopted by the General Conference of UNESCO in 1972. Its mission was to assist the conservation of the world's natural and cultural heritage by identifying sites with exceptional features or values that should be preserved and to ensure their protection through close cooperation among nations. To date, more than 750 cultural and natural heritage sites are protected worldwide under

Marine Sanctuaries

this agreement. "Natural heritage" designates outstanding physical, biological, and geological features; habitats of threatened plants or animal species; and areas of value on scientific or aesthetic grounds or from the point of view of conservation. The delicate ecosystems of the Great Barrier Reef and the Galapagos Islands are both examples of natural marine systems that have been designated as U.N. World Heritage Sites. Their listings as World Heritage Sites has led to vigorous efforts to reduce the ecological effect of factors such as human habitation, tourism, and invasions of nonindigenous species. When a site on the List is seriously endangered, it may be added the List of World Heritage in Danger, which entitles it to special attention and international assistance.

www.bioscience.jbpub.com/marinebiology
For more information on this topic, go to this book's website, http://www.bioscience.jbpub.com/marinebiology.

the Arctic. Instead, antarctic waters are seasonally occupied by numerous killer whales, large solitary sperm whales, and occasional beaked and bottle-nosed whales. Sperm, beaked, and bottle-nosed whales prey on large midwater squid, whereas killer whales express the same type of prey specialization seen in the northern hemisphere. In antarctic killer whales, an offshore population consisting of groups of 10–15 animals each prey mostly on other species of marine mammals, especially seals. The whales have been observed tipping ice floes to force resting seals or penguins into the water where they can be more easily captured. The inshore population of killer whales occurs in groups 10 times as large and feeds almost exclusively on fishes. The differences that exist between inshore and offshore antarctic killer whale populations in foraging behavior, group size, and even average body size prompted Soviet researchers to propose separate species status for the two populations.

Sperm whales are one of the most widely distributed mammals on Earth, exceeded only by humans and killer whales. Like killer whales, they also exist in matrilineal social groups. However, sexually mature male sperm whales do not remain with their natal group as do killer whale males. Instead, mature males roam all latitudes alone to feed between periods of visiting female groups for mating. Consequently, solo mature male sperm whales can be found at any latitude from the equator to polar ice edges where water depths remain greater than 1 km. Individuals are frequent visitors to the productive waters around the Antarctic where they feed on larger squid found in deeper waters. Small squid in these areas are known to feed on *E. superba*, but the diet of these larger squid is unknown, and their links to *E. superba* may contain several additional trophic levels.

STUDY GUIDE

Marine Biology On Line

Connect to this book's web site: http://www.bioscience.jbpub.com/marinebiology The site features eLearning, an online review area that provides quizzes, chapter outlines, and other tools to help you study for your class. You can also follow useful links for more in-depth information.

Summary Points

Marine Birds and Mammals in Polar Seas

- The polar ends of Earth share several environmental characteristics. Both experience long winter nights without sunlight, sea surface temperatures around 0°C, perpetual coverage by permanent fast ice, and additional ice coverage due to the seasonal freezing and thawing of pack ice.
- Yet some strong contrasts exist between the north and south poles. The Arctic is an ice-covered ocean, with an annual average productivity rate of about 25 gC/m²/yr, surrounded by continents. The Antarctic is a frozen continent surrounded by ice-covered seas with an annual production rate that is greater than 100 gC/m²/yr.

The Arctic

- Phytoplankton productivity patterns at higher latitudes are marked by large swings between long winters of essentially no productivity and short bursts of high summertime production of cells that tend to be much larger than those in lower latitudes. These in turn feed relatively large consumers and lead to short food chains supporting large aggregations of marine mammals and seabirds that cope with the seasonal variability in their food supplies via extended fasting and long migratory excursions to lower-latitude waters in winter.
- Newborn cetaceans are capable swimmers, whereas polar bear and pinniped pups are unable to swim at birth so they are born on land or ice floes. Newborn mammals are smaller than their parents, with higher surface area-to-volume ratios and poorly developed layers of insulating blubber or fur.
- The growth rates of young marine mammals can be truly astounding. These prodigious growth rates are supported by an abundant supply of high-fat milk. The energy needed to produce large offspring that are fed large quantities of fatty milk are exceedingly high; hence, marine mammals have long gestation periods, reproduce not more than once a year, and, with the exception of polar bears, produce only a single offspring each time they reproduce.
- Several species of seals, including ringed, harp, and hooded, breed on arctic fast or pack ice. These species all show a reduced level or complete absence of polygyny because males have difficulty in gaining and controlling access to multiple females due to the temporary and unpredictable nature of the location, extent, and

breakup time of typical pack ice. In these monogynous seals, the sexual dimorphism seen in polygynous males confer no selective advantage, and monomorphism or even slight reverse sexual dimorphism is typical. Mating in the water is common in these species, and some ice-breeding seals have evolved complex vocalizations to enhance their attractiveness and advertise their interest in mating.

- Polar bears have few specialized adaptations for efficient swimming and usually prey on ice seals. They endure prolonged seasonal fasts as the availability of food declines or when reproductive activities prevent foraging. Polar bears breed in late spring, with females producing two cubs about 2 months into the winter denning period. They remain with her until weaned 2 years later.

- Mysticete whales, among the largest animals on Earth, are filter feeders on planktonic crustaceans or small shoaling fishes (except gray whales, which feed on bottom invertebrates, especially amphipod crustaceans). All mysticetes show reverse sexual dimorphism, which may provide females with larger lipid stores to help offset the energetic costs associated with rapid fetal growth and lactation.

- Mysticete whales exploit these polar and sub-polar production systems with intensive summer feeding followed by long-distance migrations to low latitudes in winter months, with both mating and calving occurring in warm and often protected waters.

- Arctic waters are inhabited, or at least visited, by belugas, narwhals, and killer whales. Belugas are medium-sized whales with a very flexible neck, small flippers, thick blubber, and a stout dorsal ridge used to break through ice from below. They feed on seasonally and locally abundant fishes and invertebrates, both in the water column and on the seafloor. All narwhals possess a tusk that is actually an elongated upper tooth projecting through the lip. They feed mostly on polar and arctic cod, and sometimes dive to depths greater than 1000 m to feed on bottom fishes and midwater squid.

- Killer whales, the most easily recognized of all the whales, are found from tropical waters to both poles. Smaller social groups, or pods, occupy smaller ranges and consist of a mature female, her offspring, and her daughters' offspring (in contrast to many other social mammals, these pods include both male and female siblings that remain together all their lives). The broad ecologic and latitudinal range of killer whales reflects their wide variety of prey species, including fishes, cetaceans, pinnipeds, birds, cephalopods, sea turtles, and sea otters.

- Seabirds (only 3% of bird species) obtain their food from the sea, excreting excess salt ingested during feeding with nasal salt glands. All species of seabirds are either Charadriiformes (gulls, skuas, jaegers, terns, auks, and puffins), Procellariiformes (petrels, fulmars, and albatrosses), or Sphenisciformes (penguins).

- Most polar seabirds are found in the southern hemisphere, nesting at high latitudes during summer, often in enormous and crowded colonies, and then leaving to spend winter months at lower latitudes elsewhere. Only northern fulmars, ivory gulls, and Ross's gulls consistently depend on arctic ice-associated prey in summer.

- Almost all high-latitude sea birds must forage while raising their chicks, mandating frequent and often long feeding trips from nesting areas to the sea.

The Antarctic

- All antarctic communities depend directly or indirectly on *Euphausia superba*, a 6-cm-long, 1-g, herbivorous crustacean that occurs in enormous swarms around much of the Antarctic continent, filtering abundant pelagic diatoms and scraping ice algae from the undersides of pack ice formations. These vast shoals of krill probably support half the world's biomass of seals, and a large proportion of its baleen whales and seabirds as well.

- Adelie and emperor penguins are abundant in antarctic pack-ice habitats. Penguins are flightless birds with streamlined bodies and wings adapted as narrow flippers for efficient lift-based propulsion underwater. On land, their short legs are clumsy for walking on ice or snow, and they commonly flop down on their bellies and "toboggan" while pushing with their flippers and feet. When seeking nesting sites, Adelie and emperor penguins both penetrate pack ice to at least 77°S latitude. The antarctic prion, or whale-bird, feeds on *E. superba* and other crustaceans by running/flying along the surface of the water with outstretched wings and capturing prey by skimming and straining water with comb-like lamellae on either side of their bill, which functions in much the same manner as baleen.

- Crabeater seals, the most abundant seal on Earth, forage almost exclusively on *E. superba*, using their highly modified teeth as effective strainers. These monogamous seals spend the entire year on pack ice, migrating as the ice edge advances and retreats with the seasonal pattern of freezing and thawing. Leopard seals have teeth very much like those of crabeater seals, and about half their total diet is krill. However, they are also major predators of penguins and crabeater seals. Weddell seals are found associated with openings in the fast ice that they maintain as breathing holes by constant abrasion with their teeth. They dive to 600 m for 80 minutes or more to feed on antarctic cod and other large bottom fish and squid.

- Filter-feeding blue, fin, humpback, and minke whales occupy antarctic waters to feed on krill in summer. All are fast streamlined swimmers, with enormous baleen-filled mouths and 70–80 external grooves in the floor of the mouth and throat that expands like pleats to four times its resting size, which enables them to engulf a volume of water that is equivalent to two thirds of their body weight. After engulfing entire shoals in this manner, the lower jaw is slowly raised and the muscular tongue acts in concert with contraction of the muscles of the mouth floor to force the water out through the baleen and to assist in swallowing the trapped prey.

- Today, blue, fin, and humpback whale populations still remain depleted as a result of the intense commercial whaling during the first half of the 20th century. With the larger whale species removed from antarctic food webs, the numbers of smaller minke whales, Adelie penguins, crabeater and leopard seals, and fishes that also rely on *E. superba* have increased dramatically.

Topics for Discussion and Review

1. Compare and contrast the environmental characteristics of Arctic and Southern Oceans.
2. Explain why phytoplankton productivity patterns at higher latitudes lead to short food chains.
3. Why do the growth rates of young marine mammals need to be truly astounding?
4. Why do most arctic species of seals show a reduced level of polygyny and sexual dimorphism?
5. Why do all mysticetes show reverse sexual dimorphism?
6. How do killer whale pods differ from the groups or herds of many other social mammals?
7. All antarctic communities depend directly or indirectly on a single prey species. What is it?
8. Explain how the antarctic prion and crabeater seals forage on krill.
9. Explain how antarctic filter-feeding rorquals engulf vast volumes of water to feed on krill.
10. Why have the numbers of minke whales, Adelie penguins, crabeater and leopard seals, and fishes around Antarctica increased dramatically during the past 50–60 years?

Suggestions for Further Reading

Alonzo, S. H., P. V. Switzer, and M. Mangel. 2003. Ecological games in space and time: the distribution and abundance of antarctic krill and penguins. *Ecology* 84:1598–1607.

Ancel, A., G. L. Kooyman, P. J. Ponganis, J.-P. Gendner, J. Lignon, X. Mestre, N. Huin, P. H. Thorson, P. Robisson, and Y. Le Maho. 1993. Foraging behavior of emperor penguins as a resource detector in winter and summer. *Nature* 360:336–339.

Dunton, K. 1992. Arctic biogeography: the paradox of the marine benthic fauna and flora. *Trends in Ecology and Evolution* 7:183–189.

Eastman, J. T. 1993. *Antarctic Fish Biology*. Academic Press, San Diego.

Highsmith, R. C., and K. O. Coyle. 1993. Production of Arctic amphipods relative to gray whale energy requirements. *Marine Ecology Progress Series* 83:141–150.

Kasamatsu, F., S. Nishiwaki, and H. Ishikawa. 1995. Breeding areas and southbound migrations of southern minke whales *Balaenoptera acutorostrata*. *Marine Ecology Progress Series* 119:1–10.

Kingsley, M. C. S., and I. Stirling. 1991. Haul-out behavior of ringed and bearded seals in relation to defence against surface predators. *Canadian Journal of Zoology* 69:1857–1861.

Kooyman, G. L., T. G. Kooyman, M. Horning, and C. A. Kooyman. 1996. Penguin dispersal after fledging. *Nature* 383:397.

Kumar, N., R. F. Anderson, R. A. Mortlock, P. N. Froelich, P. Kublik, B. Dittrich-Hannen, and M. Suter. 1995. Increased biological productivity and export production in the glacial Southern Ocean. *Nature* 378:675–680.

Lydersen, C., and K. M. Kovacs. 1996. Energetics of lactation in harp seals (*Phoca groenlandica*) from the Gulf of St. Lawrence, Canada. *Journal of Comparative Physiology B* 166:295–304.

Martin, A. R., and T. G. Smith. 1992. Deep diving in wild, free-ranging beluga whales, *Delphinapterus leucas*. *Canadian Journal of Fisheries and Aquatic Sciences* 49:462–466.

McGonigal, D., L. Woodworth, and E. Hillary. 2001. *Antarctica and the Arctic: The Complete Encyclopedia*. Firefly Books, Toronto.

Moloney, C. L. 1992. Carbon and the Antarctic marine food web. *Nature* 257:259–260.

Nevitt, G. A. 2000. Olfactory foraging by Antarctic procellariiform seabirds: life at high Reynolds numbers. *Biological Bulletin* 198:245–253.

Russell, R. W. 1999. Comparative demography and life history tactics of seabirds: implications for conservation and marine monitoring. Pages 51–76 in J. A. Musick, ed. *Life in the Slow Lane: Ecology and Conservation of Long-Lived Marine Animals*. American Fisheries Society, Bethesda.

Schreer, J. F., K. K. Hastings, and J. W. Testa. 1996. Preweening mortality of Weddell seal pups. *Canadian Journal of Zoology* 74:1775–1778.

Shirihai, H., B. Jarrett, and G. M. Kirwan. 2002. *Complete Guide to the Antarctic Wildlife: Birds and Marine Mammals of the Antarctic Continent and the Southern Ocean*. Princeton University Press, NJ.

Siniff, D. B. 1991. An overview of the ecology of Antarctic seals. *American Zoologist* 31:143–149.

Sturm, M., D. K. Perovich, and M. C. Serreze. 2003. Meltdown in the north. *Scientific American* 289:60–67.

Testa, J. W. 1994. Over-winter movements and diving behavior of female Weddell seals (*Leptonychotes weddellii*) in the southwestern Ross Sea, Antarctica. *Canadian Journal of Zoology* 72:1700–1710.

CHAPTER

13

Many types of gear and vessels are used to harvest seafood. This vessel, docked in a Taiwanese harbor, is designed to capture fish larvae.

Harvesting Living Marine Resources

For much of human history, the seas have been largely isolated from the pressures created by the material needs of our land-based human population. Two thousand years ago, the human population was probably between 200 and 300 million people. The daily challenges of survival kept life expectancies short. High birth rates were balanced by high death rates, and the population grew slowly. Not until 1650, when the Industrial Revolution in Europe brought advances in medicine and technologic improvements in food production, did human mortality rates decline and the population double to 500 million. By 1850, the population had doubled again to one billion people. Declining death rates continued to accelerate the rate of population growth through the 20th century. Presently, the human population of the world is more than six billion people and is projected to surpass seven billion by 2015.

The sheer magnitude of the current human population creates enormous demands for food, fiber, minerals, and other commodities that support our modern social fabric. Today, most of the world ocean is intensively exploited for fishing, recreation, military purposes, commercial shipping, dumping of waste materials, and extraction of gas, oil, and other mineral resources. Although the impact of these activities varies geographically, no portion of the world ocean is isolated from their effects. These effects often transcend national borders as well as the boundaries of many ecologic and taxonomic groups. Where the cumulative biologic effects of these societal uses of the sea are large and measurable, they introduce a complex suite of social, political, economic, aesthetic, and biologic questions that cannot be resolved solely by scientific means.

The demand for food made by the present human population is enormous and is increasing at a rapid rate. The United Nations Food and Agriculture Organization (UNFAO) and the World Bank estimate that almost one billion people are seriously undernourished. Uncounted millions of people starve to death each year, and hundreds of millions more are deprived of good health and vigor because of inadequate diets. Many people, therefore, regard marine sources of food as a critical part of the solution to present and future food problems.

Fishing is a multibillion-dollar global industry. For our purposes, the term "fishing" is broadly defined as the capture or collection of wild marine plants and animals for food production or other commercial activities. These other activities range from collecting coral reef animals for the ornamental fish trade to killing seal pups for their skins. In 2000, the global import/export trade in fishery commodities was about 80 billion U.S. dollars. Most seafood is sold for direct human consumption, but an appreciable portion is also used for livestock fodder, pet foods, and a variety of nonfood industrial products.

The impact of fishing practices on the stability, and even the continued existence, of some marine populations has been severe, because a relatively small number of species make up the bulk of the world's fish harvest. These species have borne the brunt of the human population's demands for seafood. Some species have been completely eliminated from traditional fishing grounds, whereas others face the imminent possibility of biologic extinction.

At the beginning of the 19th century, cod in the North Atlantic were so abundant that boats often hit these fish while lowering their anchors. Haddock was a popular food fish in Europe, and salmon swam in North American coastal waters in uncounted millions. In central California, a new and promising sardine fishery was just getting started. In the Southern Hemisphere, antarctic waters teemed with large baleen whales in summer months. And very few people knew of pollack, hake, or menhaden. Now, as we begin the 20th century, the haddock populations are no more; they peaked in 1929. The cod fishery of the North Atlantic has collapsed, and we buy more farm-raised salmon than fish caught from wild populations. The sardine "Cannery Row" popularized by John Steinbeck now exists only as rows of shops and boutiques for tourists. Even the largest animal on Earth, the blue whale, currently numbers less than 10% of what it did a century earlier. And pollack, hake, and menhaden are among the most abundant marine fishes

captured. In this chapter, we look at some of the reasons for how these dramatic changes in dominant marine animal populations occurred.

13.1 A Brief Survey of Marine Food Species

The raw material of the fishing industry includes numerous species of bony and cartilaginous finfish, many mollusks and crustaceans (collectively referred to as shellfish even if, like squid, they lack shells), a variety of other aquatic animals (from worms to whales), and even some marine plants. Each year the UNFAO compiles and publishes global fishery catch statistics that are the basis of most of the figures and tables in this chapter. The units of measure used by the UNFAO (and used in this chapter) are metric tons, equivalent to 1000 kg. FIGURE 13.1 summarizes capture results and indicates capture trends at decade intervals since 1970. The annual catch size has varied somewhat; however, the relative ranking of most groups has remained reasonably constant for the past three decades.

Clupeoid fishes, including anchovies (FIGURE 13.2), anchovetas, herrings, sardines, pilchards, and menhaden, are very abundant and account for about one third of the total commercial catch. A single species, the Peruvian anchoveta (*Engraulis ringens*), provided more than 13 million tons, or nearly 19%, of the total 1970 catch, but extensive overfishing caused the fishery to collapse a few years later. The herring catch is an important part of the North Atlantic fishing industry. In the last decade, other herring fisheries in the South Atlantic and North Pacific oceans have been expanding. At its peak in the 1930s, the California sardine industry was landing over 500 thousand tons annually. The menhaden catch yields approximately one million tons annually, largely from the Atlantic Ocean and the Gulf of Mexico.

The huge size of the clupeoid fish catch is not indicative of its dollar value as an economic commodity nor its significance as a source of protein in human nutrition. Nearly all anchovies, anchovetas, and menhaden and much of the herrings, sardines, and pilchards caught are reduced to fish meal to be used as a low-cost protein supplement in livestock and poultry fodder. As fish meal, clupeoid fishes make only indirect contributions to the human diet in the form of pork or chicken.

Clupeoid fishes are found in shallow coastal waters and in upwelling regions. Their schooling behavior (Figure 13.2) simplifies catching techniques

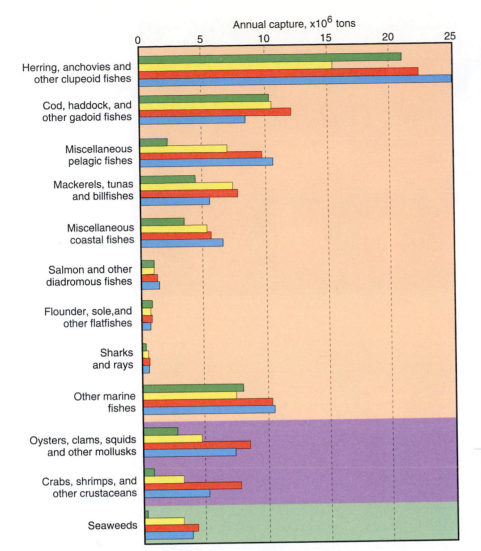

Annual capture, x10⁶ tons

- Herring, anchovies and other clupeoid fishes
- Cod, haddock, and other gadoid fishes
- Miscellaneous pelagic fishes
- Mackerels, tunas and billfishes
- Miscellaneous coastal fishes
- Salmon and other diadromous fishes
- Flounder, sole, and other flatfishes
- Sharks and rays
- Other marine fishes
- Oysters, clams, squids and other mollusks
- Crabs, shrimps, and other crustaceans
- Seaweeds

figure 13.1

Global marine and estuarine harvest by major categories for 1970 (green bars), 1980 (yellow bars), 1990 (red bars), and 2000 (blue bars). (Data complied from UNFAO statistics.)

figure 13.2

The northern anchovy, *Engraulis mordax*.

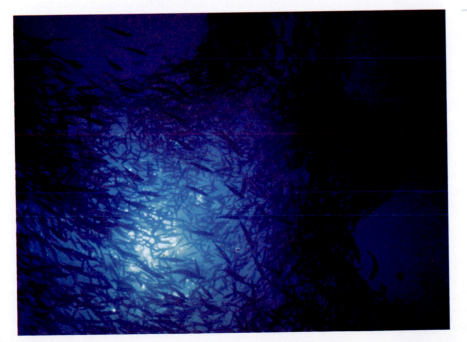

figure 13.3

Tuna being ladled out of the purse seine shown in Figure 13.4.

and reduces harvesting expenses. Large purse seines (which may be 600 m long and 200 m deep, similar to the seine shown in Figure 13.4) are used to surround and trap entire schools. Once encircled, the fish are ladled or pumped into the ship's hold. Most clupeoid fishes are small, with an average adult length of between 15 and 25 cm. Their small size, however, is compensated for by other characteristics that enhance their economic usefulness. These fish have fine gill rakers that enable them to feed on small organisms close to the base of marine food chains. Mature Peruvian anchovetas feed principally on chain-forming diatoms and other relatively large phytoplankton aggregations. Herring generally feed on herbivorous zooplankton (see Figure 2.14).

The combined catch of cod, pollack, hake, and other gadoid fishes has remained fairly constant for the past two decades, although cod catches have declined severely during that period. Gadoid fishes usually live on or near the bottom, are larger than clupeoid fishes, and feed at higher trophic levels. Fishing operations for these species are concentrated on continental shelves and other shallow areas. Cod were the foundation of coastal fisheries of many North Atlantic nations until the early 1990s. Alaskan pollack are caught by trawler fleets in shallow regions of the Gulf of Alaska and the Bering Sea. Pollack catches have increased substantially in the past two decades even as Steller sea lion populations, a major predator of pollack, experienced a serious decline. Several species of hakes abound on the continental shelves

of many oceans and are slowly gaining favor in the United States, where they are often marketed as "whitefish."

Redfish, bass, sea perch, and other miscellaneous coastal rockfishes live near the seafloor. No single species of this group dominates the catch statistics, yet the combined catch since 1980 consistently has amounted to more than five million tons. Sold fresh or frozen, most of these fish are popular fish-market items. Miscellaneous pelagic fishes such as horse mackerel and jack mackerel generally feed on smaller anchovy-sized fish and squid. These mackerels resemble tuna in form, but they seldom exceed 30 to 40 cm in length.

Tuna are individually the largest of the commercially exploited species discussed here. Tuna weighing over 100 kg are not unusual, but most range from 5 to 20 kg (FIGURE 13.3). Yellowfin, bigeye, albacore, and skipjack tuna account for most tuna catches. Tuna are often the top carnivores in complex food chains and may be separated from the primary producers by seven or more trophic levels (see Chapter 10). These fish are active predators of smaller more abundant animals, especially clupeoid fishes and sauries, and are captured in or near nutrient-rich waters where prey fish abound.

U.S. tuna fishers rely heavily on long-range purse seiners (FIGURE 13.4) to harvest schooling species of tuna, especially skipjack and yellowfin tuna. Yellowfin tuna exhibit a strong but poorly understood behavioral association with some small subtropical dolphins

A modern tuna seiner with its purse seine surrounding a school of tuna.

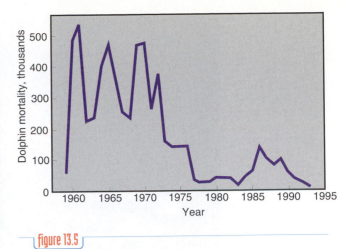

figure 13.5

Annual incidental dolphin mortality in the tropical Pacific Ocean tuna purse seine fishery.

(*Stenella* spp.). Because dolphins are easily visible at the sea surface, purse seining of yellowfin tuna is often simply a matter of setting the nets around a dolphin school with the assumption (usually correct) that a school of tuna swims just below. By the early 1970s, the global tropical tuna fishing fleet had killed several million dolphins unintentionally during capture efforts for tuna. Since then, tighter restrictions on both the design and operation of purse seining, imposed by the 1972 U.S. Marine Mammal Protection Act, have reduced annual dolphin mortality due to purse seining to about 20,000 for the U.S. tuna fleet (FIGURE 13.5).

In an effort to counter the adverse publicity associated with this high dolphin mortality, several large tuna processing companies announced policies in 1990 to buy and market only "dolphin-safe" tuna, tuna caught without setting purse seines on dolphin schools. It remains to be seen whether these policies can be effectively monitored by the tuna-consuming public.

To exploit the more widely dispersed bluefin and bigeye tuna, labor-intensive long-line fishing methods are also used. These tuna aggregate in a narrow zone of relatively high primary productivity that straddles the equator in the eastern half of the Pacific Ocean (see Figure 4.26). These large predators feed on sardines, lanternfish and other small fish, and crustaceans, which in turn graze on the smaller zooplankton. The long-line fisheries of the equatorial Pacific Ocean have developed into a valuable

oceanic fishery, with Mexico and Japan each taking about 100,000 tons of yellowfin, bigeye, and bluefin tunas. The long-line gear consists of floating mainlines that extend as far as 100 km along the surface. Hanging from each mainline are about 2000 equally spaced vertical lines that terminate with baited hooks. Once a long-line set is in place, fishers move along the mainline, remove hooked tuna, and rebait and replace the hooks.

In recent years, many fisheries have begun using gill nets. Designed to entangle fish in the net fabric, these large rectangular nets are either anchored to shore or allowed to drift. Although effective at entrapping fish, they are also invisible and indiscriminate killers of nontarget species. These nets are fast becoming major causes of mortality for some species of sharks and many seabird and marine mammal populations.

The world's commercial fishing fleet contributes more than 100,000 tons of accidentally lost nets, pots, traps, and setlines and deliberately discarded pieces of damaged fishing gear each year. The shift since 1940 from the use of natural fibers to virtually nonbiodegradable synthetic fibers for the construction of nets, lines, and other fishing gear has made commercial fisheries a major contributor to marine plastics pollution. The largest potential for lost and discarded fishing gear occurs in the North Pacific Ocean, where vessels of many nations operate under adverse climatic conditions. Over a million kilometers of gill net are now set annually by the squid fisheries. These nets constitute a large potential source of derelict gear that when lost continues to drift for thousands of kilometers. Birds, mammals, and other

Tuna entrapped in a drift net.

animals that surface frequently are especially susceptible to entanglement in the floating net fabric or other plastic debris (FIGURE 13.6).

13.2 Major Fishing Areas of the World Ocean

Shallow waters over continental shelves and near-surface banks encourages rapid regeneration of critical nutrients in the photic zone. These nutrients are prevented from escaping into deeper water and accumulating below the upper mixed layer where return to the surface requires a much longer time. Fishing activities in neritic waters are effective because the resulting animal production is "crowded" into a water column usually less than 200 m deep. In contrast, organic material produced over deep-ocean basins is shared by the many consumers thinly dispersed through several thousands of meters of water.

About 90% of the marine catch is taken from continental shelves and overlying neritic waters, a region representing less than 8% of the total oceanic area. Bottom fish and benthic invertebrates account for 10–15% of the total global catch. Halibut, flounder, sole, and other flatfishes are benthic species caught in shallow waters using bottom trawls. High prices and stable markets for these fish have resulted in heavy fishing pressures for most of the past century, and most populations are seriously overfished.

Other major fishing areas are centered in regions of upwelling where abundant supplies of critical nutrients from deeper waters are returned to the photic zone. Upwelling may, depending on the locality, occur sporadically, on a seasonal basis, or continually throughout the year. (Mechanisms of antarctic, equatorial, and coastal upwelling were described in Chapter 3.) The nutrient-rich waters surrounding the antarctic continent sustain very high summertime primary production. Yet, with the exception of the near-defunct pelagic whaling industry and a small annual harvest of krill, no large commercial fishery has yet developed in antarctic waters. Long distances to processing facilities and the absence of nearby population centers have had a restraining effect on the successful exploitation of this upwelling region.

Regions of coastal upwelling are most apparent along the west coasts of Africa and North and South America (see Figure 4.24) but occur to lesser degrees along other coastlines. High yields of commercial species result from increased rates of primary production over shallow bottoms. The greatest concentrations of clupeoid fishes are found in regions of coastal upwelling: Peruvian anchoveta from the Peru Current upwelling area, pilchard from a similar area in the Benguela Current off the west coast of South Africa, and, before a drastic decline in the 1940s, sardines from the California Current. The Peruvian and Benguelan upwelling systems also support enormous populations of seabird and pinniped predators of these small fish.

388   CHAPTER 13 Harvesting Living Marine Resources

13.3 A Perspective on Sources of Seafoods

Humans are omnivores. We obtain nourishment from a large variety of plants and animals. Yet the staples of the human diet can be narrowed down to just a few types of plants and animals. The plant staples include cereals (rice, wheat, corn, and lesser amounts of other cereal grains), vegetables, fruits, nuts, and berries. Beef, poultry (plus the milk and egg products of these animals), pork, and fish provide the major share of the food to satisfy the carnivore in us. It is important to understand just how important the present marine contribution of food is to the total human diet and how meaningful it may be in the future.

In TABLE 13.1 two categories of plant and animal foods from both terrestrial and oceanic production systems based on the state of technology used in the production of each food category are compared. Plant production is separated into *gathering*, the use of wild plants (FIGURE 13.7), and *farming*, the agricultural tending of domesticated plant species. Comparable categories for animals are used: *hunting* of wild animals and *herding* of genetically improved controlled domesticated animals. These terms are also applied to marine food items. Only a few marine plants, bivalve mollusks (FIGURE 13.8a), and salmon (Figure 13.8b) are grown in controlled conditions collectively referred to as **mariculture**. Table 13.1 lists the amount of food produced or captured in 2000 for land, fresh water, and ocean. Production statistics for wild plants and wild animals grown on land can only be approximated. The remaining figures are compiled annually by the UNFAO.

The information presented in Table 13.1 forces some uncomfortable, yet undeniable, conclusions. Most seafoods harvested are animals three or four trophic levels above the primary producers. In terrestrial agricultural systems, more plants are harvested than animals, and losses to higher trophic levels are reduced. Furthermore, most of the marine harvest consists

Table 13.1

Human Food from Land, Freshwater, and Ocean Production Systems for 2000

Food source	Production category	Food production, × 10^6 tons		
		Land	Freshwater	Ocean
Plants	Gathering	150	ND	4
	Farming	2060	ND	6
Animals	Hunting	120	8	79
	Herding	930	21	12
Totals		3260	29	101

Used for fish meal:	−25
For direct consumption:	76
Additional food from fish meal:	5
Total: (direct comsumption + additional food produced from fish meal)	81

figure 13.7

A kelp harvester in operation. As the harvester moves backward, the kelp is cut below the sea surface and then pulled up the stern loading ramp.

(a)

(b)

figure 13.8

(a) Oysters grown in mariculture. (b) Rearing cages in Canadian near-shore salmon farm.

of wild stocks that are hunted rather than controlled and domesticated. Although technologically advanced ships, nets, and fish-finding gear may be used, little of the monetary gains are reinvested in these wild fish stocks. The role of mariculture in the mix of seafood production is discussed in the next section.

Another negative trend in the use of living marine resources has developed in the past half century.

In 1955, 86% of the world fish catch was consumed directly; the remainder was reduced to fish meal for use as a protein supplement for domestic livestock. By 1970, the fish meal fraction had increased to about 40% of the world catch and has remained at 30–35% since. Assuming the reduced fish meal in 2000 was fed to pigs and chickens with trophic efficiencies of 20%, only five million additional tons of edible pork and poultry were produced (Table 13.1). So the actual amount of human food derived from the 2000 fish catch was not the total catch of 101 million tons. Instead, it was nearer 81 million tons; 76 million tons of edible fish and 5 million tons of livestock raised on fish meal derived from 25 million tons of fish.

Long food chains, unsophisticated production systems, and extensive industrial uses of marine organisms all severely limit the amount of marine food actually produced. The marine environment just does not produce very much food; for the past several decades, only 1–2% of the food consumed by the world human population came from the sea. The likelihood of greatly increasing that fraction of our marine diet depends on the magnitude of still unharvested fish stocks and how well we manage the future harvesting of these stocks.

How much fish can we continue to remove from the world ocean? The question is not easily answered, but there are several indications that we are already overharvesting most commercially important fish populations. The ultimate limit on marine sources of food is established by the rate of photosynthesis by primary producers. The marine production system begins at the first trophic level with a net production of 450 billion tonnes of biomass annually (from 45 billion tons of carbon; see Table 4.2). Neither markets nor techniques are available for economically harvesting phytoplankton, so the magnitude of the fish harvest is actually limited by the number of trophic levels in the food chain leading to harvestable fish and the efficiency with which animals at one trophic level use food derived from the previous trophic level.

The potential commercial value of a fish or other animal is, to a large degree, a function of its individual body size. These animals must be a certain minimum size before commercial exploitation is economically practical. In addition, the size of organisms at the base of marine food chains is an essential feature in establishing the number of steps in the food chain. Food chains based on smaller phytoplankton cells (such as those in subtropical gyres) generally consist of a greater number of trophic levels, as do food chains leading to large fish such as tuna (see Figure 10.13).

In general, phytoplankton cells decrease in size from greater than 100 μm in coastal and upwelled waters to less than 25 μm in the open ocean. Several moderately sized zooplankton species such as *Euphausia pacifica,* which function as herbivores in coastal North Pacific waters, must move one step up the food chain and assume a carnivorous mode of feeding in offshore waters because the phytoplankton there are too small to be captured. Other forms of oceanic zooplankton occupying the third trophic level are no more than 1 or 2 mm in length. Virtually all species of herbivorous copepods in the open ocean are preyed upon by chaetognaths that, in turn, become food for small fish.

Thus, in open ocean environments, three or four trophic levels are required to produce animals only a few centimeters in length. Food chains leading to tuna, squid, and other commercially important open ocean species consist of an average of five trophic levels; with an average number of three trophic levels for commercial species in coastal waters and one and one half for upwelling areas. The number of trophic levels for upwelling areas is low because many clupeoid fish taken from upwelling areas graze directly on phytoplankton without any intermediate trophic levels.

Accurate estimates of ecologic efficiencies in marine food chains also are difficult to achieve. Efficiency factors are based on the growth of organisms that is, in turn, a function of food assimilation minus waste and metabolic costs (such as respiration and locomotion). These factors vary widely between species, between individuals of the same species, and between various stages in the life cycles of a species. Young growing individuals often exhibit efficiencies as high as 30%, but their ecologic efficiencies decline to nearly 0% at maturity. Thus, efficiency estimates for populations composed of a variety of age groups can be little more than reasonably intelligent guesses. It is even more difficult to approximate the ecologic efficiencies across entire trophic levels consisting of a diverse group of animal types, each with its own peculiar age structure and growth rate. With that caveat, current estimates of average ecologic efficiencies are 10% for the oceanic province, 15% for coastal regions, and 20% for areas of upwelling are used here. With these estimates of net phytoplankton production (presented in Table 4.2 and repeated in part in TABLE 13.2), average numbers of trophic levels, and efficiencies of exchange between the trophic levels for each marine production province, we can estimate the total potential production of fish from the sea (Table 13.2, last column) to be about 540 million tons annually.

A potential annual production of about 540 million tons of animals is not the same as an actual harvest of that magnitude. Large portions of some commercially important species are consumed by seabirds, larger fish, and other predators. Furthermore, to maintain healthy populations available for continued exploitation, harvesting activities must allow a reasonably large fraction (generally one half to two thirds) of these populations to escape and reproduce so that harvesting can continue at that level.

Thus, using existing fishing methods to harvest presently exploited types of seafoods, the resource potential exists to double or, at best, triple our present global fish catch. But a two- or three-fold increase of the 1–2% that seafood contributes to our present diet is still a very small portion of our total food needs, and continued increases in the human population will surely offset much of those gains in future fisheries production. Other possibilities for increasing our future seafood harvests beyond the limits suggested by the numbers in Table 13.2 include expanding mariculture and moving our harvesting efforts closer to the bases of marine food webs.

Estimates of Harvestable Fish Production for Marine Production Provinces

table 13.2

Province	N.A.P.P. (×10⁹ tons)	No. trophic levels	Trophic efficiency (%)	Potential fish production (×10⁶ tons)
Open ocean	350	5	10	3
Coastal	99	3	15	334
Upwelling	4	1.5	20	200
Total				537

Net annual primary productivity (N.A.P.P.) based on values from the right column of Table 4.2 multiplied by 10 to convert carbon weight to wet (live) weight.

13.4 Fishing Down the Food Web

In theory, if we fished one trophic level down the food web and harvested what currently harvested fish eat rather than harvesting the fish themselves, a 5- to 10-fold increase could be harvested because one trophic level and its associated energy loss would be eliminated. Rather than contemplating limits of 200 or 250 million tons of seafood each year, we might anticipate harvesting one or two billion tons instead. Serious problems, however, block this path to greatly increased marine harvests. Almost without exception, animals occupying lower trophic levels are smaller and more dispersed than the animals now harvested.

Although the technologic developments necessary to harvest the zooplankton and smaller fish that comprise these lower trophic levels are not insurmountable, they may not be worthwhile. These smaller more dispersed animals are more difficult to harvest; the extra energy needed to collect these small food items may exceed the energy gained from the additional harvest. It is likely to continue to be more efficient for us to wait until the larger animals have eaten the smaller ones then to harvest the larger ones.

To provide one example, a century ago the phytoplankton crop around the antarctic continent supported a tremendous assemblage of zooplankton, particularly the krill *Euphausia superba* (see Figure 12.23). In turn, krill fed large populations of blue, fin, and humpback whales. With these whale populations now seriously reduced, they no longer serve as intermediaries to harvest the krill for us. Fisheries scientists from several countries are test-harvesting krill with an eye to expanded future production. Preliminary estimates of an annual sustained harvest of 100 million and even 200 million tons have been suggested for this one species alone. The potential of such massive harvesting is a complex issue with several ramifications unrelated to food production.

A more serious consequence of fishing down the food web is the reverse of the one just described. When intensive commercial harvesting is focused on a trophic level near the base of a food web, larger predators occupying higher trophic levels are denied adequate prey, and their populations will suffer even though they are not directly targeted by harvesting activities.

Another problem inhibiting widespread use of krill is one common to any plan to use smaller marine animals: What is done with it once it is caught? There is little demand for fresh or frozen krill. It is nutritious, but few people will buy it in its natural form. The energy costs involved in catching krill, processing it to a palatable form, and transporting it to markets in the Northern Hemisphere are enormous. Thus, the likelihood of using krill and other nontraditional food species positioned on low trophic levels in marine food webs is presently marginal at best. Our most reasonable approach to satisfy our expanding food needs must continue to rely on improving techniques for terrestrial crop production using currently available techniques to grow well-known crops already in demand by society.

13.5 Mariculture

Mariculture is the application of agricultural techniques to grow, manage, and harvest marine animals and plants. It may vary from simple enhancement of natural populations by releasing hatchery-reared juvenile fishes to intensive captive maintenance of species for their entire life span. As the numbers in Table 13.1 indicate, farming of marine plants and animals presently contributes less than 20% to the total of marine food production. For practical reasons ranging from market proximity to theft prevention, mariculturists confine their activities to coastal waters.

Salmon farming in Japan, Norway, Canada (see Figure 13.8b), Chile, and the United States has expanded rapidly in the past three decades to about one million tons annually, slightly exceeding the annual harvest of wild-caught salmon. Fish farming has been practiced for centuries in southeast Asia, Japan, and China. Mullets and milkfish are grown in shallow estuarine ponds where they graze on algae, detritus, and small animals. Estuaries, salt marshes, and other productive coastal habitats are preferred for cultivating some fish species in closed pond systems. However, intensive mariculture activities modify the nature of an estuary or salt marsh. In special circumstances, dissolved nutrients from sewer treatment plants or hot water from coastal power-generating stations can be used beneficially. Diverted into fish ponds, nutrients and the warm water could enhance the growth of primary producers to feed the fish. But these are special circumstances. More commonly, mariculture programs are, and will continue to be, restricted by adverse problems of coastal pollution and habitat destruction and by competing uses for the same land. For each fish pond installed, a portion of the native fish and shrimp populations already contributing to previously established local fisheries are displaced or denied access to these productive coastal waters.

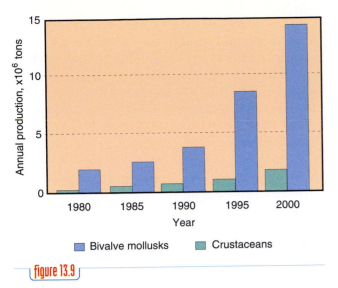

Growth of global maricultural production of marine crustaceans and bivalve mollusks. (Data complied from UNFAO statistics.)

The major expansion in global mariculture has occurred with bivalve mollusks (clams, oysters, mussels, and scallops) and, to a much smaller extent, with shrimps, crabs, and other crustaceans (FIGURE 13.9). These animals also thrive in estuaries and other shallow coastal habitats. As consumers of detritus and bacteria (see Figure 5.24) close to the bases of their respective food webs, they are ideal species for mariculture.

Mariculture is not a panacea for all our marine food production problems. When salmon are raised in crowded cage conditions, large quantities of antibiotics are used to suppress infections, and animal wastes concentrated into small areas can create serious localized water quality problems. Mariculture remains a costly business; it is and probably will continue to be restricted to expensive luxury food items that generate large returns on investments. Captive-raised oysters and salmon will add little to the diets of most people on Earth and certainly nothing to the diets of those who most need it.

13.6 The Problems of Overexploitation

Commercial and subsistence fishing represents a form of predation that has predictable effects on the prey species. When harvesting of a new population begins, initial catches are generally large and include a high proportion of large fish. Continued or increased fishing pressure tends to reduce the size of the population as well as the sizes of individual members of that population. If the fishing effort is adjusted to the growth and reproductive potential of the population, then a **maximum sustainable yield** of fish can be caught year after year without causing major upsets in the populations. Too often, however, the fishing pressure becomes much greater than the population can withstand. When losses to fishing and natural predators together exceed replacement by young animals, populations decline, and these declines are quickly reflected in reduced catches.

Numerous examples of overfished stocks can be found in most segments of the fishing industry. Most of the popular species of halibut, plaice, cod, ocean perch, herring, and salmon of the North Atlantic and Pacific Oceans are already overexploited. So are many of the warm-water tuna stocks and most of the large whales. The two examples discussed here are sufficient to demonstrate some of the problems created by overfishing activities that seem in direct conflict with the fishing industry's own best interests.

North Atlantic Cod

Cod of the North Atlantic Ocean are the inheritors of the fishing pressure that used to be concentrated on haddock early in the 20th century. When the overfished haddock populations collapsed, fishing efforts shifted to the more abundant cod. The factory trawlers introduced in the 1950s revolutionized the practice of catching these fish, hauling them in by the ton rather than by hand-lining for individual fish. By 1970, over one million tons were being harvested annually (FIGURE 13.10). In retrospect, the eventual result seems entirely predictable. As the number of mature reproducing individuals 7 years or older was reduced, the population diminished but the pressure to capture

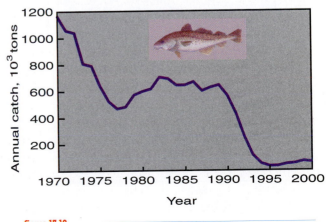

The dramatic decline in Northwestern Atlantic cod harvests, 1970–2000. (Data complied from UNFAO statistics.)

them did not. Existing fishing effort was increasingly concentrated on decreasing numbers of fish, and catches declined (Figure 13.10).

It is a sad irony that the pattern of decline, indicated by the right side of Figure 13.10, provides some of the best information about the actual sizes of cod populations. In other words, once you have eaten all the fish, you then know how many there used to be. After several years of blaming seal predation and deleterious oceanographic conditions, U.S. and Canadian regulatory agencies belatedly accepted the fact that the population had collapsed and closed nearly all cod fisheries in their waters in 2003. Several European nations followed suit for cod fishing areas on the east side of the North Atlantic. How quickly the cod populations recover is unknown, as is the type and magnitude of response to the recovery by fishers and the governmental agencies responsible for that fishery.

The Peruvian Anchoveta

The Peruvian anchoveta (*E. ringens*) has become the classic textbook example of the consequences of over-fishing. Unlike the Atlantic cod that has been exploited for centuries, the commercial harvesting of Peruvian anchoveta has been sufficiently recent that reliable data exist on the initial growth and eventual collapse of this fishery. This species is a typical clupeoid fish almost identical to the anchovy shown in Figure 13.2. It is a small fast-growing filter feeder that schools in the upwelling areas of the Peru Current. The first commercial use of the anchoveta was indirect. From the time of the Incas, droppings or guano deposits from the nesting colonies of seabirds that fed on anchoveta have been collected and used as a major source of fertilizer. These seabirds, primarily Peruvian boobies, brown pelicans, and guanay cormorants, annually converted about four million tons of anchoveta to an inexpensive fertilizer widely used by Peru's subsistence farmers.

Commercial exploitation of the Peruvian anchoveta for reduction to fish meal began in 1950. The next year, 7000 tons were landed. After 1955, the growth of the fishery was explosive (FIGURE 13.11): Over 3 million tons were landed in 1960. By 1970, the catch of this one species had surpassed 13 million tons, al-

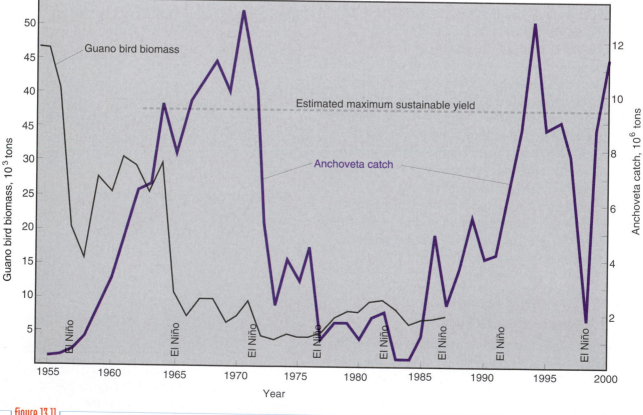

Figure 13.11

Changes in the anchoveta catch and the guano bird populations along the northwest coast of South America. El Niño years are indicated. (Adapted from Muck 1989 and UNFAO statistics.)

most one fifth of the entire world seafood harvest for that year. Nearly all the catch was taken by local fishermen and reduced to fish meal and oil for export.

Accompanying the meteoric rise in commercial anchoveta catches was a drastic drop in the number of guano birds that depended on the anchoveta for food. From 28 million in 1956, the guano bird population was reduced to six million during the El Niño year of 1957. Phytoplankton populations were dramatically reduced, anchoveta mortality increased, and the guano birds starved. After 4 years without an El Niño, the bird population had rebounded to 17 million, only to be hit by another El Niño in 1965. That time the bird population plummeted to four million, and it has not recovered.

A substantive base of biologic information was collected during the early development and growth of the Peruvian anchoveta fishery. With the advantages this information base provided, proper management procedures were expected to ensure a large and perpetual harvest from this immense population. The biologic information suggested that a maximum sustainable yield of approximately nine and a half million tons annually was possible. At the time, the reduced bird populations were taking less than one million tons each year. Tonnage reports alone, however, are inadequate and sometimes misleading. As the fishing pressure increased over the decade of the 1960s, the average size of fish being caught decreased and the number needed to make a ton increased sharply. By 1970, small anchoveta were taken with such efficiency that 95% of the juvenile fish recruited into the population were captured before their first spawning.

A glance at Figure 13.11 shows that the catches of anchoveta for 1967 through 1971 exceeded the predicted maximum sustainable yield by at least one million tons each year. The industry and the regulatory agencies responsible for managing the anchoveta stocks had ample warning of what was to come. In 1972, sampling surveys indicated the anchoveta stocks had been severely depleted and recruitment of juvenile fish was poor. As expected, the 1972 catch dropped to little more than five million tons, less than half that of the previous year. Even worse was 1973, with an estimated catch of less than three million tons. The excessive fishing pressures of the previous decade were too much even for this tremendous fish population, and it finally collapsed. Since 1980, efforts to harvest anchoveta and the population's response to that fishing effort have been extremely variable. Catches have moved between a 1994 harvest that rivaled the record 1970 catch to less than

two million tons 4 years later. In the 1990s, the huge bird populations that once relied on this species have shown no signs of recovery to pre-1970 levels.

13.7 The Tragedy of the Commons

Why have fishing enterprises and fishing nations repeatedly exploited the fish resources on which they depend to the point that returns on their fishing effort decline and too often disappear? Fishing is too often considered a right without attendant responsibilities. Even with the maze of legal and economic considerations involved, incentives for these apparently self-defeating actions are not difficult to find. Cod, tuna, anchovetas, and other oceanic species are unowned resources belonging to no single nation or individual. Many exist outside the jurisdictional limits of all nations and are therefore open to access by any nation. Historically, the concept of open access to the high seas evolved in the 16th and 17th centuries when the right to navigate freely was more crucial than the freedom to fish. But as coastal fish stocks were depleted, fishers became increasingly dependent on more distant stocks in international waters. They eventually discovered that the freedom to fish on the high seas was fundamentally different from the freedom to navigate. Unlike navigation, fishing activities remove a valuable commodity from a common resource pool at the expense of everyone, including those who do not fish.

It is commonly assumed that oceanic resources are unowned and open to access for all people. But with today's advancing pace of technology, some nations have achieved the ability to exploit these resources more quickly than others. If the fishing fleets of one nation fail to catch these unowned oceanic species, those of some other nation soon will. In species after species, such attitudes have led to inevitable and predictable results: increased competition for limited resources, duplication of effort, declining fishing efficiency, and, of course, overfishing.

Once in the net, a school of fish is no longer the property of all people; it belongs instead to those who set the net and haul the fish aboard. All people and all nations share the cost of losing the fish, the great whales, and the other marine animals that have nearly disappeared because of overfishing. Yet the short-term profits derived from overfishing are not similarly shared. This is Hardin's concept of the "tragedy of the commons." The tragedy of this situation is that it best rewards those who most heavily exploit and abuse the unprotected living resources of the open ocean.

13.8 International Regulation of Fisheries

Without controls to limit the access of fishers to fish populations in international waters, concerned nations have created a variety of multinational and international regulatory commissions for the purpose of governing the management and harvest of regional fish populations. Common strategies for management include setting quotas on the amount of fish harvested, establishing seasons and minimum size limits, defining acceptable gear such as net mesh sizes, and limiting the number of boats that are allowed to participate in a fishery. The International Commission for the Northwest Atlantic Fisheries, for instance, includes Canada, Denmark, France, Germany, Iceland, Italy, Norway, Poland, Russia, Spain, the United States, and the United Kingdom. The regulations established by this commission do not carry the weight of international law, but they are binding on member nations. Even so, they have failed to halt serious overfishing of cod and other valuable fish populations in the North Atlantic.

Other commissions, particularly those regulating the halibut and salmon fisheries in the North Pacific, have been more successful. Their very success in maintaining viable populations, however, is now creating new problems. Too many additional fishing boats from nations all over the world are attracted to these well-managed fisheries, magnifying management problems, diminishing profits of individual fishers, and increasing the possibilities for eventual overexploitation.

To counter some of the difficulties of managing fisheries by multinational commissions, several individual nations have worked to develop individual systems to regulate fish populations that are nearby yet outside their own territorial or jurisdictional limits. In 1974, the U.S. government, under pressure from its own coastal fishing interests, ended its traditional policy of opposition to extended zones of control for coastal nations. In 1976, the United States passed the Fisheries Conservation and Management Act. Under this act, the United States assumed exclusive jurisdiction over fisheries management in a zone extending from shore 200 miles to sea. Entrance of foreign fishing vessels into this exclusive economic zone is allowed on a permit basis only, and then the fishers are allowed to harvest only fish stocks with sustained yields greater than the harvesting capacity of American fishers.

In 1982, the United Nations Conference on the Law of the Sea adopted a draft Law of the Sea (LOS) Treaty that was to go into effect 12 months after ratification. Two decades later, ratification of the treaty was still seven supporting nations short of the 60 needed. However, many nations have unilaterally adopted major elements of the LOS Treaty. The LOS Treaty includes many of the key features found in the U.S. Fisheries Conservation and Management Act, in particular a 200-mile-wide exclusive economic zone granting coastal nations sovereign rights with respect to natural resources (including fishing), scientific research, and environmental preservation. Additionally, the LOS Treaty obliges nations to prevent or control marine pollution, to promote the development and transfer of marine technology to developing nations, and to settle peacefully disputes arising from competitive exploitation of marine resources.

The imposition of 200-mile-wide exclusive economic zones by essentially all coastal nations of the world has dramatically changed the concept of open access for most of the world's continental shelves, coastal upwelling areas, and major fisheries lying within 200 miles of some nations' shorelines (FIGURE 13.12). This treaty is notably silent regarding the Antarctic upwelling area, because individual national territorial claims on the Antarctic continent are not recognized. Commercial ventures there, including either whale or krill harvesting, must be consistent with the goals of the 1981 convention for the Conservation of Antarctic Marine Living Resources, namely, the "maintenance of the ecological relationships between harvested, dependent, and related populations of Antarctic marine living resources and the restoration of depleted populations." This was the first in a slowly growing trend of national and international efforts to recognize the integrity of large marine ecosystems. To date, about 50 large marine ecosystems, mostly in coastal waters, have been identified as integrated ecologic systems that respond as functional units to external stresses (such as overexploitation), environmental fluctuations (such as El Niño), or large-scale pollution problems.

The waters around the Antarctic continent represent the last relatively unspoiled large marine ecosystem on Earth. The international cooperation demonstrated so far in protecting its living resources has been unusual in the long history of our attempts to protect marine resources. The waters around this remote continent can continue to serve both as a laboratory for improving understanding of living marine systems and as a model for creating approaches to preserving those resources as a common heritage of humankind.

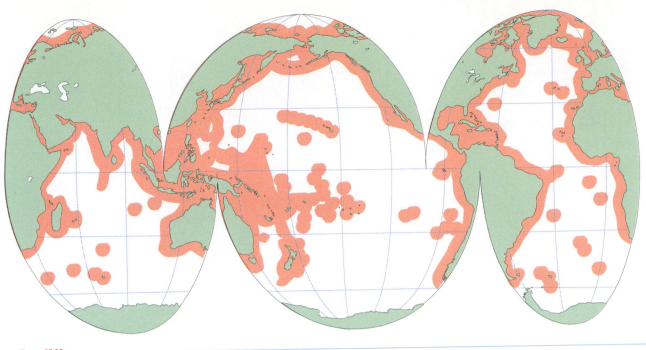

figure 13.12

Worldwide extent of the 200-mile exclusive economic zones sanctioned by the United Nations LOS Treaty.

13.9 Marine Ornamentals

It has been estimated that as many as two million people worldwide keep personal marine aquaria (600,000 households in the United States alone), supporting a trade in fishes, corals, and other invertebrates that may be worth as much as $330 million annually. Most marine ornamental animals are collected and transported from Southeast Asia, although additional Indo-Pacific Islands are becoming more important as a source of desirable species. These collected species are usually shipped to the primary markets of the United States, Europe, and Japan (80% of stony corals and 50% of fishes are exported to the United States).

This impressive trade could provide many jobs in coastal areas and significant economic incentive for coral-reef conservation because much of the Indo-Pacific region consists of low-income third-world island nations. For example, in 2000, 1 kg of Maldivian reef fishes destined for the aquarium trade was valued at nearly $500 (if used for food these fishes would be worth just $6). In addition, one metric tonne of live coral has a value of $7000 in the aquarium trade but of only $60 if used for limestone production. And 1 kg of live rock from Palau can earn as much as $4.40 if sent to aquarium hobbyists, but only two cents if used locally as construction material. Unfortunately, destructive collection techniques (such as the use of

sodium cyanide discussed in Chapter 9), overexploitation of some species, and careless handling and transport all undermine potential benefits of the trade.

In 2000, the Global Marine Aquarium Database (GMAD) was created to obtain standardized species-specific trade data from importers and exporters as a first step toward long-term conservation and sustainable use of coral reefs and their inhabitants. Today, the database contains nearly 103,000 trade records covering about 2400 species of fishes, corals, and other invertebrate species involved in the hobby. The GMAD shows that 20–24 million individual fishes in 1471 species are traded worldwide each year. Nearly half of these are damselfishes, with various angelfishes, surgeonfishes, wrasses, gobies, and butterflyfishes comprising an additional 25–30% (and just 10 extremely popular fish species accounted for 36% of all fishes traded between 1997 and 2002). GMAD also reports a total of 140 species of stony corals (11–12 million pieces), 61 species of soft corals (390,000 pieces), and 9–10 million individuals of 516 species of other invertebrates (mostly mollusks, cleaner shrimp, and sea anemones) were traded worldwide during the same time period.

Sadly, the popularity of marine ornamentals is often driven by their appearance rather than their ability to survive collection, handling, transport, or captivity. For example, although cleaner wrasses

Using Science to Detect Misleading Science

Case 1: China's Harvest

If you carefully examine the total world seafood harvest shown at decade intervals in Figure 13.1, you might reasonably conclude that the annual worldwide catch has remained more or less constant over the past three decades. According to the United Nations Food and Agriculture Organization (UNFAO) reports, China, for example, has harvested more than 15 million tons of food from the sea for each of the past 3 years, up from a total of 2.5 million tons in 1970. According to a study by two marine scientists at the University of British Columbia, China's catch has been poorer than reported and is currently declining rather than increasing each year

Only one international organization, the UNFAO, compiles global

fisheries statistics. The UNFAO must trust member countries to provide reliable statistics, even if there is some doubt about their validity. To check the reliability of the data the UNFAO uses to assess the health and viability of fish stocks, Watson and Pauly estimated fish harvests within 176,000 ocean "cells," each 0.5° of latitude by 0.5° of longitude. Included in their analysis were factors such as each cell's depth, its surface temperature, and its productivity.

Their statistical model produced reasonable year-by-year estimates for most cells around the world, but the calculated fish yields near China's shore consistently came up short. For example, Watson and Pauly estimated that the catch within China's Exclusive Economic Zone (EEZ) in 1999 was barely half as large as what was reported to the UNFAO. One explanation is that China's economic system encourages the reporting of inflated numbers because officials' promotions are partially dependent on reports of increased output. When the new estimates were substituted for China's existing UNFAO data for the past three decades, this new analysis clearly showed that the total global catch has actually been

dropping since 1988. This analysis underscores the seriousness of recent collapses in many regional fish stocks, which have increasingly forced fishers to move out of coastal waters to exploit international waters that do not compensate for depleted coastal stocks in individual nations' EEZs. These new data support the growing contention that we are seriously overfishing the ocean on a global scale. This sort of misreporting by countries with large fisheries can lead to serious errors in reported trends that impede effective management of international fisheries and misrepresents the real role of seafood in our global diet.

Case 2: Whale Meat in Japan

The North Pacific gray whale (*Eschrichtius robustus*), like most other baleen whales, is protected from commercial hunting by several multinational or international agreements dating back to 1937. Two of these agreements, the International Convention for the Regulation of Whaling (which established the International Whaling Commission, or IWC, in 1946) and the Convention on International Trade in Endangered Species of Wild Fauna and Flora

---

(*Labroides*), butterflyfishes (*Chaetodon*), and the harlequin filefish (*Oxymonacanthus longirostris*) are characterized as "truly unsuitable" for home aquaria by the GMAD (mostly because of their highly restricted diets), they are still commonly traded. In addition, the attractive appearance of carnation or strawberry coral (*Dendronephthya*) makes it one of the most commonly traded soft corals, yet it typically dies within a few weeks of captivity (because it lacks the helpful zooxanthellae described in Chapter 9).

Opponents of the aquarium trade criticize the destructive methods used to collect marine ornamentals, the potential overharvesting of some species, the wasteful mortality caused by slipshod husbandry

during shipping, and the introduction of exotic species that may disrupt local ecologic relationships. Damaging harvesting techniques are diverse and common. Sodium cyanide and other harmful chemicals, such as rotenone and quinaldine, are used to stun fishes so that they are easier to catch, yet overdoses are common, postcapture mortality of drugged fishes is very high, and incidental poisoning and killing of adjacent reef creatures and corals is common. Collection for the aquarium trade damages the reef itself in other ways, including direct collection of corals for shipment, fragmentation of the reef to provide easier access to cryptic reef dwellers, and accidental damage to neighbors of targeted individuals.

(commonly known as CITES), specifically prohibit the harvesting of gray whales or selling any of their products for commercial purposes.

Two populations of gray whales exist in the North Pacific Ocean. Since their nadir in the 1920s, the eastern Pacific population of gray whales has recovered to more than 22,000 individuals, whereas the western Pacific (or Asian) population has not recovered from commercial exploitation and remains one of the most endangered populations of whales in the world. Recent surveys of their summer feeding grounds in the Sea of Okhotsk have indicated that the total population probably numbers fewer than 100 individuals. The winter breeding area of these western Pacific gray whales is unknown but is presumed to be along the southern China coastline. The migratory corridor between feeding and breeding areas is also unknown but probably includes parts of the coasts of Korea and Japan.

As part of a continuing study using molecular genetic techniques to monitor which whales are hunted and sold for food, Baker and colleagues in 1999 purchased a total of 102 whale products from commercial markets along the Pacific coast of Japan. Using a portable laboratory for initial amplification of the mitochondrial DNA, a technique now commonly used for species identification, seven products purchased in two locations were identified as gray whale meat, although five were labeled as "minke for sashimi" and the other two were unlabeled.

There are no exceptions to the international agreements protecting gray whales in Japanese waters or to prohibiting the commercial sale of gray whale meat. The only source of gray whale products in Japan that would not be an infraction of international agreements would be fisheries bycatch or strandings, yet there are no reports to the IWC to connect any stranded or netted whale to the sold meat. With the techniques available to these researchers, it was impossible to determine whether the meat came from the eastern Pacific population or the much more endangered western Pacific population that presumably migrates through Japanese coastal waters twice each year. Even without making that population distinction, the selling of any gray whale meat in Japanese markets appears to represent a serious violation of international agreements established to protect this species.

Using Science to Detect Misleading Science

Additional Reading

Baker, C. S., M. L. Dalebout, G. M. Lento, and N. Funahashi. 2002. Gray whale products sold in commercial markets along the Pacific coast of Japan. *Marine Mammal Science* 18:295–300.

Baker, C. S., and S. R. Palumbi. 1994. Which whales are hunted? A molecular genetic approach to monitoring whaling. *Science* 265:1538–1539

Watson, R., and D. Pauly. 2001. Systematic distortions in world fisheries catch trends. *Nature* 414:534–536.

www.bioscience.jbpub.com/marinebiology
For more information on the Internet, go to this book's website http://www.bioscience.jbpub.com/marinebiology.

Studies in Sri Lanka, Kenya, the Philippines, Indonesia, and Hawaii have revealed localized depletion of several target species. Supporters of the aquarium industry criticize these studies as preliminary and point out that a study in the Cook Islands suggests that aquarium fishes there are being harvested sustainably (the catch per unit of effort was stable from 1990 to 1994). Nevertheless, the only systematic study to date assessed the impact of harvesting Hawaiian reef fishes for the aquarium trade. This study, conducted by Hawaii's Department of Land and Natural Resources, revealed that the abundance of 8 of the 10 most popular species declined at harvest sites (as opposed to control sites where collecting was not permitted). It is reasonable to assume that corals targeted for the trade are even more vulnerable to overharvesting because of their extremely slow growth rates, but a great deal more data are needed before more definitive conclusions can be reached.

The death of reef fishes after their collection is caused by many factors, some of which lead to extreme rates of mortality. One report suggests that 75% of fishes collected with narcotics, such as cyanide, die within hours of being taken from the reef. More typical mortality rates are caused by physical damage during capture, overcrowding and disease during shipping, and inattentive husbandry. A study of fishes captured in Sri Lanka reported that 15% died during and immediately after capture, 10% died during shipping, and another 5% died before being sold to consumers in the United Kingdom, and the mortality did not end there. GMAD data show that corals in home aquaria experience 76–100% mortality within 18 months. Shockingly, the most "hardy" coral species, *Plerogyra* and *Catalaphyllia,* experience 54% and 60% mortality within 18 months of purchase, yet they remain popular in the industry.

Clearly, fisheries for marine ornamentals need to be managed to ensure sustainable use, to decrease conflict with other user groups (such as sport divers who prefer that no animal be removed from natural reefs), and to keep postcapture mortalities low. Although in their infancy, such efforts already include attempts to culture popular ornamentals for trade and the implementation of typical management initiatives discussed elsewhere in this chapter, such as limiting entry into the fishery, establishment of catch quotas and size limits, creating marine reserves, and temporarily closing popular harvest sites.

13.10 Sealing and Whaling

The large body sizes and high fat and oil content of pinnipeds and whales have long made them targets of commercial harvesting efforts. Sometimes the desired product was meat, but whale and seal oil and skins with dense insulating fur were also sought.

Pinnipeds

The gregarious nature and relatively poor terrestrial locomotion of pinnipeds make them easy targets for sealers seeking skins and oil. In the past two centuries, several pinniped species have been severely decimated by commercial harvests. The northern fur seal population numbered about 2.5 million when discovered by Russian sealers in the 18th century. By 1911, the population was reduced to about 100,000 animals. With international protection, 50,000 to 100,000 fur seals are killed each year in a limited hunt that has allowed a partial recovery to about half its preexploitation population size.

Each year, nearly 100,000 ringed seal pups in the North Atlantic were killed for their skins, as were 250,000 to 370,000 harp seal pups (see Figure 12.5) for each of the 5 years from 1996 to 2000. This 5-year period had the highest number of harp seals

taken since the 1960s and was justified with the dubious assumption that harp seal predation was the major factor responsible for the recent decline in the cod harvest in the North Atlantic Ocean.

Northern elephant seals (introduced in Chapter 10) were even more seriously decimated by sealers than were the smaller fur seal and harp seal. Once distributed from central California to the southern tip of Baja California, this species came under commercial hunting pressure in the early 19th century. A scant half century later, so few survived that they were not worth hunting. No elephant seals were sighted between 1884 and 1892. In 1892, C. Townsend, a collector for natural history museums, discovered eight animals on Isla de Guadalupe, 240 km off the coast of Baja California. Seven were promptly shot for museum specimens. Early census estimates suggest that in the 1890s as few as 25 individuals survived on a single inaccessible beach on Isla de Guadalupe. Beginning with protection afforded the northern elephant seal by Mexico and the United States in the early 20th century, that remnant population slowly recovered. Since then, the northern elephant seal has again spread throughout its former breeding range. The total population has swelled to more than 300,000 animals, and the future of this species now seems secure.

Or does it? Is the present northern elephant seal population really as viable and hardy as the preexploitation population? Comparisons of blood proteins from 159 animals of the "recovered" population suggest that they are not. In marked contrast to proteins of other vertebrate species, no structural differences were demonstrated either between individuals or between separate groups of northern elephant seals breeding on different islands. The lack of structural differences in these proteins points to a complete absence of variation in the genes controlling the synthesis of these proteins. M. L. Bonnell and R. K. Selander suggest that the absence of genetic variability in the existing northern elephant seal population is the result of a genetic bottleneck when the population was at its low point in the 1890s. It is quite conceivable that on that isolated beach on Isla de Guadalupe, a lone elephant seal male dominated the breeding of all the surviving sexually mature females for several years. If so, half the pool of genetic information possessed by the surviving representatives of this species was funneled through this single animal, and much of the genetic variability presumably inherent in the predecimation population was lost. The rapid recovery of the protected population indicates that genetic variability may not be essential to the short-term survival of this species. However, this species remains vulnerable to environmental changes occurring in an extended time frame because it may lack the genetic variability necessary to cope with such changing conditions.

For some species of pinnipeds, hunting is not the cause of their population decline. Rather, as predators near the tops of their food webs, they frequently encounter intense competition from an even more efficient predator: humans. Steller sea lions provide a dramatic example of the consequence of this competition for food resources. Steller sea lions in the North Pacific numbered about 250,000 in 1960 and were important predators of pollack, salmon, herring, capelin, and rockfishes. By 1985, the population had declined to 185,000, and by 1990 to 90,000 (FIGURE 13.13). The immediate cause of the decline was determined to be reduced pup and juvenile survival due to a scarcity of food. All their usual prey species have experienced dramatic increases in commercial fishing pressure since 1975. Although it is difficult to es-

tablish a direct link between increased fishing and sea lion mortality, most biologists studying this predator–prey interaction agree that the reduction and geographic redistribution of the sea lions' prey species had a large negative effect on the survival of juvenile sea lions.

Baleen Whales

The history of the global whaling industry has been a long and tragic one. Aboriginal hunting of coastal whales for food has occurred for several thousands of years. In the 18th and 19th centuries, however, whaling took a new and ominous turn. Pelagic whales became major items of commerce as demand for their oil grew and whaling as an industry grew into a profitable commercial enterprise. Ships from a dozen nations combed the oceans for whales that could be killed with hand harpoons and lances. Right, bowhead, and gray whales were favorite targets because they swam slowly and, once killed, floated conveniently at the sea surface. In his famed whaling novel, *Moby-Dick,* Herman Melville questioned the future

(a)

(b)

figure 13.13

Steller sea lion rookery beach photographed on the same day of the year in (a) 1969 and (b) 1987.

of these great whales, faced, as they were, with "... omniscient look-outs at the mastheads of the whale-ships, now penetrating even through Behring's straits, and into the remotest secret drawers and lockers of the world; and the thousand harpoons and lances darting along all the continental coasts; the moot point is, whether Leviathan can long endure so wide a chase, and so remorseless a havoc." By the end of the 19th century the gray whale, right whale, and bowhead whale were all on the verge of extinction. Under strict international protection, the gray whale has since recovered, but the number of bowheads and right whales remains low.

The era of modern whaling was initiated in the late 19th century with the cannon-fired harpoon equipped with an explosive head. The explosive harpoon was so devastatingly effective that even the large rorquals, the blue and fin whales, were taken in large numbers. These whales had previously been ignored by whalers with hand harpoons because they were much too fast to be overtaken in sail- or oar-powered boats. The subsequent rapid decline of the whale stocks in the North Atlantic and Pacific Oceans forced ambitious whalers to seek new and untouched whaling areas. They found the Antarctic, the rich feeding grounds of the planet's largest whales. The discovery of the antarctic whale populations touched off seven decades of slaughter unparalleled in the history of whaling.

Aided by pelagic factory ships fitted with stern ramps to haul the whale carcasses aboard for processing, the kill of large rorquals rose dramatically.

From 176 blue whales in 1910, the annual take climbed to almost 30,000 in 1931 (FIGURE 13.14). After the peak year of 1931, blue whales became increasingly scarce. Blue whale catches declined steadily until they were commercially insignificant by the mid-1950s. In 1966, only 70 blue whales were killed in the entire world ocean. Only then, when substantial numbers could no longer be found to turn a profit, was the hunting of blue whales banned in the Southern Hemisphere.

The trend of increasing and then rapidly declining annual catches, shown by the blue whale curve in Figure 13.14, is distressingly similar to that of the Peruvian anchoveta fishery. But the ruthless exploitation of the great whales did not halt with the near extinction of the blue whale. As the blue whale populations gave out, whalers switched to the smaller more numerous fin whales, catches of which skyrocketed to over 25,000 whales each year for most of the 1950s. But by 1960, the fin whale catch also began to plummet, and whaling pressure was diverted to even smaller sei whales. The total sei whale population probably never exceeded 60,000. One third of the sei whale population was killed in 1965 alone. Whaling pressure quickly pushed the catches of this species far beyond its maximum sustainable yield. By the late 1960s, sei whales had followed their larger relatives to commercial extinction, and the whaling effort was shifted to the minke whale, an 8- to 9-m miniature version of a blue whale (see Figure 6.34).

As early as 1940, whaling nations were faced with undeniable evidence that some populations of pelagic

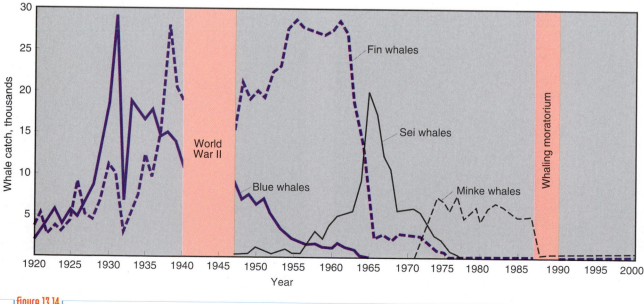

figure 13.14

Harvests of blue, fin, sei, and minke whales in the Antarctic, 1920–2000. (Data complied from UNFAO statistics.)

whales were seriously overexploited. In 1948, 20 whaling nations established the International Whaling Commission (IWC) to oversee the utilization and conservation of the world's whale resources. Unfortunately, the IWC had neither inspection nor enforcement powers. Only once during the decade from 1963 to 1973 did the whaling industry actually manage to catch the quotas established for it by the IWC. The quotas were, in effect, not quotas at all, because there were not enough whales to fill them. In 1974, the IWC, under pressure from several national governments and international conservation organizations, adopted a new set of management procedures for several geographically localized stocks of whales. Under these procedures, all species of baleen whales in the Antarctic except the minke are classified as protected stocks, and no harvesting is allowed. Four species, the blue, gray, humpback, and right whales, are protected in all oceans.

To some, the severe depletion of the large pristine stocks of krill-eating baleen whales led to the naive assumption that without these whales, a large "surplus" of krill exists for our harvesting. Such an assumption ignores the evidence that a new equilibrium has developed in krill-based food chains. Minke whales, seabirds, and seals now play larger roles than they did a century ago. This is not a unique situation, for many fish stocks depleted by excessive harvesting pressures have been replaced with other (and, from a commercial point of view, often less desirable) species. These new species assemblages, by their very existence in niches previously occupied by the exploited stocks, serve as a barrier to the recovery of those stocks.

The road to recovery for a large slowly reproducing species such as the blue whale is fraught with unanswered questions. Can the few thousand remaining blue whales scattered over the world ocean encounter each other frequently enough to mate, reproduce, and add to their decimated numbers? Will additional blue whales overcome competition from other animals for food and other resources appropriated during their tragic decline? And, most importantly, when the population begins to increase, will humans refrain from exploiting it so that it can secure a more solid grip on survival? The reproductive resiliency of the great whales is largely unknown. They may bounce back. The gray whale did, but the right whale still has not.

In 1982, the IWC approved a 5-year moratorium on the commercial killing of large whales. This moratorium was seen as a critical first step in assessing the impact of nearly a century of intense hunting of pelagic whale stocks. At the same time, it would give declining whale populations an opportunity to stabilize. Even as it went into effect, the moratorium was opposed by several nations with long whaling histories, especially Japan, Norway, and Iceland. These nations continued to take whales (particularly fin and minke whales) for so-called scientific research purposes during the moratorium.

When the 5-year moratorium against whaling expired in 1991, many IWC member nations supported its indefinite extension. Others, especially those originally opposing the moratorium, have resumed limited whaling of minke and fin whales. How the international community handles this delicate problem will be a serious test of our biologic wisdom as well as our political will.

13.11 Concluding Thoughts: Developing a Sense of Stewardship

The world ocean is a great and forgiving ecosystem, for centuries providing food, transportation, and recreation while absorbing the effluvia of human societies. Our continued use of the seas for these purposes depends on a better understanding of the consequences of our intrusion on the workings of marine ecosystems. We are the only species on Earth capable of understanding the motives and consequences of our actions. Yet the history of our economic involvement with marine populations has repeatedly demonstrated a serious lack of practical awareness of the fundamental disturbances our intervention causes in these natural systems. In our scramble for food, sport, and profit from the sea, we have repeatedly, and often with disastrous results, violated existing ecologic relationships and invented new ones. We are positioning ourselves with increasing frequency at the tops of heavily exploited marine food webs that are contaminated with the very substances we are afraid to dispose of near our homes. It would be wise to heed the plight of the cod, the anchoveta, and the great whales as sensitive indicators of what might be in store for us if we continue our unthinking and uncaring misuse of the world ocean. We hope that as we begin our journey through the 21st century, we can learn to leave some of our old and wasteful habits behind. It will not be easy or simple, but each one of us must develop a sense of stewardship toward the world ocean and its resources that is reflected in our personal choices and in our political decisions.

Summary Points

Harvesting Living Marine Resources

- The current human population creates enormous demands for food. Today, most of the world ocean is intensively exploited for fishing, recreation, military purposes, commercial shipping, dumping of waste materials, and extraction of gas, oil, and other mineral resources. These uses of the sea and their effects introduce a complex suite of social, political, economic, aesthetic, and biologic questions that cannot be resolved solely by scientific means.

A Brief Survey of Marine Food Species

- The raw material of the fishing industry includes numerous species of bony and cartilaginous finfish, shellfish, and a variety of other aquatic organisms (from worms to whales, and even some marine plants).
- Clupeoid fishes account for about one third of the world's total commercial catch. Clupeoid fishes are found in shallow coastal waters and in upwelling regions. Their schooling behavior simplifies catching techniques and reduces harvesting expenses. The magnitude of the clupeoid fish catch is not indicative of its dollar value nor its significance as a source of food (nearly all clupeoids caught are reduced to fish meal to be used in livestock and poultry fodder). Cod, pollack, hake, and other gadoid fishes, redfish, bass, sea perch, and other miscellaneous coastal rockfishes, and tuna are also important commercial targets. Bottom fishes and benthic invertebrates account for 10–15% of the total global catch

- Many fisheries use gill nets, large rectangular nets that are also invisible and indiscriminate killers of nontarget species. These nets are fast becoming major causes of mortality for some species of sharks and many seabirds and marine mammals. Moreover, these types of nets and others, as well as lost or discarded pots, traps, and setlines, all contribute to the more than 100,000 tons of gear that enter the world ocean each year.

Major Fishing Areas of the World Ocean

- About 90% of the marine catch is taken from continental shelves and overlying neritic waters, a region representing less than 8% of the total oceanic area, because shallow waters encourage rapid regeneration of crucial nutrients. Several major fishing areas are centered in regions of upwelling, such as those along the west coasts of Africa and North and South America.

A Perspective on Sources of Seafood

- An understanding of the present marine harvest forces some worrisome, yet undeniable, conclusions. First, most of the seafood harvested are animals three or four trophic levels above the primary producers (more plants than animals are harvested on land, thus losses to higher trophic levels are reduced). Second, in 1955, 14% of the world fish catch was reduced to fish meal; by 1970, the fish meal fraction had increased to about 40% of the world catch. Third, for the past several decades, only 1–2% of the food consumed by the world human population came from the sea, but there are several indications that we are already overharvesting most commercially important fish populations.
- The magnitude of the fish harvest is limited by the number of trophic levels in the food chain that lead to harvestable fishes and the efficiency with which animals at one trophic level use food derived from the previous trophic level. Because 1.5–3 trophic levels lead to most commercial species and ecologic efficiencies range from 10% to 20%, the resource potential exists to double or triple our present global fish catch. Other possibilities for increasing future

harvests include expanding mariculture and harvesting species closer to the bases of marine food webs.

Fishing Down the Food Web

- By fishing one trophic level down, a 5- to 10-fold increase could be harvested because one trophic level (and its associated energy loss) would be eliminated. Although the technologic developments necessary to harvest these lower trophic levels are not insurmountable, they may not be worthwhile because these smaller more dispersed animals are more difficult to harvest and the extra energy needed to collect them may exceed the energy gained from the harvest. Moreover, harvesting near the base of a food web may deny larger predators adequate prey, and their populations may suffer even though they are not directly targeted by commercial fisheries.

Mariculture

- Mariculture, wherein agricultural techniques are used to raise marine animals and plants, presently contributes less than 20% of our total marine food production. Estuaries, salt marshes, and other productive coastal habitats are preferred for cultivating marine species, yet for each fish pond installed, a portion of the native populations are displaced or denied access to these productive coastal waters, which often serve as essential nursery areas. In addition, when salmon are raised in crowded cage conditions, large quantities of antibiotics are used to suppress infections and animal wastes concentrated into small areas can create serious localized water quality problems.

The Problems of Overexploitation

- If our annual catch is adjusted to the growth and reproductive potential of a population, then a maximum sustainable yield can be harvested year after year without negative impact on the population. Alternatively, if the fishing pressure is greater than the population can withstand, populations decline and catches are reduced.
- Numerous examples of overfished stocks can be found in most segments of the fishing in-

dustry. Cod of the North Atlantic Ocean were among the first to be taken by factory trawlers (rather than by hand-lines for individual fish). Soon, the population diminished and catches declined, prompting U.S. and Canadian regulatory agencies to close nearly all cod fisheries in their waters in 2003. A similar fate befell the Peruvian anchoveta, a classic example of the consequences of overfishing.

The Tragedy of the Commons

- Fishing is too often considered a right without attendant responsibilities. Oceanic species exist outside the jurisdictional limits of all nations and therefore are open to access by any nation. Yet unrestricted fishing activities remove a valuable commodity from a common resource pool, and all people, including those who do not fish, share the cost of losing the fish, the great whales, and the other marine animals that have nearly disappeared because of overfishing. Yet the short-term profits derived from overfishing are not similarly shared. This is Hardin's concept of the "tragedy of the commons."

International Regulation of Fisheries

- Concerned nations have created a variety of multinational and international regulatory commissions for the purpose of governing the management and harvest of regional fish populations. Common strategies for management include setting quotas on the amount of fish harvested, establishing seasons and minimum size limits, defining acceptable gear such as net mesh sizes, and limiting the number of boats that are allowed to participate in a fishery.
- Several individual nations have developed systems to regulate nearby fish populations. In 1976, the United States passed the Fisheries Conservation and Management Act, which claimed exclusive jurisdiction over natural resources (including fishing), scientific research, and environmental preservation in an exclusive economic zone extending from shore 200 miles to sea; many other nations followed suit. This has dramatically changed the concept of open access for most of the world's continental shelves, coastal upwelling areas, and major fisheries.

STUDY GUIDE

Marine Ornamentals

- As many as two million people worldwide keep personal marine aquaria, which supports a trade that may be worth as much as $330 million annually. This impressive trade could provide many jobs in coastal areas and significant economic incentive for coral-reef conservation. Unfortunately, destructive collection techniques, overexploitation of some species, and careless handling and transport all undermine potential benefits of the trade.
- Fisheries for marine ornamentals also need to be managed to ensure sustainable use and properly regulated to decrease conflict with other user groups (such as sport divers, who prefer that no animal be removed from natural reefs).

Sealing and Whaling

- Pinnipeds and whales have long been targets of commercial harvesting for their meat, oil, and skins with dense insulating fur.
- The gregarious nature and relatively poor terrestrial locomotion of pinnipeds make them easy targets for sealers such that several pinniped species have been severely decimated by commercial harvests in the past 200 years. Hunting is not always the cause of population decline for some species. As predators near the tops of their food webs, the reduction and geographic distribution of the their prey species by human fishers can have a large negative effect on the survival of juveniles.
- In the 18th and 19th centuries, whales became major items of commerce as demand for their oil grew and whaling as an industry grew into a profitable commercial enterprise. Aided by the cannon-fired harpoon equipped with an explosive head and pelagic factory ships fitted with stern ramps to haul the whale carcasses aboard for processing, the kill of large rorquals rose dramatically, and several species were pushed to the edge of extinction.
- In 1948, 20 whaling nations established the IWC to oversee the utilization and conservation of the world's whale resources. Unfortunately, the IWC has neither inspection nor enforcement powers. How the international community handles this multinational problem will be a serious test of our biologic knowledge as well as our political wisdom.

Concluding Thoughts: Developing a Sense of Stewardship

- The world ocean has for centuries provided food, transportation, and recreation while absorbing the effluvia of human societies. Our continued use of the seas for these purposes depends on a better understanding of the consequences of our intrusion on the workings of marine ecosystems.
- Hopefully, we can learn to leave some of our old and wasteful habits behind and develop a sense of stewardship toward the sea that is reflected in our personal choices as well as our political decisions.

Topics for Discussion and Review

1. Why is it that clupeoid fishes are easily harvested in vast numbers, which accounts for them being about one third of the world's total commercial catch?
2. Regarding the trophic levels occupied by plants and animals that we consume, how does seafood and terrestrial harvests differ?
3. The magnitude of our fish harvest is limited (hypothetically) by what two ecologic factors?
4. Define "mariculture" and discuss its relative advantages and disadvantages.
5. Define "maximum sustainable yield" and discuss the biologic information required to estimate its value correctly.
6. What is the "tragedy of the commons" in relation to marine fisheries?
7. What is an exclusive economic zone? Pretend that you and your classmates are the leaders of different nations, some technologically advanced (such as the United States) and others not (such as Indonesia). Then discuss the "fairness" of establishing an exclusive economic zone around your nation.
8. In what ways has the trade in marine ornamental species been criticized?
9. What two technologic advances enabled whalers to harvest whales at previously unprecedented rates?

10. Discuss the personal choices that you can make to help ensure long-term sustainable use of marine resources.

Suggestions for Further Reading

Baker, C. S., and S. R. Palumbi. 1994. Which whales are hunted? A molecular genetic approach to monitoring whaling. *Science* 265:1538–1539.

Buschmann, A. H., J. A. Correa, R. Westermeier, M. D. C. Hernandez-Gonzalez, and R. Norambuena. 2001. Red algal farming in Chile: a review. *Aquaculture* 194:203–220.

Cato, J. C., and C. L. Brown. 2003. *Marine Ornamental Species: Collection, Culture & Conservation.* Iowa State Press, Ames.

Cipriano, F., and S. R. Palumbi. 1999. Genetic tracking of a protected whale. *Nature* 397:307–308.

Dayton, P. K., S. F. Thrush, M. T. Agardy, and R. J. Hofman. 1995. Environmental effects of marine fishing. *Aquatic Conservation: Marine and Freshwater Ecosystems* 5:205–232.

Fay, F. H., B. P. Kelley, and J. L. Sease. 1989. Managing the exploitation of Pacific walruses: a tragedy of delayed response and poor communication. *Marine Mammal Science* 5:1–16.

Frank, K. T., and W. C. Leggett. 1994. Fisheries ecology in the context of ecological and evolutionary theory. *Annual Review of Ecology and Systematics* 25:401–422.

Gubbay, S. 1995. *Marine Protected Areas: Principles and Techniques for Management.* Chapman and Hall, London.

Jackson, J. B. C., M. X. Kirby, W. H. Berger, K. A. Bjorndal, L. W. Botsford, B. J. Bourque, R. H. Bradbury, R. Cooke, J. Erlandson, J. A. Estes, T. P. Hughes, S. Kidwell, C. B. Lange, H. S. Lenihan, J. M. Pandolfi, C. H. Peterson, R. S. Steneck, M. J. Tegner, and R. R. Warner. 2001. Historical overfishing and the recent collapse of coastal ecosystems. *Science* 293:629–638.

Klimley, P. 1999. Sharks beware. *American Scientist* 87:488–491.

Kurlansky, M. 1997. *Cod: A Biography of the Fish that Changed the World.* Walker and Co., New York.

Langton, R. W., J. B. Pearce, and J. A. Gibson. 1994. *Selected Living Resources, Habitat Conditions, and Human Pertubations of the Gulf of Maine: Environmental and Ecological Considerations for Fishery Management.* U.S. DOC-NOAA-NMFS, Woods Hole, MA.

Laubier, A., and L. Laubier. 1993. Marine crustacean farming: present status and perspectives. *Aquatic Living Resources* 6:319–329.

Ludwig, D., R. Hilborn, and C. Walters. 1993. Uncertainty, resource exploitation and conservation: lessons from history. *Science* 260:17–18.

Lüning, K., and S. Pang. 2003. Mass cultivation of seaweeds: current aspects and approaches. *Journal of Applied Phycology* 15:115–119.

McDermid, K. J., and B. Stuercke. 2003. Nutritional composition of edible Hawaiian seaweeds. *Journal of Applied Phycology* 15:513–524.

Musick, J. A. 1999. *Life in the Slow Lane: Ecology and Conservation of Long-Lived Marine Animals.* American Fisheries Society, Bethesda, MD.

Myers, R. A., and B. Worm. 2003. Rapid worldwide depletion of predatory fish communities. *Nature* 423:280–283.

Naylor, R. L., R. J. Goldburg, J. H. Primavera, N. Kautsky, M. C. M. Beveridge, J. Clay, C. Folke, J. Lubchenco, H. Mooney, and M. Troell. 2000. Effect of aquaculture on world fish supplies. *Nature* 405:1017–1024.

Pauly, D., and V. Christensen. 1995. Primary production required to sustain global fisheries. *Nature* 374:255–257.

Pauly, D., V. Christensen, J. Dalsgaard, R. Froese, and F. Torres Jr. 1998. Fishing down marine food webs. *Science* 279:860–863.

Pauly, D., and J. Maclean. 2003. *In a Perfect Ocean: The State of Fisheries and Ecosystems in the North Atlantic Ocean.* Island Press, Washington, D.C.

Pauly, D., and R. Watson. 2003. Counting the last fish. *Scientific American* 289:43–47.

Punsly, R. G., P. K. Tomlinson, and A. J. Mullen. 1994. Potential tuna catches in the eastern Pacific Ocean from schools not associated with dolphins. *Fishery Bulletin* 92:132–143.

Roberts, C. M. 2002. Deep impact: the rising toll of fishing in the deep sea. *Trends in Ecology and Evolution* 17:242–245.

Thrush, S. F., J. E. Hewitt, V. J. Cummings, P. K. Dayton, M. Cryer, S. J. Turner, G. A. Funnell, R. G. Budd, C. J. Milburn, and M. R. Wilkinson. 1998. Disturbance of the marine benthic habitat by commercial fishing: impacts at the scale of the fishery. *Ecological Applications* 8:866–879.

Trippel, E. A., J. Y. Wang, M. B. Strong, L. S. Carter, and J. D. Conway. 1996. Incidental mortality of harbour porpoise (*Phocoena phocoena*) by the gill-net fishery in the lower Bay of Fundy. *Canadian Journal of Fisheries and Aquatic Sciences* 53:1294–1300.

Wabnitz, C., M. Taylor, E. Green, and T. Razak. 2003. *From Ocean to Aquarium: The Global Trade in Marine Ornamental Species*. UNEP-WCMC, Cambridge.

Wade, P. R. 1995. Revised estimates of incidental kill of dolphins (Delphinidae) by the purse-seine tuna fishery in the eastern tropical Pacific, 1959–1972. *Fishery Bulletin* 93:345–354.

Wallace, R. K., W. Hosking, and S. T. Szedlmayer. 1995. *Fisheries Management for Fishermen: A Manual for Helping Fishermen Understand the Federal Management Process*. Auburn University Marine Extension and Research Center, Mobile.

Watson, R., and D. Pauly. 2001. Systematic distortion in world fisheries catch trends. *Nature* 414:534–536.

Selected References

Chapter 1

Baker, J.J., and G.E. Allen. 1981. *Matter, Energy and Life: An Introduction to Chemical Concepts.* Reading, MA: Addison-Wesley.

Barkley, R.A. 1968. *Oceanographic Atlas of the Pacific Ocean.* Honolulu: University of Hawaii Press.

Berner, R.A., and A.C. Lasaga. 1989. Modelling the geochemical carbon cycle. *Scientific American* 260:74–81.

Bogdanov, D.V. 1963. Map of the natural zones of the ocean. *Deep-Sea Research* 10:520–23.

Bonatti, E. 1994. The Earth's mantle below the oceans. *Scientific American* 270:26–33.

Broecker, W.S., et al. 1985. Does the ocean-atmosphere system have more than one stable mode of operation? *Nature* 315:21–26.

Cloud, P. 1987. *Biogeography and Plate Tectonics.* New York: Elsevier.

Dietz, R.S., and J.C. Holden. 1970. Reconstruction of Pangaea: Breakup and dispersion of continents, Permian to present. *Journal of Geophysical Research* 75:4939–56.

Gordon, A.L. 1986. The southern ocean and global climate. *Oceanus* 26:34–44.

Hedgpeth, J.W., ed. 1957. Treatise on marine ecology and paleoecology. Volume 1: Ecology. *Geological Society of America.* Memoir 67.

Hogg, N. 1992. The Gulf Stream and its recirculation. *Oceanus* 35:28–37.

Jenkyns, H.C. 1994. Early history of the oceans. *Oceanus* 36:49–52.

Jones, P.D., and T.M.L. Wigley. 1990. Global warming trends. *Scientific American* 263:84–91.

Kunzig, R. 2000. *Mapping the Deep.* New York: W.W. Norton and Co.

Moore, III, B., and B. Bolin. 1986. The oceans, carbon cycle, and global climate change. *Oceanus* 29:16–26.

Munk, W.H. 1950. Origin and generation of waves. *Proceedings First Conference on Coastal Engineering.* Berkeley, CA: Council on Wave Research.

Parrilla, G., et al. 1994. Rising temperatures in the subtropical North Atlantic Ocean over the past 35 years. *Nature* 369:48–51.

Pickard, G.L., and W.J. Emory, eds. 1982. *Descriptive Physical Oceanography.* New York: Pergamon Press.

Pinet, P.R. 2003. *Invitation to Oceanography,* 3rd ed. Sudbury, MA: Jones and Bartlett.

Post, W.M., et al. 1990. The global carbon cycle. *American Scientist* 78:310–26.

Rand McNally Atlas of the Oceans. 1977. Chicago: Rand McNally & Company.

Rasmusson, E.M. 1985. El Niño and variations in climate. *American Scientist* 73:168–77.

Richardson, P.L. 1993. Tracking ocean eddies. *American Scientist* 81:261–71.

Southward, A.J. 1964. The relationship between temperature and rhythmic cirral activity in some *Cirripedia* considered in connection with their geographical distribution. *Helgol. wiss. Meersuntersuch.* 10:391–403.

Weller, R.A., and D.M. Farmer. 1992. Dynamics of the ocean's mixed layer. *Oceanus* 35:46–55.

Wysession, M. 1995. The inner workings of the Earth. *American Scientist* 83:134–47.

Chapter 2

Allen, T.F.H., and T.B. Starr. 1988. *Hierarchy: Perspectives for ecological complexity.* Chicago: The University of Chicago Press.

Baird, R.W., Abrams, P.A., and Dill, L.M. 1992. Possible indirect interactions between transient and resident killer whales: Implications for the evolution of foraging specializations in the genus *Orcinus. Oecologia* 89:125–32.

Caron, D.A. 1992. An introduction to biological oceanography. *Oceanus* 35:10–17.

Fuhrman, J.A. 1999. Marine viruses and their biogeochemical and ecological effects. *Nature* 399:541–48.

Hardy, A.C. 1924. The herring in relation to its animate environment, pt. 1. The food and feeding habits of the herring. *Fisheries Investigations,* London, Series II, 7:1–53.

Hedgpeth, J.W., ed. 1957. Treatise on marine ecology and paleoecology. Volume 1: Ecology. *Geological Society of America.* Memoir 67.

Hutchinson, G.E. 1961. The paradox of the plankton. *American Naturalist* 95:137–45.

Landry, M.R. 1976. The structure of marine ecosystems: An alternative. *Marine Biology* 35:1–7.

Levinton, J.S. 2001. *Marine Biology.* New York: Oxford University Press.

Lewin, R.A. 1982. Symbiosis and parasitism—definitions and evaluations. *BioScience* 32:254.

Longhurst, A.R. 2001. *Ecological Geography of the Sea,* 2nd ed. San Diego: Academic Press,

Murphy, G.I. 1968. Pattern of life history and the environment. *American Naturalist* 102:391–403.

Pomeroy, L.R. 1992. The microbial food web. *Oceanus* 35:28–35.

Raven, P.H., and G.B. Johnson. 2002. *Biology.* Boston: McGraw-Hill.

Roughgarden, J. 1972. Evolution of niche width. *American Naturalist* 106:683–718.

Russell-Hunter, W.D. 1970. *Aquatic productivity.* New York: Macmillan.

Schoener, T.W. 1974. Resource partitioning in ecological communities. *Science* 185:27–9.

Sinclair, M. 1988. *Marine populations: An essay on population regulation and speciation.* Seattle: University of Washington Press.

Smith, R.C., et al. 1999. Marine ecosystem sensitivity to climate change. *BioScience* 49:393–403.

Steele, J.H. 1991. Marine functional diversity. *BioScience* 41:470–74.

Valentine, J.W. 1973. *Evolutionary ecology of the marine biosphere.* Englewood Cliffs, NJ: Prentice-Hall.

Van Valen, L. 1974. Predation and species diversity. *Journal of Theoretical Biology* 44:19–21.

Wolfe, S.L. 1981. *Biology of the Cell.* Belmont, CA: Wadsworth.

Chapter 3

Anderson, D.M. 1994. Red tides. *Scientific American* August: 62–8.

Anderson, D. M., and D. Wall. 1978. Potential importance of benthic cysts of *Gonyaulax tamarensis* and *G. excavata* in initiating toxic dinoflagellate blooms. *Journal of Phycology* 14: 224–34.

Austin, B. 1988. *Marine microbiology.* New York: Cambridge University Press.

Bainbridge, R. 1957. Size, shape and density of marine phytoplankton concentrations. *Biological Review* 32:91–115.

Bargu, S., et al. 2002. Krill: a potential vector for domoic acid in marine food webs. Marine *Ecology Progress Series* 237:209–16.

Boney, A.D. 1992. *Phytoplankton.* London: E. Arnold.

Bougis, P. 1976. *Marine Plankton Ecology.* New York: Elsevier.

Brown, O.B., et al. 1985. Phytoplankton blooming off the U.S. east coast: A satellite description. *Science* 229:163–67.

Burkholder, J.M. 1999. The lurking peril of *Pfiesteria. Scientific American* 281:42–9.

Carmichael, W.W. 1994. *The toxins of cyanobacteria. Scientific American* January:78–86.

Carpenter, E.J.,and K. Romans. 1991. Major role of the cyanobacterium *Trichodesmium* in nutrient cycling in the North Atlantic Ocean. *Science* 254:1356–58.

Carr, N.G., and B.A. Whitton. 1983. *The Biology of Cyanobacteria.* Berkeley, CA: University of California Press.

Chisholm, S.W. 1992. What limits phytoplankton growth? *Oceanus* 35:36–46.

Coleman, G., and Coleman, W.J. 1990. How plants make oxygen. *Scientific American* February:50–8.

Cushing, D.H., and J.J. Walsh. 1976. *The Ecology of Seas.* Philadelphia: W.B. Saunders.

Dale, B., and C.M. Yentsch. 1978. Red tide and paralytic shellfish poisoning. *Oceanus* 21:41–9.

Dawes, C.J. 1981. *Marine Botany.* New York: John Wiley & Sons.

Falkowski, P.G., ed. 1980. *Primary Productivity in the Sea.* New York: Plenum Press.

Fleming, R.H. 1939. The control of diatom populations by grazing. *Journal du Conseil Permanent International pour l'Exploration de la Mer* 14:210–27.

Fryxell, G.A. 1983. New evolutionary patterns in diatoms. *BioScience* 33:92–8.

Gregg, M.C., T.B. Sanford, and D.P. Winkel. 2003. Reduced mixing from the breaking of internal waves in equatorial waters. *Nature* 422:513–15.

Griffin, D.W. 2002. The global transport of dust. *American Scientist* 90:228–35.

Hamm, C.E., et al. 2003. Architecture and material properties of diatom shells provide effective mechanical protection. *Nature* 421:841–43.

Hardy, A.H. 1971. *The Open Sea: Its Natural History. Part I: The World of Plankton. Part II: Fish and Fisheries.* Boston: Houghton Mifflin.

Hargraves, P.E., and F.W. French. 1983. Diatom resting spores: significance and strategies. In: *Survival Strategies of the Algae.* G.A. Fryxell, ed. New York: Cambridge University Press. pp. 49–68.

Harris, G.P. 1986. *Phytoplankton Ecology.* New York: Chapman and Hall.

Honhart, D.C. 1989. *Oceanography from the Space Shuttle.* Washington, DC: Office of Naval Research.

Hoppema, M., de Baar, H.J.W., Fahrbach, E., Hellmer, H.H., and Klein, B. 2003. Substantial advective iron loss diminishes phytoplankton production in the Antarctic Zone. *Global Biogeochemical Cycles* 17(1).

Humm, H.J., and S.R. Wicks. 1980. *Introduction and Guide to the Marine Blue-Green Algae.* New York: Wiley..

Jenkins, W.J., and J.C. Goldman. 1985. Seasonal oxygen cycling and primary production in the Sargasso Sea. *Journal of Marine Research* 43:465–91.

Jumars, P.A. 1993. Concepts in biological oceanography: an interdisciplinary primer. New York: Oxford University Press. .

Kaufman, P.B., et al. 1989. *Plants: Their Biology and Importance.* Reading, MA: Addison-Wesley.

Krogmann, D.W. 1981. Cyanobacteria (bluegreen algae)—Their evolution and relation to other photosynthetic organisms. *BioScience* 31:121–24.

Lalli, C.M., and T.R. Parsons. 1997. *Biological Oceanography: An Introduction.* Oxford:Butterworth Heinemann,.

Landry, M.R. 1976. The structure of marine ecosystems: An alternative. *Marine Biology* 35:1–7.

Lembi, C.A., and J.R. Waaland, eds. 1988. *Algae and Human Affairs.* New York: Cambridge University Press.

Lipps, J.H. 1970. Plankton evolution. *Evolution* 24:1–22.

Malone, T.C. 1971. The relative importance of nannoplankton and net plankton as primary producers in tropical oceanic and neritic phytoplankton communities. *Limnology and Oceanography* 16:633–39.

Margulis, L., D. Chase, and R. Guerrero. 1986. Microbial communities. *BioScience* 36:160–70.

Marshall, H.G. 1976. Phytoplankton density along the eastern coast of U.S.A. *Marine Biology* 38:81–9.

Mills, E. L. 1989. *Biological Oceanography: An Early History, 1870–1960.* Ithaca, NY: Cornell University Press.

Moll, R.A. 1977. Phytoplankton in a temperate-zone salt marsh: Net production and exchanges with coastal waters. *Marine Biology* 42:109–18.

Moore, R.E. 1977. Toxins from blue-green algae. *BioScience* 27:797–802.

Okada, H., and A. McIntyre. 1977. Modern coccolithophores of the Pacific and North Atlantic oceans. *Micropaleontology* 23:1–55.

Paasche, E. 1968. Biology and physiology of coccolithophorids. *Annual Review of Microbiology* 22:71–86.

Pace, N.R. 1997. A molecular view of microbial diversity and the biosphere. *Science* 276:734-40.

Perry, M.J. 1986. Assessing marine primary production from space. *BioScience* 36:461–66.

Peterson, M.N.A., ed. 1993. *Diversity of Oceanic Life: An Evaluative Review.* Washington, DC: Center for Strategic and International Studies.

Pickett-Heaps, J. 1976. Cell division in eucaryotic algae. *BioScience* 26:445–50.

Platt, T., and W.K.W. Li, eds. 1986. Photosynthetic picoplankton. *Canadian Bulletin of Fisheries and Aquatic Sciences.* 214:583.

Platt, T., C. Fuentes-Yaco, and K. T. Frank. 2003. Spring algal bloom and larval fish survival. *Nature* 423:398–99.

Pomeroy, L.W. 1974. The ocean's food web, a changing paradigm. *BioScience* 24:499–504.

Qasim, S.Z., P.M.A. Bhattuthiri, and V.P. Devassy. 1972. The effect of intensity and quality of illumination on the photosynthesis of some tropical marine phytoplankton. *Marine Biology* 16:22–27.

Raven, P.H., and G.B. Johnson. 2002. *Biology*. Boston: McGraw-Hill.

Raymont, J.E.G. 1983. *Plankton and Productivity in the Oceans. Volume 1: Phytoplankton*. New York: Pergamon Press.

Roman, M.R., et al. 2002. Latitudinal comparisons of equatorial Pacific zooplankton. *Deep-Sea Research (Part II, Topical Studies in Oceanography)* 49:2695–711.

Round, F.E., R.M. Crawford, and D.G. Mann. 1990. *The Diatoms*. New York: Cambridge University Press.

Rowan, K.S. 1989. *Photosynthetic Pigments of Algae*. New York: Cambridge University Press.

Russell-Hunter, W.D. 1970. *Aquatic Productivity*. New York: Macmillan.

Ryther, J.H. 1969. Photosynthesis and fish production in the sea. *Science* 166:72–6.

Ryther, J.H., and C.S. Yentsch, 1957. Estimation of phytoplankton production in the ocean from chlorophyll and light data. *Limnology and Oceanography* 2:281–86.

Saffo, M.B. 1987. New light on seaweeds. BioScience 37:654–64.

Scagel, R.F., et al. 1980. *Nonvascular Plants*. Belmont, CA: Wadsworth.

Smayda, T.J. 1970. The suspension and sinking of phytoplankton in the sea. *Oceanography and Marine Biology (Annual Review)* 8:353–414.

Smith, D.L. 1996. *A Guide to Marine Coastal Plankton and Marine Invertebrate Larvae*. Dubuque, IA: Kendall/Hunt Publishing Company.

Smith, W.O., Jr., and D.M. Nelson. 1986. Importance of ice edge phytoplankton production in the southern ocean. *BioScience* 36:251–57.

Steeman Nielsen, E. 1952. Use of radioactive carbon (C^{14}) for measuring organic production in the sea. *Journal du Conseil Permanent International pour l'Exploration de la Mer* 18:117–40.

Steeman Nielsen, E. 1975. *Marine Photosynthesis*. Amsterdam: Elsevier.

Steidinger, K.A., and K. Haddad. 1981. Biologic and hydrographic aspects of red tides. *BioScience* 31:814–19.

Sverdrup, H.U., M.W. Johnson, and R.H. Fleming. 1942. *The Oceans: Their Physics, Chemistry, and Biology*. Englewood Cliffs, NJ: Prentice-Hall.

Taylor, F.J.R., ed. 1987. *The Biology of Dinoflagellates*. London: Blackwell.

Tomas, C.R. 1997. *Identifying Marine Phytoplankton*. San Diego: Academic Press.

Vinyard, W.C. 1980. *Diatoms of North America*. Eureka, CA: Mad River Press.

Walsby, A.E. 1977. The gas vacuoles of blue-green algae. *Scientific American* August:90–97.

Walsh, J.J. 1984. The role of ocean biota in accelerated ecological cycles: A temporal view. *BioScience* 34:499–507.

Werner, D., ed. 1977. *The Biology of Diatoms*. Berkeley, CA: University of California Press.

Chapter 4

Alongi, D. M. 2002. Present state and future of the world's mangrove forests. *Environmental Conservation* 29:331–49.

Baker, J.D., and W.S. Wilson. 1986. Spaceborne observations in support of earth science. *Oceanus* 29:76–85.

Bold, H.C., and M.J. Wynne. 1985. *Introduction to the Algae*. Englewood Cliffs, NJ: Prentice-Hall.

Capone, D.G. 2001. Marine nitrogen fixation: what's the fuss? *Current Opinion in Microbiology* 4:341–48.

Chapman, A.R.O. 1979. *Biology of Seaweeds*. Baltimore: University Park Press.

Cole, K.M., and R.G. Sheath, eds. 1990. *Biology of the Red Algae*. New York: Cambridge University Press.

Correll, D.L. 1978. Estuarine productivity. *BioScience* 28:646–50.

Dawes, C.J. 1981. *Marine Botany*. New York: John Wiley & Sons.

Dawson, E.Y. 1966. *Marine Botany, an Introduction*. New York: Holt, Rinehart and Winston.

Dring, M.J. 1982. *The Biology of Marine Plants*. London: E. Arnold.

Duffy, J.E., and M.E. Hay. 1990. Seaweed adaptations to herbivory. *BioScience* 40:368–75.

Estes, J.A., and J.F. Palmisano. 1974. Sea otters: Their role in structuring nearshore communities. *Science* 185:1058–60.

Falkowski, P. G., R. T. Barber, and V. Smetacek. 1998. Biogeochemical controls and feedbacks on ocean primary production. *Science* 281:200–06.

Field, C. B., et al. 1998. Primary production of the biosphere:Integrating terrestrial and oceanic components. *Science* 281:237–40.

Foster, M.S. 1975. Algal succession in a *Macrocystis pyrifera* forest. *Marine Biology* 32:313–29.

Gaylord, B., M. W. Denny, and M. A. R. Koehl. 2003. Modulation of wave forces on kelp canopies by alongshore currents. *Limnology and Oceanography* 48:860–71.

Goering, J.J., and P.L. Parker. 1972. Nitrogen fixation by epiphytes on sea grasses. *Limnology and Oceanography* 17:320–23.

Hemminga, M. A., and C. A. Duarte. 2001. *Seagrass Ecology*. Cambridge University Press, England.

Hogarth, P. J. 2000. *The Biology of Mangroves*. Oxford University Press, Oxford.

Jacobs, W.P. 1994. *Caulerpa*. *Scientific American* 271:100–05.

Jeffries, R.L. 1981. Osmotic adjustment of the response of halophytic plants to salinity. *BioScience* 31:42–6.

Joint, I., and S. B. Groom. 2000. Estimation of phytoplankton production from space: current status and future potential of satellite remote sensing. *Journal of Experimental Marine Biology and Ecology* 250:233–55.

Kaufman, P. B., et al. 1989. *Plants: Their Biology and Importance*. Reading, MA: Addison-Wesley.

King, R.J., and W. Schramm, 1976. Photosynthetic rates of benthic marine algae in relation to light intensity and seasonal variations. *Marine Biology* 37:215–22.

Koehl, M.A.R., and S.A. Wainwright. 1977. Mechanical adaptations of a giant kelp. *Limnology and Oceanography* 22:1067–71.

Landry, M.R., et al. 2002. Seasonal dynamics of phytoplankton in the Antarctic Polar Front region at 170° W. *Deep-Sea Research (Part II, Topical Studies in Oceanography)* 49:1843–65.

Lembi, C.A., and J.R. Waaland, eds. 1988. *Algae and Human Affairs*. New York: Cambridge University Press.

Lobban, C.S., and P. J. Harrison. 1996. *Seaweed Ecology and Physiology*. New York: Cambridge University Press.

Longhurst, A., et al. 1995. An estimate of global primary production in the ocean from satellite radiometer data. *Journal of Plankton Research* 17:1245–71.

Mann, K.H. 1973. Ecological energetics of the seaweed zone in a marine bay on the Atlantic coast of Canada. *Marine Biology* 14:199–209.

McPeak, R.H., and D.A. Glantz. 1984. Harvesting California's kelp forests. *Oceanus* 27:19–26.

Meinesz, A. 1999. *Killer Algae*. Chicago: University of Chicago Press.

Pauly, D., and V. Christensen. 1995. Primary production required to sustain global fisheries. *Nature* 374:255–57.

Perry, M.J. 1986. Assessing marine primary productivity from space. *BioScience* 36:461–67.

Pettitt, J., S. Ducker, and B. Knox. 1981. Submarine pollination. *Scientific American* 244:134–43.

Philander, G. 1989. El Niño and La Niña. *American Scientist* 77:451–59.

Phillips, R.C. 1978. Sea grasses and the coastal environment. *Oceanus* 21:30–40.

Phleger, C.F. 1971. Effect of salinity on growth of a salt-marsh grass. *Ecology* 52:908–11.

Pimm S.L. 2001. *The World According to Pimm: A Scientist Audits the Earth.* New York: McGraw-Hill.

Saffo, M.B. 1987. New light on seaweeds. *BioScience* 37:654–64.

Santelices, B. 1990. Patterns of reproduction, dispersal and recruitment in seaweeds. *Oceanography and Marine Biology (Annual Review)* 28:177–276.

Segal, R.F., et al. 1980. *Nonvascular Plants.* Belmont, CA: Wadsworth.

Steele, J.H., ed. 1973. *Marine Food Chains.* Edinburgh: Oliver and Boyd.

Steele, J.H. 1974. *The Structure of Marine Ecosystems.* Cambridge, MA: Harvard University Press.

Teal, J., and M. Teal. 1975. *The Sargasso Sea.* Boston: Little Brown.

Tomlinson, P.B. 1994. *The Botany of Mangroves.* New York: Cambridge University Press.

Townsend, D.W., and Thomas, M. 2002. Springtime nutrient and phytoplankton dynamics on Georges Bank. *Marine Ecology Progress Series* 228:57–74.

Vitousek, J., et al. 1986. Human appropriation of the products of photosynthesis. *BioScience* 36:368–373.

Whittaker, R.H., and G.E. Likens. 1973. "Carbon in the biota." In G.M. Woodwell, E.V. Pecan, eds. *Carbon and the Biosphere.* Oak Ridge, TN: Technical Information Center, U.S. Atomic Energy Commission.

Chapter 5

Alexander, R.M. 1979. *The Invertebrates.* New York: Cambridge University Press.

Ayala, F.J., and A. Rzhetsky. 1998. Origin of the metazoan phyla: molecular clocks confirm palaeontological estimates. *Proceedings of the National Academy of Sciences* 95:606–11.

Barnes, R.D. 1980. *Invertebrate Zoology.* Philadelphia: Saunders College/Holt, Rinehart and Winston.

Bayne, C.J. 1990. Phagocytosis and non-self recognition in invertebrates. *BioScience* 40:723–31.

Brady, H.B. 1884. Report on the foraminifera dredged by *H.M.S. Challenger. Zoology* 9:1–814.

Brusca, R.C., and G.J. Brusca. 1990. *Invertebrates.* Sunderland, MA: Sinauer Associates.

Buzas, M.A., and S.J. Culver. 1991. Species diversity and dispersal of benthic foraminifera. *BioScience* 41:483–89.

Capriulo, G.M. 1990. *Ecology of Marine Protozoa.* New York: Oxford University Press.

Erwin, D., et al. 1997. The origin of animal body plans. *American Scientist* 85:126–37.

Grahame, J., and Branch, G.M. 1985. Reproductive Patterns of Marine Invertebrates. *Oceanography and Marine Biology Annual Review* 23:373–98.

Hickman, C. P., and L. S. Roberts. 1994. *Biology of Animals*, 6th ed. Dubuque, IA: Wm. C. Brown.

Kozloff, E.N. 1987. *Marine Invertebrates of the Pacific Northwest.* Seattle: University of Washington Press.

Levinton, J.S. 1992. The big bang of animal evolution. *Scientific American* September:84–91.

Pechenik, J. A. 2000. *Biology of the Invertebrates.* Dubuque, IA: McGraw-Hill..

Pilling, E.D., Leakey, R.J.G., and Burkill, P.H. 1992. Marine pelagic ciliates and their productivity during summer in Plymouth coastal waters. *Journal of the Marine Biology. Association* 72:265–68.

Raven, P.H., and G.B. Johnson. 2002. *Biology.* Boston: McGraw-Hill.

Richardson, J. 1986. Brachiopods. *Scientific American* September:100–06.

Russell, R. S. 1935. On the value of certain planktonic animals as indicators of water movements in the English Channel and North Sea. *Journal of the Marine Biological Association, U.K.* 20:309–32.

Russell-Hunter, W.D. 1979. *A Life of Invertebrates.* New York: Macmillan.

Sebens, K.P. 1977. Habitat suitability, reproductive ecology and the plasticity of body size in two sea anemone populations (*Anthopleura elegantissima* and *A. xanthogrammica*). Ph.D. Dissertation. Seattle: University of Washington.

Stanley, S.M. 1975. A theory of evolution above the species level. *Proceedings of the National Academy of Science U.S.A.* 72:646–50.

Valentine, W. 1978. The evolution of multicellular plants and animals. *Scientific American* September:140–58.

Villee, C.A. 1984. *Biology.* Boston: McGraw-Hill.

Willmer, P. 1990. *Invertebrate Relations.* Cambridge, UK: Cambridge University Press.

Chapter 6

Ainley, D.G., et al. 1984. *The Marine Ecology of Birds in the Ross Sea, Antarctica.* Washington, DC: American Ornithologists Union.

Alderton, D. 1988. *Turtles and Tortoises of the World.* New York: Facts on File.

Atema, J., R. R. Fay, A. N. Popper, and W. Travolga. 1988. *Sensory Biology of Aquatic Animals.* New York: Springer-Verlag.

Berta, A. 1994. What is a whale? *Science* 263:180–82.

Berta, A., C.E. Ray, and A.R. Wyss. 1989. Skeleton of the oldest known pinniped. *Enaliarctos mealsi. Science* 244:60–62.

Berta, A., and J. Sumich. 1999. *Marine Mammals: Evolutionary Biology.* San Diego: Academic Press.

Bond, C.E. 1979. *Biology of Fishes.* Philadelphia: W.B. Saunders.

Bonner, W.N. 1982. *Seals and Man: A Study of Interactions.* Seattle: University of Washington Press.

Compagno, L.J.V. 1988. *Sharks of the Order Carcharhinoformes.* Princeton, NJ: Princeton University Press.

Croxall, J.P. 1987. *Seabirds: Feeding Ecology and Role in Marine Ecosystems.* New York: Cambridge University Press.

Davoren, G.K., W.A. Montevecchi, and J.T. Anderson. 2003. Search strategies of a pursuit-diving marine bird and the persistence of prey patches. *Ecological Monographs* 73:463–81.

DeLong, R.., and B.S. Stewart. 1991. Diving patterns of northern elephant seal bulls. *Marine Mammal Science* 7:369–84.

Demski, L.S. 1987. Diversity in reproductive patterns and behavior in teleost fishes. In D. Crew, ed. *Psychobiology of Reproductive Behavior. An Evolutionary Perspective* Englewood Cliffs, NJ: Prentice Hall. pp. 1–27.

Diamond, A.W., and C.M. Devlin. 2003. Seabirds as indicators of changes in marine ecosystems: ecological monitoring on Machias Seal Island. *Environmental Monitoring and Assessment* 88:153–81.

Dunson, W.A., ed. 1975. *The Biology of Sea Snakes.* Baltimore: University Park Press.

Erdmann, M.V., and Caldwell, R.L. 2000. How new technology put a coelacanth among the heirs of Piltdown Man. *Nature* 406:343.

Ernst, C.H., and R.W. Barbour. 1989. *Turtles of the World.* Washington, DC: Smithsonian Institution Press.

Geraci, J.R. 1978. The enigma of marine mammal strandings. *Oceanus* 21:38–47.

Griffith, R.W. 1994. The life of the first vertebrates. *BioScience* 44:408-416.

Helfman, G.S., B.B. Collette, and D.E. Facey. 1997. *The Diversity of Fishes*. Malden, MA: Blackwell Science.

Kalmijn, A.J. 1982. Electric and magnetic field detection in elasmobranch fishes. *Science* 218:916–18.

Kvadsheim, P.H., and L.P. Folkow. 1997. Blubber and flipper heat transfer in harp seals. *Acta Physiologica Scandinavica* 161:385–95.

Love, J.A. 1990. *The Sea Otter*. London: Whittet Books.

Marshall, N.B. 1966. *The Life of Fishes*. New York: World Publishing.

Martin, A.P., G.J.P. Naylor, and S.R. Palumbi. 1992. Rates of mitochondrial DNA evolution in sharks are slow compared with mammals. *Nature* 357:153.

Martini, F.H. 1998. Secrets of the slim hagfish. *Scientific American* 228:70–5.

Nettleship, D.N., G.A. Sanger, and P.F. Springer, eds. 1985. *Marine Birds: Their Ecology and Commercial Fisheries Relationships.* Ottawa: Canadian Wildlife Services.

Noble, R.W., L.D. Kwiatkowski, A. De Young, B.J. Davis, R.L. Haedrich, L.T. Tam, and F.A. Riggs. 1986. Functional properties of hemoglobins from deep-sea fish: Correlations with depth distribution and presence of a swimbladder. *Biochim. Biophys. Acta.* 870:552–63.

Noren, S.R., and T.M. Williams. 2000. Body size and skeletal muscle myoglobin of cetaceans: adaptations for maximizing dive duration. Part A: Comparative biochemistry and physiology. *Molecular & Integrative Physiology* 126:181–91.

Owens, D.W. 1980. The comparative reproductive physiology of sea turtles. *American Zoologist* 20:549–63.

Popper, A.N., and Coombs, S. 1980. Auditory mechanisms in teleost fishes. *American Scientist* 68:429–40.

Raven, P.H. and G.B. Johnson. 2002. *Biology*. Boston: McGraw-Hill.

Rice, D.W., and A.A. Wolman. 1971. The life history and ecology of the gray whale (*Eschrichtius robustus*). *American Society of Mammalogy* (special publication no. 3), 142 pp.

Ridgway, S.H. 1972. *Mammals of the Sea, Biology and Medicine.* 1972. Springfield, IL: Charles C. Thomas.

Riedman, M. 1990. *The Pinnipeds: Seals, Sea Lions and Walruses.* Berkeley, CA: University of California Press.

Schreer, J.F., and K.M. Kovacs. 1997. Allometry of diving capacity in air-breathing vertebrates. *Canadian Journal of Zoology* 75:339–58.

Stirling, I., and D. Guravich. 1988. *Polar Bears*. Ann Arbor, MI: University of Michigan Press.

VanBlaricom, G.R., and J.A. Estes. 1988. *The Community Ecology of Sea Otters*. New York: Springer-Verlag.

Warner, R. R. 1973. Ecological and evolutionary aspects in the California sheephead, *Pimelometopon pulchrum*. Ph.D. dissertation. U. California, San Diego.

Webb, P.W. 1984. Form and function in fish swimming. *Scientific American* 251:72–82.

Wikelski, M., and Thom, C. 2000. Marine iguanas shrink to survive El Nino. *Nature* 403:37.

Wilson, R. P., Y. Ropert-Coudert, and A. Kato. 2002. Rush and grab strategies in foraging marine endotherms: the case for haste in penguins. *Animal Behaviour* 63:85–95.

Withers, P. C., et al 1994. Role of urea and methylamines in buoyancy of elasmobranchs. *Journal of Experimental Biology* 188:175–89.

Wrootton, R.J. 1990. *Ecology of Teleost Fishes*. New York: Chapman and Hall.

Würsig, B. 1988. The behavior of baleen whales. *Scientific American* 258:102–07.

Chapter 7

Alexander, M. 1981. Biodegradation of chemicals of environmental concern. *Science* 211:132.

Anderson, T.H., and Taylor, G.T. 2001. Nutrient pulses, plankton blooms, and seasonal hypoxia in western Long Island Sound. *Estuaries* 24:228–43.

Barnes, R.S.K. 1974. *Estuarine Biology. Studies in Biology, no. 49.* Baltimore: University Park Press.

Bertness, M.D. 1992. The ecology of a New England salt marsh. *American Scientist* 80:260–68.

Bjorndal, K.A.; A.B. Bolten; and C.J. Lagueux. 1994. Ingestion of marine debris by juvenile sea turtles in coastal florida habitats. *Marine Pollution Bulletin* 28:154–58.

Blus, L., et al. 1971. Eggshell thinning in the brown pelican: Implications of DDE. *BioScience* 21:1213–15.

Botton, M.L., and H.H. Haskin. 1984. Distribution and feeding of the horseshoe crab. *Limulus polyphemus*, on the continental shelf off New Jersey. *Fishery Bulletin* 82:383–89.

Britton, J.C. 1989. *Shore Ecology of the Gulf of Mexico*. Dallas: Texas Press.

Broecker, W.S., and G.H. Denton. 1990. What drives glacial cycles? *Scientific American* January:49–56.

Champ, M.A., and F.L. Lowenstein. 1987. The dilemma of high-technology antifouling paints. *Oceanus* 30:69–77.

Clark, R.B. 1992. *Marine Pollution*, 3rd ed. Oxford, UK: Oxford University Press.

Cloern, J., and F. Nichols. 1985. *Temporal Dynamics of an Estuary.* Boston: Kluwer Academic.

Correll, D.L. 1978. Estuarine productivity. *BioScience* 28:646–50.

Costlow, J.D., and C.G. Bookhout. 1959. The larval development of *Callinectes sapidus* Rathbun reared in the laboratory. *Biological Bulletin* 116:373–96.

Cox, J.L. 1972. DDT in marine plankton and fish in the California Current. *CalCOFI Reports* 16:103–11.

Dennison, W.C. et al. 1993. Assessing water quality with submerged aquatic vegetation. *BioScience* 43:86–93.

Epel, D., and W.L. Lee. 1970. Persistent chemicals in the marine ecosystem. *The American Biology Teacher* 207–11.

Epifanio, C. E., and R. W. Garvine. 2001. Larval transport on the Atlantic continental shelf of North America: a review. *Estuarine, Coastal and Shelf Science* 52:51–77.

Flindt, M.R. et al. 1999. Nutrient cycling and plant dynamics in estuaries: a brief review. *Acta Oecologica* 20:237–48.

Frankel, E.G. 1995. *Ocean Environmental Management: A Primer on the Role of the Oceans and How to Maintain Their Contributions to Life on Earth*. Englewood Cliffs, NJ: Prentice-Hall.

Green, E.P., 2003. *World Atlas of Seagrasses*. Berkeley: University of California Press. 320 pp.

Heinle, D.R., R.P. Harris, J.F. Ustach, and D.A. Flemer. 1977. Detritus as food for estuarine copepods. *Marine Biology* 40:341–53.

Horton, T. 2003. *Turning the Tide: Saving the Chesapeake Bay*. Baltimore: Island Press.

Jefferies, R.L. 1981. Osmotic adjustment and the Rresponse of halophytic plants to salinity. *BioScience* 31:42–6.

Johnson, D.R. 1985. Wind-forced dispersion of blue crab larvae in the Middle Atlantic Bight. *Continental Shelf Research* 4:425–37.

Johnson, D.R., B.S. Hester, and J.R. McConaugha. 1984. Studies of a wind mechanism influencing the recruitment of blue crabs in the Middle Atlantic Bight. *Continental Shelf Research* 3:425–37.

Kennish, M. J. 1999. *Estuary Restoration and Maintenance: The National Estuary Program.* Boca Raton, FL: CRC Press.

Kusler, J.A., W.J. Mitsch, and J.S. Larson. 1994. Wetlands. *Scientific American* 270:64–70.

Laws, E.A. 1993. *Aquatic Pollution: An Introductory Text,* 2nd ed. New York: John Wiley & Sons.

Lippson, J.A., and R.L. Lippson. 1984. *Life in the Chesapeake Bay.* Baltimore: The Johns Hopkins University Press.

Mangelsdorf, P. C., Jr. 1967. Salinity measurements in estuaries. In G. H. Lauff, ed. *Estuaries.* Washington, DC: American Association for the Advancement of Science. Publication no. 83, pp. 71–9.

Margulis, L., D. Chase, and R. Guerrero. 1986. Microbial communities. *BioScience* 36:160–70.

Marshall, H.G. 1980. Seasonal phytoplankton composition in the lower Chesapeake Bay and Old Plantation Creek, Cape Charles, Virginia. *Estuaries* 3:207–16.

McRoy, C.P., and C. Helfferich. 1977. *Seagrass Ecosystems.* New York: Marcel Dekker.

Miller, J.M., and M.L. Dunn. 1980. Feeding strategies and patterns of movement in juvenile estuarine fishes. In V.S. Kennedy, ed. *Estuarine Perspectives.* New York: Academic Press.

Milliman, J. 1989. Sea levels: Past, present, and future. *Oceanus* 32:40–3.

Moriarity, F. 1983. Ecotoxicology: *The Study of Pollutants in Ecosystems.* New York: Academic Press.

National Wildlife Foundation. 1989. *A Citizen's Guide to Protecting Wetlands.* Washington, DC: National Wildlife Foundation.

Nicol, J.A.C. 1967. *The Biology of Marine Animals.* London: Pitman and Sons.

Nichols, F., et al. 1986. Temporal dynamics of an estuary: San Francisco Bay. *Science* 231:567–73.

Noble, P. A., R. G. Tymowski, and M. Fletcher. 2003. Contrasting patterns of phytoplankton community pigment composition in two salt marsh estuaries in southeastern United States. *Applied and Environmental Microbiology* 69:4129–43.

Peakall, D.B. 1970. Pesticides and the reproduction of birds. *Scientific American* 222:72–8.

Peterson, C.H., and R. Black. 1987. Resource depletion by active suspension feeders on tidal flats: influence of local density and tidal elevation. *Limnology and Oceanography* 32:143–66.

Peterson, D., et al. 1995. The role of climate in estuarine variability. *American Scientist* 83:58–67.

Rabalais, N.N. 2002. Nitrogen in aquatic ecosystems. *Ambio* 31:102–12.

Remane, A. 1934. Die Brackwasserfauna. *Zoologischer Anzeiger Supplementband* 7:34–74.

Risebrough, R.W., et al. 1967. DDT residues in Pacific seabirds: A persistent insecticide in marine food chains. *Nature* 216:389–91.

Selander, R., S. Yang, R. Lewontin, and W. Johnson. 1970. Genetic variation in the horseshoe crab (*Limulus polyphemus*), a phylogenetic "relic." *Evolution* 24:402–14.

Siry, J.V. 1984. *Marshes of the Ocean Shore.* Austin: Texas A & M University Press.

Teal, J., and M. Teal. 1974. *Life and Death of the Salt Marsh.* New York: Ballantine.

Tyler, M.A., and H.H. Seliger. 1978. Annual subsurface transport of a red tide dinoflagellate to its bloom area: Water circulation patterns and organism distributions in the Chesapeake Bay. *Limnology and Oceanography* 23:227–46.

Valiela, I., and J. Teal. 1979. The nitrogen budget of a salt marsh ecosystem. *Nature* 280:652–56.

Walsh, J.P. 1981. U.S. Policy on marine pollution. *Oceanus* 24:18–24.

Warner, W.W. 1976. *Beautiful Swimmers.* Boston: Little, Brown.

Williams, A.B. 1974. The swimming crabs of the genus *Callinectes* (Decapoda: Portunidae). *Fishery Bulletin* 72:685–798.

Wurster, C.E. 1968. DDT reduces photosynthesis by marine phytoplankton. *Science* 159:1474.

Zedler, J., T. Winfield, and D. Mauriello. 1978. Primary productivity in a southern California estuary. *Coastal Zone* 3:649–62.

Chapter 8

Armstrong, R.A., and R. McGehee. 1980. Competitive exclusion. *American Naturalist* 115:151–70.

Bertness, M.D. 1989. Intraspecific competition and facilitation in a northern acorn barnacle population. *Ecology* 70:257–68.

Brafield, A.E. 1978. Life in Sandy Shores. Studies in Biology no. 89. London: Edward Arnold.

Carlton, J.T. 1985. Transoceanic and interoceanic dispersal of coastal marine organisms: The biology of ballast water. *Oceanography and Marine Biology Annual Review* 23:313–71.

Carroll, M.L., and R.C. Highsmith. 1996. Role of catastrophic disturbance in mediating *Nucella-Mytilus* interactions in the Alaskan rocky intertidal. *Marine Ecology Progress Series* 138:125–33.

Carson, R.L. 1979. The Edge of the Sea. Boston: Houghton Mifflin.

Connell, J.H. 1961. The influence of interspecific competition and other factors on the distribution of the barnacle, *Chthamalus stellatus. Ecology* 42:710–23.

Constable, A.J. 1999. Ecology of benthic macroinvertebrates in soft-sediment environments: a review of progress towards quantitative models and predictions. *Austral Ecology* 24:452–76.

Dayton, P.K. 1971. Competition, disturbance, and community organization: The provision and subsequent utilization of space in a rocky intertidal community. *Ecological Monographs* 41:351–89.

Denny, M.W. 1985. Wave forces on intertidal organisms: A case study. *Limnology and Oceanography* 30:1171–87.

Denny, M. 1995. Survival in the surf zone. *American Scientist* 83:166–73.

Eltringham, S.K. 1972. *Life in Mud and Sand.* New York: Crane, Russak.

Epel, D. 1977. The program of fertilization. *Scientific American* November:129–38.

Farmingon, J. 1985. Oil pollution: A decade of monitoring. *Oceanus* 28:2–12.

Gaines, S.D., and M.D. Bertness. 1992. Dispersal of juveniles and variable recruitment in sessile marine species. *Nature* 360:579–80.

Grahame, J., and G.M. Branch. 1985. *Reproductive Patterns of Marine Invertebrates.* Scotland: Aberdeen University Press.

Harger, J.R.E. 1972. Competitive coexistence among intertidal invertebrates. *American Scientist* 60:600–07.

Hayes, F.R. 1964. The mud-water interface. *Oceanography and Marine Biology Annual Review* 2:122–45.

Hoar, W.S. 1983. *General and Comparative Physiology.* Englewood Cliffs, NJ: Prentice-Hall.

Horn, M.H., K.L. Martin, and M.A. Chotkowski 1998. *Intertidal Fishes: Life in Two Worlds.* San Diego: Academic Press.

Jackson N.L., K.F. Nordstrom, I. Eliot, and G. Masselink. 2002. "Low energy" sandy beaches in marine and estuarine environments: a review. *Geomorphology* 48:147–62.

Johnson, C.H., and Hastings, J.W. 1986. The elusive mechanism of the circadian clock. *American Scientist* 74:29–36.

Kline, D. 1991. Activation of the egg by the sperm. BioScience, 41:89–95.

Koehl, M.A.R. 1982. The interaction of moving water and sessile organisms. *Scientific American* 247:124–34.

Kozloff, E.N. 1987. *Marine Invertebrates of the Pacific Northwest.* Seattle: University of Washington Press.

Lee, J.J. 1995. Living sands. *BioScience* 45:252–61.

Lewis, J.H. 1964. *The Ecology of Rocky Shores*. London: English Universities Press.

Little, C., and J.A. Kitching. 1996. *The Biology of Rocky Shores*. Oxford University Press, Oxford.

Loughlin, T.R., ed. 1994. *Impacts of the Exxon Valdez Oil Spill on Marine Mammals*. San Diego: Academic Press.

Lubchenco, J. 1978. Plant species diversity in a marine intertidal community: Importance of herbivore food preference and algal competitive abilities. *American Naturalist* 112:23–39.

McIntyre, A.D. 1969. Ecology of marine meibenthos. *Biological Review* 44:245–90.

Mearns, A.J. 1981. Effects of municipal discharges on open coastal ecosystems. In R.A. Geyer, ed. *Marine Environmental Pollution*. Amsterdam: Elsevier.

Menge, B.A. 1975. Brood or broadcast? The adaptive significance of different reproductive strategies in the two intertidal sea stars, *Leptasterias hexactis* and *Pisaster ochraceus*. *Marine Biology* 31:87–100.

Menge, B.A., and Lubchenco, J. 1981. Community organization in temperate and tropical rocky intertidal habitats: prey refuges in relation to consumer pressure gradients. *Ecological Monographs* 51:429–50.

Moore, P.G., and R. Seed. 1986. *The Ecology of Rock Coasts*. New York: Columbia University Press.

Morse, A.N.C. 1991. How do planktonic larvae know where to settle? *American Scientist* 79:154–67.

Newell, R.C. 1979. *Biology of Intertidal Animals*. Faversham, Kent, UK: Ecological Surveys.

Paine, R.T. 1990. Benthic macroalgal competition: complications and consequences. *Journal of Phycology* 26:12–17.

Palmer, J.D. 1974. *Biological Clocks in Marine Organisms: The Control of Physiological and Behavioral Tidal Rhythms*. New York: Interscience Publishers.

Pechenik, J.A., D.E. Wendt, and J.N. Jarrett. 1998. Metamorphosis is not a new beginning. *BioScience* 48:901–10.

Peterson, C.H. 1991. Intertidal zonation of marine invertebrates in sand and mud. *American Scientist* 79:236–48.

Ray, G.C. 1991. Coastal-zone biodiversity patterns. *BioScience* 41:490–98.

Ray, G.C., and Gregg, W.P.J. 1991. Establishing biosphere reserves for coastal barrier ecosystems. *BioScience* 41:301–309.

Reise, K. 1985. *Tidal Flat Ecology*. New York: Springer-Verlag.

Ricketts, C., J. Calvin, and J.W. Hedgpeth. 1986. *Between Pacific Tides*. Revised by D.W. Phillips. Stanford, CA: Stanford University Press.

Robles, C., R. Sherwood-Stevens, and M. Alvarado. 1995. Responses of a key intertidal predator to varying recruitment of its prey. *Ecology* 76:565–79.

Ruppert, E., and R. Fox. 1988. *Seashore Animals of the Southeast*. Columbia, SC: University of South Carolina Press.

Sanders, H.L. 1968. Marine benthic diversity: A comparative study. *American Naturalist* 102:243–82.

Scheltema, R.S. 1971. Larval dispersal as a means of genetic exchange between geographically separated populations of shallow-water benthic marine gastropods. *Biological Bulletin* 140:284–322.

Sebens, K.P. 1983. The ecology of the rocky subtidal zone. *American Scientist* 73:548–57.

Steinberger, A., and K. Schiff. 2002. *Characteristics of Effluents from Large Municipal Wastewater Treatment Facilities between 1988 and 2000*. S. Calif. Coastal Water Resources Project Special Publication.

Strathman, R. 1974. The spread of sibling larvae of sedentary marine invertebrates. *American Naturalist* 108:29–44.

Thorson, G. 1950. Reproduction and larval ecology of marine bottom invertebrates. *Biological Review* 25:1–45.

Thorson, G. 1957. Bottom communities. In J.W. Hedgpeth, ed. *Treatise on Marine Ecology and Paleoecology. Vol. I.- Ecology*. Boulder, CO: Geological Society of America. pp. 461–534.

Thorson, G. 1961. Length of pelagic life in marine bottom invertebrates as related to larval transport by ocean currents. In M. Sears, ed. *Oceanography*. Washington, DC: AAAS, pp. 455–74.

Trager, G.C., Hwang, J.S., and Strickler, J.R. 1990. Barnacle suspension-feeding in variable flow. *Marine Biology* 105:117–27.

Turner, M.H. 1990. Oil Spill: Legal strategies block ecology communications. *BioScience* 40:238–42.

Underwood, A.J. 1974. On models for reproductive strategy in marine benthic invertebrates. *American Naturalist* 108:874–78.

Underwood, AJ. 2000. Experimental ecology of rocky intertidal habitats: what are we learning? *Journal of Experimental Marine Biology and Ecology* 250:51–76.

Whitlatch, R.B. 1981. Patterns of resource utilization and coexistence in marine intertidal deposit-feeding communities. *Journal of Marine Research* 38:743–65.

Young, R.R. 1989. Prevalence and severity of shell disease among deep-sea crabs of the Middle Atlantic Bight in relation to ocean sewage sludge dumping. *Journal of Shellfish Research* 8:462.

Chapter 9

Adey, W.H. 1978. Coral reef morphogenesis: A multidimensional model. *Science* 202:831–37.

Aeby, G.S. 1991. Costs and benefits of parasitism in a coral reef system. *Pacific Science* 45:85–6.

Arvedlund, M. and L. E. Nielsen. 1996. Do the anemonefish *Amphiprion ocellaris* (Pisces: Pomacentridae) imprint themselves to their host sea anemone *Heteractis magnifica* (Anthozoa: Actinidae)? *Ethology* 102:197–211.

Babcock, R.C., P.L. Bull, A.J. Heyward, J.K. Oliver, C.C. Wallace, and B.L. Willis. 1986. Synchronous spawnings of 105 scleractinian coral species on the Great Barrier Reef. *Marine Biology* 90:379–94.

Babcock, R.C., B.L. Willis, and C.J. Simpson. 1994. Mass spawning of corals on a high latitude reef. *Coral Reefs* 13:101–9.

Baker, A. C. 2001. Reef corals bleach to survive change. *Nature* 411:765–66.

Barber, C.V. and V. Pratt. 1998. Poison and profit: Cyanide fishing in the Indo-Pacific. *Environment* 40:5–34.

Barlow, G.W. 1972. The attitude of fish eye-lines in relation to body shape and to stripes and bars. *Copeia* 1:4–12.

Bellwood, D.R. 1995. Direct estimates of bioerosion by two parrot fish species, *Chlorurus gibbus* and *C. sordidus*, on the Great Barrier Reef, Australia. *Marine Biology* 121:419–29.

Birkeland, C. 1982. Terrestrial runoff as a cause of outbreaks of *Acanthaster planci* (Echinodermata: Asteroidea). *Marine Biology* 69:175–85.

Birkeland, C. 1989. The Faustian traits of the crown of thorns starfish. *American Scientist* 77:154–63.

Birkeland, C. 1997. *Life and Death of Coral Reefs*. London: Chapman and Hall.

Bjorndal, K.A. 1995. *Biology and Conservation of Sea Turtles*. Washington, DC: Smithsonian Institution Press.

Blair, S.M., T.L. McIntosh, and B.J. Mostkoff. 1994. Impacts of Hurricane Andrew on the offshore reef systems of the central and northern Dade County, Florida. *Bulletin of Marine Science* 54:961–73.

Bolden, S.K. 2000. Long-distance movement of a Nassau grouper (*Epinephelus striatus*) to a spawning aggregation in the central Bahamas. *Fishery Bulletin* 98:642–45.

Brown, B.E., and J.C. Ogden. 1993. Coral bleaching. *Scientific American* 268:64–70.

Brown, B.E., R.P. Dunne, H. Chansang. 1996. Coral bleaching relative to elevated seawater temperature in the Andaman Sea (Indian Ocean) over the last 50 years. *Coral Reefs* 15:151–52.

Buddemeier, R.W., and Fautin, D.G. 1993. Coral bleaching as an adaptive mechanism. *BioScience* 43:320–26.

Chamberlain, J.A. 1978. Mechanical properties of coral skeleton: Compressive strength and its adaptive significance. *Paleobiology* 4:419–35.

Cheshire, A.C., C.R. Wilkinson, S. Seddon, and G. Westphalen. 1997. Bathymetric and seasonal changes in photosynthesis and respiration of the phototrophic sponge *Phyllospongia lamellosa* in comparison with respiration by the heterotrophic sponge *Ianthella basta* on Davies Reef, Great Barrier Reef. *Marine and Freshwater Research* 48:589–99.

Clifton, K.E. 1997. Mass spawning by green algae on coral reefs. *Science* 275:1116–8.

Connell, J.H. 1978. Diversity in tropical rain forests and coral reefs. *Science* 199:1302–10.

Dana, T.F. 1975. Development of contemporary Eastern Pacific coral reefs. *Marine Biology* 33:355–74.

Darwin, C. 1962. *The Structure and Distribution of Coral Reefs.* Berkeley, CA: University of California Press.

Dollar, S. J. 1982. Wave stress and coral community structure in Hawaii. *Coral Reefs* 1:71–81.

Dubinsky, Z. and P.L. Jokiel. 1994. Ratio of energy and nutrient fluxes regulates symbiosis between zooxanthellae and corals. *Pacific Science* 48:313–24.

Eakin, C.M. 1996. Where have all the carbonates gone? A model comparison of calcium carbonate budgets before and after the 1982–1983 El Niño at Uva Island in the eastern Pacific. *Coral Reefs* 15:109–19.

Earle, S.A. 1991. Sharks, squids, and horseshoe crabs—the significance of marine biodiversity. *BioScience* 41:506–09.

Edmunds, P.J. 2000. Recruitment of scleractinians onto skeletons of corals killed by black band disease. *Coral Reefs* 19:69–74.

Edmunds, P.J., and R.C. Carpenter. 2001. Recovery of *Diadema antillarum* reduces macroalgal cover and increases abundance of juvenile corals on a Caribbean reef. *Proceedings of the National Academy of Sciences* 98:5067–71.

Fadlallah, Y.H. 1983. Sexual reproduction, development and larval biology in scleractinian corals: a review. *Coral Reefs* 2:129–50.

Falkowski, PG., et al.1984. Light and the bioenergetics of a symbiotic coral. *BioScience* 34:705–09.

Falkowski, P.G., Z. Dubinsky, L. Muscatine, and L. McCloskey. 1993. Population control in symbiotic corals. *BioScience* 43:606–11.

Fankboner, P.V. 1971. Intracellular digestion of symbiotic zooxanthellae by host amoebocytes in giant clams (Bivalvia: Tridachnidae), with a note on the nutritional role of the hypertrophied siphonal epidermis. *Biological Bulletin* 141:222–34.

Fautin, D.G. 1991. The anemonefish symbiosis: what is known and what is not? *Symbiosis* 10:23–46.

Fonseca, M. S., J.C. Zieman, G.W. Thayer, and J.S. Fisher. 1983. The role of current velocity in structuring eelgrass (*Zostera marina* L.) meadows. *Estuarine and Coastal Shelf Science* 17:367–80.

Gardner, T.A. et al. 2003. Long-term region-wide declines in caribbean corals. *Science* 301:958–60.

Garrison, V.H. et al. 2003. African and Asian dust: From desert soils to coral reefs. *BioScience* 53:469–80.

Gittings, S.R., et al. 1992. Mass spawning and reproductive viability of reef corals at the East Flower Garden Bank, northwest Gulf of Mexico. *Bulletin of Marine Science* 51:420–28.

Gleason, D.F. and G.M. Wellington. 1993. Ultraviolet radiation and coral bleaching. *Science* 365:836–8.

Glynn, P.W. 1990. El Niño–Southern Oscillation 1982–1983: Nearshore population, community, and ecosystem responses. *Annual Review of Ecology and Systematics* 19:309–45.

Glynn, P. W. and L. D'Croz. 1990. Experimental evidence for high temperature stress as the cause of El Niño-coincident coral mortality. *Coral Reefs* 8:181–91.

Goreau, T. F. 1959. The ecology of Jamaican coral reefs. I. Species composition and zonation. *Ecology* 40:67–90.

Goreau, T.F. and N.I. Goreau. 1959. The physiology of skeleton formation in corals. II. Calcium deposition by hermatypic corals under various conditions in the reef. *Biological Bulletin* 117:239–50.

Goreau, T.J. 1990. Coral bleaching in Jamaica. *Nature* 343:417.

Goreau, T.F., N.I. Goreau, and C.M. Yonge. 1971. Reef corals: autotrophs or heterotrophs? *Biological Bulletin* 141:247–60.

Grigg, R.W. 1982. Darwin Point: A threshold for atoll formation. *Coral Reefs* 1:29–34.

Halstead, B.W. 1988. *Poisonous and Venomous Marine Animals of the World.* Burbank, CA: Darwin Publications.

Halstead, B.W., P.S. Auerbach, and D.R. Campbell. 1990. *A Colour Atlas of Dangerous Marine Animals.* Boca Raton, FL: CRC Press.

Harrison, P.L., R.C. Babcock, G.D. Bull, J.K. Oliver, C.C. Wallace, and B.L. Willis. 1984. Mass spawning in tropical reef corals. *Science* 223:1186–89.

Harvelt, C.D. et al. 1999. Emerging marine diseases—climate links and anthropogenic factors. *Science* 285:1505–10.

Hatcher, B.G. 1990. Coral reef primary productivity: a hierarchy of pattern and process. *Trends in Ecology and Evolution* 5:149–55.

Hawkins, J.P., C.M. Roberts, and T. Adamson. 1991. Effects of a phosphate ship grounding on a Red Sea coral reef. *Marine Pollution Bulletin* 22:538–42.

Hemminga, M.A., P.G. Harrison, and F. van Lent. 1991. The balance of nutrient loses and gains in seagrass meadows. *Marine Ecology Progress Series* 71:85–96.

Hemminga, M.A., and C.A. Duarte. 2001. *Seagrass Ecology.* Cambridge, UK: Cambridge University Press.

Highsmith, R.C. 1982. Reproduction by fragmentation in corals. *Marine Ecology Progress Series* 7:207–26.

Hillis-Colinvaux, L. 1986. Historical perspectives on algae and reefs: have reefs been misnamed? *Oceanus* 29:43–8.

Hughes, T. P. 1994. Catastrophes, phase shifts, and large-scale degradation of a Caribbean coral reef. *Science* 265:1547–50.

Hughes, T.P., et al. 2003. Climate change, human impacts, and the resilence of coral reefs. *Science* 301:929–33.

Hutchings, P. A. 1986. Biological destruction of coral reefs: a review. *Coral Reefs* 4:239–52.

Irlandi, E.A. and C.H. Peterson. 1991. Modification of animal habitat by large plants: Mechanisms by which sea grasses influence clam growth. *Oecologia* 87:307–18.

Jackson, J.B.C. 1991. Adaptation and diversity of reef corals. *BioScience*, 41:475–82.

Jackson, J.B.C. 1997. Reefs since Columbus. *Coral Reefs* 16:S23–S32.

Jackson, J.B.C., and T.P. Hughes. 1985. Adaptive strategies of coral-reef invertebrates. *American Scientist* 73:265–74.

Jackson, J. B. C. et al. 2001. Historical overfishing and the recent collapse of coastal ecosystems. *Science* 293:629–38.

Johannes, R. E. et al. 1983. Latitudinal limits of coral reef growth. *Marine Ecology Progress Series* 11:105–11.

Jones, G.P. 1990. The importance of recruitment to the dynamics of a coral reef fish population. *Ecology* 71:1691–98.

Jones, O.A., and R. Endean, eds. 1976. *Biology and Geology of Coral Reefs. Vols I and II.* New York: Academic Press.

Kinzie, R.A., III. 1993. Effects of ambient levels of solar ultraviolet radiation on zooxanthellae and photosynthesis of the reef coral *Montipora verrucosa. Marine Biology* 116:319–27.

Klumpp, D.W., B.L. Bayne, and A.J.S. Hawkins. 1992. Nutrition of the giant clam *Tridacna gigas* (L.). I. Contribution of filter feeding and photosynthesis to respiration and growth. *Journal of Experimental Marine Biology and Ecology* 155:105–22.

Knowlton, N. 2001. Sea urchin recovery from mass mortality: New hope for coral reefs? *Proceedings of the National Academy of Sciences* 98:4822–24.

Kramarsky-Winter, E., M. Fine, and Y. Loya. 1997. Coral polyp expulsion. *Nature* 387:137.

Kuta, K.G., and L.L. Richardson. 1996. Abundance and distribution of black band disease on coral reefs in the northern Florida Keys. *Coral Reefs* 15:219–23.

Lane, D.J.W. 1996. A crown-of-thorns outbreak in the eastern Indonesian Archipelago, February 1996. *Coral Reefs* 15:209–10.

Lessios, H.A. 1988. Mass mortality of *Diadema antillarum* in the Caribbean: what have we learned? *Annual Review of Ecology and Systematics* 19:371–93.

Limbaugh, C. 1961. Cleaning symbiosis. *Scientific American* August:42–9.

Lohmann, K.J., and C.M.F. Lohmann. 1998. Migratory guidance mechanisms in marine turtles. *Journal of Avian Biology* 29:585–96.

Losey, G.S., Jr. 1972. The ecological importance of cleaning symbiosis. *Copeia* 4:820–33.

Lutz, P.L., and J.A. Musick. 1996. *The Biology of Sea Turtles*. Boca Raton, FL: CRC Press.

Mariscal, R.N. 1972. Behavior of symbiotic fishes and sea anemones. In H.E. Winn and B.L. Olla, eds. *Behavior of Marine Animals*. Plenum Publishing: New York.

Marshall, A.T. 1996. Calcification in hermatypic and ahermatypic corals. *Science* 271:637–39.

Miller, M.W. and M.E. Hay. 1998. Effects of fish predation and seaweed competition on the survival and growth of corals. *Oecologia* 113:231–38.

Muscatine, L. 1980. Productivity of zooxanthellae. In P.G. Falkowski, ed. *Primary Productivity in the Sea*. New York: Plenum, pp. 381–402.

Muscatine, L., and J.W. Porter. 1977. Reef corals: Mutualistic symbiosis adapted to nutrient-poor environments. *BioScience* 27:454–60.

Nakashima, Y., T. Kuwamura, and Y. Yogo. 1995. Why be a both-ways sex changer? *Ethology* 101:301–7.

Odum, H.T., and E.P. Odum. 1955. Trophic structure and productivity of a windward coral reef community on Eniwetok Atoll. *Ecological Monographs* 25:291–320.

Ogden, J.C., et al. 1994. A long-term interdisciplinary study of the Florida Keys seascape. *Bulletin of Marine Science* 54:1059–71.

Oliver, J., and R. Babcock. 1992. Aspects of the fertilization ecology of broadcast spawning corals: sperm dilution effects and in situ measurements of fertilization. *Biological Bulletin* 183:409–17.

Pandolfi, J.M. 2003. Global trajectories of the long-term decline of coral reef ecosystems. *Science* 301:955–58.

Patterson, K.L., et al. 2002. The etiology of white pox, a lethal disease of the Caribbean elkhorn coral, *Acropora palmata*. *Proceedings of the National Academy of Sciences* 99:8725–30.

Porter, J.W. 1976. Autotrophy, heterotrophy, and resource partitioning in Caribbean reef-building corals. *American Naturalist* 110:731–42.

Porter, J.W., and O.W. Meier. 1992. Quantification of loss and change in Floridian reef coral populations. *American Zoologist* 32:625–40.

Porter, J. W., et al. 2001. Patterns of spread of coral diseases in the Florida Keys. *Hydrobiologia* 460:1–24.

Poulin, R., and A.S. Grutter. 1996. Cleaning symbioses: Proximate and adaptive explanations. *BioScience* 46:512–16

Reaka-Kudla, M.L., J.S. Feingold, and P.W. Glynn. 1996. Experimental studies of rapid bioerosion of coral reefs in the Galápagos Islands. *Coral Reefs* 15:101–07.

Richardson, L.L., et al. 1998. Florida's mystery coral-killer identified. *Nature* 392:557–58.

Richardson, L.L. 1998. Coral diseases: what is really known? *Trends in Ecology and Evolution* 13:438–43.

Richmond, R.H. 1985. Reversible metamorphosis in coral planula larvae. *Marine Ecology Progress Series* 22:181–85.

Riegl, B., C. Heine, and G.M. Branch. 1996. Function of funnel-shaped coral growth in a high-sedimentation environment. *Marine Ecology Progress Series* 145:87–93.

Robertson, D.R. 1972. Social control of sex reversal in a coral-reef fish. *Science* 177:1007–09.

Rogers, C.S. 1983. Sublethal and lethal effects of sediments applied to common Caribbean reef corals in the field. *Marine Pollution Bulletin* 14:378–82.

Rothans, T.C., and A.C. Miller. 1991. A link between biologically-imported particulate organic nutrients and the detritus food web in reef communities. *Marine Biology* 110:145–50.

Rougerie, F., and B. Wauthy. 1993. The endo-upwelling concept: from geothermal convection to reef construction. *Coral Reefs* 12:19–30.

Rowan, R.D.A.P. 1991. A molecular genetic classification of zooxanthellae and the evolution of animal-agal symbioses. *Science* 251:1348–51.

Rützler, K., D.L. Santavy, and A. Antonius. 1983. The black band disease of Atlantic reef corals. III. Distribution, ecology, and development. *Marine Ecology* 4:329–58.

Sale, P.F. 1974. Mechanisms of coexistence in a guild of territorial reef fishes. *Marine Biology* 29:89–97.

Sale, P.F. ed. 2002. *Coral Reef Fishes Dynamics and Diversity in a Complex Ecosystem*. San Diego: Academic Press. 549 pp.

Schlichter, D., H. Kampmann, and S. Conrady. 1997. Trophic potential and photoecology of endolithic algae living within coral skeletons. *Marine Ecology* 18:299–317.

Schuhmacher, H., and H. Zibrowius. 1985. What is hermatypic? *Coral Reefs* 4:1–9.

Scott, R.D., and H.R. Jitts. 1977. Photosynthesis of phytoplankton and zooxanthellae on a coral reef. *Marine Biology* 41:307–15.

Shapiro, D.Y. 1987. Differentiation and evolution of sex change in fishes. *BioScience* 37:490–97.

Shashar, N., Y. Cohen, Y. Loya, and N. Sar. 1994. Nitrogen fixation (acetylene reduction) in stony corals: evidence for coral-bacteria interactions. *Marine Ecology Progress Series* 111:259–64.

Shlesinger, Y., and Y. Loya. 1985. Coral community reproductive patterns: Red Sea versus the Great Barrier Reef. *Science* 228:1333–35.

Sorokin, Y.I. 1972. Bacteria as food for coral reef fauna. *Oceanology* 12:169–77.

Spalding, M.D., Ravilious, C., and Green, E.P. 2001. *World Atlas of Coral Reefs*. Berkeley, CA: University of California Press. 424 pp.

Stafford-Smith, M.G. and R.F.G. Ormond. 1992. Sediment-rejection methods of 42 species of Australian scleractinian corals. *Australian Journal of Marine and Freshwater Research* 43:683–705.

Stoddart, D.R. 1973. Coral reefs: The last two million years. *Geography* 58:313–23.

Swearer, S.E., et al. 1999. Larval retention and recruitment in an island population of a coral-reef fish. *Nature* 402:799–802.

Thayer, G.W., D.W. Engel, and K.A. Bjorndal. 1982. Evidence for short-circuiting of the detritus cycle of seagrass beds by the green turtle, *Chelonia mydas*. *Journal of Experimental Marine Biology and Ecology* 62:173–83.

Thresher, R.E. 1984. *Reproduction in Reef Fishes*. Neptune City, NJ: T.H.F. Publications.

Tribble, G.W., M.J. Atkinson, F.J. Sansone, and S.V. Smith. 1994. Reef metabolism and endo-upwelling in perspective. *Coral Reefs* 13:199–201.

Vernon, J.E.N. 1995. *Corals in Space and Time.* Sydney: University of New South Wales Press.

Walbran, P.D., R.A. Henderson, A.J.T. Jull, and M.J. Head. 1989. Evidence from sediments of long-term *Acanthaster planci* predation on corals of the Great Barrier Reef. *Science* 245:847–50.

Warner, R.R. 1984. Mating behavior and hermaphroditism in coral reef fishes. *American Scientist* 72:128–36.

Warner, R.R. 1990. Male versus female influences on mating-site determination in a coral reef fish. *Animal Behavior* 39:540–48.

Wilkinson, C.R. 1987. Interocean differences in size and nutrition of coral reef sponge populations. *Science* 236:1654–57.

Wilson, J.R., and P.L. Harrison. 1998. Settlement-competency periods of larvae of three species of scleractinians. *Marine Biology* 131:339–45.

Wolanski, E., et al. 2003. Mud, marine snow, and coral reefs. *American Scientist* 91:44–51.

Yamamuro, M., H. Kayanne, M. Minagawa. 1995. Carbon and nitrogen stable isotopes of primary producers in coral reef ecosystems. *Limnology and Oceanography* 40:617–21.

Chapter 10

Alexander, R. McNeill. 1970. *Functional Design in Fishes.* London: Hutchinson.

Alexander, R. McNeill. 1988. *Elastic Mechanisms in Animal Movement.* New York: Cambridge University Press.

Allan, J.D. 1976. Life history patterns in zooplankton. *American Naturalist* 110:165–80.

Alldredge, A. 1976. Appendicularians. *Scientific American* July: 94–102.

Alldredge, A.L., and L.P. Madin. 1982. Pelagic tunicates: Unique herbivores in the marine plankton. *BioScience* 32:655–63.

Bainbridge, R. 1960. Speed and stamina in three fish. *Journal of Experimental Biology* 37:129–53.

Barham, E.G. 1966. Deep scattering layer migration and composition: Observations from a diving saucer. *Science* 151:1399–1403.

Blake, R.W. 1983. *Fish Locomotion.* New York: Cambridge University Press.

Block, B.A., and E.D. Stevens. 2001. *Tuna: Physiological Ecology and Evolution.* San Diego: Academic Press.

Boden, B.P., and E.M. Kampa. 1967. The influence of natural light on the vertical migrations of an animal community in the sea. *Symposium of the Zoological Society of London* 19:15–26.

Bond, C.E. 1983. *Biology of Fishes.* Philadelphia: W.B. Saunders.

Boyd, C.M. 1976. Selection of particle sizes by filter-feeding copepods: A plea for reason. *Limnology and Oceanography* 21:175–79.

Bright, T., et al. 1972. Effects of a total solar eclipse on the vertical distribution of certain oceanic zooplankters. *Limnology and Oceanography* 17:296–301.

Brinton, E. 1962. The distribution of Pacific euphausiids. *Bulletin. Scripps Institution of Oceanography* 8:51–270.

Bundy, M.H., and H.A. Vanderploeg. 2002. Detection and capture of inert particles by calanoid copepods: the role of the feeding current. *Journal of Plankton Research* 24:215–23.

Carey, F.G. 1973. Fishes with warm bodies. *Scientific American* February: 36–44.

Carr, A. 1965. The navigation of the green turtle. *Scientific American* May:79–86.

Clinton, W.L. 1994. Sexual selection and growth in male northern elephant seals. In B.J. Le Boeuf, and R.M. Laws, eds. *Elephant Seals.* Berkeley, CA: U. California Press. pp. 154–168.

Compagno, L.J.V. 1988. *Sharks of the Order Carcharhinoformes.* Princeton, NJ: Princeton University Press.

Cushing, D.H. 1968. *Fisheries Biology.* Madison, WI: U. Wisconsin Press.

DeLong, R.L.; and B.S. Stewart. 1991. Diving patterns of northern elephant seal bulls. *Marine Mammal Science* 7:369–84.

Denton, E.J., and J.P. Gilpin-Brown. 1973. Flotation mechanisms in modern and fossil cephalopods. *Advances in Marine Biology* 11:197–268.

Eastman, J.T., and A.L. DeVries. 1986. Antarctic fishes. *Scientific American* 255:106–14.

Emlet, R.B., and R.R. Strathman. 1985. Gravity, drag, and feeding currents of small zooplankton. *Science* 228:1016–17.

Ege, V. 1939. A revision of the genus *Anguilla* Shaw, a systematic, phylogenetic, and geographical study. *Dana Reports* 3:1–256.

Elsner, R., and B. Gooden. 1983. *Diving and Asphyxia: A Comparative Study of Animals and Men.* New York: Cambridge University Press.

Fierstine, H.L., and V. Walters. 1968. Studies in locomotion and anatomy of scombroid fishes. *Southern California Academy of Science, Memoirs* 6:1–34.

Fish, J.F., J.L. Sumich, and G.L. Lingle. 1974. Sounds produced by the gray whale, *Eschrichtius robustus. Marine Fisheries Review* 36:38–45.

Fish, F.E. 1999. Energetics of swimming and flying in formation. *Comments on Theoretical Biology* 5:283–304.

Frost, B.W. 1972. Effects of size and concentration of food particles on the feeding behavior of the marine planktonic copepod *Calanus pacificus. Limnology and Oceanography* 17:805–15.

Gilmer, R.W. 1972. Free-floating mucus webs: A novel feeding adaptation for the open ocean. *Science* 176:1239–40.

Giorgio, P.A., and Duarte, C.M. 2002. Respiration in the open ocean. *Nature* 420:379–84.

Hardy, A.H. 1971. *The Open Sea: Its Natural History.* Part I: "The World of Plankton." Part II: "Fish and Fisheries." Boston: Houghton Mifflin.

Hasler, A.D., A.T. Sholz, and R.M. Horrall. 1978. Olfactory imprinting and homing in salmon. *American Scientist* 66:347–54.

Hawryshyn, C.W. 1992. Polarization vision in fish. *American Scientist* 80:164–75.

Herman, L.M., ed. 1980. *Cetacean Behavior: Mechanisms and Functions.* New York: John Wiley & Sons.

Hoar, W.W. 1983. *General and Comparative Physiology.* Englewood Cliffs, NJ: Prentice-Hall.

Jackson, G.A. 1990. A model of the formation of marine algal flocs by physical coagulation processes. *Deep-Sea Research* 37:1197–211.

Jumper, G.Y., Jr., and R.C. Baird. 1991. Location by olfaction: A model and application to the mating problem in the deep-sea hachetfish *Argyropelecus hemigymnus. American Naturalist* 138:1431–58.

Kalmijn, A.J. 1977. The electric and magnetic sense of sharks, skates, and rays. *Oceanus* 20:45–52.

Kanwisher, J., and A. Ebling. 1957. Composition of swim bladder gas in bathypelagic fishes. *Deep-Sea Research* 4:211–17.

Keenleyside, M.H.A. 1979. *Diversity and Adaptation in Fish Behavior.* New York: Springer-Verlag.

Klimley, A.P. 1994. The predatory behavior of the white shark. *American Scientist* 82:122–33.

Koehn, R.K. 1972. Genetic variation in the eel, a critique. *Marine Biology* (Berlin) 14:179–81.

Kooyman, G. 1989. *Diverse Divers: Physiology and Behavior.* New York: Springer-Verlag.

Kooyman, G.L., M.A. Castellini, and R.W. Davis. 1981. Physiology of diving in marine mammals. *Annual Review of Physiology* 43:343–57.

Lam, R.K., and B.W. Frost. 1976. Model of copepod filtering response to changes in size and concentration of food. *Limnology and Oceanography* 21:490–500.

Lockyer, C. 1981. Estimates of growth and energy budget for the sperm whale, *Physeter catodon*. *Mammals in the Sea* 3:489–504.

Lohmann, K.J., and C.M.F. Lohmann. 1998. Migratory guidance mechanisms in marine turtles. *Journal of Avian Biology* 29:585–96.

Nicol, J.A.C. 1989. *The Eyes of Fishes*. New York: Oxford University Press.

Norris, K.S. 1968. Evolution of acoustic mechanisms in odontocete cetaceans. *Evolution and Environment* 297–324.

Norris, K.S., and G.W. Harvey. 1972. A theory for the function of the spermaceti organ of the sperm whale (*Physeter catodon*). In K.S. Norris ed. *Animal Orientation and Navigation*. Washington, DC: NASA. pp. 397–417.

O'brien, W.J., H.I. Browman, and B.I. Evans. 1990. Search strategies in foraging animals. *American Scientist* 78:152–60.

Ohman, M.D. 1990. The demographic benefits of diel vertical migration by zooplankton. *Ecological Monographs* 60:257–81.

Partridge, B.L. 1982. The structure and function of fish schools. *Scientific American* 246:114–23.

Pennisi, E. 1989. Much ado about eels. *BioScience* 39:594–98.

Perutz, M.F. 1978. Hemoglobin structure and respiratory transport. *Scientific American* 239:92–125.

Porter, K.G., and J.W. Porter. 1979. Bioluminescence in marine plankton: A coevolved antipredation system. *American Naturalist* 114:458–61.

Quetin, L.B., and Ross, R.M. 1991. Behavioral and physiological characteristics of the antarctic krill, *Euphausia superba*. *American Zoologist* 31:49–63.

Richman, S., D.R. Heinle, and R. Huff. 1977. Grazing by adult estuarine calanoid copepods of the Chesapeake Bay. *Marine Biology* 42:69–84.

Rodin, E.Y., and S. Jacques. 1989. Countercurrent oxygen exchange in the swim bladders of deep-sea fish: A mathematical model. *Mathematical and Computer Modelling* 12:389–93.

Rome, L.C., D. Swank, and D. Corda. 1993. How fish power swimming. *Science* 261:340–43.

Rommel, S.A., Jr., and J.D. McCleave. 1972. Oceanic electric fields: Perception by American eels? *Science* 176:1233–35.

Royce, W., L.S. Smith, and A.C. Hartt. 1968. Models of oceanic migrations of Pacific salmon and comments on guidance mechanisms. *Fishery Bulletin* 66:441–62.

Rubenstein, D.I., and M.A.R. Koehl. 1977. The mechanisms of filter feeding: Some theoretical considerations. *American Naturalist* 111:981–94.

Russel, R.S. 1935. On the value of certain planktonic animals as indicators of water movements in the English Channel and North Sea. *Journal of the Marine Biological Association, U.K.* 20:309–32.

Sanderson, J.L., and R. Wassersug. 1990. Suspension-feeding vertebrates. *Scientific American* 262:96–101.

Scholander, P.F. 1957. The wonderful net. *Scientific American* 196:96–107.

Sheldon, R.W., et al. 1972. The size distribution of particles in the ocean. *Limnology and Oceanography* 17:327–40.

Silver, M.W., A.L. Shanks, and J.D. Trent. 1978. Marine snow: Microplankton habitat and source of small-scale patchiness in pelagic populations. *Science* 201:371–73.

Smith, D.L. 1996. *A Guide to Marine Coastal Plankton and Marine Invertebrate Larvae*. Dubuque, IA: Kendall/Hunt.

Smith, O.L., et al. 1971. Resource competition and an analytical model of zooplankton feeding on phytoplankton. *American Naturalist* 109:571–91.

Steele, J.H., ed. 1973. *Marine Food Chains*. Edinburgh: Oliver and Boyd.

Steele, J.H. 1980. Patterns in plankton. *Oceanus* 25:3–8.

Steele, J.H. 1976. Patchiness. In D.H. Cushing and J.J. Walsh, eds. *Ecology of the Seas*. Oxford, UK: Blackwell Scientific.

Stewart, B.S., and R.L. Delong. 1993. Seasonal dispersion and habitat use of foraging northern elephant seals. *Symposium of the Zoological Society of London* 66:179–94.

Stoecker, D.K., and J.M. Capuzzo. 1990. Predation on protozoa: Its importance to zooplankton. *Journal of Plankton Research* 12:891–908.

Strickler, J.R. 1985. Feeding currents in calanoids: two new hypotheses. In M.S. Laverack, ed. *Physiological Adaptations of Marine Animals. Symposium, Society of Experimental Biology* 39:459–85.

Sverdrup, H.U., M.W. Johnson, and R.H. Fleming. 1970. *The Oceans: Their Physics, Chemistry, and Biology*. Englewood Cliffs, NJ: Prentice-Hall.

Triantafyllou, M.S., and Triantafyllou, G.S. 1995. An efficient swimming machine. *Scientific American* 272:64–70.

Tucker, D.W. 1959. A new solution to the Atlantic eel problem. *Nature (London)* 183:495–501.

Turner, J.T., P.A. Tester, and W.F. Hettler. 1985. Zooplankton feeding ecology. *Marine Biology* 90:1–8.

Vlymen, W.J. 1970. Energy expenditure of swimming copepods. *Limnology and Oceanography* 15:348–56.

Vourinen, I., M. Rajasilta, and J. Salo. 1983. Selective predation and habitat shift in a copepod species—support for the predation hypothesis. *Oecologia* 59:62–4.

Waickstead, J.H. 1976. *Marine Zooplankton*. London: E. Arnold.

Webb, P.W. 1984. Form and function in fish swimming. *Scientific American* 251:72–82.

Webb, P.W. 1988. Simple physical principles and vertebrate aquatic locomotion. *American Zoologist* 28:709–725.

Wilbur, R.J. 1987. Plastic in the North Atlantic. *Oceanus* 30:61–68.

Zaret, T.M., and J.S. Suffern. 1976. Vertical migration in zooplankton as a predator avoidance mechanism. *Limnology and Oceanography* 21:804–13.

Chapter 11

Arp, A.J., and J.J. Childress. 1983. Sulfide binding by the blood of the hydrothermal vent tube worm *Riftia pachyptila*. *Science* 219:295–97.

Baco, A.R., and C.R. Smith. 2003. High species richness in deep-sea chemoautotrophic whale skeleton communities. *Marine Ecology Press Series* 260:109–14

Baker, E.T. 1996. Extensive distribution of hydrothermal plumes along the superfast-spreading East Pacific Rise. 13 degrees 50' - 18 degrees 40'S. *Journal of Geophysics* 101:8685–95.

Cary, S.C., et al. 1998. Worms bask in extreme temperatures. *Nature* 391:345–46.

Childress, J. J. 1995. Are there physiological and biochemical adaptations of metabolism in deep-sea animals? *Trends in Ecology and Evolution* 10:30–6.

Childress, J., H. Felback, and G. Somero. 1987. Symbiosis in the deep sea. *Scientific American* 256:114–20.

Corliss, J.B., et al. 1979. Submarine thermal springs on the Galapagos Rift. *Science* 203:1073–83.

Dayton, P.K., and R.R. Hessler. 1972. Role of biological disturbance in maintaining diversity in the deep sea. *Deep Sea Research* 19:199–208.

Distel, D.L. 1998. Evolution of chemoautotophic endosymbioses in bivalves. *BioScience* 48:277–86.

Fisher, C.R. 1996. Ecophysiology of primary produciton at deep-sea vents and seeps. *Biosystematics and Ecology Series* 11:313–36.

Gage, J.D. 1996. Why are there so many species in deep-sea sediments? *Journal of Experimental Marine Biology and Ecology* 200:257–86.

Grassle, J.F. 1991. Deep-sea benthic biodiversity. *BioScience* 41:464–69.

Gray, J.S. 1974. Animal-sediment relationships. *Oceanography and Marine Biology (Annual Review)* 12:223–61.

Gage, J.D., and P.A. Tyler.1991.*Deep-Sea Biology: A Natural History of Organisms of the Deep-Sea Floor.* Cambridge, UK: Cambridge University Press.

Haymon, R.M., and Macdonald, K.C. 1985. The geology of deep-sea hot springs. *American Scientist* 73:441–49.

Herring, P. 2002. *The Biology of the Deep Ocean.* Oxford: Oxford University Press.

Hessler, R.R., J.D. Isaacs, and E.L. Mills. 1972. Giant amphipod from the abyssal Pacific Ocean. *Science* 175:636–37.

Higgins, R.P., and H. Thiel, eds. 1988. *Introduction to the Study of Meiofauna.* Washington, DC: Smithsonian Institution Press.

Isaacs, J.D., and R.A. Schwartzlose. 1975. Active animals of the deep-sea floor. *Scientific American* 233:84–91.

Jannasch, H.W. 1984. Chemosynthesis: The nutritional basis for life at deep-sea vents. *Oceanus* 27:73–8.

Jannasch, H.W. 1989. Serendipity in deep-sea microbiology: lessons from the *Alvin* lunch. *Oceanus* 31:28–33.

Kim, S.L. and L.S. Mullineaux 1998. Distribution and near-bottom transport of larvae and other plankton at hydrothermal vents. *Deep-Sea Research* 45:423–40.

Lampitt. R. S., et al. 2001. Material supply to the abyssal seafloor in the Northeast Atlantic. *Progress in Oceanography* 50:27–63.

Levin, L.A. 2002. Deep-ocean life where oxygen is scarce. *American Scientist* 90:436–44.

Lutz, R.A., et al.1994. Rapid growth at deep-sea vents. *Nature* 371:663–64.

Lutz, R.A., T.M. Shank, and R. Evans. 2001. Life after death in the deep sea. *American Scientist* 89:422–31.

Malahoff, A. 1985. Hydrothermal vents and polymetallic sulfides of the Galapagos and Gorda/Juan De Fuca ridge systems and of submarine volcanoes. *Biological Society of Washington Bulletin* 6:19–41.

Margulis, L., Chase, D., and Guerrero, R. 1986. Microbial communities. *BioScience* 36:160–70.

Marshall, N.B. 1980. *Deep Sea Biology: Developments and Perspectives.* New York: Garland S.T.P.M.

Menzies, R.J., R.Y. George, and G.T. Rowe. 1973. *Abyssal Environment and Ecology of the World Oceans.* New York: John Wiley.

Mullineaux, L. S., P. H. Wiebe, and E. T. Baker. 1991. Hydrothermal vent plumes: larval highways in the deep sea? *Oceanus* 34:64–8.

Page, H.M., C.R. Fisher, and J.J. Hildress. 1990. Role of filter-feeding in the nutritional biology of a deep-sea mussel with methanotrophic symbionts. *Marine Biology* 104:251–57.

Prieur, D. 1997. Microbiolology of deep-sea hydrothermal vents. *Trends in Biotechnology* 15:242–44.

Rex, M.A. 1973. Deep-sea species diversity: Decreased gastropod diversity at abyssal depths. *Science* 181:1051–53.

Rex, M. A., et al. 1993. Global-scale latitudinal patterns of species diversity in the deep-sea benthos. *Nature* 365:636–39.

Rokop, F.J. 1974. Reproductive patterns in the deep-sea benthos. *Science* 186:743–45.

Sanders, H.L. 1968. Marine benthic diversity: A comparative study. *American Naturalist* 102:243–82.

Sanders, H.L., and R.R. Hessler. 1969. Ecology of the deep-sea benthos. *Science* 163:1419–24.

Sanders, N.K. and J.J. Childress. 1993. Plume surface are and blood volume in the hydrothermal vent worm *Riftia pachyptila.* *Amercan Zoologist* 33:95A.

Sebens, K.P. 1985. The ecology of the rocky subtidal zone. *American Scientist* 73:548–57.

Snelgrove, P.V.R. 1999. Getting to the bottom of marine biodiversity: Sedimentary habitats. *BioScience* 49:1129–38.

Sokolova, M.N. 1970. Weight characteristics of meiobenthos in different regions of the deep-sea trophic areas of the Pacific Ocean. *Okeanologia* (in Russian) 10:348–56.

Sverdrup, H.U., Johnson, M.W., and Fleming, R.H. 1942. The Oceans: Their Physics, Chemistry, and General Biology. Englewood Cliffs, NJ: Prentice-Hall.

Tait, R.V. 1968. *Elements of Marine Ecology.* London: Butterworths.

Thorson, G. 1957. Bottom communities. In J.W. Hedgepeth, ed. *Treatise on Marine Ecology and Paleoecology. Geological Society of America, Vol. 1, Ecology* pp. 461–534.

Thorson, G. 1966. Some factors influencing the recruitment and establishment of marine benthic communities. *Netherlands Journal of Sea Research* 3:241–67.

Tunnicliffe, V. 1992. Hydrothermal-vent communities of the deep sea. *American Scientist* 80:336–49.

Tunnicliffe, V.R., et al. 1997. Biological colonization of new hydrothermal vents following an eruption on Juan de Fuca Ridge. *Deep-Sea Research* 44:1627–44.

Van Dover, C.L. 2000. *The Ecology of Deep-Sea Hydrothermal Vents.* Princeton, NJ: Princeton University Press.

Chapter 12

Alonzo, Suzanne H., Switzer, Paul V., Mangel, M. 2003. Ecological games In space and time: The distribution and abundance of antarctic krill and penguins. *Ecology* 84:1598–607

Ancel, A., et al. 1993. Foraging behavior of emperor penguins as a resource detector in winter and summer. *Nature* 360:336–39.

Bargu, S., Powell, C.L., Coale, S.L., Busman, M., Doucette, G.J., and Silver, M.W. 2002. Krill: a potential vector for domoic acid in marine food webs. *Marine Ecology Progress Series* 237: 209–16.

Bartholomew, G.A. 1970. A model for the evolution of pinniped polygyny. *Evolution* 24:546–59.

Berta, A., and J. Sumich. 1999. *Marine Mammals: Evolutionary Biology.* San Diego: Academic Press.

Bonnell, M.L., and R.K. Selander. 1974. Elephant seals: Genetic variation and near extinction. *Science* 184:908–9.

Bonner, W.N. 1982. *Seals and Man: A Study of Interactions.* Seattle: University of Washington Press.

Bonner, W.N. 1989. *Whales of the World.* New York: Facts on File.

Charney, J.I., ed. 1982. *The New Nationalism and the Use of Common Spaces: Issues in Marine Pollution and the Exploitation of Antarctica.* Lanham, MD: Rowman and Littlefield.

Clarke, A. and C.M. Harris. 2003 Polar marine ecosystems: major threats and future change. *Review Environmental Conservation* 30:1–25

Croxall, J.P. 1987. *Seabirds: Feeding Ecology and Role in Marine Ecosystems.* New York: Cambridge University Press.

Garrison, D.L. 1991. Antarctic sea ice biota. *American Zoologist* 31:17–33.

Hempel, G. 1991. Life in the Antarctic sea ice zone. *Polar Record* 27:249–54.

Hickie, B.E., et al. 2000. A modelling-based perspective on the past, present, and future polychlorinated biphenyl contamination of the St. Lawrence beluga whale (*Delphinapterus leucas*) population. *Canadian Journal of Fisheries and Aquatic Sciences* 57:101–12.

Hunt, G.L.J. 1991. Marine ecology of seabirds in polar oceans. *American Zoologist* 31:131–42.

Kingsley, M.C.S., and I. Stirling. 1991. Haul-out behavior of ringed and bearded seals in relation to defence against surface predators. *Canadian Journal of Zoology* 69:1857–61.

Laws, R.M. 1961. Reproduction, age and growth of southern fin whales. *Discovery Reports* 31:327–486.

Laws, R.M. 1985. The Ecology of the Southern Ocean. *American Scientist* 73:26–40.

Lydersen, C., and K. M. Kovacs. 1996. Energetics of lactation in harp seals (*Phoca groenlandica*) from the Gulf of St. Lawrence, Canada. *Journal of Comparative Physiology* 166:295–304.

Mackintosh, N.A. 1966. The distribution of southern blue and fin whales. In Norris, K.S., ed. *Whales, Dolphins and Porpoises.* Berkeley, CA: University of California Press.

Nelson, C., and K. Johnson. 1987. Whales and walruses as tillers of the seafloor. *Scientific American* 256:112–17.

Nemoto, T., Okiyama, M., Iwasaki, N., and Kikuchi, T. 1988. Squid as predators on krill (*Euphausia superba*) and prey for sperm whales in the Southern Ocean. *Antarctic Ocean and Resources Variability* 292–96.

Nerini,M. 1984. A review of gray whale feeding ecology. In Jones, M.L., S.L. Swartz, and S. Leatherwood, eds. *The Gray Whale, Eschrichtius robustus (Lilljeborg, 1861).* New York: Academic Press, pp. 423–50.

Nevitt, G. 1999. Foraging by seabirds on an olfactory landscape. *American Scientist* 87:46–53.

Nicol, S., and de la Mare, W. 1993. Ecosystem management and the antarctic krill. *American Scientist* 81:36–47.

Pierotti, R., and C.A. Annett. 1990. Diet and reproductive output in seabirds. *BioScience* 40:568–74.

Pivorunas, A. 1979. The feeding mechanisms of baleen whales. *American Scientist* 67:432–40.

Pollard, R.T., Lucas, M.I., and Read, J.F. 2002. Physical controls on biogeochemical zonation in the Southern Ocean. *Deep-Sea Research (Part II, Topical Studies in Oceanography)* 49:3289–305.

Quetin, L.B., and Ross, R.M. 1991. Behavioral and physiological characteristics of the antarctic krill, *Euphausia superba. American Zoologist* 31:49–63.

Rice, D.W., and A.A. Wolman. 1971. The life history and ecology of the gray whale (*Eschrichtius robustus*). *American Society of Mammalogists*, special publication no. 3.

Stevens, J.E. 1995. The Antarctic pack-ice ecosystem. *BioScience* 45:128–221.

Siniff, D. B. 1991. An overview of the ecology of Antarctic seals. *American Zoologist* 31:143–49.

Sturm, M., D.K. Perovich, and M.C. Serreze. 2003. Meltdown in the north. *Scientific American* 289:60–7.

Sumich, J.L. 1986. Latitudinal distribution, calf growth and metabolism, and reproductive energetics of gray whales, *Eschrichtius robustus.* Unpublished Ph.D. dissertation. Corvallis, Oregon: Oregon State University.

Tovar, H., Guillen, V., and Nakama, M.E. Monthly population size of three guano bird species off Peru, 1953 to 1982. In D. Pauly and I. Tsukayama eds. *The Peruvian Anchoveta and Its Upwelling System: Three Decades of Change. ICLARM Studies and Reviews* 15:208–18.

Wilson, R.P., et al. 1991. To slide or stride: when should Adelie penguins (*Pygoscelis adeliae*) toboggan? *Canadian Journal of Zoology* 69:221–25.

Zopal, W. 1987. Diving adaptations of the Weddell seal. *Scientific American* 256:100–05.

Chapter 13

Adey, W.H. 1987. Food production in low-nutrient seas. *BioScience* 37:340–48.

Alexander, L.M. 1993. Large marine ecosystems. *Marine Policy* 17:186–98.

Bardach, J. 1987. Aquaculture. *BioScience* 37:318–19.

Beddington, J.R., R.J.H. Beverton, and D.M. Lavigne. 1985. *Marine Mammals and Fisheries.* Boston: George Allen and Unwin.

Beddington, J.R., and R.M. May. 1982. The harvesting of interacting species in a natural ecosystem. *Scientific American* November:62–9.

Bell, F.W. 1978. Food from the sea: *The Economics and Politics of Ocean Fisheries.* Denver: Westview Press.

Borgese, E.M. 1983. The law of the sea. *Scientific American* March:42–9.

Buschmann, A. H., et al. 2001. Red algal farming in Chile: a review. *Aquaculture* 194:203–20.

Caddy, J.F., ed. 1988. *Marine Invertebrate Fisheries: Their Assessment and Management.* New York: Wiley-Interscience.

Castanza, R., et al. 1998. Principles for sustainable governance of the oceans. *Science* 281:198–99.

Cushing, D.H. 1981. *Fisheries Biology.* Madison, WI: University of Wisconsin Press.

Cushing, D.H., and R.R. Dickson. 1976. The biological response in the sea to climatic changes. *Advances in Marine Biology* 14:1–122.

Dahlberg, K.A. 1991. Sustainable agriculture—fad or harbinger? *BioScience,* 41:337–40.

Emery, K.O., and C.O. Iselin. 1967. Human food from ocean and land. *Science* 157:1279–81.

Falkowski, P.G., ed. 1980. *Primary Productivity in the Sea.* New York: Plenum Press.

Field, C. B., et al. 1998. Primary production of the biosphere: Integrating terrestrial and oceanic components. *Science* 281:237–40.

Food and Agricultural Organization of the United Nations. 1982. *Atlas of the Living Resources of the Seas.* Rome: Department of Fisheries (FAO).

Food and Agricultural Organization of the United Nations. 2002. *Yearbook of Fisheries Statistics, Catches, and Landings.* Rome: Department of Fisheries (FAO).

Frank, K.T., and W.C. Leggett. 1994. Fisheries ecology in the context of ecological and evolutionary theory. *Annual Review of Ecology and Systematics* 25:401–22.

Fye, P.M. 1982. The law of the sea. *Oceanus* 25:7–12.

Gulland, J.A. 1971. *The Fish Resources of the Ocean.* Surrey, England: Fishing News (Books).

Hardin, G.J. 1993. *Living within Limits: Ecology, Economics, and Population Taboos.* New York: Oxford University Press.

Hardy, A.H. 1971. *The Open Sea: Its Natural History.* Part I: The World of Plankton.Part II: Fish and fisheries. Boston: Houghton Mifflin.

Hempel, G., and K. Sherman,eds. 2003. *Large Marine Ecosystems of the World Change and Sustainability.* Amsterdam: Elsevier Science. 450 pp.

Heslinga, G.A., and Fitt, W.K. 1987. The domestication of reef-dwelling clams. *BioScience* 37:332–39.

Horn, M.H., and R.N. Gibson. 1988. Intertidal fisheries. *Scientific American* 258:64–70.

Jackson, J.B.C., et al. 2001. Historical overfishing and the recent collapse of coastal ecosystems. *Science* 293:629–38.

Jarre, A., Muck, P., and Pauly, D. 1991. Two approaches for modelling fish stock interactions in the Peruvian upwelling ecosystem. In N. Daan and M. Sissenwine, eds. *Multispecies Models Relevant to Management of Living Resources. ICES Marine Science Symposium* 193:171–84.

Johnson, S.W. 1994. Deposition of trawl web on an Alaska beach after implementation of MARPOL Annex V legislation. *Marine Pollution Bulletin* 28:477–81.

Koslow, J.A. 1997.Seamounts and the ecology of deep-sea fisheries. *American Scientist* 85:168–76.

Laws, R.M. 1985. The ecology of the southern ocean. *American Scientist* 73:26–40.

Levin, P.S., and M.H. Schiewe. 2001. Preserving salmon diversity. *American Scientist* 89:220–27.

Longhurst, A., et al. 1995. An estimate of global primary production in the ocean from satellite radiometer data. *Journal of Plankton Research* 17:1245–71.

Lüning, K., and S. Pang. 2003. Mass cultivation of seaweeds: current aspects and approaches. *Journal of Applied Phycology* 15:115–19.

Marshall, N.B. 1980. *Deep-Sea Biology: Developments and Perspectives.* New York: Garland S.T.P.M. Press.

Myers, R.A., and B. Worm. 2003. Rapid worldwide depletion of predatory fish communities. *Nature* 423:280–83.

Muck, P. 1989. Major trends in the pelagic ecosystem off peru and their implications for management. *PROCOPA Contribution #90.* 386–403.

Murphy, R.C. 1925. *Bird Islands of Peru.* New York: G.P. Putnam and Sons.

Nicol, S., and W. de la Mare. 1993. Ecosystem management and the antarctic krill. *American Scientist* 81:36–47.

Norse,E., ed. 1993. *Global Marine Biological Diversity: A Strategy for Building Conservation into Decision Making.* Washington, DC: Island Press.

Pauly, D. and V. Christensen. 1995. Primary production required to sustain global fisheries. *Nature* 374:255–57.

Pauly, D., et al. 2000. Fishing dowm aquatic food webs. *American Scientist* 88:46–51.

Pauly, D. and J. Maclean. 2003. *In a Perfect Ocean: The State of Fisheries and Ecosystems in the North Atlantic Ocean.* Washington, DC: Island Press.

Radmer, R.J. 1996. Algal diversity and commercial algal products. *BioScience,* 46:263–70.

Roberts, C.M. 2002. Deep impact: the rising toll of fishing in the deep sea. *Trends in Ecology and Evolution* 17:242–45.

Robinson, M.A., and A. Crispoldi. 1975. Trends in world fisheries. *BioScience* 18:23–9.

Ross, R.M., and L.B. Quetin. 1986. How productive are Antarctic krill? *BioScience* 36:264–69.

Rothschild, B.J., ed. 1983. *Global Fisheries: Perspectives for the 1980s.* New York: Springer-Verlag.

Rudloe, J., and A. Rudloe. 1989. Shrimpers and sea turtles: A conservation impasse. *Smithsonian* 29:45–55.

Ryther, J.H. 1969. Photosynthesis and fish production in the sea. *Science* 166:72–6.

Ryther, J.H., et al. 1972. Controlled eutrophication—increasing food production from the sea by recycling human wastes. *BioScience* 22:144–52.

Ryther, J.H. 1981. Mariculture, ocean ranching, and other culture-based fisheries. *BioScience* 31:223–30.

Scarff, J.E. 1980. Ethical issues in whale and small cetacean management. *Environmental Ethics* 3:241–79.

Schaefer, M.B. 1970. Men, birds, and anchovies in the Peru Current—dynamic interactions. *Transactions of the American Fisheries Institute* 99:461–67.

Shivji, M., et al. 2002. Genetic identification of pelagic shark body parts for conservation and trade monitoring. *Conservation Biology* 16:1036–47.

Sissenwine, M.P., and A.A. Rosenberg. 1993. U. S. Fisheries: Status, long-term potential yields, and stock management ideas. *Oceanus* Summer:48–54.

Smith, W.O., Jr., and D.M. Nelson. 1986. Importance of ice edge phytoplankton production in the southern ocean. *BioScience* 36:251–57.

Vitousek, P.M., et al. 1997. Human appropiation of Earth's ecosystems. *Science* 277:494–99.

Wabnitz, C., M. Taylor, E. Green, and T. Razak. 2003. *From Ocean to Aquarium: The Global Trade in Marine Ornamental Species.* Cambridge, UK: UNEPWCMC.

Wade, P.R. 1995. Revised estimates of incidental kill of dolphins (Delphinidae) by the purse-seine tuna fishery in the eastern tropical Pacific, 1959–1972. *Fishery Bulletin* 93:345–54.

Walsh, J.J. 1984. The role of ocean biota in accelerated ecological cycles: A temporal view. *BioScience* 34:499–507.

Watson, R., and D. Pauly. 2001. Systematic distortion in world fisheries catch trends. *Nature* 414:534–36.

Zabel, R. W., et al. 2003. Ecologically sustainable yield. *American Scientist* 91:150–57.

Glossary

aboral relating to side of the body opposite the mouth of radially symmetrical animals

absorptive feeding a means of taking up dissolved food material through specialized organs or across the body wall

abyssal plains the flat, sediment-covered floor of the ocean basin, usually 3000–5000 m deep

accessory pigments nongreen photosynthetic pigments found in marine plants that absorb light energy from the center of the visible light spectrum and transfers it to the green pigment chlorophyll

acrorhagi nematocyst-armed defensive structures of anemones

adenosine triphosphate (ATP) a complex organic compound composed of the molecule adenosine and three phosphate groups, which serves in short-term energy storage and conversion in all organisms

aerobic dive limit (ADL) the longest breath-hold dive possible that does not lead to an increase in blood lactate concentration (i.e., a shift to anaerobic respiration)

aerobic respiration cellular respiration occurring in the presence of oxygen

air sacs lateral branches of the nasal passages of smaller toothed whales; sources of echolocation sounds

alcoholic fermentation a form of anaerobic respiration in which sugar is degraded to alcohol and CO_2 and energy is released

algal grazer an animal that consumes either large algae or thin algal films

alveoli minute air sacs in the lungs of vertebrates where gas exchange occurs (singular = alveolus)

amniotic egg egg of reptiles, birds, and mammals containing an embryo that develops within an amniotic membrane

amoebocyte motile cells within sponges that have the shape or movements of an amoeba

anadromous an animal (such as a salmon) that spends much of its life at sea and then returns to a freshwater stream or lake to spawn

anaerobic respiration cellular respiration occurring in the absence of oxygen

ancestral character any character that remains unchanged in the descendant from its condition or appearance in the ancestor

androgen a male sexual hormone in vertebrates

anoxic without oxygen

antennae elongated sensory structures projecting from the heads of arthropods

anterior the front end of a bilaterally symmetrical animal

antitropical distribution a type of distribution in which similar climatic regimes in both the northern and southern hemispheres are inhabited by the same species

aorta large artery (or arteries) carrying blood away from the heart to the body of vertebrates

aphotic zone the portion of the water column, usually deeper than 1000 m, where sunlight is absent

apneustic breathing breathing pattern exhibited by marine mammals in which several rapid breaths alternate with a prolonged cessation of breathing

areolus a structural unit of diatom frustules that is usually hexagonal

asexual reproduction reproduction by a single individual involving fission, budding, or fragmentation that does not involve the union of gametes

aspect ratio index of propulsive efficiency in fishes obtained by dividing the square of its caudal fin height by the caudal fin area

atmosphere a unit of pressure (atm) equal to 14.7 lbs/in^2 (or 760 mmHg) and equivalent to the pressure created by a 1-inch-square, 10-meter-tall column of seawater

atoll a ring-shaped chain of coral reefs from which a few low islands project above the sea surface

autosome a chromosome not designated as a sex chromosome

autotroph any organism that synthesizes its own organic nutrients from inorganic raw materials

auxospore the naked cell of a diatom after the frustule has been shed

baleen rows of comb-like material that project from the outer edges of the upper jaws of filter-feeding whales

bar-built estuary a type of estuary formed behind a coastal barrier or bar

barrier island a long narrow island of sand or mud running parallel to a coastal area

barrier reef a coral reef separated from the shore by a lagoon

benthic pertaining to the seafloor and the organisms that live there

benthos marine organisms that live in or on the sea bottom

bilateral symmetry animal body plan with mirror image left and right sides

bioaccumulation increasingly concentrated accumulation of substances, especially pollutants, at successively higher trophic levels via food-chains amplification

biochemical oxygen demand (BOD) the quantity of dissolved O_2 needed to sustain a given community

biogeochemical cycle pathway of biologically important nutrients between living organisms and its nonliving reservoirs

bioluminescence production of visible light by living organisms

black-band disease a disease of reef-forming corals caused by a sulfate-reducing cyanobacterium that results in a band of dead, blackened tissue surrounding a coral colony

blade the flattened, usually broad, leaf-like structure of seaweeds

blastocyst in mammals, a small ball of cells representing an early stage of embryonic development

blastula early embryonic stage of many animals, consisting of a hollow ball of cells

bleaching expulsion of pigmented zooxanthellae by reef-forming corals when stressed

bloom the sudden appearance of dense concentrations of phytoplankton that occur in response to optimum growth conditions

blubber the sheath of insulating fat and connective tissue of cetaceans and other marine mammals that has been used as a source of oil and food by humans

bradycardia marked slowing of the heart rate

broadcast spawners marine animals that reproduce by releasing eggs and sperm into the water

bronchi the paired ventilatory tubes of a vertebrate that branch into each lung at the lower end of the trachea (singular = bronchus)

bubble-net feeding a unique method of cooperative feeding used by humpback whales in which one member of a group produces a curtain of ascending bubbles of exhaled air that cause prey to clump for easier capture

buffer a chemical substance that tends to limit changes in pH levels

byssal thread a strong elastic fiber produced by mussels to attach themselves to a solid substrate

calorie a unit of heat energy equivalent to the amount of energy required to change the temperature of 1 g of pure water 1°C

carnivore an animal that preys on other animals

carpospore diploid spores produced by the carposporophyte form of red algae

carposporophyte a generation of plants, unique to red algae, that produces carpospores

catadromous fishes that migrate from fresh water to spawn in the ocean

caudal fin an enlarged fin at the posterior end of most fishes

caudal peduncle the area where the caudal fin joins the rest of a fish's body

cell wall a supportive structure that encloses the cells of most plants, bacteria, and fungi

cenosarc a thin layer of soft tissue serving to interconnect adjacent polyps of a coral colony

centimeter a unit of length equal to one-hundredth of a meter

centric relating to diatoms that possess shapes that can be described as circular, triangular, or modified squares

cephalization the evolutionary process of increasing specialization of the head, especially the brain and sense organs

cerata fingerlike branching external gills that project from the dorsal surface (mantle) of nudibranchs

character any countable or measurable attribute of an organism

chemosynthesis bacterial synthesis of organic material from inorganic substances using chemical energy

chitin a flexible, impermeable material found in the exoskeletons of arthropods

chloride cell a specialized cell or "ionocyte" in the gills of bony fishes that excretes excess ions

chlorophyll the green photosynthetic pigment of plants, protists, and bacteria

chloroplast a chlorophyll-containing organelle found in photosynthetic eukaryotes

choanocytes flagellated cells lining the spongocoel that extract food and oxygen from passing water currents

chromatophore surface pigment cell found in many animals that expands and contracts to produce changes in color and appearance

chromosome a subcellular structure that contains the genetic information of the cell

cilia numerous short hairline cellular projections used for locomotion or transport

circadian rhythm a cycle of activity or behavior that recurs about once per day

circalunadian rhythm a cycle of repeating activity each lunar day, or 24.8 hours

clade see **lineage**

cladistics see **phylogenetic taxonomy**

cladogram see **phylogenetic tree**

cleaning symbiosis a form of mutualism in which one partner picks external parasites and damaged tissue from the other partner

clones genetically identical group of individuals derived from a single individual

cnidocyte cell that contains the stinging nematocyst of cnidarians

coastal plain estuary an estuary formed as rising sea level floods a low-lying coastal river valley

coccolith a small calcareous plate imbedded in the cell wall of coccolithophores, a type of phytoplankton

coelom the internal body cavity of most animals, lined with a peritoneum

columella the central supporting structure of gastropod mollusk shells and of the corallites of corals

commensalism a symbiotic relationship in which the symbiont benefits without affecting the host in any measurable or demonstrable way

community an assemblage of interacting populations living in a particular locale

conduction the molecular transfer of heat through a medium

consumer an organism that consumes and digests other organisms to satisfy its energy and material needs

continental boundary current a surface ocean current flowing generally north or south along a continental edge

continental drift the gradual movement of continents in response to seafloor spreading processes

continental shelf the relatively smooth underwater extension of the edge of a continent that slopes gently seaward to a depth of 120–200 m

continental slope the relatively steep portion of the sea bottom between the outer edge of the continental shelf (the shelf break) and the deep ocean basin

convection the transfer of energy via the flow or mixing within a volume of liquid or gas with regions of different temperature and different density

convective mixing the vertical mixing of water masses driven by wind stresses or density changes at the sea surface

corallite the calcareous skeletal cup in which a coral polyp lives

Coriolis effect the apparent change in direction of a moving object (to the left in the Southern Hemisphere and to the right in the Northern Hemisphere) due to the rotation of the Earth

cost of transport (COT) the energy costs associated with an animal's locomotion

countercurrent an ocean current that flows directly back into another current; also, in some animals, relating to paired blood vessels containing blood flowing in opposite directions or streams of blood and seawater flowing through a gill in opposite directions

countershading the coloration pattern exhibited in pelagic animals for camouflage, with the upper surfaces darkly pigmented and the sides and ventral surfaces silvery or only lightly pigmented

covalent bond a chemical bond between two atoms created by the sharing of two electrons

critical depth the depth at which ambient light intensity is so diminished that the rate of photosynthesis (gross primary production) equals the rate of cellular respiration such that no net primary production occurs

cropper a deep-sea animal in which the roles of predator and deposit feeder have merged

crossover speed the velocity at which it becomes more efficient for a swimming animal to leap above the surface, or porpoise, rather than to remain submerged

ctene bands of cilia found on the body surfaces of ctenophores

cypris the final bivalved larval stage of a barnacle

cyst a resistant and photosynthetically inactive diatom

cytoplasm the internal fluid environment of a cell

dark reaction that part of the photosynthetic process that, in the absence of light, utilizes the preformed high-energy molecules ATP and NADPH to synthesize complex organic molecules

decimeter a unit of length equal to one-tenth of a meter

decomposer an organism that consumes and breaks down dead organic material

deep sound-scattering layer (DSSL) one or more layers of midwater marine animals that reflect and scatter the sound pulses of echo sounders

delta a low-lying sediment deposit often found at the mouth of a river

density ratio of the mass of a substance to its volume

deposit feeding a process in which an animal engulfs masses of sediments and processes them through its digestive tract to extract nourishment

derived character any character whose condition or appearance is changed in the descendant or is different from that of its ancestor

detritus particulate dead organic matter, including excrement and other waste products, shed body parts (such as exoskeletons, skin, hair, or leaves), and minute dead organisms

diffusion the transfer of substances along a gradient from regions of high concentrations to regions of low concentrations

diploid cells that contain two of each type of chromosome characteristic of its species

diurnal tide a tidal pattern with one high tide and one low tide each lunar day

dive reflex the suite of internal responses, including bradycardia and peripheral circulation shutdown, that occurs during dives by an air-breathing vertebrate

dorsal the back or upper surface of a bilaterally symmetrical animal

echolocation a means of navigation and environmental assessment that includes producing high-frequency sounds and listening for reflected echoes as the sounds bounce off target objects

ecologic adaptation short-term response expressed by an individual organism to its changing environment

ectotherm an animal whose body temperature is governed by external ambient temperatures

elver a newly pigmented juvenile eel that swims upriver to take up residence in fresh water until sexual maturity

endoplasmic reticulum a system of folded membranes within the cytosol of eucaryotic cells

endotherm an animal whose body temperature is established by internal sources of metabolic heat

enzyme protein catalyst that regulates a particular chemical reaction

epibiont a species that live on another species, using the host as a form of hard substrate

epifauna benthic animals that crawl about on the seafloor or are firmly attached to it

epipelagic zone the upper 200 m of the oceanic province

epiphyte a plant that attaches itself to other organisms

epitheca the larger portion of a diatom frustule

estrogen a female sexual hormone in vertebrates

estuary the portion of the mouth of a river in which there is substantial mixing of fresh water and seawater

eucaryotes cells characterized by an organized nucleus and other membrane-bound subcellular structures

euryhaline an organism capable of withstanding a wide range of environmental salinities

eurythermal the ability to tolerate wide variations in environmental temperature

evolutionary adaptation the changes occurring in a population of individuals over many generations by processes of natural selection

exoskeleton an external supporting skeleton, commonly found in arthropods and mollusks

external auditory canal the sound-transmitting channel connecting the external and middle ears

fast ice permanent sea ice that exists year-round

fecundity the number of offspring an organism can produce in a given time span

feedback mechanisms control mechanisms in organisms and communities in which a change in a given factor either inhibits or stimulates processes controlling the production, release, or use of that factor

fertilization the process wherein two haploid gametes unite to produce one diploid zygote

fetch the extent of the ocean over which winds blow to create waves

fineness ratio ratio of an animal's body length to its maximum body diameter

finlet small median fin on the dorsal and ventral sides of the rear parts of tuna and similar fishes

fjord a deep coastal embayment caused by glacial erosion

flagellum whiplike structure used by cells for locomotion (plural = flagella)

flushing time the time required for all of the water of an estuary to be completely exchanged

food chain a diagrammatic representation of trophic relationships

food web a diagrammatic representation of the complete set of trophic relationships of an organism

form drag see **pressure drag**

frictional drag the resistance created by an animal's body surface when it moves through a fluid medium

fringing reef a large coral-reef formation that closely borders the shoreline

frustule the siliceous wall of a diatom; consists of two halves

fucoxanthin a golden or brown pigment characteristic of Phaeophyta, Chrysophyta, and Dinophyta

gamete an egg or sperm cell

gametophyte a gamete-producing haploid plant

gastrovascular cavity digestive cavity of some lower invertebrates, with a single opening for both mouth and anus, that also plays a circulatory role

gestation period the portion of the reproductive cycle in a female mammal extending from fertilization to birth of its offspring

gill arch the skeletal supports inside the gill tissues of fishes

globigerina ooze thick blankets of foraminiferan tests that cover the seafloor

gram the basic unit of mass in the metric system (g)

grana flattened saclike structures inside chloroplasts containing chlorophyll and other photosynthetic enzymes

gross primary production the total amount of photosynthesis accomplished in a given period of time

guano the droppings from seabird nesting colonies

guyot a sunken volcanic island topped by a dead, usually flat coral reef

gyre the large loop of interconnected surface currents within a single ocean basin, usually spanning 20 to 30° in latitude

halophyte flowering plants that are tolerant to seawater exposure, including complete submergence

haploid cell a cell containing only one of each type of chromosome characteristic of its species

haptera short, sturdy root-like structures that form the holdfast of seaweeds

heat capacity the measure of heat energy required to raise the temperature of 1 g of a substance 1°C

hemocyanin an oxygen-binding pigment found in the blood of several kinds of invertebrates

hemoglobin an oxygen-binding red blood pigment found in vertebrates and some invertebrates

herbivore an animal adapted to feeding on plants

hermaphrodite an animal that has the sex organs of both sexes at some point during its lifetime, sometimes simultaneously

hermatypic a coral that contributes to the building of massive carbonate reefs and contains mutualistic dinophytes

heterocercal a type of caudal fin found in sharks and rays, as well as some primitive bony fishes (such as sturgeons), wherein the vertebral column extends into the upper lobe of the tail

heterotroph an organism that is unable to synthesize its own food from inorganic substances and must consume other organisms and organic compounds for nourishment

high tide the highest level reached by the rising tide

holdfast a structure that attaches seaweeds to the sea bottom or to other substrates

holoplankton species of zooplankton that are part of the plankton community throughout their lives

homeostasis the condition in which the internal environment of an organism (salinity, temperature, blood sugar) remains relatively stable, within physiological limits

homeotherm an animal, such as a bird or mammal, that maintains precisely controlled internal body temperatures using its own physiologically controlled heating and cooling mechanisms

homocercal a type of caudal fin found in advanced bony fishes wherein the supporting skeletal rays are arranged symmetrically around the end of the vertebral column

homology similarity of characteristics resulting from a shared ancestry

host one member of the host-symbiont pairing characteristic of all symbiotic relationships

hydrogen bond (H-bond) a weak electrostatic bond formed by the attractive force between the charged ends of water molecules and other charged molecules or ions

hydrostatic skeleton the body fluid contained within some animals, against which muscles work to provide shape changes

hyperosmotic relating to a water solution with a higher concentration of ions than that of an adjacent solution separated by a selectively permeable membrane

hyphae elongated and multinucleated threadlike structures making up the body of a fungus

hypoosmotic relating to a water solution with a lower concentration of ions than that of an adjacent solution separated by a selectively permeable membrane

hypotheca the smaller portion of a diatom frustule

induced drag the resistive force that is produced via the generation of lift

infauna animals that live within the sediments of the seafloor

inquilinism a protection benefit of symbiosis resulting from the proximity of a large and imposing host

interstitial animal an animal that occupies the spaces (interstices) between sediment particles

intertidal zone the vertical extent of the shoreline between the high and low tide lines

ion an electrically charged atom or molecule formed by gaining or losing one or more electrons

ionic bond in crystalline structures, an atomic bond formed between adjacent oppositely charged ions

iridocyte a fish skin cell that contains reflecting crystals made of guanine

island a area of land, much smaller than a continent, that is completely surrounded by water

isohaline having the same salinity, or a line on a chart or map connecting points of equal salinity

isosmotic relating to a water solution with the same concentration of ions (osmotic pressure) as an adjacent solution separated by a selectively permeable membrane

junk a fat-filled organ of similar size to the spermaceti organ of sperm whales that is hypothesized to be homologous with the melon of smaller toothed whales

kelp a group of large brown seaweeds

labyrinth organ one of a pair of equilibrium and balance organs in the inner ears of vertebrates that contains three fluid-filled semicircular canals (only one in hagfishes and two in lampreys)

lagoon a coastal body of water surrounded by rock, sand, mud, or coral that presents a partial barrier to the open sea

Langmuir cells parallel pairs of surface ocean convection cells driven by winds

last glacial maximum (LGM) the time at which the last major continental glacial advance in the Northern Hemisphere reached its maximum extent (= about 18,000 years ago)

latent heat of fusion the heat that must be extracted from a liquid to freeze it to a solid at the same temperature; for water, it is 80 cal/g

latent heat of vaporization the heat energy required to convert a liquid to a gas at the same temperature, for water, it is 540 cal/g

latitude the angular distance north or south of the equator of a position on the earth's surface; measured in degrees (°)

lek a communal area used by adult males during the breeding season as a stage for the competitive attraction of females

leptocephalus larva leaf-shaped, transparent larva of some eels, bonefish, and tarpon

light reaction that part of the photosynthetic process that, in the presence of light, captures energy to form ATP and NADPH to be used to synthesize complex organic molecules in the dark reaction

limiting factor any factor necessary for the growth or health of an organism that limits further growth if in insufficient supply

lineage the line of descent from an ancestor (or group of ancestors) to a descendent (or group of descendents)

liter the basic unit of volume in the metric system; one liter (L) is slightly more than 1 quart

littoral see **intertidal zone**

longitude the angular distance of a position on the earth's surface east or west of the Greenwich Prime Meridian; measured in degrees (°)

lophophore tentacle-bearing feeding structure of bryozoans, brachiopods, and phoronids

low tide the lowest level reached by a falling tide

lunar day the time between one high tide and the same high tide on the next day, a period of about 24 hours and 50 minutes

macrofauna benthic animals larger than about 1 mm

mangal a tropical community of mangrove plants and associated organisms

mariculture the collective techniques applied to grow marine organisms in captive, controlled situations

matrilineal used to describe the line of descent that follows the female side of the family

maximum sustainable yield the maximum level of fishing effort that a fish stock can withstand without causing major upsets in the abundance of its stock

medusa free-swimming, or jellyfish, generation of cnidarian life cycles

meiofauna benthic infaunal organisms that are 100–1000 μm in size

meiosis a process of cellular division that reduces the chromosome number by half

melon fat-filled structure in the foreheads of toothed whales

meristematic tissue within some seaweeds, specific tissue sites where most cell division for growth occurs

meroplankton larval forms of benthic and nektonic adults that are temporary members of the plankton community

mesoglea the jellylike layer in the bodies of cnidarians

mesopelagic zone the portion of the pelagic division that extends from the bottom of the epipelagic zone to about 1000 m

metabolism the sum total of all biochemical processes occurring in a living organism

metamere a repeated body unit, or segment, along the long axis of some bilateral animals

meter the basic unit of length in the metric system; the meter (m) is slightly longer than 1 yard

microbial loop the portion of a planktonic food web composed of bacteria and other decomposers.

microfauna benthic infaunal organisms that are less than 100 μm in size

mimicry the ability of an organism to disguise itself by assuming the behavior or appearance of another species.

mitochondria in eucaryotes, subcellular organelles (singular = mitochondrion) that conduct cellular respiration

mitosis a process of cell division resulting in two descendant cells genetically identical to their parent cell

mixed semidiurnal tides tidal pattern during a lunar day with unequal high tides and unequal low tides

molt a process of punctuated growth in arthropods wherein they shed their old exoskeleton and replace it with a large version

mortality the rate at which individuals of a population die

mutualism a type of symbiotic relationship in which both the symbiont and the host benefit from the association

myoglobin a red muscle pigment with a strong chemical affinity for oxygen (similar to that of hemoglobin)

myomere one of a series of muscle segments along the trunk of vertebrates, especially fishes

nanoplankton unicellular plankton with cells 5–20 μm in width

nasal gland salt-excreting gland in the nostrils of seabirds

nauplius microscopic free-swimming planktonic stage of barnacles and some other crustaceans (plural = nauplii)

neap tide tide that recurs every two weeks (near quarter moons) that is characterized by a less than average tidal range

nekton active, usually large, marine animals that are able to swim up current

nematocyst organelle within the cnidocytes of cnidarians that is often venomous, used in prey capture

neritic pertaining to the portion of the marine environment that overlies the continental shelves

net primary production the measure of primary production remaining after loses due to cellular respiration have been subtracted from gross primary production

neuromast a mechanosensory cell, similar to sensory hair cells, found in the lateral line system that detect water movements thereby aiding in prey capture, schooling behavior, and avoidance of obstacles and predators

neuston planktonic organisms living (usually floating) at or on the sea surface

niche the functional role of an organism in the ecology of an environment, as well as the suite of physical and chemical factors that limit its range of existence

nitrogen fixation conversion by bacteria and cyanobacteria of atmospheric N_2 to other forms of nitrogen that can be directly used by eucaryotic plants

node the branching point of a phylogenetic tree (or cladogram)

non-point source a source of pollution originating from a broad and vaguely defined source, such as fertilizers washed off fields into streams

notochord an elongated, cartilaginous rod that forms the central skeletal support of chordate embryos

nuclear membrane the membrane surrounding the nucleus of eucaryotic cells

nucleus the membrane-bound central structure of eucaryotic cells that contains the chromosomes

oceanic pertaining to the portion of the marine environment that overlies the deep ocean basins, between continental shelf breaks

oceanic reef coral reefs growing around oceanic islands lacking continental shelves

olfaction the sense of smell, or the ability to detect and identify chemicals dissolved in air or water by using nasal sensory cells

ooze the layer of fine sediment found on the seafloor that is composed of the mineralized skeletal remains of planktonic organisms that have settled there from the photic zone

open access the concept of international law that permits free access by any nation to marine resources existing outside national jurisdictions

oral relating to the side of the body containing the mouth of radially symmetrical animals

osmoregulator an organism that spends ATP to regulate the salt content of its internal fluid environment such that it remains relatively constant regardless of ambient salinity

osmosis diffusion of water from a more concentrated solution to an adjacent solution of lower concentration through a selectively permeable membrane

osmotic conformers organisms that tolerate large variations of internal ionic concentrations without serious damage by remaining isosmotic with the water around them

osmotic pressure in hypoosmotic conditions, the internal fluid pressure that develops from the osmotic inflow of water

oviparous relating to the reproductive strategy of some vertebrates that reproduce via eggs that develop and hatch outside of the mother

ovoviviparity an intermediate condition between viviparity and oviparity in which the eggs develop and hatch while still inside the mother, with subsequent live birth of the young

oxygen minimum zone the layer of the water column below the photic zone where dissolved oxygen concentration is lowest

ozone O_3, formed by the action of ultraviolet light on atmospheric O_2

pack ice seasonal sea ice that thaws and disappears each summer

Pangaea the supercontinent that consisted of all the present land masses prior to their breakup and subsequent drift to their present positions

parasitism a type of symbiotic relationship in which the symbiont (parasite) lives on or in the host and benefits from the relationship at the expense of the host

pelagic pertaining to the ocean's water column and the organisms that live there

pelagic larval duration the length of time during which a particular larva remains a part of the plankton community as meroplankton prior to metamorphosis

pen in squids, a thin, chitinous structure extending the length of the mantle tissue

pennate relating to diatoms that are elongated and display various degrees of bilateral symmetry

pentamerous five-sided radial body symmetry commonly displayed by echinoderms

period in ocean waves, the time required for two successive waves to pass a reference point

pH a numerical scale from 0 to 14 that is used to represent the H ion concentration of a water solution

pheromone a chemical substance that is released from an animal that influences the behavior and development of other members of the same species

phonic lips enlarged folds of tissue associated with the nasal sacs of cetaceans that vibrate rapidly to produce the characteristic clicks and whistles of these animals

photic zone the portion of the ocean where light intensity is sufficient to accommodate gross primary production (photosynthesis)

photoinhibition reduction of photosynthetic rates due to too much light

photophores light-producing organs found in animals

photosynthesis the biological synthesis of organic compounds from inorganic substances using light as an energy source

phototrophic the reliance on the photosynthetic products of symbiotic zooxanthellae to obtain nutrition; found in hermatypic corals, some sponges and other reef invertebrates

phycobilin a type of pink or blue accessory photosynthetic pigment found in cyanobacteria and red algae

phylogenetic relating to the evolutionary history of a taxon or group of taxa

phylogenetic taxonomy the science of classifying organisms according to their branching pattern along a cladogram, within considering the degree of morphological divergence

phylogenetic tree a branching diagram depicting hypothesized evolutionary relationships between groups of organisms

physoclistous swim bladder a swim bladder lacking an air passage to the esophagus

physostomous swim bladder a swim bladder with an air passage or duct to the esophagus

phytoplankton photosynthetic members of the plankton community

picoplankton unicellular plankton with cells less than 2 μm in width

plankton free-floating, usually minute, marine organisms that drift with the currents

planula the planktonic larval form of cnidarians that develops into the polyp generation

plasma membrane the selectively permeable outer membrane of a cell

plate tectonics the collective geologic processes that move the crustal plates of the Earth and cause continental drifting and seafloor spreading

pneumatic duct the connection between the esophagus and swim bladder of fishes

pneumatocyst gas-filled float present in several types of kelp plants

pneumatophore a gas-filled float used by some siphonophores to maintain buoyancy in the water

pod a small group of marine mammals

poikilotherm an animal that lacks the ability to control its body temperature physiologically

point source a source of pollution originating at a definable point, such as a storm drain

polar easterlies winds that blow from east to west at very high latitudes

pollen immature male gametophytes of flowering plants

polygyny a type of social and breeding organization in which a male is dominant over and mates with several females

polyp the attached, benthic generation of many types of cnidarians

population a group of freely interbreeding organisms of the same species

porpoising a type of locomotion, characteristic of porpoises and other small cetaceans, in which subsurface swimming is punctuated by leaping above the water–air interface and gliding airborne for a few body lengths

posterior the rear end of a bilaterally symmetrical animal

predator a carnivorous animal that feeds by killing other animals

pressure drag hydrodynamic drag on an organism caused by its cross-sectional area

primary producer autotrophic organisms that synthesize organic compounds via photosynthesis or chemosynthesis

primary production the synthesis of organic compounds from inorganic substances

procaryotes bacteria and cyanobacteria that lack the structural complexity and defined nucleus found in eucaryotes

producer an organism, usually photosynthetic, that contributes to the production of organic compounds for a community

protandry a form of sequential hermaphroditism in which the individual matures first as a male, then transforms into an adult female later in life

protogyny a form of sequential hermaphroditism in which the individual matures first as a female, then changes sex to an adult male later in life

pseudopodia temporary and changing extensions of the cells of some protozoans, used for locomotion

pycnocline the ocean layer, usually near the bottom of the photic zone, marked by a sharp change in density that separates the less dense surface waters from denser deep waters

radula rasping tonguelike structure of most mollusks

raphe a groove in the frustules of pennate diatoms through which portions of the cell extend for locomotion

red tide a bloom of toxic dinophytes that may cause serious morality to other forms of marine life

reef massive near-shore deposits of coral skeletal material

rete mirabile a complex network of capillary-sized blood vessels, usually in a countercurrent arrangement

retina the light-sensitive layer of nerve cells in the eyes of most vertebrates and some invertebrates

rhizome the horizontal underground stem of seagrasses and some macroalgae

ridge and rise system the interconnecting chain of seafloor mountains that trace the edges of crustal plates and the sites of production of new oceanic crust

salinity a measure of the total amount of dissolved ions in seawater

salt marsh a low-lying intertidal grass community found along shores of temperate estuaries

saxitoxin a paralytic toxic produced by dinophytes that accumulates in the butter clam (*Saxadoma*)

scavenger an animal that feeds on the dead remains of other animals and plants

school a well-defined social organization of marine animals consisting of a single species with all members of a similar size all swimming in the same direction at the same speed

seafloor spreading a global process of oceanic crust moving away from ridge and rise systems where it formed

seamount an isolated undersea volcano that rises well above the seafloor but falls well beneath the sea's surface

seasonal delayed implantation a pattern in the reproductive cycle of some mammals causing the blastocyst to remain dormant in the female's uterus for some time before implantation on the uterine wall

secondary lamella a small extension of a gill filament containing blood capillaries for gas exchange

selectively permeable membrane a membrane that allows small atoms, molecules, compounds, and ions (such as H_2O, O_2, CO_2, Na^+) to pass through but is not permeable to larger atoms, molecules, compounds, and ions

semidiurnal tides tidal patterns with two high tides and two low tides each lunar day

sensory hair cell specialized cells found in the inner ear of vertebrates that detect movements and vibrations resulting in the senses of equilibrium, balance and hearing

septa the thin structures separating internal parts of organisms (singular = septum)

sex chromosome one of a pair of chromosomes whose composition determines gender

sexual dimorphism an obvious and often dramatic difference in the size and appearance (and behavior) of adult males and females of the same species

sexual reproduction a mode of reproduction involving the production of gametes by meiosis, followed by a fusion of the gametes in fertilization

shelf break the point at which the outer edge of the continental shelf meets the continental slope, typically at a depth of 120–200 m

shelf reef coral reefs growing on continental shelves

siphon tube-like structure of mollusks used to take in and expel water from the mantle cavity

siphuncle a central tube-like tissue connecting the chambers of shelled cephalopods such as *Nautilus*

smolt a young salmon just before it migrates downstream and out to sea

speciation event the origin of a new species via evolution

species a particular kind of organism; members share a common ancestry, similar morphology, and, if the species reproduces sexually, reproductive isolation from other species

spermaceti organ a large organ in the forehead of sperm whales that is filled with a fine-quality liquid or waxy spermaceti oil

sperm competition a strategy used by copulating males competing with previous copulators by attempting to displace or dilute their sperm within a female's reproductive tract, thus increasing the probability of being the male to fertilize the female's single ovum

spicules small, mineralized skeletal structures of sponges

spongin fibrous material making up the flexible skeleton of many sponges

spongocoel the internal cavity of sponges

sporangia special plant cells or structures that produce spores

spore in plants, the single-celled haploid structure produced by sporophytes

sporophyte a spore-producing diploid plant

spring tide tide that recurs every two weeks (near new and full moons) that is characterized by a greater than average tidal range

standing crop total amount of plant or animal material in an area at any one time

statocyst a gravitationally sensitive vesicle lined with sensory cells and containing dense bodies; found in many invertebrates

stenohaline an organism that tolerates exposure to only slight variations in salinity

stenothermal the tendency to tolerate only narrow variations in environmental temperature.

stigma a structure on the female part of a flower on which pollen grains are received and germinate

stipe the flexible, stem-like structure found in the large seaweeds

stroma the fluid of the chloroplast containing the enzymes for the dark reactions of photosynthesis

succession the gradual replacement, through time, of one group of species in a community by other groups

surface current long-term, unidirectional horizontal movement of water at the sea surface

surface tension the mutual attraction of water molecules at the surface of a water mass that creates a flexible molecular "skin" over the water surface

suspension feeder an animal that uses a filtering device or sticky mucus to obtain plankton or detritus from the water

swim bladder gas-filled buoyancy organ of bony fishes

symbiont the beneficiary of a symbiotic relationship

symbiosis an intimate and prolonged association between two (or more) organisms in which at least one partner obtains some benefit from the relationship

taxon any group or category in a taxonomic system of classification

taxonomy the science of naming and grouping organisms according to their evolutionary relationships

tectonic estuary an estuary that fills in a depression created by one or more faults in the nearshore crust

temperature a relative intensity measure of the condition caused by heat

testosterone a male sexual hormone in vertebrates

thermocline the ocean layer, usually near the bottom of the photic zone, marked by a sharp change in temperature that separates the warmer surface waters from the colder deep waters

thermohaline circulation mixing of the world ocean on a global scale that is initiated by the sinking of very cold, very dense seawater at the poles

tidal range vertical distance between high and low tides

tide a long-period wave noticeable as a periodic rise and fall of the sea surface along coastlines

trace element an element needed for normal metabolism but available only in minute amounts from the environment

trachea the windpipe of vertebrates

trade winds subtropical winds that blow from northeast to southwest in the Northern Hemisphere and from southeast to northwest in the Southern Hemisphere

trench deep area in the ocean floor, generally deeper than 6000 m, that is the site of subduction of oceanic crust into the Earth's core

trochophore early free-swimming, ciliated larval stage of many marine mollusks, annelid worms, bryozoans, and brachiopods

trophic pertaining to feeding or nutrition

trophic level the position of an organism or species in a food chain or web

trophosome an internal, symbiotic bacteria-filled organ of the giant tube worm, *Riftia*

turbulence random, nonlaminar flow of a fluid

turnover rate the rate at which members of a population or community replace themselves

tympanic bulla bony case in the middle ear that encloses the sound-processing structures of mammals

ultraplankton unicellular plankton with cells 2–5 μm in width

ultrasonic frequencies of sound that are above the normal range of human hearing (usually greater than 18 kHz

upwelling the process that carries nutrient-rich deep waters upward to the photic zone

urea a nitrogen-containing waste product excreted in the urine of many vertebrates

uric acid the main nitrogenous excretory product in birds, reptiles, some invertebrates, and insects

vacuole a liquid or food-filled cavity within a cell

vasoconstriction narrowing of blood vessel diameter via contraction of smooth muscles in the vessel wall with resulting reduction in blood flow and increase in blood pressure

veliger early larval form of some mollusks

vena cava the major vein returning blood to the heart of vertebrates; the superior vena cava drains blood from the head and upper extremities, the inferior vena cava drains blood from the torso and lower extremities

ventral the underside of a bilaterally symmetrical animal

vertebrae the series of articulated bones that make up the backbone of vertebrates (singular = vertebra)

vertical migration daily or seasonal movement of small marine animals between the photic zone and midwater depths

viscosity the resistance of water molecules to external forces that would separate them

visible light that portion of electromagnetic radiation to which the human eye is sensitive, usually represented as a spectrum of colors from shorter-wavelength violet to longer-wavelength red

viviparity a condition describing the reproductive strategy wherein embryos develop within the mother while receiving nutrients via a placenta with subsequent live birth

water-vascular system the unique internal circulatory system of echinoderms that simultaneously operated numerous tube feet hydraulically for locomotion

wave a periodic, traveling undulation of the sea surface

wave drag a force resistive to movement that is created by the production of surface waves when swimming at or near the sea surface

westerlies winds that blow primarily from the west in the mid-latitudes of both hemispheres

white-band disease an epidemic coral disease that most often infects elkhorn, staghorn, and other corals in genus *Acropora*

white plague a wasting disease of corals caused by a bacterium common to untreated human sewage

xanthophyll a group of yellow or golden photosynthetic pigments

zoea an early larval stage of many crustaceans

zooplankton heterotrophic members of the plankton community

zooxanthellae mutualistic unicellular dinophytes found in corals, sea anemones, mollusks, and several other types of marine animals

zygote a single diploid cell that is the product of the fusion of two haploid gametes

Index

Note: page numbers in *italics* indicate figures.

blooms, 300
  in English Channel, 90
  in polar seas, 119
  in temperate seas, 114, *116*
chains of, *75*
chloroplasts of, 71
as deep-sea ooze, 336
in estuaries, 73
habitat, 67
light requirements, 86
limiting factors, 84
locomotion, 71
loss of in Chesapeake Bay, 214
on mudflats, 204
photosynthetic pigments, 67, 88
as prey of copepods, 301–303
reproduction, 72–73
resting spore, *74, 79, 116*
shapes, 71
size, 67, 71, 77
skeleton, 71–72, 89
storage products, 67
with toxins, 76
Dictyocha, 71
Diffusion, 19
Dinoflagellates, 129
Dinophytes, 73, 75–76
  biodiversity, 67
  blooms of toxic species, 208
  in Chesapeake Bay, 214
  as coral symbionts, 257–258
  diurnal rhythms, 75
  habitat, 67
  as heterotrophs, 129
  light limitations, 86
  limiting factors, 84
  photosynthetic pigments, 67, 88
  population fluctuations, 115–116
  reproduction, 75
  resting stages, 79
  size, 67, 75, 77
  storage products, 67
  symbiosis, 76
Diodon, 269
Dioxins, 208–210
Diploid cells, 46
Diseases, of corals, 266–268
Disruptive coloration, in reef fishes,
  273–274
Distribution
  of coelacanths, 165–166, 184–185
  of corals, 256–257
  of estuarine species, 202
  of kelp, 114
  of killer whales, 366–367
  of mangals, *103*
  of red algae, 113
  of salt marshes, *103*
  of sea turtles, 172
  of walrus, 360

Ditylum, 78
Diurnal tides, *36*
Dive response
  in mammals, 182
  in Weddell seals, 183
Diving physiology, 181–184
Division, definition, 50
Dogfish sharks, development, *165*
Dolphinfishes, *49–50*
Dolphins, common, *49*
Dolphins
  cost of swimming in, 310–311
  countercurrent heat exchangers in, *176*
  echolocation in, 327–328
  mortality in tuna fishery, *315–316,*
    386–387
  porpoising behavior, 309–311
  schooling behavior in, 314–316
  speed in water, 309
  taxonomy, 50
Domoic acid poisoning, 76
Donax, 238
Dorado, *49–50*
Drag, as a resistance to swimming, 309
Dredging
  in Chesapeake Bay, 211
  in estuaries, 196
Dugongs
  feeding in, 177
  phylogeny, *49*
Dynamite
  role in coral mortality, 399
  use in marine ornamental trade,
    266, 399

Earth size, 5
East Australia Current, *33*
East Pacific Rise, 13–14, *344*
Echinodermata, *147–149*
  in the deep sea, 340, *342*
  feeding, 222
  meroplankton of, *223*
  as reef builders, 256
  salt and water balance, 19, *21*
  settling of larvae, *225*
Echinometra, 264
Echiura, 143
Echolocation, 326–329
  in killer whales, 368
Ecologic efficiency, in the sea, 391
  of a food web, 57
Ecologic succession, 73
  on rocky shores, 233–234
Ectoprocta, as reef builders, 256
Ectotherms, 24
Eelgrass, 100–*101*
  epiphytes of, 204
  grazing by sea turtles, 213
  loss of coverage, 214
  on mudflats, 204–205

Eels, *167*
  migrations of, 317–319, 325
  olfaction in, 324
Eggs of fishes
  benthic, 275–276
  broadcast spawning, 276–277
Egregia, 107, 245
Ekman spiral, 32
El Niño, 117–119, 244, 396
  role in coral bleaching, 268
Elacatinus, 272
Elatomma pinetum, 130
Electromagnetic spectrum, 22–23
Electroreception, in vertebrates, 186
Elephant seals, 318–324
  diving in, 183–184
  genetic bottleneck after harvest,
    400–401
  harvest of, 400
  maximum dive depths in, 180, 318
  maximum dive duration in, 180, 318
  predator avoidance, 183
  as prey of great white sharks, 323–*325*
Elkhorn coral, diseases of, 267–268
Elkhorn kelp, 106, *245*
Elminius modestus, 222
Elvers, 317
Emerita, 236, 238
Emiliania huxleyi, 70
Endangered species, in estuaries,
  196–197
Endoplasmic reticulum, *52*
Endotherms, 25
Enewetok Atoll, 260
Engraulis mordax, 385
Engulfment feeding of rorquals,
  373–374
Enhalus, 286
Enhydra, 244–246
  diving in, 180
ENSO, 117–119, 244, 396
  role in coral bleaching, 268
Entoprocta, *137*
Epidemics
  in corals, 265–268
  in sea urchins, 266
Epifauna, 44, 220–222
  in deep sea, 341–342
  of mudflats, 206
Epinephelus striatus, 276
Epipelagic zone, 291–294, 298–329
Epiphytes, on seagrasses, 109, 204
Epitheca, 71
Equatorial countercurrents, role in
    migrations, 317–318
Equatorial currents, *33*
Equilibrium, in vertebrates, 187–188
Eratosthenes, 5
Eschrichtius, 179
Essential Fish Habitat, 197

Pneumatocysts of seaweeds, 106, *108*
Pneumatophores of zooplankton, *304*
Pocasset estuary, *201*
Pocillopora damicornis, 262
Podon, 59
Pods
  of killer whales, 367
  of sperm whales, 378
Pogonophora, 143–144
  biodiversity in deep sea, 340–341
Poikilotherms, 24, 178
Point sources of pollution, 208
Polar bears, 361–362
  reproduction in, 175–176
Polar easterlies, 32–33
Polar seas, 255–378
  diatom bloom in, 119
  net primary production in, 120–121
  primary production in, 119–120
  thermocline, 119
Pollen of seagrasses, 101
Pollution
  in atmosphere, 5
  in estuaries, 196, 200
  from mariculture, 393
  in salt marshes, 203
  in the sea, 207–211
  sewage outfalls, 246–247
Polychaetes, 144
  as reef builders, 256
  as reef grazers and scrapers, 256
  interstitial, *238*
  in the deep sea, 340
  meroplankton of, *223*
Polygyny
  in cleaner wrasses, 171
  in elephant seals, 321
  in reef fishes, 279
Polyps
  in Cnidaria, 133–*135*
  of corals, *254*–255
Pomacanthus, 269
Porbeagle sharks
  reproduction, 164
  warm body of, 312–313
Porcupinefish, *269*
Porifera, 131–*132*
Porites, 258, 265
Porolithon, 264
Portuguese man-of-war, 134, *304*
Postelsia, 106, 112
Predators of corals, 266–267
Pressure drag in swimming, 309
Pressure
  effect on buoyancy, 304–307
  in the sea, 28, 180–*181*, 336–338
Prestige, 240–241
Prey shifts during ontogeny, of great
  white sharks, 323
Priapulida, *143*
Primary producers, definition, 53

Primary production, 67, 79–94
  Antarctic, 356
  Arctic Ocean, 356
  in Atlantic Ocean, *81*
  on coral reefs, 254, 257–258
  in estuaries, 196
  factors affecting the rate of, 84–94,
    114–120
  in mudflats, 204
  measurement of, 80–84
Proboscis
  of echiurans, 143
  of elephant seals, 321
  of innkeeper worms, 143
  of ribbon worms, 136
Procaryotes, 68, 69
  biodiversity, 67
  bioluminescence, 75
  cell wall, *51*
  DNA, *51*
  flagellum, *51*
  in food web, *60*
  habitat, 67
  photosynthetic pigments, 67
  reproduction, 68
  size, 67, 77
  storage products, 67
  taxonomy, 50
  ultrastructure, *51*
Producers, definition, 53
  on coral reefs, 254
Prop roots of mangroves, 102, *227*
Protandry, in reef fishes, 277–280
Protista, 52, 67, 126–131
  reproduction in, 128
Protogyny, in reef fishes, 277–280
Protoperidinium, 76
Protostomes, 126, 137–147
Protozoea larva, *147*
Psammodrilus, 238
Pseudocoelomates, 136
Pseudonitzschia, 76
Pseudopods
  of protists, 129
  of silicoflagellates, *71*
Pterogophora, 112, 245
Ptychodiscus, 75–76
Purple sail, *304*
Pycnocline, 27–28, *30,* 62
  in tropics, 91–92, *116*
Pycnogonida, 145–*146*
  in the deep sea, 340
Pytheas, 5

Radial symmetry, 126, *133*
  of echinoderms, secondary, 147–148
Radioactive wastes, 208
Radiolarians, 129–*130*
  as deep-sea ooze, 336
Radula, 140–*141*
Rays, 161–165, *269*

  of coral reefs, 269–270
  electroreception in, 186
  magnetoreception in, 186, 325
  and mangrove trees, 104–105
  osmoregulation, 161
  phylogeny of, *159*
  reproduction, 163–*165*
  skeletons, 161
  swimming in, 308
Red blood cells, of vertebrates, 157
Red mangrove, 102
Red muscles vs. white muscles, 312–314
Red Sea, 68
  coral reefs, 259
Red tides, 75
Redshank, bill shape of, *176*
Reef flat, *263*
Reef shark, *269*
Remoras, symbiosis in, 270–*271*
Reproduction
  in animals, 130–131
  in Antarctic penguins, 371–372
  in blue crabs, 213–214
  in chondrichthyes, 163–165
  in cnidarians, 133–*134*
  in corals, 261–263
  in crabeater seals, 372
  in cyanobacteria, 68
  in the deep sea, 342–343, 348–349
  in diatoms, 72–73
  in dinophytes, 75
  in eels, 317–319
  in elephant seals, 318–324
  in gray whales, 363–366
  in green sea turtles, 213
  in grunion, 238–239
  in horseshoe crabs, 213
  in labyrinthomorpha, 130
  in marine birds, effects of DDT, 209
  in marine mammals, 175–176
  in menhaden, 213
  in oysters, 213
  in placozoans, 132
  in polar bears, 362
  in procaryotes, 68
  in protists, 128
  in red mangroves, 102–*103*
  in reef fishes, 275–280
  in ringed seals, 358–359
  in salmon, 316–317
  in sea anemones, 231
  in sea snakes, 173, 282
  in sea turtles, 172, *174,* 283–286
  in seagrasses, 101
  in seaweeds, 110–112
  in silicoflagellates, 70
  in stingrays, 104
  in tuna 316–317
  in walruses, 360–361
  in Weddell seals, 372–373
  in whales, *179*

442   Index

Photo Credits

(*Note:* Photos for which there are no credits are owned and supplied by the authors.)

Chapter 1

Opener: Courtesy NASA. 1.4 (inset): Courtesy NOAA. 1.5: Courtesy of National Geophysical Data Center. 1.7: Scripps Institution of Oceanography, University of California San Diego. 1.11: Courtesy National Geophysical Data Center. 1.12: Courtesy NASA. 1.15: Hermann Eisenbeiss/Photo Researchers. 1.25: Michael Van Wort/ORA/NESDIS/NOAA. 1.39: Courtesy NOAA.

Chapter 3

3.B1: Courtesy NASA/JPL Ocean Surface Topography Team. 3.3: Courtesy C. Osolin, Lawrence Berkeley Lab. 3.4: Photo by D. Doubilet. 3.6: Courtesy F. Reid, Scripps Institution of Oceanography. 3.8: Courtesy F. Reid, Scripps Institution of Oceanography. 3.9: Courtesy G. Fryxell, Scripps Institution of Oceanography. 3.10: Courtesy E. Venrick, Scripps Institution of Oceanography. 3.13: Courtesy F. Reid and K. Lang, Scripps Institution of Oceanography. 3.15: Courtesy F. Reid, Scripps Institution of Oceanography. 3.16: Courtesy E. Venrick and F. Reid, Scripps Institution of Oceanography. 3.21: Courtesy NASA. 3.22: Provided by the SeaWiFS Project, NASA/Goddard Space Flight Center and ORBIMAGE. 3.25: Courtesy Mobil Oil Corp. 3.28: Courtesy Herbert W. Israel, Cornell University. 3.33: Provided by the SeaWiFS Project, NASA/Goddard Space Flight Center and ORBIMAGE. 3.39: Provided by the SeaWiFS Project, NASA/Goddard Space Flight Center and ORBIMAGE.

Chapter 4

4.3: Ben Mieremet/OSD/NOAA. 4.12: Courtesy G. Dudley. 4.14: Courtesy G. Dudley. 4.19: Photo by T. Phillipp. 4.22: Provided by the SeaWiFS Project, NASA/Goddard Space Flight Center and ORBIMAGE. 4.24: Provided by the SeaWiFS Project, NASA/Goddard Space Flight Center and ORBIMAGE. 4.26: Courtesy Gene Felderman, NASA/GSFC Space Data Computing Division.

Chapter 5

5.3: Courtesy Kuzo Takahashi, Scripps Institution of Oceanography. 5.11: Photo by T. Phillipp. 5.12: Photo by T. Phillipp. 5.19: Courtesy J. Richardson, Victoria Museum, Melbourne. 5.26: Courtesy R. Brusca. 5.28: Photo by T. Phillipp. 5.30: Courtesy A. Kuzirian, Marine Biological Laboratory. 5.38: Photo by T. Phillipp.

Chapter 6

6.9: Courtesy M. Snyderman. 6.10: Courtesy R. E. Williams. 6.12: Courtesy Scripps Institution of Oceanography (UCSD). 6.13: Anchovy: Courtesy of OAR/NURP/NOAA. Flying fish:

Courtesy of Shannon Rankin, NMFS, SWFSC/NOAA. Seahorse: Courtesy of Mr. Mohammed Al Momany, Aqaba, Jordan/NOAA. Triggerfish: Courtesy of Georgia Department of Natural Resources/NOAA. Yellowtail snapper: Courtesy of Florida Keys National Marine Sanctuary/NOAA. Striped Bass: Courtesy of USFWS. Damselfish: Photo by Hereon Verbruggen. Grouper: Photodisc. Scorpionfish: Courtesy of D. Weaver, FGBNMS/NOAA. Eel: Courtesy of J. Moore, OAR/NURP/NOAA. Salmon: Courtesy of William W. Hartley, USFS. Flounder: Photo by John Morrissey. 6.18: Courtesy NOAA. 6.22: Courtesy Corel Professional Photos. 6.23: Courtesy J. Harvey. 6.25: Photo by Scott A. Eckert, WIDECAST. 6.26a: Michael Van Woert/ORA/NESDIS/NOAA. 6.29: S.H. Ridgway. 1972. *Mammals of the Sea.* Courtesy of Charles C. Thomas, Publisher, Springfield, IL. 6.30: Commander John Bortniak, NOAA Corps (ret.). 6.33: Courtesy Sirenia Project, U.S. Fish and Wildlife Service. 6.35: From Rice and Wolman,1971. *The Life History and Ecology of the Gray Whale* (*Eschrichtius robustus*). Spec. Publ. No. 3, Lawrence KS: American Society of Mammalogists. Courtesy D. Rice. 6.37: Courtesy New England Aquarium. 6.39: S.H. Ridgway. 1972. *Mammals of the Sea.* Courtesy of Charles C. Thomas, Publisher, Springfield, IL. 6.42: Louise Kane/NOAA Restoration Center.

Chapter 7

Opener: NOAA National Estuarine Research Reserve Collection. 7.2: Courtesy NASA. 7.9b: Louise Kane/NOAA Restoration Center. 7.15: Image courtesy the SeaWiFS Project, NASA/Goddard Space Flight Center, and ORBIMAGE.

Chapter 8

Opener: Russel Illig/PhotoDisc, Inc. 8.B1: NOAA. 8.5: Photo by T. Phillipp. 8.7: Photo by T. Phillipp. 8.12: Courtesy M. Tegner, Scripps Institution of Oceanography. 8.14: Courtesy U.S. Geological Survey. 8.15a: NOAA. 8.15b: OAR/NURP/NOAA. 8.31: Photo by C. Birkland, courtesy P. Dayton, Scripps Institution of Oceanography. 8.33: Sea cucumber: Courtesy of Dr. James P. McVey, NOAA Sea Grant Program. Olive shell: Courtesy of Image*After. Blue crab: Courtesy of Mary Hollinger, NODC biologist, NOAA. Flounder: Courtesy of NW Fisheries Science Center/NOAA. Sand dollar and Cockle: Photodisc. Clam: Courtesy of Dr. Roger Mann, VIMS/NOAA. Bristle worm and Land crab: Courtesy of NOAA. Amphipod: Courtesy of M. Quigley, GLERL/NOAA, Great Lakes Environmental Research Laboratory. 8.35: Courtesy D. Garcia. 8.39: Courtesy J. Harvey. 8.40: Courtesy J. Harvey. 8.41: Courtesy J. Harvey. 8.44: Photo by T. Phillipp.

Chapter 9

9.1: Photo by T. Phillips. 9.B1: David Bowman, US Fish and Wildlife Service. 9.4: Photo by Heroen Verbruggen. 9.7: NOAA. 9.9: Courtesy of National Geophysical Data Center. 9.10: P. Harrison/Animals Animals/OSF. 9.11: Photo by P. Parks, Oxford Scientific Films. 9.13: Courtesy of T. Ebert. 9.14: Courtesy of T. Phillipp. 9.15: Photos by T. Phillipp. 9.16: NOAA/AIMS. 9.19: Angelfish photo from Joyce and Frank Burek/NOAA; all other photos from John Morrissey. 9.24: Photo by T. Phillipp. 9.25: NOAA. 9.26: Courtesy C. Farwell, Scripps Institution of Oceanography. 9.27a: Photo by T. Phillipp. 9.27b: Courtesy C. Stepien, Scripps Institution of Oceanography. 9.28a: Photo by T. Phillipp. 9.30: Courtesy of Michael P. Robinson. 9.31: Courtesy of Jack Randall. 9.32: © F. Pacorel/Peter Arnold, Inc. 9.34: © Carlos Villoch/Fragile Ocean Stock Photography.

Chapter 10

10.2: Courtesy of J. Trent, Scripps Institution of Oceanography. 10.5: Courtesy P. Azam, Scripps Institution of Oceanography. 10.10: Courtesy E. Kampa, Scripps Institution of Oceanography and the National Institute of Oceanography. 10.11: Courtesy A. Alldredge. 10.15: Courtesy R. Strickler. 10.19: Courtesy J. Trent, Scripps Institution of Oceanography. 10.29: Courtesy NOAA/NMFS/SWFC. 10.30: Courtesy Honolulu Laboratory, National Marine Fisheries Service, NOAA, Department of Commerce. 10.44: Courtesy J. Harvey. 10.45a: From T. Cranford.

Chapter 11

11.B1: NOAA. 11.2: Courtesy Scripps Institution of Oceanography. 11.4: Courtesy Scripps Institution of Oceanography, Marine Life Research Group. 11.6: Courtesy R. Lampitt. 11.9: Courtesy Scripps Institution of Oceanography, Marine Life Research Group. 11.10: Courtesy E. Barham, National Marine Fisheries Service. 11.12: Courtesy Paul Johnson, HMRG. 11.13: Courtesy Scripps Institution of Oceanography. 11.15: Photo by D. Foster. 11.17: Courtesy K. Van Damm, UNH. 11.19: Courtesy A. Dittel, U. Delaware.

Chapter 12

Opener: Bill Hickey/US Fish and Wildlife Service. 12.1: Provided by the SeaWiFS Project, NASA/Goddard Space Flight Center, and ORBIMAGE. 12.3: Photo by Kit and Christian, A.R.C. 12.4: Photo by Kit and Christian, A.R.C. 12.5: © AbleStock. 12.6: Photos by Kit and Christian, A.R.C. 12.8: NOAA. 12.10a: NOAA. 12.12: Courtesy B. Reitherman. 12.14: NASA. 12.18b: Photo by Kit and Christian, A.R.C. 12.19: NOAA. 12.20: Courtesy R. Baird. 12.22: Courtesy Sea World, San Diego. 12.23 (inset): Frank T. Awbrey/Visuals Unlimited. 12.25a: Courtesy of Tony Palliser. 12.25b: Courtesy of Janet Hinshaw, University of Michigan, Museum of Zoology. 12.30: Courtesy F. Sharpe.

Chapter 13

13.2: Photo by T. Phillipp. 13.3: Courtesy W. Perryman, National Marine Fisheries Service. 13.4: Courtesy W. Perryman, National Marine Fisheries Service. 13.6: Photo by Greenpeace. 13.7: Courtesy Kelco, a division of Merck Pharmaceuticals. 13.8a: Doug Plummer/Photo Researchers. 13.8b: Jim Wark/Peter Arnold, Inc. 13.13a,b: Courtesy National Marine Mammal Laboratory, NMFS, NOAA.